PLANT PHYSIOLOGY

PLANT PHYSIOLOGY

With Reference to the Green Plant

2nd Edition

– Volume 1 –

EDWIN CYRUS MILLER
Professor, of Plant Physiology,
Kansas State College of Agriculture and Applied Science, and Plant Physiological, Kansas,
Agricultural Experimental Station

BIOTECH
2005
Biotech Books
Delhi - 110 035

ISBN 81-7622-154-6 (SET)
ISBN 81-7622-155-4 (Volume 1)
ISBN 81-7622-156-2 (Volume 2)
ISBN 81-7622-157-0 (Volume 3)

Published by : **BIOTECH BOOKS**
1123/74, Tri Nagar,
DELHI – 110 035
Phone: 27382765
e-mail: biotechbooks@yahoo.co.in

Showroom : 4762-63/23, Ansari Road, Darya Ganj,
NEW DELHI - 110 002
Phone: 23245578, 23244987

Printed at : **Chawla Offset Printers**
New Delhi – 110 052

PRINTED IN INDIA

To my wife

DELLA SLYE MILLER

PREFACE TO THE SECOND EDITION

In the preparation of this edition, various topics have been enlarged or revised to include the investigations and findings that have been made since the publication of the first edition. Certain topics that were overlooked in the preparation of that edition have been added. New material has been presented under appropriate headings.

Although no claim to completeness is made, it has been the effort of the author to present the more important citations relating to each subject. He has tried to present impartially the data that have been reported. He does not feel qualified to speak with authority on a majority of the topics, and his own ideas, if presented, thus would be of little consequence.

In order to conserve space, the questions that concluded each chapter in the first edition have been omitted.

The author is indebted in various ways to several of his associates at the Kansas State College of Agriculture and Applied Science, and he takes this occasion to thank them. He wishes to thank especially Hugh G. Gauch, for his invaluable aid in assembling and typing the bibliographies and other portions of the manuscript, and for aid in editing it; John C. Frazier, for aid in editing the manuscript, for helpful advice, and for unselfish service in many ways; Dr. Fritz Moore, head of the Department of Modern Languages, and Mrs. Irene Moore, for checking, respectively, the German and the French references; and Miss Esther Lewis, for clerical help in preparing the last half of the manuscript.

The author is especially indebted to Professor L. E. Melchers, head of the Department of Botany and Plant Pathology, Kansas State College of Agriculture and Applied Science, for his kindness in allotting clerical and stenographic help for the revision of this text.

EDWIN C. MILLER.

KANSAS STATE COLLEGE OF AGRICULTURE AND APPLIED SCIENCE,

PREFACE TO THE FIRST EDITION

There has existed for some time a need for an advanced text in plant physiology. The various texts by European investigators and teachers, although summarizing the work that has been done on the Continent, have failed to cover adequately the contributions of American and English plant physiologists. These contributions during the past two decades have been outstanding and dominate in many cases the work along certain lines. A summary of these contributions should be available to students, teachers, and investigators in the field of plant physiology.

It has been the intention of the author to bridge this gap in the literature and to summarize in this book the more important findings of English, American, and Continental plant physiologists. The material has been assembled in such form as to be available as a text for upperclassmen and graduate students and at the same time to be sufficiently comprehensive as a reference book for investigators in plant physiology.

The work has been confined entirely to the physiology of the green plant, as it has been thought that the processes and phenomena of the fungi belong, at present at least, to the province of plant pathology and allied fields. The discussions and summaries of the various investigators have been used freely, and in many cases their phraseology has been altered but little. The author has tried, however, in each instance to indicate the source of the information presented. Some topics are discussed more thoroughly than others. This, however, is to be expected as an author naturally is more familiar with those subjects which are in the field of his research.

With but few exceptions the references cited at the end of each chapter have been mentioned or discussed in the text. It is considered that these references will be of value to those who desire to familiarize themselves more thoroughly with any of the topics than they are herein discussed. The questions following each chapter cover all the topics discussed in that portion of the text. It has been the experience of the author that these are valuable to the student as an aid in guiding him to a mastery of the subject matter.

This text is the outgrowth of twenty years' experience as a teacher of plant physiology in the Kansas State Agricultural College and as plant physiologist of the Kansas Agricultural Experiment Station. The material included herein has been presented to classes composed for the most part of juniors, seniors, and graduate students. The favorable manner in which the subject matter and its method of presentation have been received in the classroom has prompted the author to present this volume to the public.

I am indebted to many of my associates for assistance in the preparation of this text. Most of the drawings were made by Mr. S. Fred Prince, formerly biological artist at the Kansas State Agricultural College. I am indebted to Miss Nora E. Dalbey, associate professor of botany in the Kansas State Agricultural College, for certain drawings and for helpful suggestions in the arrangement of the illustrations. I wish to thank Dr. Frank M. Schertz of the U. S. Bureau of Chemistry and Soils, Dr. John W. Shive of the New Jersey Agricultural Experiment Station, Dr. W. E. Tottingham of the University of Wisconsin, Dr. F. F. Blackman of Botany School, Cambridge, England, and Professor W. E. Davis of the Kansas State Agricultural College for permission to use certain figures as indicated in the text.

I am grateful to Mr. J. C. Frazier, formerly of the Department of Botany and Plant Pathology, Kansas State Agricultural College, for helpful advice in the preparation of the manuscript and to Dr. R. P. White, formerly of the same department, for criticism and advice in the preparation of Chap. II. I wish to thank Dr. J. V. Cortelyou, professor of modern languages at Kansas State Agricultural College, and Miss Madge Wardell, instructor in French at the same institution, for checking respectively the German and French references. I wish also to acknowledge the most valuable aid and advice of Miss Nellie Jacobs, clerk in the Department of Botany and Plant Pathology, Kansas State Agricultural College, in the preparation of the manuscript.

I am especially indebted to the following of my colleagues in the Kansas State Agricultural College for their constructive criticism in the preparation of the text: S. C. Salmon, professor of farm crops; Dr. P. L. Gainey, professor of bacteriology; R. J. Barnett, professor of horticulture; Dr. H. H. Haymaker, professor of plant pathology; W. E. Davis, professor of plant physiology; Dr. F. C. Gates, professor of taxonomy and ecology; Dr. F. L. Duley, professor of soils; Dr. C. W. Colver, professor of organic chemistry; and Dr. E. L. Tague, associate professor of chemistry. The author alone, however, is responsible for the statements in the text.

EDWIN C. MILLER.

KANSAS STATE AGRICULTURAL COLLEGE,

CONTENTS

CHAPTER III

CHAPTER IV

CHAPTER V

CHAPTER VI

CHAPTER VII

CHAPTER VIII

CHAPTER IX

CHAPTER XI

CHAPTER XII

PAGE

CHAPTER XIII

PLANT PHYSIOLOGY

CHAPTER I

THE PLANT CELL

The cell is not only the structural unit but also the functional unit of the plant and animal body. Since plant physiology is the study of the functions of plants, a general knowledge of the morphological parts of the plant cell is necessary for a clear understanding of the various life activities of the plant. In this chapter the intention is to discuss the parts of the plant cell only in a general way, since a consideration of the more minute details belongs to the province of plant cytology. The parts of a living plant cell are generally classified as follows:

- Parts of the plant cell
  - Cell wall
    - Middle lamella
    - Primary wall
    - Secondary wall
  - Protoplasm
    - Cytoplasm
    - Nucleus
    - Plastids
      - Leucoplasts
      - Chloroplasts
      - Chromoplasts
    - Chondriosomes or mitochondria
  - Vacuole or cell sap

I. THE CELL WALL

A. Origin

The protoplasm of plant cells is with but few exceptions enclosed by a more or less solid membrane which has been secreted by it. This membrane, which is called the "cell wall," is relatively very thin when it is first formed but increases in thickness through the deposition by the protoplasm of new particles upon those already present. The wall may increase in thickness but little over its original measurements, or it may become so much thickened that it almost fills the cell cavity. Thus, for example, the amount of thickening in the cell wall of the flax fiber may amount to as much as 90 per cent of the area of the cell in cross section (Anderson, 1927). (See Fig. 2.) The cell wall in many cases is thickened irregularly, leaving thin places in it which are termed "pits," which facilitate the interchange of materials from one cell to another (Fig. 2p). The manner in which the original cell wall becomes thickened is little understood and may be different in different types of plant cells. Aldaba (1927) observed in the case of the development of the cell walls of the bast fibers of *Boehmeria* and *Linum* that the cell wall is formed by a series of hyaline membranes that gradually become transformed into cell-wall lamellae. The modified hyaline membranes resemble plasma membranes and react similarly towards stains and during plasmolysis. The unmodified hyaline membranes are in direct contact with the protoplasm, but the thickening and differentiating portions are separated from it by one or more subsequently formed layers. The mode of formation is suggestive of a process of transformations. It cannot be satisfactorily accounted for by the theories of apposition and intussusception.

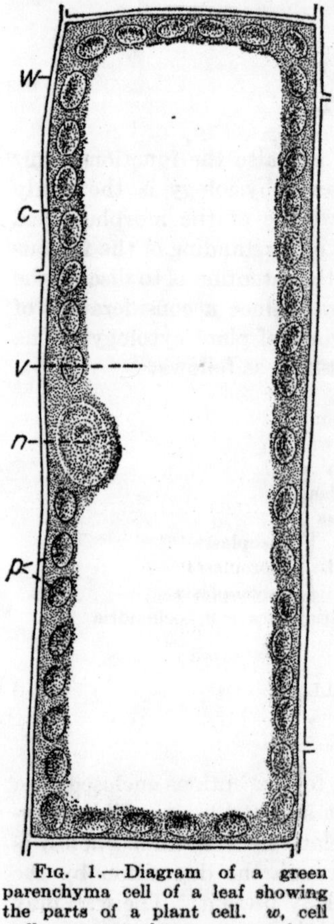

FIG. 1.—Diagram of a green parenchyma cell of a leaf showing the parts of a plant cell. *w*, cell wall. *c*, cytoplasm. *n*, nucleus. *p*, plastids. *v*, vacuole.

The formation of the cell wall of the cotton fiber has been studied intensely by Farr and Eckerson (1933, 1934) and Farr and Sisson (1934). They noted that, when a cotton fiber is only 5 to 6 days old, a thin layer of fibrils that give the cellulose reaction is closely affixed to the thin original wall. From the very beginning of the elongation of the fiber, there are present in the protoplasm ellipsoidal particles of cellulose covered with a thin layer of pectic substance. Toward the center of the cell these particles are separate or in short chains while in the extreme outer regions of the protoplasm next to the cell wall the chains are longer and in some instances have entered into the formation of fibrils. Regardless of the ultimate structure of these particles, they apparently represent standard units which build up the fibrils that compose the cell wall of the cotton fiber. This conclusion was reached because by various microscopic and submicroscopic procedures, including X-ray examinations, a relationship was definitely established between these more or less free cellulose particles in the cytoplasm and the compact cellulose membrane of the mature wall of the fiber. This behavior in cell-wall formation was found to be similar in various other plants and is considered to be of general occurrence.

Sponsler (1934) made observations of the dividing cells of *Rhizoclonium* in which the cross wall grows from the side walls toward the center of the cell. The growth rate of this cross cell wall is approximately 0.1μ per minute, and in the process monosaccharoses are apparently condensed into polysaccharose chains. During a 15-min. interval of growth there were deposited about 3,200 layers, or between 3 and 4 per second, each of the thickness of a monosaccharose unit. The bordering protoplasm evidently takes an active part in the wall formation, for a change in environment may practically stop the process.

Bonner (1935) believed that the young cell wall must be considered as living. According to him the growth of the cell wall cannot be regarded as a simple plastic stretching, because the alignment effect of the cell-wall micelles is partially or entirely lacking. The increase in the plasticity of the cell wall through the addition of growth substances can be traced back to a loosening up of the contacts of the cellulose micelles. He considered that growth may be a simultaneous action of stretching and stratification of cellulose micelles. Bonner and Heyn (1935) thought that the increase in cell-wall plasticity by growth substance might be due to the indirect effect of this substance upon the charges of the micelles of the cell wall. This might be brought about by certain ions entering the wall from the protoplasm thus increasing the micellar charges in the cell wall. These might cause a greater repulsion between the micelles and an increase of micellar hydration, thus increasing the plasticity of the wall. Investigations, however, indicated that growth substance had no effect upon the charges of the micelles.

That portion of the cell wall which is laid down during the early life of the cell is different in optical, chemical, and staining properties from that part of the wall which is later deposited. This central layer or plate is readily seen in those cell walls that have been greatly thickened, but it can be observed after proper treatment in nearly all cell-wall membranes, even when they are quite young. This central portion of the cell wall is called the "middle lamella" and is generally defined as the cell wall that was originally deposited by the protoplasm, although there is considerable uncertainty as to the length of the period during which this portion of the wall was formed. Allen (1901) studied the origin of the middle lamella in the cells of a considerable number of plants and concluded that the middle lamella consists of the layers first deposited by the protoplasm plus a certain amount of material, subsequently in contact with these layers, which is generally rich in pectic compounds as compared with the still later deposited strata.

The general terms "middle lamella," "primary wall" and "secondary wall" have been proposed to clarify the designation of the cell-wall parts. This terminology, however, is open to criticism (Kerr and Bailey, 1934).

Middle Lamella.—The amorphous isotropic material, largely, if not entirely, pectic compounds, first deposited by the cytoplasm.

Primary Wall or Cambial Wall.—The first anisotropic layer of the wall composed largely of cellulose and some pectic materials. Cells with walls of this type are capable of growth and extension. This wall has the capacity for undergoing reversible changes in thickness.

Secondary Wall.—The additional wall layers formed on the primary wall. Cells having secondary walls lose their potentiality for growth.

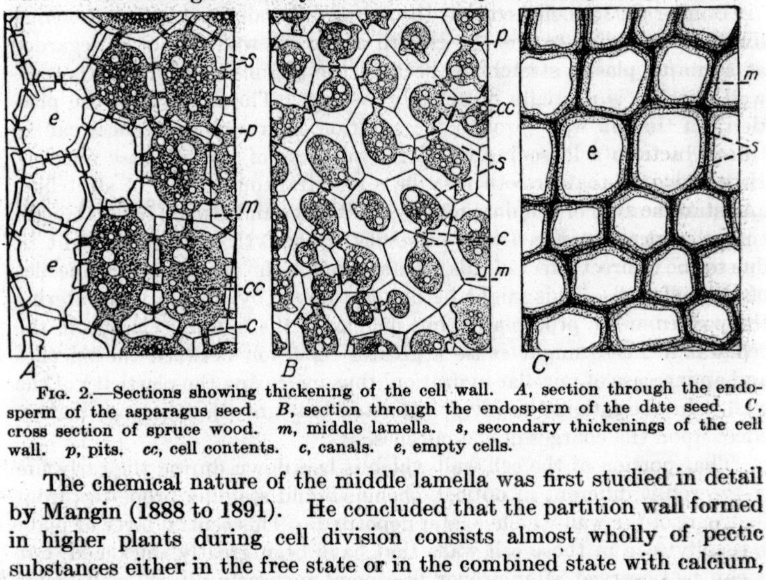

FIG. 2.—Sections showing thickening of the cell wall. *A*, section through the endosperm of the asparagus seed. *B*, section through the endosperm of the date seed. *C*, cross section of spruce wood. *m*, middle lamella. *s*, secondary thickenings of the cell wall. *p*, pits. *cc*, cell contents. *c*, canals. *e*, empty cells.

The chemical nature of the middle lamella was first studied in detail by Mangin (1888 to 1891). He concluded that the partition wall formed in higher plants during cell division consists almost wholly of pectic substances either in the free state or in the combined state with calcium, in which form it was considered the bulk of the middle lamella in mature cells.

Tupper-Carey and Priestley (1923) investigated the cell walls of the apical meristem of the stem and root and found the existence of cellulose in the walls of the meristem, although its presence is marked by association with other substances. Protein closely linked to it is considered to be the factor which prevents its detection with iodine and sulphuric acid. Pectin is present, but the middle lamella in the meristem is never of calcium pectate but is probably a mixture of pectin and protein. Fat is present in the cell walls of the meristem closely linked to cellulose and probably is responsible for the failure of chloriodide of zinc to give the cellulose reaction. Wood (1924, 1926), however, considered that there is not enough protein present in the cell wall to interfere with the cellulose-pectin reaction, since according to his tests not more than 0.001 per cent protein occurs in the cellulose cell walls of any of the plants examined.

Lüdtke (1931), in an examination of young and old winter wheat plants, found that a whole series of intermediate products is present in the younger cells from the simple saccharoses, as xylose and glucose, up to cellulose, thus indicating that the formation of this cell-wall constituent occurs in a series of transitional steps. Ritter (1925) considered that in woody tissue the middle lamella is not pectic in nature but is strongly lignified. Harlow (1927) observed in the young cortex unlignified pith and cambium region that the middle lamella is composed of pectin in one or more of its forms, while in woody tissue this layer is heavily lignified. There are two explanations for these observed changes in the older cell walls: (1) that the pectins are metamorphosed to form part of the lignin of the mature cell wall and (2) that the initial pectic structure may be completely covered with lignin substances which render it inert to the action of the usual pectic stains and solvents. A combination of these two processes might also occur. The evidence indicates that, at least in the softer tissues of plants, the middle lamella is composed, for the most part, of pectic substances that may become lignified in woody tissues. The middle lamella in most tissues dissolves readily under the action of potassium chlorate and nitric acid, which do not affect the later deposited portions of the cell wall. It resists the action of strong sulphuric and other mineral acids that cause the other portions of the cell wall to swell and finally dissolve.

B. General Physical and Chemical Properties

The chemical and physical structure of plant cell walls varies widely in different plant groups and in the same individuals. Thus the cell walls of root hairs, wood fibers, bast cells, cork cells, or epidermal cells have but few characteristics in common. All plant cell walls, however, have two general characteristics in common: (1) They are not chemically homogeneous but are composed of two or more chemically distinct substances. (2) They are not physically homogeneous but show stratification, striation, and other indications of structural variation.

1. Methods of Studying.—The present conception of the general structure of the cell wall is really a composite pattern built from the different insights that have been gained by the following methods (Scarth, Gibbs, and Spier, 1929; and Anderson, 1935):

a. Direct Observation of Untreated Walls.—This method reveals the presence of definite layers, surface markings, striations, radial lines, pits, and irregular thickenings.

b. Differential Staining of Walls.—The presence of the various compounds in the wall can be determined after this treatment. In addition some idea can be gained concerning the spatial distribution of the submicroscopic units in the wall (Czaja, 1930, 1934).

c. Differential Solubilities.—Extraction of the different compounds from the walls is accomplished by this method.

d. Ash Analysis.—Kind and amount of the various minerals in the walls are determined in this manner.

e. Use of the Spierer Lens.—Images produced by this lens have been supposed to indicate the presence of definite structural units of ultramicroscopic dimensions. It has been contended that the use of this lens provides further evidence for the micellar structure of the cellulose cell wall (Seifriz, 1931, 1934, and Thiessen, 1932). Seifriz (1936) considered that these microscopic units revealed in a Spierer picture are not the micelles the chemist has in mind, so he has termed these units "super micelles."

f. Measurement of Refractive Indexes.—The different lamellae of the wall have different refractive indexes so that by this method the prominence of certain portions of the wall is greatly emphasized.

g. Use of Fluorescent Light.—This detects the degree of lignification of the walls. They are irradiated with ultraviolet rays which on passing through lignified walls become visible due to fluorescence. The intensity of fluorescence varies directly with the degree of lignification.

h. Hydration.—By the use of suitable reagents the cell wall may be greatly swollen. In this swollen condition, structural relations appear that are invisible in the unswollen state. By this method Balls and Hancock (1919) and Balls (1923) were able to count the layers present in the cotton fiber.

i. X-ray Analysis.—This method has been used by numerous investigators and has probably contributed more to our knowledge of the ultimate structure of the cell wall than has any other single procedure. The actual size of the molecular aggregates in the wall, their distance apart, and their spatial relationships are shown.

j. Polarized Light.—This method is used to distinguish isotropic substances, those which show no double refraction, from anisotropic substances, which show this property. It also furnishes information regarding the spatial orientation of the submicroscopic units of the wall.

2. Theories of Cell-wall Structure.—In 1864 Nägeli suggested that the cell wall was composed of submicroscopic units, or micelles, arranged in definite layers. Each micelle was believed to be surrounded by a layer of water and thus separated from the adjacent micelles. The modern methods of the investigation have only served to emphasize its general accuracy and today it is probably more widely accepted than any other theory of cell-wall structure. The present theories have certain features in common and differ mainly in regard to the supposed size of the structural units concerned. These theories may be grouped under three headings:

a. The work of Frey (1926–1928), Frey-Wyssling (1930), and Astbury, Marwick, and Bernal (1932) with polarized light, and of Herzog (1925), Meyer and Mark (1928), Meyer (1929, 1931), and Clark (1930) with X-ray examinations indicates that the cell wall is composed of submicroscopic units or micelles of a crystalline nature. The crystalline micelles are separated from each other by a noncellulose hydrophilic colloidal material. The micelle is considered to be composed of a bundle of basic cellulose groups or molecules. The basic cellulose molecule appears to be a long chain of glucose residues (glucose minus a molecule of water). The length of the chain is variously estimated to be from 50 to 1,000 of these glucose residues (Meyer, 1929; Sponsler, 1929, 1930, 1933; Freudenberg, 1932; Seifriz, 1934; and Gibbs, 1935). These chainlike cellulose "molecules" are grouped in parallel rows into a bundle or micelle (Davidson and Richardson, 1936). The number of parallel chains present in a single micelle has been variously estimated to be from 40 to 60. Its length has been estimated to be from 15 to 100mμ and its diameter from 2.0 to 6.6mμ depending upon the source of the material. The size may vary in different plants, in different plant parts, or even in the same cell wall. The forces that hold the glucose residues together in long chains are the primary valence forces, while the forces holding the chains together in the micelle are those of molecular cohesion. In fiber cells the micelles are grouped into fibrils that collectively make up the cell wall. These fibrillar units may be parallel (Preston, 1934) or inclined to the long axis of the cell wall.

b. It was considered by Sponsler (1926–1933), and Sponsler and Dore (1936), that the evidence for the existence of micelles as definite entities in the cell wall is inadequate. They considered the wall to be composed of long parallel chains of glucose residues running lengthwise of the fiber. These chains are oriented in such a way as to form a spatial lattice of three dimensions. The cellulose wall is considered to be a complex system built up of very thin molecular layers each composed of long chains of glucose residues. Certain layers extend lengthwise of the wall, some extend across the wall, and other layers form definite angles with the lengthwise layers. A single fiber cell wall may be some 40,000 of these lattice units in thickness. The wall of a fiber cell like cotton consists then, according to Sponsler, of an enormous number of parallel molecular layers and not of a system of bricklike units as suggested by the micellar hypothesis.

c. According to Hess (1928), Lüdtke (1931), Farr and Clark (1932), Farr and Eckerson (1933, 1934), and Farr and Sisson (1934), the cell wall is composed of definite cellulose units microscopically visible, each unit being enclosed in a thin film of noncellulose material that serves to cement the units together. Farr and Eckerson (1933, 1934) considered this noncellulose constituent to be pectic material with its nature apparently varying in different walls. This theory of cell-wall structure finds support in the behavior of the walls during swelling, in dissolution by various reagents, and in the reactions to certain dyes. Certain cellulose-staining dyes frequently fail to impart color to cellulose walls except when these walls are subjected to mechanical pressure. It is presumed that this pressure ruptures the noncellulose constituent and permits the stain to reach the cellulose particles. Bailey and Kerr (1935), however, considered that there is no reliable evidence to indicate that the matrix of the cell wall is composed of entities of visible size bound together by a noncellulose material. They believed that such putative entities are actually heterogeneous fragments that are shredded or disrupted from an originally continuous and coherent matrix. If these are discontinuous in the structural pattern, they are confined to the submicroscopic field—to the realm of micelles and molecular chains.

C. Chemical Composition

The chemical nature of the cell wall is but little understood. The substances that compose it are complex, and their constitution is seldom known definitely. Furthermore, it is seldom known whether the cell wall is composed of a mixture of substances or whether these substances are chemically combined with each other. Any discussion of the chemical composition of the cell wall must, therefore, be rather meager and superficial. In general, however, it may be stated that the cell wall of the plant cell is composed of a mixture of celluloses, together with certain other compounds depending upon the nature of the cells in question. The following diagram shows the substances that may enter into the composition of the cell wall:

- Materials that may enter into the composition of the cell wall
  - Celluloses
    - Normal, typical, or true celluloses — α, β, γ types
    - Hydrocelluloses
    - Oxycelluloses
    - Compound celluloses—Lignocelluloses
    - Hemicelluloses or Pseudo-celluloses
      - Skeletal hemicelluloses — Pentosans (Xylan, Araban), Galactosans — In cell walls of wood and seed coats
      - Reserve hemicelluloses — Pentosans (Xylan, Araban), Mannosans, Galactosans — In cell walls of endosperm and young fibers of wood and bark
  - Suberin and cutin
  - Pectic substances
    - Pectic acid
    - Pectates
    - Pectose
    - Pectin
  - Other constituents — Resins, gums, tannins, minerals, coloring matters, proteins, fats, phospholipides, ethereal oils, and callose

The celluloses considered collectively may be defined as a group of relatively inert substances composed of polysaccharose units that constitute the major portion of the cell walls of plant cells.

1. Normal, Typical, or True Celluloses.—Normal cellulose is apparently the first cellulose formed by the protoplasm and is the one from which all the other celluloses are derived. It is a white hygroscopic substance that absorbs 6 to 12 per cent of water but loses it on being heated to 100°C. When heated with water under pressure to 260°C., it dissolves completely without disintegrating. Its general formula is $(C_6H_{10}O_5)_n$, so that it is chemically similar to but not identical with starch. Normal cellulose is insoluble in ordinary solvents but dissolves in ammoniacal cupric oxide (Schweitzer's reagent), or in zinc chloride in two times its weight of concentrated hydrochloric acid in the cold. It also goes into solution by heating 1 part of cellulose with 6 parts of zinc chloride in 10 parts of water. Concentrated sulphuric acid dissolves normal cellulose gradually, converting it into dextrin and then into glucose. If this solution as soon as made is diluted with water, a gelatinous precipitate is formed that gives a blue color with iodine. The same substance is formed by the action of chloriodide of zinc, which stains the compound at first a rose-red and later a violet color. When cellulose is acted upon with acetic anhydride and concentrated sulphuric acid, or is hydrolyzed under the influence of the enzyme cytase, which is secreted by bacteria and various fungi, it breaks down into cellobiose ($C_{12}H_{22}O_{11}$), an isomer of maltose. By further hydrolysis it is converted into glucose. Cellobiose stands in the same relation to cellulose as does maltose to starch. According to Haworth (1929) the sole constitutional difference between cellobiose and maltose is a difference in configuration, at the linkage between the glucose residues, similar to that which distinguishes the α- and β-stereoisomeric forms of methyl glucoside. Maltose is the alpha form, and cellobiose is the beta form.

Cotton fiber is the best example of almost pure normal cellulose occurring in nature. An analysis of American cotton fiber showed it to be composed of 91 per cent normal cellulose, 8 per cent water, and only 1 per cent of other materials, principally waxes, gums, minerals, and pectose derivatives. The cell walls of the younger plant cells are composed, for the most part, of normal cellulose or some of its derivatives. Wood (1924) studied the nature of the cellulose constituents of the cell wall of the cells of the young root tips of *Vicia faba* and *Helianthus annuus* as well as the parenchymatous tissues of the stems of these plants. He found that the cellulose constituents of the cell wall consist of normal cellulose, oxycellulose, and hydrocellulose, the proportion of each constituent varying with the age and kind of cell.

An illustration of the similarity of normal celluloses obtained from different sources was reported by Sanders and Cameron (1933). They found the cellulose of the cotton stalk and cusps to be the same type of cellulose found in cotton lint and spruce, pine, and poplar wood. However, the compounds termed "normal celluloses" are composed frequently of three different types or forms of cellulose, which have been designated as alpha, beta, and gamma celluloses. These differ from each other especially in regard to the action of alkalies and possess different degrees of dispersion, dehydration, and corresponding polymerization (Gibbs, 1935).

2. Hydrocelluloses and Oxycelluloses.—These celluloses are degradation products that are frequently found associated with normal celluloses in the cell wall. Hydrocellulose is apparently produced by partial hydrolysis and is a compound that possesses active carbonyl groups and shows a very strong absorptive power for different organic dyes. It may be prepared, according to Mehta (1925), by soaking cotton with as small a quantity as possible of cold 72 per cent sulphuric acid and treating almost immediately with 95 per cent alcohol. The hydrocellulose thus formed is washed free from acid and dried in the air. It has the formula $C_{12}H_{22}O_{11}$ but retains the fibrous structure of the original cellulose, can be rubbed into a powder, and stains blue with iodine. The formation of hydrocellulose is marked by an increase in reducing power, a fall in tensile strength and in viscosity in Schweitzer's reagent, and an increase in solubility in cold sodium hydroxide solution. The interrelations of these properties are always the same regardless of the acid treatments (Davidson and Richardson, 1936). The action of acid thus apparently is a hydrolysis of glucosidic linkages in the chain molecules, with the production of shorter chains and the liberation of reducing groups.

Oxycelluloses are mentioned by many authors to designate various compounds of an indefinite character which are produced by the action of various oxidizing agents on cellulose. The compounds of this type possess acidic properties, contain active carbonyl groups, and reduce Fehling's solution. The oxidation process always leads to a decrease of tensile strength, a fall in viscosity, and an increase in solubility showing that chain molecules are broken. Although in any given series of oxycelluloses these properties show definite relations to each other and to the chemical properties, the relations vary from one type of oxycellulose to another (Davidson and Richardson, 1936).

3. Compound Celluloses.—This is a general term that includes those compounds of the cell wall which consist of some form of cellulose combined or closely associated with a noncellulose constituent. The most common and widespread compound cellulose in the plant kingdom is lignocellulose. Lignocellulose is the chief constituent of wood, jute, and cereal grasses and straw. It is generally considered to consist of cellulose chemically combined with lignin, but some authorities are of the opinion that the lignin is simply an encrusting substance and consider its combination with the cellulose to be only a physical one.

The manner in which the lignin might be combined with the cellulose is a matter for speculation. Some have assumed that an esterlike union exists between an acidic group in the lignin and a hydroxyl of the carbohydrate, while others consider that an etherlike linkage exists between the lignin and the cellulose or other carbohydrates. It has also been shown that some of the lignin is combined in an acetal-like manner. It is very probable that all three of these combinations may occur, depending upon the kind of plant or the conditions under which the lignocellulose is formed (Phillips, 1934). It was also considered by Mehta (1925) that lignocellulose consists of a chemical combination of lignin with cellulose and related polysaccharoses as an aromatic glucoside. According to Candlin and Schryver (1928), and others, decarboxylation occurs when plant tissues lignify. There is also evidence that the formation of lignocellulose is an infiltration process. Fischer and Lieske (1928), Freudenberg and his workers (1931), and Harlow (1932) have noted that sections of wood still retain their structural features when treated with various reagents which remove either the cellulose or the lignin. According to Anderson (1935) the lignification of cellulose walls does not alter the X-ray patterns of cellulose. The process thus does not involve a modification of the cellulose units themselves but indicates that the lignin is deposited between the cellulose micelles, hence lignification involves a change in the intermicellar material and not in the micelles generally considered to compose the cellulose.

The lignocelluloses cannot be crystallized without decomposition and cannot be obtained in the form of true solutions, although colloidal solutions are possible. They are among the most inert plant compounds and probably have no role other than that of adding stiffness to the plant tissues. Lodged wheat stalks (Davidson and Phillips, 1930) and brash specimens of fir (Dadswell and Hawley, 1929) show a higher lignin content than the tougher ones. Lignification is not to be considered, however, as an indication of senescence, for Mehta (1925) found the lignin content of the buds of Oregon pine and spruce relatively high, and Norman (1933) noted that young cereal plants often contain appreciable amounts of this compound.

The location of lignocellulose in the cell wall is somewhat in dispute. Schellenberg (1896) seems to have been the first to note that the middle lamella can become lignified. Ritter (1925) considered that the greater percentage of lignocellulose in wood is located in the middle lamella of the cell wall. According to Harlow (1928) the secondary cell walls of soft wood are appreciably lignified, while those of the hardwoods with but few exceptions contain practically no lignin. The lignin of the secondary cell wall may be removed by mild chlorination without affecting to any great extent the lignin that comprises the middle lamella. In 1932 Harlow obtained additional evidence that most of the lignin in wood is located in the middle lamella, while Norman (1933) considered that at least 50 per cent of this compound is located therein. Kerr and Bailey (1933) considered that the middle lamella of lignified tissue is not composed entirely of lignin.

Lignocellulose is resolved completely into its constituents by a process of chlorination or by heating for 1 hr. with a 4 per cent sodium hydroxide solution at a pressure of 10 atmospheres. The lignin can then be precipitated by acidifying the alkaline liquid and isolated in the pure condition quantitatively by extraction with alcohol. Lignin as isolated from wood by the above method is a brown, amorphous, faintly acidic substance having an aromatic odor and melting at 170°C. It is insoluble in water at ordinary pressure but partially soluble under pressure and soluble in dilute alkalies and alcohol and is more readily attacked by oxidizing agents than cellulose. The lignins contain methyl, methoxyl, formyl, and acetyl groups and consequently have a higher carbon content than cellulose.

Lignin from different sources varies greatly in its methoxyl and acetyl contents. In barley (Phillips and Goss, 1935) the lignin from the young plants differs from that of mature plants in that the lignin from the former contains a much smaller percentage of methoxyl. As maturity is approached, the percentage of lignin increases and there is a rapid methylation of the hydroxyl groups. The general opinion is that lignin contains condensed, unsaturated aromatic nuclei perhaps with an aldehyde or ketone group and is substituted by hydroxyl, methyl, and possibly acetyl groups. Some consider lignin to be a polymerized and substituted ether. There is considerable discussion, however, whether lignin belongs to the aliphatic, aromatic, or heterocyclic series. It is a disputed question as to whether lignin contains a grouping related to coniferyl alcohol or aldehyde, although some investigators (Klason, 1910) support this hypothesis.

It should be kept in mind as Harlow (1927, 1928) has pointed out that the lignin of botanists may not be the same as the lignin of chemists. It is customary for the former to regard walls as "lignified" when they give color reactions with certain staining reagents, while the latter regard as lignin that material which is removable from the cell walls by certain chemical treatments. It has long been suspected that the color reactions are due to small amounts of substances of an aldehyde nature occurring with lignin but not necessarily a part of it.

The source of lignin is purely a matter of speculation. Lignified tissue contains lignin and hemicelluloses in relatively large amounts with only traces of pectin, while nonlignified tissues contain relatively large amounts of pectin, small amounts of hemicellulose and no lignin. It is thus considered by some that pectins are the substances from which lignins are formed. Phillips and Goss (1935) noted in the development of the barley plant that the percentage of the lignin, cellulose, and pentosans increased progressively with the age of the plant, but no indications were found that lignin increased at the expense of cellulose or pentosans. Smith (1935), on the other hand, considered that in the Kieffer pear the sugar that the leaves supply to the fruit is converted to pectins, to hemicelluloses, and finally to lignocellulose. There was a relative decrease in sugars during the lignification of this fruit. As shown by Crist and Batjer (1931), the first clear microscopic evidence of lignification appeared in 15 and 20 days after blooming in Bartlett and Kieffer pears, respectively. This lignification reached a maximum in from 3 to 5 weeks after its first appearance.

There is some evidence (Norman, 1933) of a relationship between lignins, certain resins, and the phenolic residues of tannins (Menon, 1935). The whole question of the formation of aromatic nuclei in nature, however, is a very perplexing one. The assumption is that they are not direct products of photosynthesis and must therefore be formed from pyran or furan rings of sugar.

There is at present no known commercial use of lignin, and it is estimated that 1,500,000 tons are wasted annually, mostly in the manufacture of paper pulp. In 1936 the production of a commercial compound known as "ralig" was announced. This compound is a 25 per cent solution of the lignin of hemlock, and it was stated that it was to be utilized in highway construction.

4. Hemicelluloses or Pseudocelluloses.—The term "hemicellulose" was first used by E. Schulze in 1892 to denote a group of substances in the cell wall which he considered very closely related to cellulose in chemical constitution and as intermediate substances in its development. The outstanding characteristics of this group of substances are their relative ease of hydrolysis by hot dilute mineral acids, under ordinary pressure, and their solubility in dilute alkali. They are commonly extracted from the cell wall by treatment with dilute alkali. This solution is then acidified and the hemicelluloses precipitated by the addition of ethyl alcohol. The properties of the hemicelluloses prepared after this manner are determined to a large extent by the concentration of the alkali used, by the conditions maintained during the extraction, and by the general technique used to precipitate and purify them (Burkhart, 1936). Schulze prepared these compounds from a large number of plant parts and obtained upon their hydrolysis the sugars, arabinose, xylose, galactose, and mannose. Most of his preparations contained more than one type of sugar unit. The term "hemicellulose" thus came to include a rather heterogeneous collection of substances in the cell wall, including galactosans, pentosans, and mannosans, which are the anhydrides of the abovementioned sugars. Waksman and Diehm (1931) suggested that hemicelluloses be defined as the noncellulose polysaccharose constituents of the cell wall.

From the work of numerous investigators including O'Dwyer (1923), Candlin and Schryver (1928), Norman (1929), Norris and Preece (1930), Preece (1930, 1931), and Anderson (1931), it has been found that many of the preparations of hemicellulose contain uronic acids in addition to the polysaccharoses that have been mentioned. The uronic acids that have thus been found are generally glucuronic and galacturonic, while mannuronic is less common. These acids possess a formaldehyde radical at

Investigators	Plant	Plant part	Compounds obtained by hydrolysis
Schulze (1893).....	Corn	Seed coat	Xylose and galactose
	Lupine	Seed coat	Pentose and galactose
Grüss (1897)......	Date	Seed	Mannose
Castoro (1906–1907)...........	*Ruscus aculeatus*	Seed	Mannose and arabinose
	Lupinus angustifolius	Seed coat	Galactose
	Lupinus albus	Seed coat	Galactose
	Cucurbita pepo	Seed coat	Xylose and galactose
Schulze and Pfenniger (1910).....	*Pisum sativum*	Pod	Galactose and arabinose
	Phaseolus vulgaris	Pod	Galactose and arabinose
Tottingham, Roberts, and Lepkovsky (1921)......	Apple	Twig	Xylose, glucose, and galactose
Cake and Bartlett (1922)..........	Asparagus	Seed	Mannose, glucose, fructose, and galactose
Schryver and Thomas (1923)..	Various starches	Starch grains	Glucose
O'Dwyer (1923)...	White oak	Wood	Xylose and arabinose
Link (1929).......	Corn	Seedling	Xylose and glucose
Norris and Preece (1930)...........	Wheat	Wheat bran	Uronic acid, glucose, and pentose
Preece, (1930).....	Corn	Cob	Arabinose, xylose, methylpentose, and uronic acid
Anderson and Kinsman (1931).....	Cotton	Seed hull	1-Xylose and 1-glycuronic acid
Widdowson (1932).	Apple	Developing fruit	Arabinose
Tottingham (1933).	Cherry, apple, peach, plum, pear	Wood	Glucose and xylose
Burkhart (1936)...	Alfalfa	Xylem and phloem of roots	*d*-Xylose, *d*-glucose, and *d*-glucuronic acid

one end of the molecule and a carboxyl group at the other. The carboxyl group has been derived from the oxidation of the primary alcohol radical of the sugar molecule. The uronic acids associated with the hemicelluloses are apparently all in a combined form, being linked to anhydrous sugar units. Such compounds may rightfully be termed polyuronides. The hemicelluloses thus comprise a wide variety of compounds with pure polysaccharoses related to starch and inulin at the one extreme and pure uronic complexes at the other. Thus by the hydrolysis of these cell-wall constituents there may be obtained pure monosaccharoses, a mixture of monosaccharoses, or a mixture of monosaccharoses and uronic acids. A general list of the hydrolytic products obtained from various plant parts by different investigators is shown in the table on page 13.

Of all the various hemicelluloses, the pentosans have received the most consideration owing primarily to their greater abundance and to the ease of their determination by the furfuraldehyde method. The hemicelluloses may make up the entire cell wall or they may simply form incrustations on the cellulose framework of the wall. The former may occur in the cell walls of the endosperm or embryo in certain seeds, while the latter prevail in the walls of wood and bark fibers. Norman and Shrinkhande (1935) and Sands and Nutter (1935) noted in the cell walls of wheat and barley straw, hay, and woods of oak, fir, and mesquite that a large proportion of the hemicellulose is combined or closely associated with lignin.

The components of the cell wall designated as "hemicelluloses" rank next to cellulose in quantitative importance. They are found in the seed coats of beans, peas, and other legumes. They occur in the cell walls of the seeds of coffee, soybean, nasturtium, white lupine, date, onion, asparagus, vegetable-ivory palm, and many others (Mitchell, 1930). They are also found in pods of legumes, stony fruits, shells of nuts, hay, straw, and fodder; to some extent in the walls of the xylem elements of wood: and in certain fibers of the bark. In many of the cases mentioned the hemicelluloses undoubtedly function as reserve carbohydrates that are transformed into sugars when the need arises. Thus in the seed of the date, vegetable-ivory palm, and others, the hemicelluloses are the only reserve carbohydrates present and are transformed into sugars that are utilized by the young seedling. Tottingham, Roberts, and Lepkovsky (1921) and Murneek (1929) emphasized the importance of reserve hemicelluloses in the fruiting branches of apple twigs as reserve carbohydrates. The latter investigator found in the new growth of the fruit-bearing spur of the apple that starch constitutes only 1 to 4 per cent of the total dry matter, while the hemicelluloses represent 17 to 22 per cent. Similar proportions of these two groups of substances were found also in the leaves, flowers, and fruits. Murneek found that hemicelluloses were formed in relatively large quantities in developing fruits and considered that they were later hydrolyzed to form some of the sugar of the ripening fruit. Widdowson (1932), however, noted in developing apples a constant increase of hemicelluloses, but could find no time when they decreased in amount. He thus considered that they do not act as a reserve carbohydrate supply for the fruit.

The origin of the hemicelluloses is not definitely known. Malhotra (1930) claimed to have found them in young cell walls and in the protoplasm. Norman (1933) stated that there is apparently a close generic relationship between pectins and hemicelluloses and on purely chemical grounds it is possible to account for their formation by the protracted mild oxidation of linked hexose units. Norman and Norris (1930) considered that there is some evidence that hemicelluloses may be formed by the mild oxidation of pectin. Buston (1935) considered that by the oxidation of the free primary alcoholic groups of the hexoses, after the condensation of the sugars themselves to hexosans, there would be formed uronic anhydride units. By the decarboxylation of these units, pentosans would be formed, and stereochemical considerations show that by such a process arabans would arise from galactosans, while glucosans would pass through glucuronic anhydrides to xylans. It is evident that from a parent hexosan, even if based on a single sugar, a very wide variety of intermediate products could be formed, all of which might be classed as hemicelluloses. Sands and Gary (1933) showed there is no relationship between the hemicelluloses and gum of the mesquite that would indicate that one was the precursor of the other.

5. Suberin and Cutin.—These terms are used to designate those substances which render certain cell walls of the plant more or less impermeable to water. These two substances have frequently been called "adipo-" or "cutocelluloses" and were classified originally as compound celluloses, since it was assumed that they consisted of a cellulose base chemically combined or closely associated with compounds of a fatty or waxlike nature. The work of Gilson (1890), Van Wisselingh (1888 to 1892), von Schmidt (1903 to 1910), and Priestley and his coworkers (1921 to 1925), however, seems to show conclusively that suberin and cutin are not combined with a cellulose nucleus but are a collection of distinct compounds, although in some cases they may infiltrate a cellulose matrix.

The general properties of suberin and cutin may be stated as insolubility in and impermeability to water, considerable insolubility in fatty solvents, great resistance to concentrated sulphuric acid, but ready oxidizability by nitric or chromic acid and ready solubility in warm alkali. They are stained by fat stains such as Sudan III or scarlet red.

a. Suberin.—This is the substance that renders the cell walls of cork cells impervious to water, and on chemical grounds it may be regarded as an aggregate of variously modified forms (condensation products or anhydrides) of the suberogenic acids of which phellonic, suberic, and phloionic are representatives. Phellonic acid gives color reactions with iodine reagents, and this fact may account for the impression that cellulose is present in the suberin lamella. According to Priestley (1921) and Rhodes (1925), the distribution of suberin in the wall of the cork cell is as follows: Within each cell there is formed an inner layer which possesses fat-storing properties and which is called the "suberin lamella." Similarly, fatty substances impregnate the middle lamella of the cell wall. The continuous suberin lamella within the cell is laid down simultaneously in all parts of the cell and is undoubtedly formed from the contents of the cell itself which have accumulated at the surface of the protoplast. Between this suberin lamella and the protoplasm is located a layer of cellulose impregnated to some extent with fat but quite recognizable when all fat bodies are removed. The fatty substances in the walls of cork cells are thus not in a continuous sheet but are for the most part concentrated within a series of isolated lamellae lining individual cells and to some extent diffused throughout a carbohydrate basis of cellulose or of pectates. Rhodes (1925) found that neutral fat solvents remove only a small portion of the fatty substances of the cork cells of the potato and that of the fats so extracted only a small portion were glycerides. He removed the remainder of the fatty substances from the residual material by saponification with 3 per cent alcoholic sodium hydroxide. The fatty material thus removed is regarded as consisting largely of the salts of various organic acids, many of which have the characteristics of hydroxy-fatty acids. It seems evident that suberin owes its peculiar properties to the presence of normal and hydroxy-fatty acids in a state of combination and condensation and that the hydroxy acids are derived from the normal acids by some process of oxidation. Rhodes found that the chloroform-soluble material in the cork cells of the potato remained constant as the thickness of the cork cells increased, and he considered that this chloroform-soluble material is responsible for the staining of the cork tissue with Sudan III.

The formation of suberin in the healing of the cut or injured surfaces of the potato, sweet potatoes, bulbs, sugar beets, and various other plant parts, has received the attention of numerous investigators. In the white potato Priestley and Woffenden (1923), and Artschwager (1927) found that the first process in the healing of the cut tuber is the deposition of a fatty "suberin" layer formed by the oxidation and condensation of fatty substances that remain as a residue when the exuded sap dries at the cut surface. A supply of oxygen and a proper relative humidity of the air are necessary for the formation of this layer. Under suitable conditions this outer layer of suberin is formed within 24 to 48 hr. If the humidity is too low, the cut surface will dry and crack while the tuber will shrink from the loss of water. After a period of time varying with the prevailing temperature, relative humidity, solar radiation, and variety of tuber, a periderm is developed at varying depths beneath the cut surface. Peacock (1931) observed that skinned potatoes which were exposed during the early morning or late afternoon were not affected in their normal healing, while those exposed during the middle of the day were hindered in this process. When the injured portion of the tuber was placed on freshly stirred soil, a normal periderm developed regardless of the time of day. The relative humidity, light relations, and temperature under this condition are apparently optimum for the formation of periderm. Under optimum conditions of humidity Artschwager (1927) noted that the periderm cells appeared about the tenth day at 7°C., while at 21°C. they appeared after 2 days. Smith (1933) noted that in wounded potatoes suberization made its appearance on the second day in tubers stored at 68°F. but at 46°F. was not detected until the sixth day.

The sweet potato and the corms of gladiolus respond to wounding by a process of surface suberization and a subsequent formation of wound periderm in a manner similar to that of the potato tuber. Artschwager and Starrett (1931) observed in the sweet potato that temperatures of 22 to 35°C. and a relative humidity of 90 per cent or above were the best conditions for the formation of periderm. According to Weimer and Harter (1921) the production of a cork layer in a wound of the sweet potato is preceded by the formation of a layer of starch-free cells usually 3 to 10 cells deep, beneath the injured surface. Cross walls begin to appear in these cells from the second to the third day, and by the fourth to the sixth day a distinct layer of cork cells has formed a covering over the wound. No well-developed cork layer was produced over wounds under conditions existing in the storage house. Artschwager and Starrett (1933) noted that sugar beets respond to wounding by a suberization of the cut surface but show only a belated development of shallow wound periderm. Woodhead (1934) found that wounded plants of *Kleinia articulata* showed in their healing the same general sequence of events that occur in wounded potato tubers. Little suberization of the phelloderm layer occurs, however, and some of the phelloderm layers become collenchymatous. This formation of collenchyma is associated with the deposition of calcium oxalate crystals. Bloch (1935) noticed that the wound reaction in *Tradescantia fluminensis* varies according to the location of the injury so that the formation of suberized layers occurs only in certain regions of the plant.

b. Cutin.—Cutin is a substance that is more or less impervious to water and is found in a continuous external layer or lamella on the outer cellulose wall of the epidermal cells of leaves and stems. Lee and Priestley (1924) have shown that in some cases there may be a cellulose cutinized lamella impregnated with fatty deposits beneath this external layer of cutin or cuticle. The structure of the outer epidermal wall of the leaves of *Clivia nobilis* has been carefully studied by Anderson (1928, 1934). From the protoplasm outward the epidermal cell wall presents the following layers: (1) cellulose and pectic materials, (2) pectic substances, (3) cellulose and pectic material infiltrated with cutin, (4) cellulose infiltrated with cutin, and (5) an outer layer of pure cutin. This last layer is more resistant to saponification than other cutinized areas of the wall.

The thickness of the cuticle varies greatly for different plants and for different plant parts. Stevens (1932) noted in his study of 33 varieties of cranberries grown in Massachusetts during 1929 that the thickness of the cuticle on the fruits ranged from 9.9 to 13.7μ. This thickness was greater than that of the two following years and was found not to be correlated with the size or keeping quality of the fruit.

The cuticle of the leaves of many plants permits the loss of a considerable amount of water, and Crafts (1933) noted that acids and various plant poisons can readily penetrate it. The rate of penetration of these substances generally varies inversely as the thickness of the cuticle, although its composition may enter into the problem. Lee and Priestley (1924) considered that the cuticle is thickened by fatty substances migrating along the walls from interior tissues and that the mobility of the fatty substances in the walls may be modified by external conditions, which thus affect the thickness of the developing cuticle. Thoday (1933) believed that the formation of the cuticle depends primarily on the distinctive physiological activity of the epidermal cells and that it is not the relatively simple process of condensation of fatty materials. He fortified his opinion with the observation of Damm (1901) who noted that where the stem of *Viscum album* was cracked under the strain of growth a cuticle was developed by the outer cells.

Lee (1925) made a detailed macrochemical study of the cuticle of the petals of chrysanthemum and rose and of the petioles of rhubarb leaves. From her work it appears that cutin is a complex mixture of fatty substances, consisting of free fatty acids and condensation products of fatty acids, fatty acid combined with alcohol, soaps, unsaponifiable material, and resinous substances. A large amount of the fatty acids present are of the hydroxy-fatty type. These hydroxy-fatty acids are less soluble in chloroform than the normal fatty acids and are extracted from the cuticle only after a lengthy saponification. In the petals of chrysanthemum the chloroform extract of the cuticle contains 45 per cent of normal fatty acids and 16 per cent of hydroxy-fatty acids, while the saponified extract contains 5 per cent of the normal fatty acids and 80 per cent of hydroxy-fatty acids. Lee considered that the preponderance of the hydroxy-fatty acids is the result of oxidation processes taking place during the deposition of the cuticle. Since no glycerol has been identified, it seems that the fatty constituents of the cuticle have the characteristics of waxes rather than fats and have as their base monohydric alcohols rather than the trihydric alcohol glycerin. Legg and Wheeler (1925) extracted the cuticle of *Agave americana* and obtained four classes of compounds with the solvents used: (1) water-soluble material; (2) waxy material soluble in alcohol, benzene, or chloroform; (3) cellulose soluble in cuprammonia; and (4) the residue insoluble in the solvents used. This residue was considered as the characteristic constituent of the cuticle and was termed "cutin" by these investigators. More than 90 per cent of this cutin dissolved in alcoholic potash. The residue consisted for the most part of two semiliquid organic acids not hitherto reported as occurring in nature. One of these was termed "cutic acid" ($C_{26}H_{50}O_6$), and the other "cutinic acid" ($C_{13}H_{22}O_3$).

These investigators in 1929 reported that the cutin from *Agave rigida* on oxidation with dilute nitric acid gives suberic acid, a mixture of azelaic and sebacic acids, an acid ($C_{11}H_{20}O_3$) insoluble in water but soluble in cold alkali, and an acid ($C_{22}H_{12}O_3$) insoluble in both cold water and alkali. Chibnall (1931) and his workers found in the wax of apple peel that the amount of fatty acid obtained on saponification of the original crude wax was very small. The fatty acids found may be present as wax esters with long-chain alcohols or be derived from the residual amount of glycerides present in the crude wax. The major portion of the alcohols present is wax-free. The main general distinction between cutin and suberin that is now definitely known is the absence from the cutin of phellonic acid and glycerin.

6. Pectic Substances.—There are present in the cell wall—especially in the parenchymatous tissue of fruit, as pears, apples, currants, leaf stem of rhubarb, and fleshy roots, as carrots, beets, and turnips, as well as many others—a group of compounds that have the property of forming gels under certain conditions. These substances play an important part in the jam- and fruit-preserving processes, and it is to their peculiar properties that the "setting" of jams and jellies is due. It was formerly supposed that these substances consisted of a chemical combination of cellulose and certain mucilaginous compounds, so that they were classified as compound celluloses under the name of "pectocelluloses." It is now fairly well established, however, that these substances are not chemically combined with cellulose, although they are closely associated with it, so that they are classified under the general name of "pectic compounds" or "pectins."

These pectic substances are relatively unstable in contrast to the other constituents of the cell wall and their linkages are easily ruptured by both acids and alkalies. The formation of pectic substances in plants appears to occur when metabolism and growth are at their maxima. As growth slows and as maturity is reached the production of pectic materials decreases and those present in the wall are slowly converted into other substances. According to Anderson (1935) the pectic compounds at times apparently permeate the intermicellar spaces of the cellulose portions of the wall. Thus the removal of cellulose with Schweitzer's reagent leaves the cell wall in many cases unchanged in thickness indicating that the pectic compounds permeated the cellulose structure. Sometimes the removal of pectic materials brings about a striking lamellation of the wall, which suggests that definite lamellae composed mostly of pectic materials are present between the cellulose layers. All pectic compounds stain red with ruthenium red. With methylene blue they stain violet while the other cell-wall constituents stain blue or green. The membranes composed of pectic substances do not exhibit double refraction—they are isotropic—while cellulose membranes exhibit double refraction or are anisotropic. The pectic compounds are of variable composition but their principal components are galacturonic acid, galactose, arabinose, and methyl alcohol with the galacturonic acid forming in general from 65 to 85 per cent of the whole.

It has been considered that pectic substances, hemicelluloses, and lignin are interrelated and that the passage from pectic substances to lignin takes place through an intervening formation of certain hemicelluloses. No scheme, however, has been proposed concerning the method of this transformation. The common possession of a high methoxyl content by both soluble pectin and insoluble lignin has been regarded as evidence of their interrelation. There is apparently a close relationship between the pectic substances and the hemicelluloses. The pectins represent a type of highly oxidized hexosan witu but little of the unchanged parent hexose and a small amount of the ultimate pentose constituent. The pectic substances are thus mainly distinguished from the hemicelluloses by containing approximately 70 per cent of uronic anhydride, while in the hemicelluloses the amount rarely exceeds 25 per cent. Support has been found for the suggestion that the pectins are the precursors of the hemicelluloses from the fact that the former predominate in young fleshy or unlignified tissues, while in the older lignified tissues the relative amount is reversed.

The literature concerning the pectic substances is voluminous and contradictory. Over no other constituent of the cell wall is there such lack of agreement among investigators in regard to terminology as over this aggregation of pectic compounds, so that it is impossible to summarize the literature on the pectic substances with any degree of satisfaction. The different terms applied to these compounds have been tabulated by Ahmann and Hooker (1925). The lack of agreement of the various investigators on this subject can probably be traced to two causes. In the first place, the term "pectins" includes, no doubt, a large aggregation of compounds rather than a few substances, so that the reactions obtained by one individual may be due to one substance and those by another individual to another substance. In the second place, the existence of many pectic compounds, which have been described by numerous investigators, must be regarded as uncertain. These pectic compounds are undoubtedly of a complex colloidal nature, and changes of physical properties and state of aggregation may occur without any definite change of chemical composition. On that account, solubility, the loss of solubility, or gelatinization—the tests that have so frequently been applied as determiners of pectic compounds—cannot be regarded as evidence in themselves of chemical change.

The real foundation of our knowledge of the pectic substances was laid by the researches of Fremy (1840 to 1848), and his general classification of these compounds is the one that has been in general use until the present time.

A study of the general properties of the pectic substances indicates that there are three distinct types of compounds as shown in the following outline:

Pectic substances	Pectic acid		
	Salts of pectic acid or pectates	Soluble in water	Potassium pectate Sodium pectate
		Insoluble in water	Calcium pectate Magnesium pectate
	Esters of pectic acid	Insoluble in water	Pectose
		Soluble in water	Pectin

a. Pectic Acid and Pectates.—The key number of the group is pectic acid from which all the other pectic compounds are derived (Norman, 1933). Pectic acid from different sources contains galacturonic acid, galactose, and arabinose in different proportions. According to one formula, 4 molecules of galacturonic acid are linked to 1 molecule each of galactose and arabinose, the whole forming a closed, six-membered ring. In some material its composition seems to be 1 arabinose molecule to 4 of galacturonic acid. Pectic acid is only slightly soluble in water even in the presence of a considerable amount of electrolytes. It forms soluble pectates with alkalies and insoluble ones with metals of the alkaline earth, of which calcium pectate is the most widely distributed. The pectic materials in the middle lamella of the cell wall are for the most part calcium pectates. The relatively large amount of pectates in some tissues indicates that they may occur in other parts of the cell than the cell wall (Norman, 1929). Pectic acid is found especially in overripe and decaying fruits. It may be obtained by the saponification of pectin with dilute alkali, pectinic acids, with decreasing methyl alcohol content, occurring as intermediate or transition compounds in this reaction. Pectic acid and its salts possess the property of forming gels in the presence of calcium. Conrad (1926) has described a method for the simultaneous removal and quantitative separation and determination of protopectin and pectic acid and its salts. Protopectin or its hydrolytic product, pectin, was found in all the tissues studied, but pectic acid or pectates were found only in radishes. He considered that the many cases of pectates reported in plant tissue are probably the results of enzyme action during the preparation of the material for analysis.

b. Pectose.—The term "pectose" has been generally used since the time of Fremy to denote an insoluble pectic substance that occurs in the cell walls of green fruits and other immature plant parts and is considered to be the parent substance of pectin. The terms "protopectin" and "pectinogen" have been used by von Fellenberg (1918) and Clayson, Norris, and Schryver (1921), respectively, to denote pectose. It has been claimed by the majority of investigators that during the ripening of the fruit, pectose decomposes into a soluble compound generally known as "pectin," which consequently appears in increasing amounts in the expressed juice as the fruit matures.

Protopectin is considered by some as a glucosidelike pectin-cellulose or pectin-hemicellulose compound, although it has not been isolated, and there is no substantial evidence as to its existence. The extraction of protopectin-containing materials with boiling water or dilute acid yields pectin. Protopectin has been defined as the water-insoluble, acid-soluble, and acid-hydrolyzable parent substance of the pectins.

Tutin (1923) concluded from his investigations with the apple that pectose or protopectin does not exist and that pectin is present in the green fruits as well as the ripe ones, but it has not previously been extracted, since the tissues are extremely hard to reduce to a fine powder and also because certain insoluble compounds by their presence prevent the pectin from going through the filter. He claimed that he could with certain precautions obtain pectin from the green fruit. Tutin's work has been severely criticised by Carre (1925) who introduced evidence to show that his conclusions are in error.

c. Pectin.—The compound or aggregate of compounds termed "pectin" has the following characteristics: It is soluble in water and may be precipitated from this solution by alcohol. It cannot be crystallized, and its only known method of purification is treatment in aqueous solution with animal charcoal and reprecipitation with alcohol. The chemical composition of pectin has not been determined with any certainty. von Fellenberg (1918) established the fact that pectin is the methyl ester of pectic acid. Tutin (1921) considered pectin to be the dimethyl isoprophenyl ester of pectic acid. Schryver and Haynes (1916) considered from the analysis of pectin from turnip, apple, strawberry, and rhubarb stems that it has the formula $C_{17}H_{24}O_{16}$ with strong evidence that it contains one pentose group.

Pectin constitutes a great part of the pectic material in the cell wall and is located for the most part in the primary and secondary portions of the wall. According to Norman (1933) it may also occur in the cell sap. Pectin is the only one of the pectic substances that is of commercial value. The commercial uses of pectin are based on three of its characteristics—its jellying properties, its colloidal stabilizing properties, and its imbibing power for water. Under suitable conditions pectin has the property of giving a viscous solution when boiled with sugar and dilute acid. Upon cooling, this solution sets to a firm gel. The process of jelly or jam making involves boiling to concentrate the pectin and fruit acids present so that with added sugar a satisfactory jelly will be obtained. Prolonged or over-heating of the solution destroys the power of the pectin to gel. It is believed that the degree of esterification of the pectin is intimately connected with jelly formation (Norman, 1933). The general investigations concerning the practical application of the pectin-sugar gel formation are difficult of interpretation because the substances used have been neither ash-free nor well-defined salts.

Fremy also discovered the action of the enzyme pectase on pectin. This enzyme causes a solution of pectin to set to a stiff gel and this change occurs more readily in the presence of chalk. The production of the gel has been considered to be due to the formation of calcium pectate. Tutin (1921) found that the action of cold dilute alkali and of pectase on pectin are essentially identical. In both cases a salt of pectic acid is formed and methyl alcohol and acetone are eliminated. Kopaczewski (1925) found that when pectin from carrots and pectase from alfalfa leaves which had been purified by long dialysis were mixed, coagulation of the pectin no longer occurs. Coagulation, however, occurs upon the addition of salts of calcium, barium, strontium, magnesium, iron, or copper to the mixture. All these salts show power to coagulate pectin in the absence of pectase. According to this author, a neutral medium gives the best results from the action of pectase. He concluded that any purely chemical explanation of the coagulation of pectin is at present impossible.

Considerable work has been done on the transformation of the pectic substances in fruits during their ripening and storage. Haller (1929) stated that the softening of apples on the tree as maturity approaches is to some extent associated with a decrease in the percentage of protopectin and a corresponding decrease in total pectic substances. The amount of pectin, however, remains constant. The softening of apples in maturing cannot be accounted for entirely by the changes in the pectic constituents. In storage the softening of apples is apparently due to the conversion of pectose into pectin. In the ripening of grapes, Hopkins and Gourley (1930) found that the total amount of pectic materials falls off in a regular manner and that there is a conversion of the insoluble pectic materials to the soluble form. The pears stored at different temperatures the development of soluble pectin is the chief factor concerned in their softening (Emmett, 1929). The normal ripening processes in citrus fruits, according to Gaddum (1934), are accompanied by a gradual transition from the insoluble protopectins to the soluble pectins with a consequent decrease in cell-wall rigidity. This transition terminates finally in demethylated pectic acids so that overmature fruits contain no pectins.

In the ripening of bananas, peaches, strawberries, and dewberries, Conrad (1930) showed that pectin increases at the expense of pectose. Concurrently with this behavior there is liberated an unidentified furfural-yielding substance, soluble in 70 per cent alcohol, which increases progressively with the pectin. This substance is also liberated when material containing pectose is heated with dilute acid. Gerritz (1935) found that the insoluble pectic substances in June-dropped apples could be converted to pectin of good jellying qualities by the treatment of this immature fruit with 0.5 per cent hydrochloric, sulphuric, or tartaric acid.

7. Other Constituents of the Cell Wall.—The cell walls of the plant cell have infiltrations more or less of resins, gums, tannins, callose, fats, oils, coloring matters, ethereal oils, proteins, phospholipides, and the various inorganic salts of which potassium, sodium, silicon, and calcium are the more prominent. The kind and amount of these substances will depend upon the kind of plant and the conditions under which it was grown. The presence of phospholipides in the cell wall has been especially emphasized by Hansteen-Cranner (1914 to 1926). He considered that the plasma membrane bordering the cell wall is largely composed of these substances and that they are in direct and intimate connection with the phospholipides of the cell wall. The plasma membrane is thus considered not sharply delimited from the cell wall but is continuous with the phospholipides that saturate it. This, however, has been little considered by workers on the subject of plasmolysis, absorption, and general cell response. The occurrence of phospholipides in plants will be discussed in detail in Chap. X.

Uber and Goodspeed (1935) reported on preliminary methods for microincineration studies to determine the localization of the inorganic substances in the cell wall.

Callose is sometimes found in the cell walls of root hairs, in that portion of the wall bordering sieve pits, and in calcified cell layers as in cystoliths. The chemical nature of callose is not known, but its staining properties and its solubility distinguish it from other cell-wall substances. It dissolves in a solution of hydroxide, in calcium chloride, and in stannous chloride but will not dissolve in Schweitzer's reagent or in ammoniacal oxalate. It stains blue with a dilute solution of resorcin blue or aniline blue and takes a brilliant red stain in rosolic acid.

The gums are not true polysaccharoses but consist of a number of sugar molecules united to a central nucleus of a sugar-acid nature. The attachment seems to be of an ethereal or glucosidic type since the gums retain acidic properties and are capable of forming salts. They are usually found as salts of calcium, magnesium, and potassium. The classification of a substance as a gum is as yet mainly an arbitrary one and is not based on any constitutional knowledge but is decided by its origin and physical behavior together with the fact that it is of a carbohydrate nature. Judged from the products of hydrolysis, there is apparently little or no essential difference in structure between gums and hemicelluloses (Norman, 1929; Bailey and Norris, 1932).

8. Commercial Uses of Cell Walls.—The most important commercial component of the cell wall is cellulose, and it constitutes one of the most important raw materials of industry. The finished products of cellulose enter into the daily life and routine of every individual. The sources of cellulose may be classified according to the parts of the plant from which it is derived (Skinner, 1932).

a. Fibers of Seeds or Fruits.—The fibers are borne on seeds or on the inner walls of the fruit. Cotton is the principal and by far the most important of this class, which also includes the silk cottons.

b. Fibers from Stems Other than Trees.—In some cases these are from the inner bark while in certain other cases they are obtained from the sclerenchyma of the rind and that surrounding the bundles. In this group are found flax, hemp, jute, ramie, the cereal straws, esparto grass, and the stalks of corn and sugar cane.

c. Fibers from Leaves.—Examples of these are Manila hemp, sunn hemp, and sisal hemp.

d. The Wood of Trees.—From this source the useful elements vary greatly according to the species used.

The materials from these various sources are used in the manufacture of textiles, paper, cordage, and for the production of pure cellulose, from which are produced rayon, cellophane, and explosives.

II. THE PROTOPLASM

The protoplasm is the name given to that aggregate of matter in the cell which displays those series of phenomena that we term "life." Its perceptible characters show that it has a viscous slimy consistency, that it is neither a true solid nor a true liquid, and that it possesses considerable power of cohesion. It is heavier than water and shows no tendency to mix with the same when placed therein. Its viscosity ranges from a degree slightly more than that of water to the firmness of a fairly rigid gel (Seifriz, 1920). Although a certain degree of viscosity may characterize the protoplasm as a whole, it is always divided into regions more or less definite, *e.g.*, the plasma membrane, nucleus, or plastids which differ in viscosity from its general mass (Beams and King, 1935). Its general appearance seems to support the view that protoplasm is essentially a colloidal solution of the emulsion type. It consists of a water phase containing dissolved minerals, a fat phase, a protein phase, and numerous minor phases, so that protoplasm may be considered as a polyphase colloidal system. The colloidal nature of protoplasm is manifested in its power of adsorption, in its staining properties, and in the changes in its physical state under varied external conditions.

Protoplasm as viewed through the microscope has the appearance of an emulsion (Seifriz, 1935). It has also been considered an emulsion because in certain cases it behaves like an emulsion. Thus an electric current will cause an emulsion to become more permeable to certain ions, and electric currents have the same effect on living tissue (Dixon and Clark, 1932). This observation, however, may only mean that two rather diverse types of systems—an emulsion and a living jelly—show similar responses to the same environmental changes and not that the one is the same type of colloidal mixture as the other. Protoplasm is characteristically elastic, and pure emulsions are not. Protoplasm coagulates and emulsions do not. Scarth (1927) says that the impression of fluidity in protoplasm is illusory. According to Seifriz (1935) the properties of protoplasm that force us to recognize it to be essentially a jelly or a lyophilic colloidal system, whether firm or fluid, are elasticity, rigidity, extensibility, imbibition, water immiscibility, thixotropy, synaeresis, and coagulation. Many of these properties are greatly accelerated in protoplasm by the addition of alkalies and are reduced or often wholly eliminated by the addition of certain acids. This effect is produced regardless of whether these substances are added separately or whether they originate at the cathode or anode of an electric current (Seifriz, 1935). The fact that protoplasm appears to the naked eye or under the microscope as an emulsion is thus not proof that its ultimate arrangement is emulsive.

Concerning the chemical constitution of living protoplasm, but little is known. Strictly speaking, living protoplasms cannot be analyzed, since under any known method of chemical analysis the protoplasm is killed and dead matter only is being analyzed. The chemical analysis of protoplasm shows that it consists of from 80 to 90 per cent of water. The residue that remains after driving off the water consists of both inorganic and organic compounds. The inorganic substances make up from 5 to 7 per cent of the dry weight and consist chiefly of the chlorides, phosphates, sulphates, and carbonates of magnesium, potassium, sodium, calcium, and iron. The organic constituents are chiefly protein, carbohydrates, and fatty substances. The nitrogenous compounds make up from 40 to 50 per cent, and the carbohydrates and fats each from 12 to 14 per cent, of the dry weight of the protoplasm. The principal component part of protoplasm thus consists of proteins and of fats and phospholipides collectively termed "lipoids" (Lepeschkin, 1928). The carbohydrates and decomposition products of protein substances are considered only as admixtures because in some cases they are absent. The various salts, however, are always present, although in some cases in very small amounts. The compounds that enter into the composition of protoplasm are thus no different from those found elsewhere, which are always considered as nonliving material. Since protoplasm does not lose weight when it is killed, it is inferred that no elements or compounds escape from it at death and that it is the manner in which these various compounds are combined that gives the protoplasm the outstanding characteristics which distinguish it from dead matter. Death of living matter seems to be due to disorganization.

The most outstanding characteristic of protoplasm is its power of organization. Seifriz (1935) states that, although this characteristic cannot be physically interpreted, structure is the most fundamental requisite for its manifestation, and this structure must be both continuous and exceedingly labile. The protoplasm is able to carry on many different processes simultaneously without one interfering with another. To account for this it must be assumed that within the protoplasm there are delicate membranes consisting of nothing more than firm protoplasm traversing it in all directions. The fact that protoplasm does not mix with water implies structural continuity. On account of its colloidal nature, protoplasm presents a multitude of surfaces, and it is at these surfaces that the numerous chemical reactions occur.

It is of interest at this point to note the age that the living cells of plants may attain. MacDougal (1926), MacDougal and Long (1927), MacDougal and Smith (1928), MacDougal and Brown (1928), and Faul (1928) have made observations along this line. Living cells were observed in the medulla and cortex of the tree cactus (*Carnegiea*) and in the massive cactus *Ferocactus wislizenii* which had an age of at least 100 years. Ray cells of *Sequoia sempervirens* were found in the heartwood that was over 100 years of age, while living ray cells and tracheids 250 years old were observed in *Parkinsonia microphylla*. It was observed that the pentosans or mucilages which are abundant in the young cells of the cacti decrease with age. The glucose content increases, while the lipoid and protein content undergoes the least change.

Bristol (1916) claimed to have found dormant but living moss protonema in air-dry soils that had been stored in sealed bottles for nearly 50 years. Lipman (1934) believed that he observed unicellular blue-

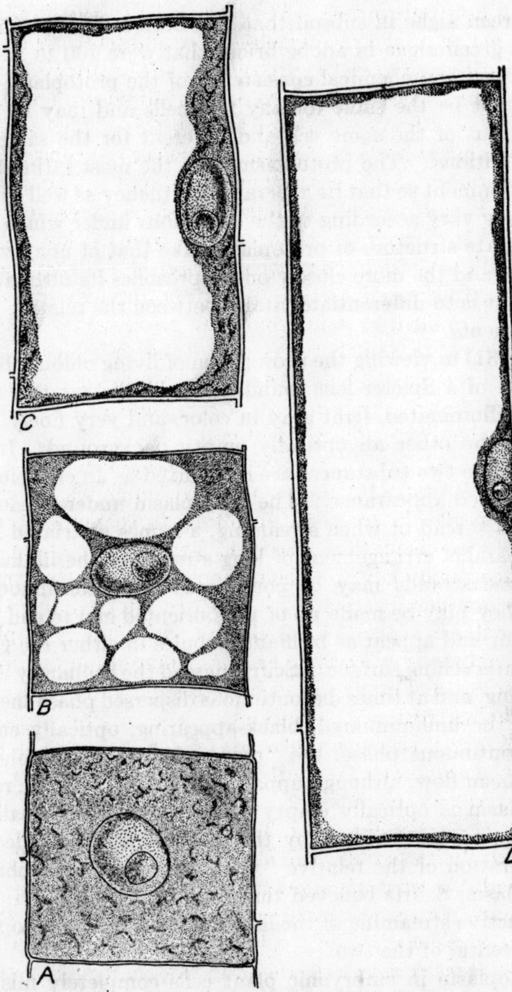

FIG. 3.—Stages in the development of the vacuole and protoplasmic membrane as shown by the cells of the growing region of a root of corn. *A*, an embryonic cell. *D*, a cell that has reached its full growth in length and in which the vacuole is fully formed, the protoplasm being in a thin membrane along the cell wall. *B* and *C*, intermediate stages in the enlargement of the cell, showing the formation of numerous vacuoles which finally coalesce to form a single large central one.

green and green algae in subsoil that had been sealed for 65 years, and single-celled green algae in adobe bricks that were 100 to 150 years old.

The physical and chemical consistency of the protoplasm is variable, and it may not be the same for any two cells and may be different in different regions of the same cell and different for the same cell under different conditions. The protoplasm is in the most intimate exchange with its environment so that its general constituency as well as its position in the cell may vary according to the conditions under which it is found.

The ultimate structure of protoplasm, like that of non-living matter, is not visible, and the more closely one approaches its ultimate structure the less easy it is to differentiate vitally between the relative importance of its constituents.

Seifriz (1931) in viewing the protoplasm of living onion cells under the oil immersion of a Spierer lens found he could observe two substances: one brightly illuminated, light gray in color, and very finely granular in texture, and the other an optically empty background. In quiescent protoplasm these two substances are intermixed as an emulsion and then present a mottled appearance. The protoplasm under tension, as when formed into a thread or when streaming, assumes a striated appearance due to the parallel arrangement of long strands of the illuminated substance. These strands may be continuous or discontinuous. In the latter case they may be made up of rods oriented end to end. Included particles occur and appear as brilliant globules in either the gray matter or the dark intervening surface. Seifriz named the brilliantly illuminated, gray-appearing, and at times discontinuous dispersed phase the "phaneroplasm" and the unilluminated, black-appearing, optically empty background or continuous phase, the "cryptoplasm." Both phaneroplasm and cryptoplasm flow, although apparently not at the same rate. Since the cryptoplasm is optically empty and thus cannot actually be seen, its streaming is made evident by the movement of included particles. In a consideration of the relative "vital" significance of phaneroplasm and cryptoplasm, Seifriz believed that the frequent discontinuity in the former and active streaming of the latter suggest that cryptoplasm is the more fundamental of the two.

The protoplasm in embryonic plant cells completely fills the entire cavity within the cell wall, but as the cells increase in size the protoplasmic content is altered by the appearance within the protoplasm of cavities filled with cell sap and called "vacuoles." These vacuoles are in most cases ultimately combined to form a single large sap cavity so that the protoplasm is spread out in a thin film along the cell wall (Fig. 3). The parts of the protoplasm of plant cells that are at this time definitely recognized under the ordinary microscope are the cytoplasm, nucleus, plastids, and chondriosomes.

A. The Cytoplasm

The cytoplasm may be considered as the less specialized portion of the protoplasm of the cell, since its functions are not so clearly defined as those of the other parts of the living substance. The cytoplasm makes up the general groundwork of the protoplasm and the greater portion of its bulk. Its physical and chemical characteristics are practically the same as those that have been mentioned in the discussion of the protoplasm. In general, the cytoplasm is the more fluid portion of the protoplasm in which the more highly differentiated morphological portions of the protoplasm are embedded and is thus that portion of the protoplasm that comes into the most intimate relationship with the environment of the cell. The surfaces of the cytoplasm that border on the cell wall and on the vacuole are generally modified in a greater or less degree in texture and consistency from its general mass. These modifications seem to consist of special layers or membranes of nongranular hyaline cytoplasm. Whether these membranes are the result of inherent specialization of the cytoplasm or simply the resultant of the physical forces of surface tension is not definitely known. The outside membrane next to the cell wall is generally called the "ectoplast" and the inside membrane bordering the vacuole the "tonoplast," while the more granular turbid bulk of the cytoplasm lying between these two membranes has been termed the "endoplasm." Any matter that enters the vacuole from the exterior of the cell or passes from the vacuole to the outside must pass through the cytoplasmic layer. The ectoplast and the tonoplast have been called by some authors the "plasma membranes" and have been considered by some authors to exercise a selective influence on the passage of materials into or out of the regions on which they border, but whether they are more specialized in this regard than the general mass of the cytoplasm has not been proved by any adequate experimental evidence. Concerning the physiological function of the cytoplasm but little is known. What little experimental work has been done seems to be in harmony with the hypothesis advanced almost 50 years ago by Bernard (1878) who maintained that the cytoplasm is the seat of destructive metabolism. If it is not thus involved, it would at least be safe to assume that it is a neutral region which translocates materials toward and away from the nucleus and plastids. It has been shown by Poirault (1893), Gardiner and Hill (1901), and Livingston (1933, 1935) that in many cases the cytoplasm of one cell is connected with that of the adjoining cell by delicate strands of cytoplasm that extend through the cell wall. These strands or fibrils are termed "plasmodesms" and lie in minute passageways through the cell wall. The plasmodesms are characteristic at least of those living cells which have cellulose or unlignified walls. These strands are the largest in the cells of mosses and the smallest in the cells of seed plants. They have been observed in the thick-walled cells of endosperms, in storage tissue, phloem, meristem, and in the medullary rays. The plasmodesms bear a close resemblance to the connecting strands of sieve tubes, and these strands are considered by some to be merely enlarged plasmodesms. Although the plasmodesms have been little recognized in plant physiology, it seems probable that they may play an important role in the transmission of stimuli and in translocation of materials from cell to cell.

B. The Nucleus

The nucleus is the most conspicuous organ of the protoplasm, and, owing largely to its behavior in the process of reproduction and cell division, cytologists have largely concentrated their attention upon it with the result that the nucleus is the best known body in the protoplasm in regard to its structure and morphology. A detailed discussion of the nucleus is beyond the scope of this work, and it is the intention to make only a few general statements concerning its structure and function. It is generally a rounded or ellipsoidal portion of the protoplasm whose structural basis is essentially similar to that of the cytoplasm. In its physical nature, the nucleus consists in part at least of a gel of higher viscosity than the cytoplasm. It is bounded by a distinct membrane within which is the nuclear sap or karyolymph. This substance is generally highly transparent and homogeneous and may occur as either a sol or a gel. In the karyolymph is embedded a network or reticulum that may be relatively uniform throughout or only fragmentary. In its chemical properties the nucleus is distinguished from the cytoplasm by the abundance of nucleoprotein that it contains. The nucleoproteins under proper procedure can be split into nucleic acid and a form of protein, the two existing in the nucleus in chemical combination. In its physiological functions the nucleus seems to be especially the center of the synthetic activities of the cell. The limited experimental evidence here again tallies with the opinion of Bernard (1878) that the nucleus is the organ of constructive metabolism.

C. The Plastids

The plastids are differentiated portions of the protoplasm of plant cells which perform certain physiological functions and are classed as protoplasmic organs on the same morphological and physiological plane as the nucleus of the cell. According to our present knowledge, the plastids or their primordia are present in every plant cell of the green plant, and the number of them formed in any given cell and the degree or form of differentiation that they assume will depend upon varied internal and external factors. Since the time of Schimper (1885) the plastids have been classified into three more or less distinct groups: the chloroplasts, the chromoplasts, and the leucoplasts.

1. The Chloroplasts.—The chloroplasts are the green plastids and owe their characteristic color to the presence of the pigment chlorophyll. They occur for the most part in those cells of the plant that are exposed to the sunlight. They vary much in shape and size in the algae, but for the most part, in the bryophytes, pteridophytes, and spermatophytes, they are small round or discoid bodies that more or less completely fill the peripheral cytoplasm of the plant cell. According to Eyster (1929), Mobius, in the examination of 215 species of plants belonging to many families of the bryophytes, pteridophytes, and spermatophytes, found 105 to have chloroplasts 5μ, 70 to have chloroplasts ranging from 3 to 5μ, 31 to have chloroplasts ranging from 5 to 7μ, and 9 to have chloroplasts from 7 to 10μ in their longest dimensions. In certain genetic strains of corn, Eyster (1929) found the chloroplasts in one strain to be only 3 to 4μ in length, while in other strains they ranged from 10 to 25μ in their longest dimensions, as compared with 6 to 8μ in most strains. He observed a direct correlation between the number and size of the chloroplasts in individual cells. The smaller the chloroplasts the greater the number in the cell. Our meager information concerning the internal structure of the chloroplast seems to indicate that it consists of a colorless protoplasmic groundwork in which are embedded the chlorophyll granules. Priestley and Irving (1907) considered that the chlorophyll is restricted to the peripheral ring of the chloroplast where it is held in the meshes of a network of protoplasm in two species of *Selaginella* examined by them. Zirkle (1926) examined the chloroplasts under normal conditions in a large number of higher plants in monochromatic light of a known wave length without any fixing or staining. He considered that the ground substance or stroma of the chloroplast is in the form of a hollow, flattened prolate spheroid surrounding a large central "vacuole." The stroma has a granular appearance that is due to numerous pores which connect the central vacuole with the cytoplasm surrounding the chloroplast. The pigments of the chloroplast are intimately mixed and evenly distributed throughout the protein ground substance. The starch inclusions of the chloroplasts are contained within the central vacuole. Zirkle also observed that in many leaves a constant differentiation of chloroplasts occurs. Some contain little starch and are mainly concerned in photosynthesis, while others contain much included starch and little green tissue and apparently function mainly as storage organs. It has not been adequately shown that the chloroplast or any of the other plastids have a specific membrane of their own like the plasma membrane, or that of the nucleus, although Zirkle (1926) determined that the chloroplast is surrounded by a more or less permanent sheath of nongranular cytoplasm. Harper (1919), however, considered that when the chlorophyll is dissolved from the chloroplast it has little to distinguish it from the adjacent cytoplasm. He considered that the plastid is to be regarded as a region of the protoplasmic complex rather than as a differentiated and definitely delineated body and that cytologically the function of the chloroplast is perhaps little more than an area of the cytoplasm impregnated or infiltered with chlorophyll. Weier (1933) defined the plastid of *Anthoceros* as a localized region of chlorophyll-impregnated cytoplasm.

The chloroplasts in the cells of the higher plants increase in number by the process of simple fission, which resembles the amitotic division of nuclei, but, so far as is known, the structure of the chloroplast undergoes no change in the process. The outstanding function of the chloroplasts is the formation of carbohydrates from carbon dioxide and water in the presence of sunlight. It is thus due primarily to the presence of the chloroplasts in certain of their cells that enable plants to manufacture their own food, while the members of the animal kingdom in whose cells the chloroplasts do not occur are totally unable to do so.

2. The Chromoplasts.—The chromoplasts are plastids that contain red, yellow, or orange pigments. They occur principally in the petals of flowers and in ripe fruits and may originate from chloroplasts or leucoplasts, as will subsequently be described. They have no special physiological function so far as is now known.

3. The Leucoplasts.—The leucoplasts are the colorless plastids and occur for the most part in those regions of the plant where light does not penetrate. They are of a denser consistency than the chloroplasts and are generally spherical in shape, although they may be greatly modified by the inclusions of starch and other compounds. Although the leucoplasts are colorless, they have the inherent capacity of forming coloring matter under certain conditions and thus being transformed into chloroplasts or chromoplasts. The term "leucoplasts" is generally used in a broad sense and is made to include not only the colorless plastids in the fully developed cells of the plant but also the colorless undifferentiated plastids of the embryonic cells which may eventually develop into chloroplasts or chromoplasts. The term also includes the colorless plastids of the epidermal cells and hairs which are apparently abortive chloroplasts. One of the chief functions performed by the leucoplasts, especially in the storage cells, is the formation of definite starch grains from the soluble carbohydrates that are present in the cell. The capacity for forming definite starch grains is thus possessed only by plant cells and does not occur at all in the animal kingdom, whose cells in no case possess leucoplasts. Mottier (1921) investigated the formation of the protein granules in the aleurone layer of the endosperm of corn and the formation of oil in the castor bean. He was of the opinion that leucoplasts may elaborate bodies other than starch grains and concluded from his cytological investigations that the protein and oil bodies found in the above storage cells owe their origin to plastids whose primordia are permanent organs of the cell.

The three different types of plastids seem to be homologous structures which in many cases display direct relationship. The primordia from which the chloroplasts and leucoplasts develop are identical, as far as our present knowledge goes, since they cannot be differentiated from each other by any staining or chemical reactions at present at our disposal. The more mature plastids under certain conditions change apparently from one into the other. The plastids of the tomato at a very early age may undoubtedly be classified as leucoplasts but soon develop into chloroplasts, which, upon the ripening of the fruit, become the chromoplasts that give the fruit its characteristic color.

4. The Origin of the Plastids.—There are two prevalent opinions in regard to the origin of the plastids in the cell. One group of investigators, including Gris (1857), Sachs (1875), Mikosch (1885), Belzung (1887 to 1891), and Stone (1932), holds the view that the plastids originate directly from the cytoplasm of the cell. According to them, the cells of the embryo of the mature seed contain no plastids. Their opinion is that, if any plastids were present in the young embryo, they lose their identity and disintegrate at the ripening of the seed and that at germination the cytoplasm in the cells gives rise to new plastids that function during the period of activity of the plant.

Stone (1932) studied the development of the chloroplasts in the young leaves of the potato plant. The young leaves observed contained green coloring matter that was diffused throughout the cell, and no definite bodies could be observed. The chlorophyll gathers into specific, delimited regions that eventually give rise to the chloroplasts. During their rounding off, these masses are often held together for a time by cytoplasmic connections that disappear as the plastids become fully formed. The process of chloroplast development is one of concentration and contraction rather than of expansion and growth since these original patches of chlorophyll may be larger than the resulting fully developed chloroplasts. The chloroplasts thus are considered to arise directly from cytoplasmic regions impregnated with chlorophyll.

The other group of investigators is inclined to the opinion first set forth by Meyers (1883), Schimper (1885), Bredow (1890), and Weier (1930 to 1933), that plastids never originate from the protoplasm in the cell but have an individual existence in the cell and multiply by division and are handed down from generation to generation. They consider that the fertilized egg contains plastids that have been derived from the parent plant. During the development of the egg into the embryo the plastids multiply and in this manner provide every cell of the embryo with plastids. The plastids go through the resting stage of the embryo intact and at germination increase in number and thus supply each cell of the new plant. In certain *Conjugatae*, *Chlorophyceae*, and *Archegoniatae*, the chloroplasts pass through the entire life cycle apparently unchanged. They are reproduced exclusively by the division of preexisting plastids and are apparently permanent cell organs. In the higher plants, however, the life history of the chloroplasts is more difficult and different, since in certain stages of the cycle no chloroplasts apparently occur but later reappear. They must then have originated from the cytoplasm or have passed through the preceding stage in an inconspicuous form. Although plastids in both the undifferentiated and differentiated state have been observed in the spores and fertilized eggs of several species of plants, the number of cases observed has been small owing to the difficulty encountered in proper differentiation by staining, so that, because of this lack of evidence in this link of the theory, it must remain more or less hypothetical. The work of later investigators has tended to strengthen this theory rather than to weaken it. Miller (1911) repeated the work of Famintzin (1893) on the origin of the chloroplasts in the cotyledons of the sunflower (*Helianthus annuus*) and showed that the plastids are present in their usual place in the cells of the seed but are very minute and that as the seed begins to germinate these plastids increase in size and divide by simple fission and develop into the chloroplasts of the green cotyledons of the seedling.

Mottier (1918) investigated the young tissues of *Pisum sativum*, *Zea mays*, and *Elodea canadensis* among others and concluded that the primordia of the leucoplasts and chloroplasts are permanent organs of the cell and that the egg and sperm have sufficient cytoplasm to carry these over into the new individual. Twiss (1919) traced in the root-tip cells of *Zea mays* from the embryonic cells backward an unbroken series of bodies varying in size from the most minute forms to plastids. She did not consider, however, that the observation of a series of gradations in size from mitochondria to the differentiated or mature chloroplasts is sufficient evidence that the mitochondria are the rudimentary plastids, since there is no way of showing that the rudimentary plastids are actually mitochondria and not young plastids.

Zirkle (1927, 1929) studied the origin of the plastids in the apical meristem of *Elodea canadensis, Lunularia vulgaris,* and *Zea mays.* He considered that the plastids develop from primordia in the meristematic tissue and that they are not reproduced *de novo* but reproduce by division, the division occurring in one plane only. The primordia are rod shaped, frequently found in strands end to end, and are often grouped around the nucleus. These enlarge, become green, and develop into chloroplasts. The immature plastids frequently contain starch inclusions. In the vascular bundles of corn many transitional stages can be observed in the course of development of mitochondria into plastids.

The origin of the plastids has been intensively and extensively studied in the leaves with various chlorophyll types including those that are entirely devoid of green pigment, variegated ones, and those in which the chlorophyll is evenly distributed throughout the leaf but lacks the intensity of greenness that is commonly observed.

Randolph (1922) made a cytological study of various chlorophyll types of maize in order to determine the status of the plastids in the cells of these strains. All the chlorophyll types examined were found to contain the same initial, minute primordia of plastids (protoplastids) of the same size and general appearance. In normal green plants the protoplastid first appears in the cell as a minute granule at the limit of visibility, gradually enlarging and developing chlorophyll until it becomes a mature chloroplast. In plants of the white virescent and other inheritance strains the unusual characters of the plants are due to the failure of the primordia initially present to develop into plastids with the normal size, color, or both. Randolph thought that the behavior of the primordia of the chloroplasts may in part at least be under the control of the nucleus of the cell. In the case of *Coleus,* which has variegated leaves, Stout (1915) observed that plastids are present in both green and yellow cells but that in yellow cells they are fewer in number, smaller in size, and somewhat distorted in shape and, further, that in extreme cases of yellow development nearly all the cells fail in the production of chlorophyll. Hein (1926) cut sections of the leaves of certain variegated plants so that they included both normal green and abnormal white and yellow areas and the marginal cells lying between. He found in the leaves of *Dieffenbachia seguine* (Schott) in going from the normal green cells through the critical zone to the pure white areas that the plastids decreased in number, size, and color until the cells showing no green were reached and that in these cells no evidence of chlorophyll bodies could be observed. In contrast to this, the nuclei of the various cells showed no modification in shape, size, or appearance. In the leaves of *Dracaena godseffiana* (Sanders) a more abrupt change from the normal green tissue to the colorless tissue was shown than in *Dieffenbachia,* but it is not absolutely sharp and definite, since the marginal cells show both normal and abnormal chloroplasts.

In the aberrant types of leaves that have been studied, it appears that the primordia of the plastids are normal but that these divergent forms are due (*a*) to a delayed development of the plastids, (*b*) to an arrestment of their development, (*c*) to an arrestment of development followed by degeneration, and (*d*) to a normal development of only a portion of the plastids (Küster, 1919; Zirkle, 1929; Eyster, 1933).

A study of chlorophyll inheritance has thrown considerable light on the development and subsequent behavior of the plastids. A summary of these observations has been made by Lubimenko (1926) and Priestley (1929). In some variegated plants the evidence indicates that the inheritance of the chloroplast mechanism can be interpreted in the terms of Mendelian segregation of certain pairs of unit characters. To the geneticist this indicates that at least part of the mechanism associated with the chloroplast inheritance is in the nucleus of the germ cell. Certain plants, however, afford evidence that the chloroplast mechanism is passed on by only one parent, and it is suggested that in these cases plastid gives rise to plastid through the germinal tract from this one parent to that of the opposite sex. In 15 experiments of crossing two pure lines of a species that differed markedly in chlorophyll content, there was found in 6 cases with the F_1 and succeeding generations a chlorophyll distribution that could be interpreted on Mendelian lines, while in the other 9 observations no such interpretation seemed possible. This suggests that both methods of chlorophyll inheritance may be involved. Chloroplasts may persist as units throughout the life cycle in the cytoplasm of the germ cell but factors segregating in the nucleus along Mendelian lines may determine their behavior during somatic development in succeeding generations.

D. The Mitochondria or Chondriosomes

In the cytoplasm of the cells of many plants representing all the great groups of the plant kingdom there have been described by numerous investigators small rod-shaped or granular bodies under the name of "chondriosomes" or "mitochondria." These minute bodies can seldom be observed in living fresh tissue and are, in general, not preserved by the usual methods of fixation, so that they are not detected by the ordinary methods of staining. It is due to these facts that their discovery was not made until comparatively recent times in botanical history. A complete and thorough review of the literature on chondriosomes is given by Cavers (1914). The literature on the chondriosomes is very confusing. Many of the investigators do not distinguish between the chondriosomes and the primordia of the plastids, and some (Guilliermond, 1911; Lewitsky, 1910; and others) consider that the bodies which they call chondriosomes give rise to the leucoplasts and chloroplasts. Rudolph (1912), Schmidt (1912), Mottier (1918), and others considered that the chondriosomes and the primordia of the plastids are entirely different structures and that in no case are chondriosomes changed into plastids. The evidence seems to indicate that there are present in the cytoplasm of plant cells certain definite minute bodies other than the primordia of the plastids. Guilliermond (1911 *et seq.*) and Mottier (1918) assigned to these bodies the same morphological rank as the nucleus of the cell. The latter considered that they are permanent organs of the cell and are transmitted from individual to individual through the cytoplasm of the sperm and egg. The mitochondria are considered by physiologists (Politis, 1921; Marston, 1923; Cowdry, 1926; and Horning and Petrie, 1927) to be the seat of both synthetic and hydrolytic degradation processes in plants. The phase boundary of the mitochondria and the surrounding cytoplasm is regarded as the seat of these processes. Marston (1923) from the reaction of the mitochondria to azine dyes concluded that they contain proteolytic enzymes and developed the theory that the mitochondria are the seat of enzymatic protein synthesis in the cell. He considered that the "water-poor" phases which exist at the surface of the lipoidal constituents of the mitochondria provide suitable conditions for synthetic activity. He considered that not only protein but also other foods and complex compounds may be synthesized in the same regions of the mitochondria. The function of the mitochondria in hydrolytic processes has been studied by Horning and Petrie (1927) in the germination of corn, wheat, and barley. During germination the mitochondria become more numerous apparently by simple fission in the scutellum and are secreted in large numbers from the epithelial cells into the adjacent starch-containing cells of the endosperm. These secreted mitochondria then aggregate around the starch grains prior to their corrosion. Their behavior appears to lend considerable support to the enzymatic conception of mitochondrial activity, that the starch-splitting enzyme is located within them or at their surfaces and is liberated therefrom when the bodies have reached the surface of the starch grains.

III. THE VACUOLE OR CELL SAP

A. FORMATION

As an embryonic plant cell begins to increase in size, the protoplasm soon develops cavities of varying sizes which become filled with water and materials in solution and suspension. These cavities with their contents are termed "vacuoles," and as the cell increases in size, these vacuoles usually coalesce to form a single large vacuole or cell cavity. The formation of the vacuole is apparently a much more complicated procedure than had formerly been considered. According to Guilliermond (1929), the vacuoles in the meristematic cells of flowering plants, as a result of the hydration and swelling of numerous and tiny mitochondria-like bodies, contain a thick colloidal solution. As the cell differentiates, these bodies swell into small, round vacuoles which fuse into one single large vacuole, making the bulk of the cell. The large liquid vacuoles originating from the mitochondria-like bodies by hydration of their contents ultimately contain a very dilute colloidal solution which stains faintly as a whole. The whole system of vacuoles in any given cell at any stage is termed the "vacuome." The colloidal substances in the vacuoles of flowering plants differ widely in different cases. According to Guilliermond, the vacuolar contents are most often composed of alcohol-soluble protein mixed with phenol compounds of the tannin group and closely related anthocyanin compounds, all of which absorb intra-vitam dyes.

The vacuome is considered to be always present in any cell, and it may appear in two altogether different aspects according to the age of the cell: (1) It may be distributed throughout the cytoplasm as numerous tiny thick bodies composed of a concentrated solution of colloids and showing as mitochondria-like grains or threads. (2) It may be in the form of one or a few large liquid vacuoles, where colloids are in weak solution, these vacuoles developing from the small, mitochondria-like bodies by hydration and coalescence. Either of these aspects may be converted into the other. The vacuoles may be made to give up water when the cell is dehydrated. They shrink and break down into thread-like thick drops, as is shown when a seed begins to lose water preparatory to maturing and as can be experimentally induced by plasmolyzing cells.

B. Composition

The vacuole or cell sap may contain the following materials:

1. Water.—As much as 98 per cent of the cell sap may consist of water.

2. Inorganic Salts.—Any of the inorganic salts that are absorbed by the plant from the soil may be found in the cell sap, but those which occur in the greatest abundance are the salts of sodium, potassium, magnesium and calcium.

3. Carbohydrates.—The carbohydrates that are generally found in the cell sap include cane sugar, maltose, glucose, fructose, dextrins, and inulin.

4. Nitrogenous Compounds.—The principal nitrogenous compounds are proteins, amino acids, amides, proteoses, and peptones.

5. Soluble Alkaloids and Glucosides.—These products, which appear to be waste products of the cell, include the tannins, amygdalin, anthocyanin, and the mustard-oil glucosides.

6. Gases.—Oxygen and carbon dioxide are the gases in the cell sap that are of the most physiological importance. The oxygen may come from the exterior, or if it is a green cell a portion of it may arise as a by-product in photosynthesis. Carbon dioxide occurs in all plant cells as an end product of respiration. It may also enter the cells from the outside, especially in green cells where it is used in photosynthesis.

7. Enzymes.—The principal enzymes that may be found in the cell sap are diastase, invertase, lipase, inulase, and proteolytic enzymes.

8. Organic Acids.—These compounds are present either as acids or in combination with mineral bases. The principal organic acids present in the vacuole are oxalic, malic, citric, tartaric, tannic, acetic, and formic. The last two seem to be very widespread in their distribution and are probably present in every living cell.

According to their origin, the materials found in the vacuole fall into three general groups: (*a*) the materials that have been absorbed by the cell from its exterior, (*b*) the materials that have been synthesized by the protoplasm, and (*c*) the waste products or excretions that arise in the protoplasm as a result of its life functions. It is evident, therefore, that the composition of the vacuole will be an extremely variable one, since the functions that give rise to the materials therein are dependent upon and influenced by a great many factors.

C. Concentration

The concentration of the cell sap is generally expressed in terms of osmotic pressure. By the osmotic pressure of the cell sap is meant the pressure that the water in the vacuole would be capable of exerting upon the protoplasm and the cell wall if the cell were placed in pure water. This definition of osmotic pressure is based on the assumption that the cell wall is absolutely rigid and that the protoplasm will allow water to enter freely and that neither will allow dissolved materials to escape from the vacuole. The osmotic pressure of the cell sap can also be defined by stating that it is the pull that the cell sap is capable of exerting upon pure water if it were separated from it by a perfectly semipermeable membrane. The magnitude of the osmotic pressure of the cell sap is commonly expressed in terms of atmospheric pressure. The osmotic pressure or pull of the cell sap varies directly with its concentration and is due to certain relationships of membranes and solutions that will subsequently be discussed in Chap. II.

1. Factors That Influence the Concentration of the Cell Sap.—The factors that influence the concentration of the cell sap are (*a*) the habitat of the plant, (*b*) the kind of plant, (*c*) the portion of the plant, (*d*) the age of the tissues, and (*e*) the time of sampling, *i.e.*, season, day or night.

a. The Habitat of the Plant.—It was shown by Stange (1892) with plants of wheat, beans, and peas, Hill (1908) with the root hairs of *Salicornia* and *Suaeda*, Roberts (1916) with the root hairs of numerous crop and vegetable plants, and Ohga (1926) with wheat and buckwheat seedlings that the osmotic pressure of the cell sap can be varied by changing the concentration of the medium in which the plants are growing. Roberts (1916) found that the concentration of the cell sap of the root hairs of numerous plants grown in moist air was equivalent to 5.1 to 7.7 atmospheres so that under this condition there was a minimum difference of four atmospheres in concentration between the inside and outside of the root-hair cell. By growing the roots of radish in a series of increasing concentrations of cane sugar from 0.02 *M* to 0.65 *M*, she showed that the cell sap of the root hairs increased in concentration directly with that of the external medium and maintained an osmotic pressure of from 4 to 6 atmospheres above it. Ohga (1916), studying wheat and buckwheat seedlings, observed that the maximum increase in osmotic pressure in the roots, due to placing them in sucrose solution, was about 0.8 *M* or 18.7 atmospheres.

McCool and Millar (1917) found that the osmotic pressure in the roots and tops of Canada pea and wheat and corn seedlings is indicative but not directly proportional to the density of the solution in which they are grown. The concentration of the soil solution was varied by adding solutions of $Ca(NO_3)_2$, $MgSO_4$, and K_3PO_4. The following figures show the variation of the osmotic pressure of the soil solution and the osmotic pressure of the corn roots growing therein:

Osmotic pressure of the soil solution, atmospheres	Osmotic pressure of sap of root cells, atmospheres
1.21	4.59
1.99	5.48
3.38	6.61
4.96	7.51
7.22	8.19

The application of the various mineral nutrients to the soil as fertilizers generally results in increased concentration of these elements in the cell sap (McCool and Weldon, 1928). It appears, however, that unless there is considerable contrast between two environments the osmotic value of the expressed sap cannot be taken as a safe criterion of the type of environment (Meyer, 1927).

Harris and Pascoe (1930) observed a general positive correlation between the salinity of the soil and the osmotic concentration of the cell sap in Prima Egyptian cotton and Lone Star upland cotton. This fact was indicated when measured in terms of freezing-point depression, chloride content, sulfate content, or electrical conductivity.

Magistad and Truog (1925) found that the application of fertilizer in the hill increases the osmotic pressure of the sap of corn plants which in turn lowers the freezing temperature of the plants from one to two degrees centigrade. This is often sufficient to prevent plants from being frozen by late spring frosts if they are not too severe. The greatest benefits in this regard would be derived from the application of fertilizers to peat, muck, or other soils of low-soluble salt content.

In a changing medium, plants differ widely in the extent to which they take up salts and increase the concentration. Thus Eaton (1927) found that the cell sap of wheat plants was 98 per cent more concentrated when growing on a soil with a higher salt content than on a nonsaline soil, while under the same conditions a difference of but 42 per cent was shown for the saltbush plant. Drabble and Drabble (1907) examined 50 species of plants growing under various conditions of water supply and found the osmotic pressure of the cells varied from 4 to 20 atmospheres. Their conclusions concerning the relation between environment and the osmotic pressure of the cell sap of plants have been verified by other investigators and can now be stated under the general law, that "In plants of the same species growing under the same conditions of water supply, the osmotic strength of the cell sap is generally the same, and in any area the osmotic strength of the cell sap varies indirectly with the physiological scarcity of water."

The osmotic pressure in the cell sap is more rapidly changed by the fluctuations in the moisture conditions of the site than by any other environmental factors (Korstian, 1924). The highest densities of cell sap occur on the sites with the lowest moisture supply regardless of whether this low supply of moisture is due to a direct lack of water or to a high concentration of solutes in the soil. Stoddart (1936) noted as the soil dried with the progress of the season that in all the prairie plants examined the osmotic pressure increased in their cell sap. This increase amounted to as much as 30 atmospheres or more in certain upland plants, while in plants growing in low moist habitats the increase in osmotic pressure was usually only 2 to 10 atmospheres. The water content of a plant is a fair indicator of osmotic pressure—plants with high pressures usually having a low water content and plants with low osmotic pressures usually having a high water content.

Beck (1929) observed that the cell sap of the assimilatory tissues manifests the greatest regularity, speed, and degree of response to drought of any of the plant tissues. The osmotic values of the epidermal cells show the least variations.

Stanfield (1932) found a correlation between the osmotic pressure and rate of evaporation in the leaves of *Pinus scopulorum*, the higher pressures occurring with the higher rates of evaporation. A difference of approximately 3 atmospheres was noted between the pressures occurring with the lowest and highest rates of evaporation.

Iljin (1929) noted that the osmotic pressure of the roots is closely related to the degree of moisture in the soil as shown by the following concentrations of the cell sap of plants from different habitats:

In swamps, 0.13 to 0.35 M — In steppe plants, 0.40 to 0.48 M
In meadow, 0.25 to 0.35 M — In exposed declivities, 0.55 to 0.60 M

Hibbard and Harrington (1916) found that the osmotic pressure of the cell sap of the root cells of corn was highest in those roots growing in the soil with the lowest moisture content (see also Whitfield, 1932; and McKay, 1935).

Fitting (1911) has given us the most complete record of the osmotic pressures in the cell sap of desert plants. He worked in the Sahara desert and examined 46 species of plants growing in oases, saline soils, rocky wastes, and sand dunes. His observations may be summarized briefly as follows:

Number of plants	Osmotic pressure	
	Salt equivalent	Atmospheres
10	3.0 N (KNO_3) or above	100 or above
16	1.5 N (KNO_3) or above	53 or above
24	1.0 N (KNO_3) or above	34 or above
5	0.3 to 0.6 N (KNO_3) or above	10 to 20 or above

The highest osmotic pressures were found in the plants growing in dry soil and no extremely high pressures were met with in moist nonsaline soil. The work of Fitting has been verified by Ilin and Nazarova (1915) and by Keller (1920) working in Russia.

Harris, Lawrence, and Gortner (1915 to 1921) and Harris (1934) studied the concentration of the cell sap of a large number of different plants growing under a wide range of conditions. Their investigations include not only plants growing in widely distributed geographical areas of North America but also those growing in different habitats of the same area. In an investigation of the spring flora of the mesophytic region of Cold Spring Harbor, N. Y., and that of the desert region of Tucson, Ariz., they reported the following: At Coid Spring Harbor 50 per cent of the pressures were equal to 10.5 atmospheres or lower, while at Tucson in the desert 50 per cent of the pressures equaled 15.7 atmospheres or above. At Cold Spring Harbor only 3 per cent of the plants exceeded a pressure of 20 atmospheres, while at Tucson 30 per cent of the pressures exceeded this figure. The maximum pressure observed at Cold Spring Harbor was 30 atmospheres, while the highest pressure at Tucson was 65 atmospheres. The average pressure of plants growing at Cold Spring Harbor was 14.4 atmospheres as compared to an average pressure of 38.8 atmospheres of those growing on the coastal slopes of Jamaica, B.W.I.

b. The Kind of Plant.—Fitting (1911) observed that the concentration of the cell sap is highest in trees and shrubs and lowest in the annuals. Harris and Lawrence (1916) in their work at Tucson, Ariz., reported the following average determinations of the cell-sap concentration:

Kind of plants	Osmotic pressure of cell sap, atmospheres
Trees and shrubs	28.10
Dwarfs and half shrubs	21.45
Perennial herbs	16.35
Winter annuals	14.73

The osmotic pressure of the sap of trees and shrubs growing at Tucson, Ariz., Jamaica, B.W.I., and Cold Spring Harbor, N. Y., was found by Harris and others (1921) to be 30.5 per cent higher than that of the herbaceous plants growing in the same region. Harris (1918) observed that the cell sap of epiphytic plants is only one-third to two-thirds as concentrated as that of the terrestrial plants. It was observed by him and is considered a general biological law that the conductivity of the sap of herbaceous plants is distinctly higher than in woody species. The concentration of ionized electrolytes is thus lower in woody forms than in herbaceous forms, although the reverse is true in regard to the total solutes. It was also observed by Korstian (1924) that the concentration of the cell sap of woody species is much higher than that of herbaceous species (see also Meyer, 1927).

The crowns, stalks, and roots of sorghums have a higher osmotic concentration than those of corn, while the leaf juices of sorghums have a lower freezing-point depression and a lower conductivity than corn.

Harris and Lawrence (1916) studied the osmotic pressure of the cell sap of the mistletoes growing on 20 different hosts. They found the average pressure for the parasites to be 14.43 atmospheres and that of the hosts to be 13.59 atmospheres. The osmotic pressure of the leafless species of mistletoe is distinctly lower than that of the leafy species.

The cell-sap concentration of the cells of *Phoradendron juniperinum*, parasitic on *Juniperus utahensis*, showed according to Harris, Pascoe, and Jones (1930) a pressure of 3.71 atmospheres in excess of the host plant. The chloride content of the cell sap of the parasite exceeds by 1.5 times that of the host. The almost universal absence of *Phoradendron californicum* as a parasite on *Covillea tridentata* is considered to be due to a higher osmotic pressure in the latter plant (Harris, Harrison, and Pascoe, 1930).

Harris and others (1925) found that the sulphate content of the leaves of the upland varieties of cotton is higher than that of the Egyptian variety, the differences ranging from 3 to 4 g. per liter. This is just the opposite from the chloride content, the Egyptian types taking up larger quantities of chlorides than the upland types. All of the Egyptian varieties considered have a higher osmotic pressure than the upland varieties.

c. The Portion of the Plant.—Hannig (1912) determined the relative concentration of the cell sap in leaves and roots of a large number of different plants. His observations are expressed as follows:

Percentage of plants examined	Ratio of the concentration of the sap of the leaves to that of the roots
14	1
51	1.25
12	1.5
23	2

Dixon and Atkins (1916) found that the osmotic strength of the cell sap of the leaves was greater than that of the roots. The concentration of the sap of the leaves of

Ilex aquifolium varied from 13 to 15 atmospheres, while that of the roots varied from 7 to 10 atmospheres. The leaves of *Iris* had a pressure of 13 atmospheres and the roots a pressure of 9 atmospheres. Lutman (1919) made an extended study of the osmotic pressures in the cells of the various organs of the potato at various stages of growth. He also observed the osmotic pressures in several plants closely related to the potato or closely resembling it in the manner of storing food. He found that the normal pressure in the seed tubers as they are taken from storage is between 7 and 10.3 atmospheres and that the sprouts which come from these tubers not in the soil exhibit a pressure slightly in excess of that of the tubers themselves. He found that the juice of the leaves of the young plant showed a higher osmotic pressure than that of the stalk and that the tubers as they develop maintain an almost constant pressure until maturity. The osmotic pressure of the sap of the growing tubers is always intermediate between that of the sap of the stalks and the roots whose osmotic pressure is always the least of any of the organs of the potato plant. Garden beets examined on Sept. 28 showed a pressure in the leaf sap of 8.83 atmospheres and a pressure in the root sap of 11.81 atmospheres. The sugar beets examined at the same time showed a pressure of 12.21 atmospheres for the sap of the leaves and of 17.18 atmospheres for the roots. According to Chandler (1914), the molecular concentration of the leaves of fruit trees is generally considerably greater than that of the fruit except in the case of some ripe fruits. The concentration of the cell sap of the leaves at different heights of insertion on the stem has been studied by Dixon (1914), Harris, Gortner, and Lawrence (1917), Korstian (1924), and Hurd-Karrer (1925). Dixon found that the leaves of wistaria at a height of 3 ft. above the ground had a concentration of 5.25 atmospheres, while the leaves at a height of 27 ft. had a sap concentration of 6.61 atmospheres. The leaves of the elm 1 ft. from the ground showed a sap concentration of 9.2 atmospheres, while those at a height of 10 ft. showed a pressure of 12.4 and 18.6 atmospheres in the shade and sun, respectively. Harris and others (1917) made determinations of the osmotic concentration of the leaf sap at different levels on 26 trees belonging to 12 species and concluded that the concentration of the leaves as measured by the freezing-point lowering method increases from lower to higher levels almost without exception. The following are the results that they obtained with *Betula lutea* and *Fagus grandifolia:*

Name of plant	Date	Height of leaves, feet	Osmotic pressure of cell sap, atmospheres
Betula lutea	July, 1914	11	12.6
Betula lutea	July, 1914	25	14.1
Betula lutea	July, 1914	39	15.1
Betula lutea	July, 1914	52	16.1
Betula lutea	July, 1914	66	15.6
Fagus grandifolia	August, 1913	19	17.33
Fagus grandifolia	August, 1913	64	21.92

Korstian also (1924) demonstrated an osmotic gradient in the cells of plants. In the case of the yellow pine the 2-year-old needles at the base of the leaves had a sap concentration equivalent to 17.3 atmospheres while at the top the osmotic pressure was equivalent to 18.5 atmospheres. Hurd-Karrer (1925) determined the concentration gradient in the juice of corn stalks as measured by the specific gravity of the expressed sap. The plants used were from 8 to 11 ft. high with large, but immature, ears. The sap from each internode showed a gradual gradation upward, the specific gravity of the sap of the first internode being 1.0347, while that of the thirteenth and fourteenth internode had a value of 1.0497. The electrical conductivity is always less at the higher levels than at the lower ones, which indicates that the greater concentration of the uppermost leaves is due to the increase of nonelectrolytes as sugar.

It was shown by Martin, Harris, and Jones (1931) that in the sorghums the highest osmotic concentration of the cell sap is in the terminal leaf and that there is a progressive decrease in the leaf juices from the top of the plant downward. The osmotic pressure of the leaves of the date palm in the Algerian desert is frequently 10 atmospheres higher than that of the roots (Killian and Faurel, 1934). It was stated by Iljin (1929) that the osmotic pressure of the roots is more closely correlated with the habitat than is that of the cell sap of the leaves.

d. The Age of the Tissues.—Dixon (1912) observed in the two evergreens, *Ilex aquifolium* and *Hedera helix*, that the osmotic pressure of the cell sap of the older leaves was from 2 to 4 atmospheres higher than that of the younger leaves.

Korstian (1924) in the case of the pine found that there was a difference of from 1 to 5 atmospheres in the osmotic pressure of the young and old leaves, the young having the lower concentration. Gail and Cone (1929) also observed this fact. In the mulberry it was observed by Lombardi (1932) that the older the leaves the higher the concentration of their cell sap. The conductivity of the sap of the older leaves was higher than that of the younger leaves and indicates that their greater concentration is due to the accumulation of electrolytes.

e. The Time of Sampling.—Since the sugars contribute to the solutes in the cell sap, the conditions that would affect their formation would modify its concentration. Shedd (1915) stated large differences in the composition of the sap of the wild grape were found when it was collected at the same point on the vine at different times during the same season. The mineral constituents were generally higher during the day and the sap had a more uniform condition during the night. Harris, Hoffman, and Lawrence (1925) noted an increase in the magnitude of the chloride concentration in Egyptian cotton plants as the season advanced. McCool and Millar (1917) found that the osmotic pressure of corn tops was lowest in the morning at sunrise and greatest at 1 P.M. The concentration of the cell sap of the leaves at sunrise was 6.06 atmospheres and at 1 P.M. it was 7.52. The osmotic pressure decreased after that period so that at 10 P.M. it amounted to only 6.72 atmospheres. The roots showed an osmotic pressure of 4.52 atmospheres at sunrise and a pressure of 6.05 atmospheres at 1 P.M. and then a decrease to 4.33 atmospheres at 10 P.M. These maximum pressures observed correspond in a general way to the maximum point of sugar production, and this fact apparently accounts for the increase in the osmotic pressures (Gail, 1926; and Greathouse, 1932). Korstian (1924) observed that the density of the cell sap changes during the day, increasing as the day advances and decreasing toward evening. The daily minimum occurs in the early morning, indicating a relationship between concentration and photosynthesis. The same fact apparently accounts for the differences observed by Dixon and Atkins (1912) on the concentration of the cells of leaves of a north exposure and those of a south exposure. In *Hedera helix* they found that the concentration of leaves of a north exposure averaged 8.7 atmospheres, while that of a south exposure averaged 9.3 atmospheres. They also observed that shaded or covered leaves always had a lower concentration than the ones exposed to the sunlight. This fact has also been observed by Lutman (1919) for the stem and leaves of potato, and by Halma and Haas (1928) for the leaves of Eureka lemons.

2. Cell-sap Concentration and Growth.—The relationship of cell-sap concentration and the rate of growth has received considerable attention.

It was observed by Chandler (1914) that the molecular concentration of the cortex sap is the lowest during the period of rapid growth and that the molecular concentration of the cortex sap is not so great during early summer with trees that have been forced into vigorous growth by heavy pruning as with trees not so pruned.

Reed (1921) investigated the relationship of growth and cell-sap concentration in the young trees of walnuts, apricot, and orange. The weekly increment in concentration, including the leaves, was determined. The more important facts observed by him are as follows: The rate of growth and sap concentration have a tendency to vary in opposite directions so that rapid growth is associated with a generally lower concentration of sap in the shoot while slower growth is accompanied by higher concentration of sap. There was a gradual increase in sap concentration as the season advanced. In the apricot tree the concentration of the sap continued to increase for some time after active growth had ceased. The average concentration of the cell sap of the apricot expressed in atmospheres by months was as follows: May, 11.84; June, 13.66; July, 14.34; August, 15.04; September, 15.18; and October, 16.48. Of several environmental factors measured, soil moisture was the only one having an obvious effect upon sap concentration. The addition of water to the soil usually diminished the concentration of the plant sap. A concentration gradient appears to exist in the shoot. The concentration of the sap in the apical portion of the stem was greater than that in the basal region. The sap concentration of shoots in trees heavily pruned was lower than that of shoots on trees not pruned because of the more rapid growth of the former. In the case of the lemon, however, Reed and Halma (1926) found that neither severe nor moderate pruning affects the sap concentration of leaves and shoots of lemon trees for more than a few weeks after the operation. In view of these results, these authors concluded that there is something about the lemon tree which operates to maintain osmotic pressure in the face of a growth rate undergoing extreme fluctuations.

D. The Acidity of the Cell Sap

1. General Discussion.—The cell sap as a rule has an acid reaction and only rarely is it alkaline and then only slightly so. A pH value on the alkaline side of 7 to 8 has been observed, while on the acid side the highest acid value so far observed has been 0.9 or a value equivalent to about 0.1 *N* sulphuric acid. The pH value of the cell sap of plants generally lies somewhere between 4.6 and 7.0 (Atkins, 1922; and Dunne, 1932). Oxalic, malic, and citric acids are generally the cause of this acid reaction. These acids are normally present for the most part in the form of their acid salts or of their neutral salts. The concentration of free acids in the cell sap is usually rather low.

The concentration of free malic acid in plant tissue rarely exceeds 0.05 per cent of the fresh weight and is usually much lower, although the total malate content may be as high as 2 per cent of the fresh weight. According to Jodidi (1933) citric acid in the Alaska pea equals 0.36 per cent of the oven-dried material. In the expressed sap from *Pelargonium*, Armstrong (1929) found oxalic acid 0.0573 *M*, malic acid 0.0337 *M*, citric acid 0.0066 *M*, and tartaric acid 0.003 *M*. Clague and Fellers (1934) reported that the benzoic acid content of 24 varieties of cranberries averaged 0.065 per cent.

In studying the acidity of the cell sap, four quantities must be considered (Bennet-Clark, 1933): (*a*) The hydrogen-ion concentration, which is very frequently termed the "actual acidity." This quantity refers to the number of free hydrogen ions in the cell sap. (*b*) The concentration of the free, undissociated acid. This term refers to the acids in the sap which are not ionized. (*c*) The "titratable acidity." This quantity includes the combined hydrogen ions of the acids and the acid salts in the cell sap as well as the free hydrogen ions therein. (*d*) "The total acidity." This is the total quantity of anions and molecules and is therefore best expressed as the number of mols or equivalents of the acid. This quantity is obtained by neutralizing and precipitating the acid as an insoluble salt. It is termed by some the "quantitative acidity." The difference between the total acidity and the titratable acidity may be termed the "nontitratable acidity." Ordinarily the quantity of acid formed or used during metabolism is obtained by determining the increase or decrease in the titratable acidity and under most circumstances this is equal to the change in quantity of total acid.

The oxalates and oxalic acid are probably present in almost all plants. The genera *Rumex, Oxyria, Geranium, Tropaeolum,* and *Oxalis* are especially rich in these two compounds. Most plant saps contain malic acid or malates (Bennet-Clark and Woodruff, 1935). The water extract of wheat, barley, rye, and corn plants at 2 weeks preceding heading contained aconitic, citric, malic, and malonic acids, with a trace of oxalic acid (Nelson and Hasselbring, 1931; and Nelson and Mattern, 1931).

The striking changes in the acidity of plant tissues that occur under natural conditions indicate that the organic acids play some important role in metabolism and that they are by no means waste products. It has been variously suggested that they may be converted into proteins, alkaloids, fats, or carbohydrates.

Plant	Actual acidity pH value estimate in normality	Total acidity by titration
Cranberry	0.004*N**	0.917*N*
Grapefruit	0.001*N*	0.192*N*
Apple	0.0004*N*	0.071*N*
Rhubarb (leaf stalk)	0.0006*N*	0.157*N*
Orange	0.00016*N*	0.094*N*
Pineapple	0.000035*N*	0.137*N*

* This means that a 0.004*N* acid if completely dissociated would give a pH reading equivalent to that observed.

The amount of titratable acidity and the value of the actual acidity of the cell sap of plants have been studied by a large number of investigators including Haas (1917); Atkins (1922); Smith and Quirk (1926); Martin, Rea, and Small (1926); and Dunne (1932).

Haas, who examined the more acid fruits, found the values of acidity for the cell sap as shown in the table on page 46.

Smith and Quirk (1926) determined the acidity of the cell sap of the shoots and leaves of *Begonia lucerna*. This plant is the most acid on record, the leaves being the most acid and the acidity decreasing from top to base. The juice of the leaves was twice as acid as lemon juice and practically as acid as 0.1 *N* hydrochloric acid. A considerable part of the acid is oxalic acid, although other acids are present. The hydrogen-ion concentration and total acidity of the sap of this plant ranged as follows: pH 0.9 to 1.36, 1.22 to 2.23, and 3.30 to 3.42 for the leaves, top, and base of shoots, respectively. The total acidity (undissociated salts) for the leaves, top, and base of shoots was estimated at +250 to 263, +99 to 140, and +41 to 56, respectively. The total acidity as here expressed in terms of Fuller's scale denotes the number of cubic centimeters of a *N* NaOH solution required per liter of juice to bring it from the given pH to pH 8.2 (phenolphthalein neutrality). Smith and Quirk also determined the acidity of the sap of other plants according to the following table:

Plant	Tissues	pH value	Total acidity	Plant	Tissues	pH value	Total acidity
Oxalis	Leaves	1.70	+224	Sugar cane	Soft top	4.88	+ 37
Rumex crispus	Leaves	3.53	+ 66	Bananas	Young petiole of midrib	4.74	+ 12
Wandering Jew	Leaves	5.47	+ 10				
Pineapple	Leaves	4.72	+ 40	White garlic	Bulbs	3.46	+165
Boston fern	Leaves	5.20	+ 38	Olives	Young shoots	5.22	+ 52
				Onions	Leaves	4.33	+ 30

The buffer systems in the cell sap have been studied by various workers. Small (1929) considered that inorganic phosphates, malates, oxalates, asparagine, and probably bicarbonates, act as buffers in the cell saps examined by him. During the growth of seedling wheat plants, Hurd-Karrer (1930) believed that asparagine was the substance in the sap of the etiolated seedlings that was responsible for the inflection of the titration curve near pH 8.0. The phosphates appeared to be the principal buffers between pH 6.0 and 7.5, while above pH 6.0 the buffering seemed largely due to the presence of an organic constituent. This was a solution of sodium malate alone or a combination of tartrate and citrate. Dunne (1932) considered that organic acids, amides, amino acids, phosphates, and sugars were, in the cases examined by him, the principal types of substances responsible for the buffering of cell saps.

2. Factors Influencing the Acidity of the Cell Sap.—Some of the factors that may influence the acidity of the cell sap are: (*a*) the reaction of the medium in which the plant grows, (*b*) the time of sampling, (*c*) previous treatment, (*d*) kind of tissue, and (*e*) age and vigor.

a. The Reaction of the Medium.—The effect of the changes in the soil solution upon the acidity of the cell sap is not yet definitely known, since the data which have been presented are very conflicting. Thus Truog and Meacham (1919) grew alfalfa, clover, lupine, sorghum, buckwheat, corn, and rape on unlimed acid soil and on limed acid soil. In 12 of the 16 cases the plant juices are more acid on the unlimed soils than on the limed ones and in the opposite cases the differences are quite small. Some of their data are shown in the following table:

Plant	pH value of juice from plants grown on	
	Unlimed soil	Limed soil
Alfalfa	5.78	6.03
Clover (*Trifolium pratense*)	5.69	5.83
Soybean	5.81	5.94
Corn	5.21	5.31
Buckwheat	4.02	4.05

Haas (1920) found that in most cases the effect of the limed soil on the sap of the tops of alfalfa, barley, garden beans, alsike clover, mustard, timothy, and wheat was to decrease the actual acidity. He considered that his data are in favor of the suggestion of Truog (1919) that the main specific harmful influence of soil acidity on certain plants is due to its influence in preventing the plant from securing rapidly enough the bases that are needed to neutralize and precipitate acids within the plant.

Bennet-Clark (1930) found that plants of *Crassula lactea* from cuttings of the same parent plant had different titratable acidities when grown on soils of different pH reactions. Thus the sap of plants which were grown in soils with a pH of 6.8 and 9.0, had titratable acidities respectively of 8.3 and 2.2 mg. equivalents per 100 grams fresh weight. Loehwing (1930) noted that the acidity of wheat plants grown on acid soils was much greater than that of plants grown on the same soils after the acidity was corrected by an application of calcium carbonate. Clevenger (1919), however, found that the acidity of the tops of oats, buckwheat, soybeans, and cowpeas grown in limed and unlimed acid sand was higher in those plants grown in the former medium, and his conclusions are the opposite of those obtained by Truog and Meacham (1919) and Haas (1920). Miyake and Masashi (1924) found that either phosphoric acid or potassium applied to the soil tends to raise the hydrogen-ion concentration of the cell sap of oats and flax, while nitrogen tends to lower it. More acid materials as determined by electrodialysis were found in tomato plants grown on acid soil and more basic materials in tomato plants grown in alkaline soils.

Hoagland and Davis (1925) found that the hydrogen-ion concentration of the cell sap of normal cells remained at a practically constant pH value of 5.2 when the cells were immersed in solutions of widely different reactions. They considered that the preservation of a definite reaction in the vacuolar sap may be essential to the normal functioning of the cell. Different types of cells may have different reactions so that the reactions of the cell sap expressed from a mass of tissue may not represent the reactions of any particular cell. Crozier (1919) noted that the cell sap of *Valonia* had a pH of 5.0 to 6.7, the average being about 6.0, while the pH value of sea water in which the plant grew ranged from 8.2 to 8.3. Masoni (1920) observed that juices of plants maintained an acid reaction several days after contact with calcareous soil or when in contact with an excess of pure calcium carbonate. Robertson and Smith (1931) reported that the acidity of the potato tuber is independent of environment.

b. The Time of Sampling.—According to Bennet-Clark (1933), the changes in the acidity of plant tissues are associated with diurnal and seasonal fluctuations in light intensity and temperature. The majority of succulent plants and some nonsucculent plants increase markedly in acidity during the night and decrease during the day. Kraus (1883) showed in *Bryophyllum calycinum* that this fluctuation in acidity was associated with a fluctuation of the carbohydrate content in the opposite direction. It appears that the nightly increase in acidity may be brought about by the formation of malic acid from sugar. Wolf (1931) showed, however, that the decrease in sugar content during the night was not parallel in most cases to the increase in the production of acid. Truog and Meacham (1919) found the juice of the alfalfa plant to be more acid in the morning than that of plants examined later in the day. In one case at 6 A.M. the pH value of the juice was 5.97 and at 2 P.M. 6.06. In another case at 8 A.M. the pH reading was 6.04 and at 2:30 P.M. it amounted to 6.11.

Clevenger (1919) followed the diurnal changes in the acidities of the leaves, stems, and roots of the cowpea during a 24-hr. period. The acidity of the leaves and stems was found to be highest in the morning and decreased during the day, while the acidity of the roots was found to be highest during the day.

Loehwing (1930) observed that the tissue fluids from entire wheat tops showed a diurnal acid periodicity in that the free acidity reached a minimum in the evening and a maximum in the early morning.

According to Reed and Halma (1926), the hydrogen-ion concentration of both lemon and orange trees shows a rather small amount of seasonal fluctuation. There is some increase in acidity, however, during the period of rapid vegetative growth when carbohydrates are presumably being extensively oxidized.

According to Rea and Small (1927) the pH of the pith and xylem of stem tissues tends to remain constant or to fluctuate only slightly throughout the year. The pH of the cortex, endodermis, and phloem, however, in most species tends to increase during the winter and decrease during the summer, the degree of variation differing with the species.

c. The Previous Treatment.—Gray and Ryan (1921) observed that the acidity of both navel and Valencia oranges was greatly reduced when a spray composed of soap, sodium carbonate, lead arsenate, and sulphur was applied to trees even for one season. This reduction in acidity amounted to more than 50 per cent in some cases and the arsenic compound seemed to be chiefly responsible for the effect, but the facts do not correlate the reduction of acidity with the local absorption of the spray by the fruit. The action of the spray is thought to be systemic, affecting the whole tree.

It was shown by Ruhland and Wetzel (1927) that a rich nitrogen manuring causes a considerable rise in the acidity of *Begonia* plants. Miller (1931) noted that the treatment of dormant potatoes with ethylene chlorohydrin invariably results in an increase in the pH value of the expressed juice. With the use of phosphate buffers, the addition of various amounts of ethylene chlorohydrin directly to the potato juice produces no change in the pH value until a concentration of 4 per cent is reached. Orange trees sprayed with lead arsenate produce a fruit juice which is less acid than that of the unsprayed fruit (Nelson and Mattern, 1932). According to Böning and Böning-Seubert (1932) the nutritional conditions have only a limited influence on the actual acidity of the sap of the tobacco leaf.

d. Kind of Tissue.—Gustafson (1924) showed that there was a hydrogen-ion concentration gradient in corn, squash, pole bean, pumpkin, and sunflower plants. This gradient is not always in the same direction in different species or in the leaves and stems of the same species. In corn, squash, and pole bean the older leaves had a higher hydrogen-ion concentration than the younger leaves, while in pumpkin and sunflower the reverse was the case. The bases of the stems of corn, sunflower, and pumpkin had a lower hydrogen-ion concentration than the tops.

Martin, Rea, and Small (1926) made a survey of 166 species of plants with especial regard to the young flowering stems and noted that the tissue reaction varied, in general, between pH 6.0 and 4.0. The tissue of the xylem and epidermis tended to be more acid than the others. Their work indicates that some tissues may remain constant in their reaction, while the reaction of others varies with the age of the plant.

Martin (1927) found in the sunflower that the cortex, root hairs, flower buds, and epidermis had pH values of 5.7, 4.2, 5.0, and 4.4, respectively. The epidermal hairs showed an alkaline reaction with a pH of 9.0 and 10.0. After examining almost 200 species of plants, Small (1929) concluded that in general the xylem tissue is more acid than the phloem tissue. Robertson and Smith (1931) noted that the acidity is not uniform throughout the potato tuber. In the early stages of development the underground stem and the basal end of the tuber are more acid than the other portions. When sprouting begins, the active buds have the most acid reaction.

e. Age and Vigor.—Hurd (1923) investigated the acidity of corn and its relation to vegetative vigor. She found that the concentration of titratable acid was always higher in the juice of the leaves than in that of the stalk regardless of the plant's vigor. The hydrogen-ion concentration was higher in the leaves than in the stalk in the vigorous plants only; in the stunted plants, it was greater in the stalks than in the leaves. The hydrogen-ion concentration of the tops of corn plants ranged from pH 5.0 to 5.6 and was inversely correlated with the degree of vegetative vigor induced by the environmental conditions affecting the different plots. The titratable acidity of these tops varied in the same manner, the values ranging from an average of 10 cc. of 0.05 *N* sodium hydroxide solution required to neutralize 10 cc. of juice from plants of the most stunted plot to 5 cc. required to neutralize the same quantity of juice from the most rapidly growing plants. The same author (1924) studied the course of the acidity changes in the cell sap of the wheat plant during different stages of growth. The acidity determinations were made on the sap at frequent intervals from the time the plants were 3 in. high until they were in the flowering stage at the age of about 6 months. The titratable acidity of the juice decreases progressively sometimes to one-half the initial concentration between the ages of 2 and 6 weeks. This period is followed by one of relatively low acidity extending up to the visible approach of maturity, when the acid concentration rises. The final acidity at maturity may be twice the highest seeding concentration and almost three times that of the least acid stage.

The hydrogen-ion concentration of the juice of the wheat plant does not decrease to any extent between the ages of 2 and 6 weeks. It is greatly increased during the preripening period and reaches a relatively high value at the flowering stage and later. This increasing acid concentration during the final stages of growth seems to be correlated primarily with the rate of drying rather than with head formation or kernel development.

Hurd (1924) observed that the electrometric titration curve of juice expressed from wheat plants changes progressively during the seedling stage and during the maturation period. Only minor changes, correlated with environmental factors, occur during the tillering stage and most of the shooting stage. The difference in the form of the titration curve for juice of seedlings in successive stages of development indicates differences in composition associated with increasing photosynthetic activity. The end of this period of change in the curve may be interpreted as the end of the seedling stage.

Gustafson (1924) showed that in *Zea mays*, *Cucurbita maxima*, *Helianthus sp.*, and *Bryophyllum calycinum* the total acid of the plant juice is not responsible for the hydrogen-ion concentration gradient found in plants and that there is no constant relation between total and actual acidity in these plants. The juice from young parts of plants requires more sodium hydroxide to neutralize it than does juice from older parts of the same plant, even when the former is nearer to the neutral point at the beginning than the latter.

Böning and Böning-Seubert (1932) stated that the actual acidity of the leaves of tobacco increases with age. Within a pH range of 3.0 to 6.0 the young leaves are more poorly buffered than are the fully developed older leaves.

Bibliography

AHMANN, C. F., and H. D. HOOKER, 1925: The estimation of pectin and a study of the constitution of pectin, *Mo. Sta. Res. Bull.* 77.

ALDABA, V. C., 1927: The structure and development of the cell wall in plants. I, Bast fibers of *Boehmeria* and *Linum*, *Am. J. Bot.*, **14**: 16–24.

ALLEN, C. E., 1901: On the origin and nature of the middle lamella, *Bot. Gaz.*, **32**: 1–34.

ALVARADO, S., 1923: Die Entstehung der Plastiden aus Chondriosomen in den Paraphysen von *Mnium cuspidatum*, *Ber. bot. Ges.*, **41**: 85–96.

ANDERSON, D. B., 1927: A microchemical study of the structure and development of flax fibers, *Am. J. Bot.*, **14**: 187–212.

———, 1927: Über die Struktur der Kollenchymzellwand auf Grund mikrochemischer Untersuchungen, *Sitzungsber. Akad. Wiss. Wien.*, Abt. I, **136**: 429–439.

———, 1928: Some recent work on the structure of the plant cell wall, *Ohio J. Sci.*, **28**: 299–314.

———, 1928: Struktur und Chemismus der Epidermus-Aussenwand von *Clivia nobilis*, *Jahrb. wiss. Bot.*, **69**: 501–515.

———, 1934: The distribution of cutin in the outer epidermal wall of *Clivia nobilis*, *Ohio J. Sci.*, **34**: 9–18.

———, 1935: The structure of the walls of the higher plants, *Bot. Rev.*, **1**: 52–75.

ANDERSON, E., 1931: Studies on the hemicelluloses. I, The evolution of carbon dioxide by plant materials and some hemicelluloses under the action of boiling 12 per cent hydrochloric acid, *J. Biol. Chem.*, **91**: 559–568.

——— and S. KINSMAN, 1931: Studies on the hemicelluloses. II, The composition of the hemicellulose from cotton seed hulls, *J. Biol. Chem.*, **94**: 39–49.

———, 1936: The isolation of pectic substances from wood, *J. Biol. Chem.*, **112**: 531–539.

Appel, O., 1906: Zur Kenntnis des Wundverschlusses bei den Kartoffeln, *Ber. bot. Ges.*, **24**: 118–122.

Armstrong, J. I., 1929: Hydrogen ion phenomena in plants. II, An investigation of the buffer complex of sap from the stems of *Pelargonium sp.*, *Protoplasma*, **8**: 313–343.

Artschwager, E., 1927: Wound-periderm formation in the potato as affected by temperature and humidity, *J. Agr. Res.*, **35**: 995–1000.

——— and R. C. Starrett, 1931: Suberization and wound-periderm formation in sweet potato and gladiolus as affected by temperature and relative humidity, *J. Agr. Res.*, **43**: 353–364.

——— and ———, 1933: Suberization and wound-cork formation in the sugar beet as affected by temperature and relative humidity, *J. Agr. Res.*, **47**: 669–674.

Astbury, W. T., T. C. Marwick, and J. D. Bernal, 1932: X-ray analysis of the structure of the wall of *Valonia*, *Proc. Roy. Soc.*, B, **109**: 443–450.

Astruc, A., 1903: Recherches sur l'acidité végétale, *Ann. Sci. Nat. Bot.*, **17**: 1–64.

Atkins, W. R. G., 1922: The hydrogen-ion concentration of plant cells, *Sci. Proc. Roy. Dublin Soc.*, **16**: 414–426.

———, 1922: The hydrogen-ion concentrations of some Indian soils and plant juices, *Agr. Res. Inst. Pusa Bull.* **136**: 1–12.

Bailey, I. W., and T. Kerr, 1935: The visible structure of the secondary wall and its significance in physical and chemical investigations of tracheary cells and fibers, *J. Arnold Arboretum*, **16**: 273–300.

Bailey, K., and F. W. Norris, 1932: The nature and composition of the mucilage of the seed of white mustard (*Brassica alba*), *Biochem. J.*, **26**: 1609–1623.

Baker, G. L., and R. Kneeland, 1935: Viscosity of apple pectin. Effect of commercial diastatic enzymes at 40°C., *Ind. and Eng. Chem.*, **27**: 92–94.

Balls, W. L., and H. A. Hancock, 1919: The existence of daily growth rings in the cell wall of cotton hairs, *Proc. Roy. Soc.*, B, **90**: 542–555.

———, 1923: Further observations on the cell wall structure as seen in cotton hairs, *Proc. Roy. Soc.*, B, **93**: 426–440.

Bamford, K. F., and W. G. Campbell, 1936: The determination of lignin in the analysis of woods, *Biochem. J.*, **30**: 419–428.

Beams, H. W., and R. L. King, 1935: The effect of ultracentrifuging on the cells of the root tip of the bean *Phaseolus vulgaris*, *Proc. Roy. Soc.*, B, **118**: 264–276.

Beck, W. A., 1929: The effect of drought on the osmotic value of plant tissue, *Protoplasma*, **8**: 70–126.

Beer, R., 1904: The present position of cell wall research, *New Phytol.*, **3**: 159–164.

Belzung, E., 1887: Recherches morphologiques et physiologiques sur l'amidon et les grains de chlorophylle, *Ann. Sci. Nat. Bot.*, VII, **5–6**: 179–311.

———, 1891: Nouvelles recherches sur l'origine des grains d'amidon et des grains chlorophylliens, *Ann. Sci. Nat. Bot.*, VII, **13**: 1–22.

Bennet-Clark, T. A., 1930: The metabolism of succulent plants, *thesis*, *Cambridge Univ.*

———, 1933: The role of organic acids in plant metabolism. I, *New Phytol.*, **32**: 37–71.

———, 1933: The role of organic acids in plant metabolism. III, *New Phytol.*, **32**: 197–230.

——— and W. M. Woodruff, 1935: Seasonal changes in acidity of the rhubarb (*Rheum hybridum*), *New Phytol.*, **34**: 77–91.

Bernard, C., 1878: "Leçons sur les phénomènes de la vie," 1st ed., 1878; 2d ed., 1885, Paris.

Bloch, R., 1935: Wound healing in *Tradescantia fluminensis, Ann. Bot.*, **49**: 651–670.

Böning, K., and E. Böning-Seubert, 1932: Wasserstoffionenkonzentration und Pufferung im Pressaft von Tobakblättern in ihrer Abhängigkeit von der Ernährung und Entwicklung der Pflanze. I, *Biochem. Z.*, **247**: 35–67.

Bonner, J., 1935: Zum Mechanismus der Zellstreckung auf Grund der Micellarlehre, *Jahrb. wiss. Bot.*, **82**: 377–412.

———, 1935: Some colloidal properties of the pectins, *Proc. Akad. Amsterdam*, **38**, No. 3: 1–11.

——— and A. N. J. Heyn, 1935: Cell elongation and the electrical properties of the cell wall, *Protoplasma*, **24**: 466–469.

———, 1936: The chemistry and physiology of the pectins, *Bot. Rev.*, **2**: 475–497.

Bowen, R. H., 1927: A preliminary report on the structural elements of cytoplasm in plant cells, *Biol. Bull.*, **53**: 179–195.

Brady, J., 1934: Some factors influencing lodging in cereals, *J. Agr. Sci.*, **24**: 209–232.

Bragg, W., 1930: Cellulose in the light of the X-ray, *Nature*, **125**: 315–322.

Brannon, M. A., 1913: Osmotic pressure in potatoes, *Bot. Gaz.*, **56**: 433–439.

Bredow, H., 1890: Beiträge zur Kenntniss der Chromatophoren, *Jahrb. wiss. Bot.*, **22**: 349–414.

Bristol, B. M., 1916: On the remarkable retention of vitality of moss protonema, *New Phytol.*, **15**: 137–143.

Brunstetter, B. C., and C. A. Magoon, 1930: A microelectrode for the rapid determination of the hydrion concentration of expressed juices from small amounts of plant tissue, *Plant Physiol.*, **5**: 249–256.

Burkhart, B. A., 1936: Hemicellulose constituents in alfalfa roots, *Plant Physiol.*, **11**: 421–428.

Buston, H. W., and V. H. Chambers, 1933: Some cell-wall constituents of *Cetraria islandica* ("Iceland moss"), *Biochem. J.*, **27**: 1691–1702.

———, 1935: Observations on the nature, distribution, and development of certain cell-wall constituents of plants, *Biochem. J.*, **29**: 196–218.

Cake, W. E., and H. H. Bartlett, 1922: The carbohydrate content of the seed of *Asparagus officinalis, J. Biol. Chem.*, **51**: 93–102.

Candlin, E. J., and S. B. Schryver, 1928: Investigations of the cell-wall substances of plants, with special reference to the chemical changes taking place during lignification, *Proc. Roy. Soc.*, B, **103**: 365–376.

Carre, M. H., and D. Haynes, 1922: The estimation of pectin as calcium pectate and the application of this method to the determination of the soluble pectin in apples, *Biochem. J.*, **16**: 60–70.

———, 1925: Chemical studies in the physiology of apples. IV, Investigations on the pectic constituents of apples, *Ann. Bot.*, **39**: 811–839.

———, 1925: The relation of pectose and pectin in apple tissue, *Biochem. J.*, **19**: 257–266.

Castoro, N., 1906: Beiträge zur Kenntnis der Hemicellulosen, *Z. physiol. Chem.*, **49**: 96–107.

———, 1907: Ueber die in den Samenschalen von *Cucurbita pepo* enthaltenen Hemicellulosen, *Z. physiol. Chem.*, **52**: 521–525.

Cavers, F., 1914: Chondriosomes (mitochondria) and their significance, *New Phytol.*, **13**: 96–106, 170–180.

Chandler, W. H., 1914: Sap studies with horticultural plants, *Mo. Sta. Res. Bull.* 14.

Chibnall, A. C., S. H. Piper, A. Pollard, J. A. B. Smith, and E. F. Williams, 1931: The wax constituents of the apple cuticle, *Biochem. J.*, **25**: 2095–2110.

Chrysler, H. L., 1934: Amounts of bound and free water in an organic colloid at different degrees of hydration, *Plant Physiol.*, **9**: 143–155.

Clague, J. A., and C. R. Fellers, 1934: Relation of benzoic acid content and other constituents of cranberries to keeping quality, *Plant Physiol.*, **9**: 631–636.

Clark, G. L., 1930: Cellulose as it is completely revealed by X rays, *Ind. and Eng. Chem.*, **22**: 474–487.

Clayson, D. H. F., F. W. Norris, and S. B. Schryver, 1921: The pectic substances of plants. Part II, A preliminary investigation of the chemistry of the cell walls of plants, *Biochem. J.*, **15**: 643–654.

———, and S. B. Schryver, 1923: The hemicelluloses. I, The hemicellulose of wheat flour, *Biochem. J.*, **17**: 493–496.

Clevenger, C. B., 1919: Hydrogen-ion concentration of plant juices. I, The accurate determination of the hydrogen-ion concentration of plant juices by means of the hydrogen electrode, *Soil Sci.*, **8**: 217–226.

———, 1919: Hydrogen-ion concentration of plant juices. II, Factors affecting the acidity or hydrogen-ion concentration of plant juices, *Soil Sci.*, **8**: 227–242.

Conrad, C. M., 1926: A biochemical study of the insoluble pectic substances in vegetables, *Am. J. Bot.*, **13**: 531–548.

———, 1930: A furfural-yielding substance as a splitting product of protopectin during the ripening of fruits, *Plant Physiol.*, **5**: 93–103.

———, 1931: Decarboxylation studies on pectins and calcium pectates, *J. Am. Chem. Soc.*, **53**: 1999–2003.

———, 1931: The decarboxylation of *d*-galacturonic acid with special reference to the hypothetical formation of *l*-arabinose, *J. Am. Chem. Soc.*, **53**: 2282–2287.

Cowdry, E. V., 1924: "General Cytology," University of Chicago Press.

———, 1926: Surface film theory of the function of mitochondria, *Am. Nat.*, **60**: 157–165.

Crafts, A. S., 1933: Sulfuric acid as a penetrating agent in arsenical sprays for weed control, *Hilgardia*, **8**: 125–147.

Crist, J. W., and L. P. Batjer, 1931: The stone cells of pear fruits, especially the Kieffer pear, *Mich. Sta. Tech. Bull.* 113.

Cross, C. F., and C. Dorée, 1923: "Researches on Cellulose," Longmans, Green & Company, New York.

Crozier, W. J., 1919: Intracellular acidity in *Valonia*, *J. Gen. Physiol.*, **1**: 581–585.

Culpepper, C. W., and H. H. Moon, 1933: Composition of rhubarb at different stages of maturity in relation to its use in cooking and canning, *J. Agr. Res.*, **46**: 387–402.

Czaja, A. T., 1930, 1934: Untersuchungen über metachromatische Farbungen von Pflanzengeweben. I, *Planta*, **11**: 582*ff.*; II, **21**: 531–601.

Dadswell, H. E., and L. F. Hawley, 1929: Chemical composition of wood in relation to physical characteristics. A preliminary study, *Ind. and Eng. Chem.*, **21**: 973–975.

Damm, O., 1901: Ueber den Bau, die Entwicklungsgeschichte und die mechanischen Eigenschaften mehrjähriger Epidermen bei den Dicotyledonen, *Beih. bot. Cent.*, **11**: 219–260.

Davidson, G. F., and W. A. Richardson, 1936: The molecular structure of cellulose and starch, *Sci. Prog.*, **31**: 68–77.

Davidson, J., and M. Phillips, 1930: Lignin as a possible factor in lodging of cereals, *Sci.*, **72**: 401–402.

Dixon, H. H., and W. R. G. Atkins, 1910: On osmotic pressure in plants, and on a thermo-electric method of determining freezing points, *Sci. Proc. Roy. Dublin Soc.*, n. s., **12**: 275–311.

——— and ———, 1912: Changes in the osmotic pressure of the sap of the developing leaves of *Syringa vulgaris*, *Notes Bot. School Trinity Col. Dublin*, **2**: 99–102.

——— and ———, 1912: Variations in the osmotic pressure of the leaves of *Hedera helix*, *Notes Bot. School Trinity Col. Dublin*, **2**: 103–110.

——— and ———, 1912: Variations in the osmotic pressure of the sap of *Ilex aquifolium*, *Notes Bot. School Trinity Col. Dublin*, **2**: 111–120.

——— and ———, 1913: Osmotic pressures in plants. I, Methods of extracting sap from plant organs, *Notes Bot. School Trinity Col. Dublin*, **2**: 154–165.

——— and ———, 1913: Osmotic pressures in plants. II, Cryoscopic and conductivity measurements on some vegetable saps, *Notes Bot. School Trinity Col. Dublin*, **2**: 166–172.

——— and ———, 1916: Osmotic pressure in plants. VI, On the composition of the sap in the conducting tracts of trees at different levels and at different seasons of the year, *Sci. Proc. Roy. Dublin Soc.*, **15**: 51–62.

———, and T. A. Bennet-Clark, 1932: Electrical properties of oil water emulsions, with special reference to the structure of the plasmatic membrane. II, *Sci. Proc. Roy. Dublin Soc.*, **20**: 211*ff*.

Dore, W. S., 1924: Recent progress in the chemistry of pectin and its industrial applications, *Ind. and Eng. Chem.*, **16**: 1042–1044.

Dorée, C. and E. C. Barton-Wright, 1927: Contributions to the study of lignin. I, Meta-lignin a new type of alkali lignin, *Biochem. J.*, **21**: 290–300.

Drabble, E., and H. Lake, 1905: The osmotic strength of cell sap in plants growing under different conditions, *New Phytol.*, **4**: 189–192.

———, and H. Drabble, 1907: The relation between the osmotic strength of cell sap in plants and their physical environment, *Biochem. J.*, **2**: 117–132.

Dunne, T. C., 1932: Plant buffer systems in relation to the absorption of bases by plants, *Hilgardia*, **7**: 207–234.

Eaton, F. M., 1927: The water requirement and cell-sap concentration of Australian salt brush and wheat as related to the salinity of the soil, *Am. J. Bot.*, **14**: 212–226.

Emmett, A. M., 1929: An investigation on the changes which take place in the chemical composition of pears stored at different temperatures, with special reference to the pectic changes, *Ann. Bot.*, **43**: 269–308.

Esselen, G. J., 1923: A few high spots in the chemistry of cellulose, *Ind. and Eng. Chem.*, **15**: 306–308.

Eyster, W. H., 1929: Variation in the size of plastids in genetic strains of *Zea mays*, *Sci.*, **69**: 48.

———, 1933: Plastid studies in genetic types of maize: argentia chlorophyll, *Plant Physiol.*, **8**: 105–121.

Faber, O., and B. Tollens, 1899: Untersuchungen über die Oxycellulose, *Ber. chem. Ges.*, **32**: 2589–2601.

Famintzin, A., 1893: Sur les grains de chlorophylle dans les graines et les plantes germeantes, *Bull. de l'Acad. imp. de Sciences de St. Petersbourg*, IV, **36**: 75–85.

Farr, W. K., and G. L. Clark, 1932: Cotton fibers. II, Structural features of the wall suggested by X-ray diffraction analyses and observations in ordinary and plane polarized light, *Contr. Boyce Thompson Inst.*, **4**: 273–296.

——— and S. H. Eckerson, 1933, 1934: The formation of cellulose membranes by microscopical particles of uniform size in linear arrangement, abst., *Am. J. Bot.*, **20**: 672; *Contr. Boyce Thompson Inst.*, **6**: 189–204.

——— and ———, 1934: Separation of cellulose particles in membranes of cotton fibers by treatment with hydrochloric acid, *Contr. Boyce Thompson Inst.*, **6**: 309–313.

——— and W. A. Sisson, 1934: X-ray diffraction patterns of cellulose particles and interpretation of cellulose diffraction data, *Contr. Boyce Thompson Inst.*, **6**: 315–321.

———, 1934: Cotton fibers. IV, Fiber abnormalities and density of the fiber mass within the boll, *Contr. Boyce Thompson Inst.*, **6**: 471–478.

Faul, J. H., 1928: Living cells in heartwood of trees, *Sci.*, **67**: 296.

Fink, B., 1925: Some considerations of protoplasm, *Ohio J. Sci.*, **25**: 99–113.

Fischer, F., and R. Lieske, 1928: Untersuchungen über das Verhalten des Lignins bei der natürlichen Zersetzung von Pflanzen, *Biochem. Z.*, **203**: 351–362.

Fitting, H., 1911: Die Wasserversorgung und die osmotischen Druckverhältnisse der Wüstenpflanzen, *Zeits. f. Bot.*, **3**: 209–275.

Franzen, H., and E. Keyssner, 1923: Kritisches über das Vorkommen der Äpfelsäure in der Pflanzen, *Biochem. Z.*, **135**: 183–216.

——— and F. Helwert, 1923: Kritisches über das Vorkommen der Citronensäure in der Pflanzen, *Biochem. Z.*, **135**: 384–415.

——— and ———, 1923: Kritisches über das Vorkommen der Weinsäure, *Biochem. Z.*, **136**: 291–305.

——— and R. Ostertag, 1923: Kritisches über das Vorkommen der Bernsteinsäure, *Biochem. Z.*, **136**: 327–335.

Fremy, M. E., 1840: Recherches sur la pectine et l'acide pectique, *J. de Pharm.*, **26**: 368–393.

———, 1859: Recherches chimiques sur la composition des cellules végétales, *Ann. d. Sci. Nat.*, IV, t. 12.

Freudenberg, K., F. Sohns, W. Durr, and C. Niemann, 1931: Ueber Lignin, Coniferylalcohol und Saligenin, *Cellulosechemie*, **12**: 263–276. Quoted by Harlow, 1932.

———, 1932: The relation of cellulose to lignin in wood, *J. Chem. Ed.*, **9**: 1171–1180.

Frey, A., 1926: Der heutige Stand der Micellartheorie, *Ber. bot. Ges.*, **44**: 564–570.

———, 1926: Die submikroskopische Struktur der Zellmembranen, *Jahrb. wiss. Bot.*, **65**: 195–223.

———, 1927: Der submikroskopische Feinbau der Zellmembran, *Die Natur. Wiss.*, **15**: 760–765.

———, 1928: La structure micellaire des membranes cellulaires, *Bull. Soc. Bot. France*, **75**: 950–956.

———, 1928: Über die intermicellar-Räume der Zellmembranen, *Ber. bot. Ges.*, **46**: 444–456.

Frey-Wyssling, A., 1930: Mikroskopische Technik der Micellaruntersuchung von Zellmembranen, *Z. wiss. Mikrosk.*, **47**: 1–46.

Gaddum, L. W., 1934: The pectic constituents of citrus fruits, *Fla. Sta. Bull.* 268.

Gail, F. W., 1926: Osmotic pressure of cell sap and its possible relation to winter killing and leaf fall, *Bot. Gaz.*, **81**: 434–445.

——— and W. H. Cone, 1929: Osmotic pressure and pH measurements on cell sap of *Pinus ponderosa*, *Bot. Gaz.*, **88**: 437–441.

Gardiner, W., and A. W. Hill, 1901: II, The histology of the cell wall with special reference to the mode of connection of cells, *Trans. Phil. Roy. Soc.*, B, **194**: 83–125.

Gerhardt, F., 1929: Effect of acid and alkaline hydrolysis on the estimation of hemicellulose and associated groups in young apple wood, *Plant Physiol.*, **4**: 373–383.

Gerritz, H. W., 1935: Extraction of pectin from apple thinnings, *Ind. and Eng. Chem.*, **27**: 1458–1459.

Gibbs, R. D., 1935: Studies of wood. I, The cell wall, *Can. J. Res.*, **12**: 715–726.

GILSON, E., 1890: La suberine et les cellules du liège, *La Cellule*, **6**: 63.
———, 1892: La cristallisation de la cellulose et la composition chimique de la membrane cellulaire végétale, *La Cellule*, **9**: 397–441.
GRAY, G. P., and H. J. RYAN, 1921: Reduced acidity in oranges caused by certain sprays, *Month. Bull. Dept. Agr. Calif.*, **10**: 11–33.
GREATHOUSE, G. A., 1932: Effects of the physical environment on the physicochemical properties of plant saps and the relation of these properties to leaf temperature, *Plant Physiol.*, **7**: 349–390.
GRIS, A., 1857: Recherches microscopiques sur la chlorophylle, *Ann. Sci. Nat. Bot.*, **7**: 179–219.
GRÜSS, J., 1897: Studien über Reservecellulose, *Bot. Cent.*, **70**: 242–261.
GUILLIERMOND, A., 1911: Sur la formation des chloroleucites aux dépens des mitochondries, *C. R. Acad. Sci. Paris*, **153**: 290–292.
———, 1911: Sur les mitochondries des cellules végétales, *C. R. Acad. Sci. Paris*, **153**: 199–202.
———, 1911: Sur l'origine des leucoplastes et sur le processus cytologique de l'élaboration de l'amidon dans le tubercule de la pomme de terre, *C. R. Acad. Sci. Paris*, **153**: 1492–1494.
———, 1912: Sur les différentes modes de la formation des leucoplastes, *C. R. Soc. Biol. Paris*, **73**: 110–112.
———, 1914: Bemerkungen über die Mitochondrien der vegetativen Zellen und ihre Verwandlung in Plastiden. Eine Antwort auf einige Einwürfe, *Ber. bot. Ges.*, **32**: 282–301.
———, 1929: The recent development of our idea of the vacuome of plant cells, *Am. J. Bot.*, **16**: 1–23.
———, 1932: La structure de la cellule végétale: les inclusions du cytoplasme et en particulier les chondriosomes et les plastes, *Protoplasma*, **16**: 291–337.
———, 1932: La structure des cellules végétales à l'ultramicroscope, *Protoplasma*, **16**: 454–477.
GUSTAFSON, F. G., 1922: Studies of the hydrogen-ion concentration of plant juices, *Mich. Acad. Sci.*, **1922**: 49–53.
———, 1924: Hydrogen-ion concentration gradient in plants, *Am. J. Bot.*, **11**: 1–6.
———, 1924: Total acidity compared with actual acidity of plant juices, *Am. J. Bot.*, **11**: 365–370.
HAAS, A. R. C., 1916: The acidity of plant cells as shown by natural indicators, *J. Biol. Chem.*, **27**: 233–241.
———, 1917: The reaction of plant protoplasm, *Bot. Gaz.*, **63**: 232–235.
———, 1920: Studies on the reaction of plant juices, *Soil Sci.*, **9**: 341–370.
———, and F. F. HALMA, 1931: Sap concentration and inorganic constituents of mature citrus leaves, *Hilgardia*, **5**: 407–424.
HABER, E. S., 1928: The influence of soil reaction on the ionizable constituents of the tomato as determined by electro-dialysis, *J. Agr. Res.*, **37**: 101–114.
HABERLANDT, G., 1914: "Physiological Plant Anatomy," English Trans., 4th ed. Macmillan & Company, Ltd., London.
HALLER, M. H., 1929: Changes in the pectic constituents of apples in relation to softening, *J. Agr. Res.*, **39**: 739–747.
HALMA, F. F., and A. R. C. HAAS, 1928: Effect of sunlight on sap concentration of citrus leaves, *Bot. Gaz.*, **86**: 102–107.
HANSTEEN-CRANNER, B., 1914: Ueber das Verhalten der Kulturpflanzen zu den Bodensalzen. III, Beiträge zur Biochemie und Physiologie der Zellwand lebender Zellen, *Jahrb. wiss. Bot.*, **35**: 536–598.

———, 1919: Beiträge zur Biochemie und Physiologie der Zellwand und der plasmatischen Grenzschichten, *Ber. bot. Ges.*, **37**: 380–391.

———, 1922: "Zur Biochemie und Physiologie der Grenzschichten lebender Pflanzenzellen," Kristiania, 1922.

———, 1926: Untersuchungen über die Frage ob in der Zellwand lösliche Phosphatide vorkommen, *Planta Archiv. wiss. Bot.*, **2**: 438–445.

HARLOW, W. M., 1927: The chemical nature of the middle lamella, *N. Y. State Col. Forestry Syracuse Univ. Tech. Bull.* 21.

———, 1928: Contributions to the chemistry of the plant cell wall. II, Lignification in the secondary and tertiary layers of the cell wall of wood, *N. Y. State Col. Forestry Syracuse Univ. Tech. Bull.* 24.

———, 1932: Contributions to the chemistry of the plant cell wall. VI, Further studies on the location of lignin in the cell walls of wood, *Am. J. Bot.*, **19**: 729–739.

HARPER, R. A., 1919: The structure of protoplasm, *Am. J. Bot.*, **6**: 273–301.

HARRIS, J. A., J. V. LAWRENCE, and R. A. GORTNER, 1915: The osmotic pressure of the juices of desert plants, *Sci.*, **41**: 656–658.

——— and W. POPENOE, 1916: Freezing point lowering of the leaf sap of the horticultural types of *Persea americana*, *J. Agr. Res.*, **7**: 261–269.

——— and J. V. LAWRENCE, 1916: The cryoscopic constants of expressed vegetable saps as related to local environmental conditions in the Arizona deserts, *Physiol. Res.*, **2**: 1–49.

——— and ———, 1916: On the osmotic pressure of the tissue fluids of Jamaican Loranthaceae parasitic on various hosts, *Am. J. Bot.*, **3**: 438–455.

——— and ———, 1917: Cryoscopic determinations on tissue fluids of plants of Jamaican coastal deserts, *Bot. Gaz.*, **64**: 285–305.

——— and ———, 1917: The osmotic concentration of the sap of the leaves of mangrove trees, *Biol. Bull. Marine Biol. Lab. Woods Hole*, **32**: 202–211.

——— and ———, 1917: The osmotic concentration of the tissue fluids of Jamaican montane rain forest vegetation, *Am. J. Bot.*, **4**: 268–299.

——— and ———, 1917: Vegetable saps, *Carnegie Inst. Wash. Yearbook*, **16**: 131–132.

———, 1917: Osmotic concentration of tissue fluids in relation to geographical distribution, *Carnegie Inst. Wash. Yearbook*, **16**: 88–89.

———, R. A. GORTNER, and J. V. LAWRENCE, 1917: The relationship between the osmotic concentration of leaf sap and height of leaf insertion in trees, *Bull. Torrey Bot. Club*, **44**: 267–286.

———, 1918: On the osmotic concentration of the tissue fluids of phanerogamic epiphytes, *Am. J. Bot.*, **5**: 490–506.

———, R. A. GORTNER, and J. V. LAWRENCE, 1921: On the differentiation of the leaf-tissue fluids of ligneous and herbaceous plants with respect to osmotic concentration and electrical conductivity, *J. Gen. Physiol.*, **3**: 343–347.

———, J. V. LAWRENCE, and R. A. GORTNER, 1921: The osmotic concentration and electrical conductivity of the tissue fluids of ligneous and herbaceous plants, *J. Phys. Chem.*, **25**: 122–146.

———, R. A. GORTNER, W. F. HOFFMAN, J. V. LAWRENCE, and A. T. VALENTINE, 1924: The osmotic concentration, specific electrical conductivity, and chlorid content of the tissue fluids of the indicator plants of Tooele Valley, Utah, *J. Agr. Res.*, **27**: 893–924.

———, J. V. LAWRENCE, and Z. W. LAWRENCE, 1924: The chloride content of the leaf tissue fluids of Egyptian and upland cotton, *J. Agr. Res.*, **28**: 695–705.

———, 1925: The accumulation of chlorides in the leaf-tissue fluids of Egyptian cotton with the march of the season, *Proc. Soc. Exp. Biol. Med.*, **22**: 415–417.

———, C. T. Hoffman, and W. F. Hoffman, 1925: Sulfate content of the leaf-tissue fluids of Egyptian and upland cotton, *J. Agr. Res.*, **31**: 653–663.

———, W. F. Hoffman and others, 1925: The leaf-tissue fluids of Egyptian cotton, *J. Agr. Res.*, **31**: 1027–1035.

———, 1926: The relationship between the concentration of the soil solution and the physico-chemical properties of the leaf-tissue fluids of Egyptian and upland cotton, *J. Agr. Res.*, **32**: 605–649.

———, G. J. Harrison, and T. A. Pascoe, 1930: Osmotic concentration and water relations in the mistletoes, with special reference to the occurrence of *Phoradendron californicum* on *Covillea tridentata*, *Ecology*, **11**: 687–702.

——— and T. A. Pascoe, 1930: Further studies on the relationship between the concentration of the soil solution and the physico-chemical properties of the leaf-tissue fluids of cotton, *J. Agr. Res.*, **41**: 767–789.

———, ———, and I. D. Jones, 1930: Note on the tissue fluids of *Phoradendron juniperinum* parasitic on *Juniperus utahensis*, *Bull. Torrey Bot. Club*, **57**: 113–116.

———, 1934: "The Physico-chemical Properties of Plant Saps in Relation to Phytogeography. Data on Native Vegetation in its Natural Environment," 339 pp., University of Minnesota Press.

Hawley, L. F., and A. G. Norman, 1932: The differentiation of hemicelluloses, *Ind. and Eng. Chem.*, **24**: 1190–1194.

Haworth, W. N., 1929: "The Constitution of Sugars," London.

——— and E. L. Hirst, 1936: The chemistry of the carbohydrates and the glucosides, *Ann. Rev. Biochem.*, **5**: 81–100.

Haynes, D., 1914: The gelatinisation of pectin in solutions of the alkalies and the alkaline earths, *Biochem. J.*, **8**: 553–583.

Hein, I., 1926: Changes in plastids in variegated plants, *Bull. Torrey Bot. Club*, **53**: 411–419.

Herzog, R. O., 1925: Zur Erkenntnis der Cellulose Faser, *Ber. chem. Ges.*, **58**: 1254–1262.

Hess, K., 1928: Zur Frage des Aufbaues pflanzlicher Membrane, *Biochem. Z.*, **203**: 409–420.

Hibbard, R. P., and O. E. Harrington, 1916: The depression of the freezing point in triturated plant tissue and the magnitude of this depression as related to soil moisture, *Physiol. Res.*, **1**: 441–454.

Hill, T. G., 1908: Observations on the osmotic properties of the root hairs of certain salt marsh plants, *New Phytol.*, **7**: 133–142.

Hoagland, D. R., and A. R. Davis, 1925: Suggestions concerning the absorption of ions by plants, *New Phytol.*, **24**: 99–111.

Hooper, F. E., 1931: Studies on parenchymatous and vascular plant tissues, some analytical and specific gravity data, *Plant Physiol.*, **6**: 531–540.

Hopkins, E. F., and J. H. Gourley, 1930: Nitrogen fertilization and pectic materials in grapes, *Proc. Am. Soc. Hort. Sci.*, **27**: 164–169.

Horning, E. S., and A. H. K. Petrie, 1927: The enzymatic function of mitochondria in the germination of cereals, *Proc. Roy. Soc.*, B, **102**: 188–206.

Hurd, A. M., 1923: Acidity of corn and its relation to vegetative vigor, *J. Agr. Res.*, **25**: 457–469.

———, 1924: The course of acidity changes during the growth period of wheat with special reference to stem rust resistance, *J. Agr. Res.*, **27**: 725–735.

Hurd-Karrer, A. M., 1925: A concentration gradient in corn stalks, *J. Gen. Physiol.*, **9**: 341–345.

———, 1928: Changes in the buffer system of the wheat plant during its development, *Plant Physiol.*, **3**: 131–153.

———, 1929: Relation of leaf acidity to vigor in wheat grown at different temperatures, *J. Agr. Res.*, **39**: 341–350.

———, 1930: Titration curves of etiolated and of green wheat seedlings reproduced with buffer mixtures, *Plant Physiol.*, **5**: 307–328.

ILIN, V. S., and P. S. NAZAROVA, 1915: Osmotic pressure in roots and leaves of plants with respect to the moisture content of their habitats, *Bull. Acad. Imp. Sci. Petrograd.*, 6 ser., **9**: 749–768; *Exp. Sta. Rec.*, **36**: 733.

ILJIN, W. S., 1929: Der Einfluss der Standortsfeuchtigkeit auf den osmotischen Wert bei Pflanzen, *Planta*, **7**: 45–58.

JODIDI, S. L., 1933: The presence of citric acid in the Alaska pea, *J. Am. Chem. Soc.*, **55**: 4663–4664.

JUNGERS, V., 1930: Recherches sur les plasmodesmes chez les végétaux, *La Cellule*, **40**: 7–81.

KAMERLING, Z., 1897: Zur Biologie und Physiologie der Zellmembrane, *Bot. Cent.*, **72**: 49–54, 85–91.

KELLER, B. A., 1920: The osmotic pressure of the cell sap of plants of different ecological types and different habitats, *Proc. Russ. Bot. Soc.*, **5**: 84–92.

KERR, T., and I. W. BAILEY, 1933: Composition of the so-called middle lamella in the cambium and its lignified derivatives, abst. *Am. J. Bot.*, **20**: 672.

——— and ———, 1934: Structure, optical properties, and chemical composition of the so-called middle lamella, *J. Arnold Arboretum*, **15**: 327–349.

KILLIAN, C., and L. FAUREL, 1934: Observations sur la pression osmotique des végétaux désertique et subdésertiques de l'Algérie, *Bull. Soc. Bot. France*, **80**: 775–778; *Biol. Abst.*, **9**: 1710.

KLASON, P., 1911: "Beiträge zur Kenntniss der chemischen Zusammensetzung des Fichtenholzes," Berlin.

KOPACZEWSKI, W., 1925: Sur la coagulation de la pectine, *Bull. Soc. Chim. Biol.*, **7**: 419–428.

KORSTIAN, C. F., 1924: Density of cell sap in relation to environmental conditions in the Wasatch Mountains of Utah, *J. Agr. Res.*, **28**: 845–909.

KOZLOWSKI, A., 1922: Critique de l'hypothèse des chondriosomes, *Rev. Gén. Bot.*, **34**: 641–658.

KRAUS, G., 1883: Über die Wasserverteilung der Pflanzen. IV, Die Acidität des Zellsaftes, *Abh. naturf. Ges. zu Halle*, **16**: 141–205.

KÜSTER, E., 1919: Über weissrandige Blätter und andere Formen der Buntblättrigkeit, *Biol. Cent.*, **39**: 212–251.

LAMBRECHT, E., 1926: Beitrag zur Kenntnis der osmotischen Zustandsgrössen einiger Pflanzen des Flachlandes, *Beitr. Biol. Pflanzen*, **17**: 87–136; *Biol. Abst.*, **5**: 394.

LEE, B., and J. H. PRIESTLEY, 1924: The plant cuticle. I, Its structure, distribution and function, *Ann. Bot.*, **38**: 525–545.

———, 1925: The plant cuticle. II, A macrochemical study, *Ann. Bot.*, **39**: 755–768.

LEGG, V. H., and R. V. WHEELER, 1925: Plant cuticles. I, Modern plant cuticles, *J. Chem. Soc.*, **127**: 1412–1421.

——— and ———, 1929: Plant cuticles. Part I (continued), Modern plant cuticles, *J. Chem. Soc.*, **1929**: 2444–2449.

——— and V. W. RICHARD, 1929: Plant Cuticles. II, Fossil plant cuticles, *J. Chem. Soc.*, **1929**: 2449–2458.

LEONTJEW, H., 1927: Über das spezifische Gewicht des Protoplasmas. III, *Protoplasma*, **2**: 59–64.

LEPESCHKIN, W. W., 1928: The chemical and physical composition of protoplasm, *Sci.*, **68**: 45–48.

———, 1930: My opinion about protoplasm, *Protoplasma*, **9**: 269–297.

———, 1932: Influence of visible and ultraviolet rays on the stability of protoplasm, *Am. J. Bot.*, **19**: 547–558.

LEWITSKY, G., 1910: Über die Chondriosomen in pflanzlichen Zellen, *Ber. bot. Ges.*, **28**: 538–546.

———, 1911: Die Chloroplastenanlagen in lebenden und fixierten Zellen von *Elodea canadensis*, *Ber. bot. Ges.*, **29**: 697–703.

LINK, K. P., 1929: Chemical composition of corn (*Zea mays*) seedlings. I, The isolation of xylan and cellulose from the cell walls, *J. Am. Chem. Soc.*, **51**: 2506–2516.

———, 1929: The chemical composition of corn (*Zea mays*) seedlings. II, The isolation of a dextrin similar to the trihexosan obtained by the thermal depolymerization of potato starch, *J. Am. Chem. Soc.*, **51**: 2516–2522.

LIPMAN, C. B., 1934: Further evidence on the amazing longevity of bacteria, *Sci.*, **79**: 230–231.

LITVINOV, L. S., 1927: Chemical nature of the sap of plants, *Bull. Inst. Rech. Biol. and Sta. Biol. Univ. Perm.*, **5**: 311–329; *Biol. Abst.*, **4**: 45.

LIVINGSTON, L. G., 1933: The nature and distribution of plasmodesma in the tobacco plant, abst. *Am. J. Bot.*, **20**: 673.

———, 1935: The nature and distribution of plasmodesmata in the tobacco plant, *Am. J. Bot.*, **22**: 75–87.

LOEHWING, W. F., 1930: Effects of insolation and soil characteristics of tissue-fluid reaction in wheat, *Plant Physiol.*, **5**: 293–305.

LOMBARDI, L., 1932: Ricerche sulla pressione osmotica nelle pianta di *Morus alba* L., *Boll. Staz. Bachi-Gesilcolt. Ascoli-Piceno*, **10**: 11–28; *Biol. Abst.*, **8**: 872.

LUBIMENKO, V., 1926: Recherches sur les pigments des plastes sur la photosynthèse. I, La physiologie des plastes, *Rev. Gén. Bot.*, **38**: 307–328, 381–400.

———, 1927: Recherches sur les pigments des plastes sur la photosynthèse. II, Les pigments des plastes et leur transformation dans les tissues vivants de la plante, *Rev. Gén. Bot.*, **39**: 1–115.

———, 1928: Recherches sur les pigments des plastes sur la photosynthèse. III, La biologie de la photosynthèse, *Rev. Gén. Bot.*, **40**: 1–59.

LÜDTKE, M., 1931: Untersuchungen über Aufbau und Bildung der pflanzlichen Zellmembran und ihre stofflichen Komponenten, *Biochem. Z.*, **233**: 1–57.

———, 1931: Intermediate products of cellulose formation, *Cellulosechemie*, **12**: 307–310; *Chem. Abst.*, **26**: 4939.

———, 1935: Werden und Organisation der pflanzlichen Zellmembran, *Protoplasma*, **22**: 457–488.

LUTMAN, B. F., 1919: Osmotic pressures in the potato plant at various stages of growth, *Am. J. Bot.*, **6**: 181–202.

MA, R. M., 1930: Starch deposition in the sporogenous cells of certain mosses, *Bull. Torrey Bot. Club*, **57**: 525–533.

———, 1931: The chloroplasts of *Selaginella*, *Bull. Torrey Bot. Club*, **57**: 277–284.

MACDOUGAL, D. T., and H. A. SPOEHR, 1920: The components and colloidal behavior of plant protoplasm, *Proc. Am. Phil. Soc.*, **59**: 150–170.

———, 1926: Growth and permeability of century old cells, *Am. Nat.*, **60**: 393–415.

——— and F. L. LONG, 1927: Character of cells attaining a great age, *Am. Nat.*, **61**: 385–406.

——— and J. G. BROWN, 1928: Living cells two and one half centuries old, *Sci.*, **67**: 447–448.

——— and G. M. Smith, 1927: Long lived cells of the redwood, *Sci.*, **66**: 456–457.

Magistad, O. C. and E. Truog, 1925: The influence of fertilizers in protecting corn against freezing, *J. Am. Soc. Agron.*, **17**: 517–526.

Magness, J. R., 1920: Investigations in the ripening and storage of Bartlett pears, *J. Agr. Res.*, **19**: 473–500.

Mahood, S. A., and D. E. Cable, 1922: The chemistry of wood. IV, The analysis of the wood of *Eucalyptus globulus* and *Pinus monticola*, *J. Ind. Eng. Chem.*, **14**: 933–935.

——— and ———, 1922: Comparison of wood cellulose and cotton cellulose, *J. Ind. Eng. Chem.*, **14**: 727–731.

Malhotra, R. C., 1930: Microchemical study of hemicellulose in some plant cells with special reference to its distribution in the protoplasm, *J. Biochem.* (*Tokyo*), **12**: 341–349.

Mangin, M. L., 1888: Sur la constitution de la membrane de végétaux, *C. R. Acad. Sci. Paris*, **107**: 144–146.

———, 1889: Sur la présence des composés pectiques dans les végétaux, *C. R. Acad. Sci. Paris*, **109**: 579–582.

———, 1890: Sur la substance intercellulaire, *C. R. Acad. Sci. Paris*, **110**: 295–297.

———, 1890: Sur les réactifs colorants des substances fondamentales de la membrane, *C. R. Acad. Sci. Paris*, **111**: 120–123.

———, 1891: Observations sur la membrane cellulosique, *C. R. Acad. Sci. Paris*, **113**: 1069–1072.

———, 1893: Sur l'emploi du rouge de ruthénium en anatomie végétale, *C. R. Acad. Sci. Paris*, **116**: 653–656.

Marston, H. R., 1923: The azine and azonium compounds of the proteolytic enzymes, *Biochem. J.*, **17**: 851–859.

Martin, J. H., J. A. Harris, and I. D. Jones, 1931: Freezing-point depression and specific conductivity of sorghum tissue fluids, *J. Agr. Res.*, **42**: 57–70.

Martin, S. H., M. W. Rea, and J. Small, 1926: The reaction of plant tissues, *Brit. Assoc. Adv. Sci. Rept.*, **94**: 410.

———, 1927: The hydrion concentration of plant tissues. III, The tissues of *Helianthus annuus*, *Protoplasma*, **1**: 497–521.

Mason, T. G., 1920: On some factors affecting the concentration of electrolytes in the leaf sap of *Syringa vulgaris*, *Notes Bot. School Trinity Col. Dublin.*, **3**: 67–82.

Masoni, G., 1920: Acidity of plant roots and soil lime, *Staz. Sper. Agr. Ital.*, **53**: 421–437, *Exp. Sta. Rec.*, **49**: 121.

McBain, J. W., and D. A. Scott, 1936: Micellar structure as related to cellulose, *Ind. and Eng. Chem.*, **28**: 470–473.

McCool, M. M., and C. E. Millar, 1917: The water content of the soil and the composition and concentration of the soil solution as indicated by the freezing point lowerings of the roots and tops of plants, *Soil Sci.*, **3**: 113–138.

——— and M. D. Weldon, 1928: The effect of soil type and fertilization on the composition of the expressed sap of plants, *J. Am. Soc. Agron.*, **20**: 778–793.

McKay, E., 1935: Salt tolerance of *Ruppia maritima* in lakes of high magnesium sulfate content, *Plant Physiol.*, **10**: 425–446.

Mehta, M. M., 1925: Biochemical and histological studies of lignification. I, The nature of lignin; its physiological significance and its estimation in timbers, *Biochem. J.*, **19**: 958–978.

———, 1925: Biochemical and histological studies of lignification. II, Histological studies on the polysaccharides and aromatic constituents of the cell wall, *Biochem. J.*, **19**: 979–997.

MENKE, W., 1934: Chloroplasten-studien, *Protoplasma*, **21**: 279–298.
———, 1934: Chloroplasten-studien. 2, *Protoplasma*, **22**: 56–62.
MENON, S. R. K., 1935: The variation in the methoxyl and cellulose values of the fibre of the fruit of *Cocos nucifera* during its growth from infancy of the nut to its ultimate drying on the tree, *Biochem. J.*, **29**: 282–284.
MEYER, B. S., 1927: Studies on the physical properties of leaves and leaf saps, *Ohio J. Sci.*, **27**: 263–288.
———, 1929: Some critical comments on the methods employed in the expression of leaf saps, *Plant Physiol.*, **4**: 103–112.
MEYER, K. H., and H. MARK, 1928: Über den Bau der kristallinen Anteile der Zellulose, *Ber. chem. Ges.*, **61**: 593–614.
———, 1929: Die Chemie der Micelle und ihre Anwendung auf biochemische und biologische Probleme, *Biochem. Z.*, **208**: 1–31.
———, 1931: The molecular structure of the cell wall, *New Phytol.*, **30**: 1–11.
MEYERS, A., 1883: "Das Chlorophyllkorn in chemischer, morphologischer und biologischer Beziehung," Leipzig.
MIKOSCH, C., 1885: Über die Entstehung der Chlorophyllkörner, *Sitzungsber. Akad. Wiss. Wien.*, **92**: 168.
MILLER, E. C., 1911: The origin of the chloroplasts in the cotyledons of *Helianthus annuus*, *Bot. Gaz.*, **51**: 378–384.
MILLER, L. P., 1931: The effect of treatments with ethylene chlorhydrin on the pH of the expressed juice of potato tubers, *Contr. Boyce Thompson Inst.*, **3**: 321–335.
MITCHELL, E. M., 1930: A microchemical study of hemicelluloses of endosperms and cotyledons, *Am. J. Bot.*, **17**: 117–138.
MIYAKE, K., and A. MASASHI, 1924: On the influence of three manurial ingredients upon the hydrogen-ion concentration of the cell sap of oats and flax, *J. Biochem. (Tokyo)*, **4**: 317–321.
——— and Y. IMAI, 1935: Maternal transmission of mutated plastids in the Japanese morning glory, *Bot. Gaz.*, **96**: 571–574.
MOTTIER, D. M., 1918: Chondriosomes and the primordia of chloroplasts and leucoplasts, *Ann. Bot.*, **32**: 91–114.
———, 1921: On certain plastids with special reference to the protein bodies of *Zea*, *Ricinus* and *Conopholis*, *Ann. Bot.*, **35**: 349–365.
MURNEEK, A. E., 1929: Hemicellulose as a storage carbohydrate in woody plants with special reference to the apple, *Plant Physiol.*, **4**: 251–265.
NÄGELI, C., 1864: Über den inneren Bau der vegetabilischen Zellmembranen, *Sitzber. Bay. Akad. Wiss. München*, **1**: 282–323; **2**: 114–171.
NAYLOR, G. L., and B. RUSSELL-WELLS, 1934: On the presence of cellulose and its distribution in the cell walls of brown and red algae, *Ann. Bot.*, **48**: 635–641.
NELSON, E. K., and H. HASSELBRING, 1931: Some organic acids of wheat plants, *J. Am. Chem. Soc.*, **53**: 1040–1043.
———, and H. H. MATTERN, 1931: Some organic acids in barley, maize, and rye plants, *J. Am. Chem. Soc.*, **53**: 3046–3048.
NORMAN, A. G., 1929: The chemical constitution of the gums. I, The nature of gum arabic and the biochemical classification of the gums, *Biochem. J.*, **23**: 524–535.
———, 1929: The biochemistry of pectins, *Sci. Prog.*, **24**: 263–279.
———, and F. W. NORRIS, 1930: Studies on pectin. IV, The oxidation of pectin by Fenton's reagent and its bearing on the genesis of the hemicelluloses, *Biochem. J.*, **24**: 402–409.
———, 1933: Some aspects of the chemistry of the plant cell wall, *Sci. Prog.*, **28**: 229–245.

———, and S. H. JENKINS, 1933: A new method for the determination of cellulose based upon observations of the removal of lignin and other encrusting materials, *Biochem. J.*, **27**: 818–831.

———, 1934: The biological decomposition of pectin, *Ann. Bot.*, **43**: 233–243.

———, 1935: The hemicelluloses. I, Alcoholic sodium hydroxide as a pretreatment to extraction, *Biochem. J.*, **29**: 945–952.

——— and J. G. SHRINKHANDE, 1935: The hemicelluloses. II, The association of hemicellulose with lignin, *Biochem. J.*, **29**: 2259–2266.

———, 1936: The composition of some vegetable fibers with particular reference to jute, *Biochem. J.*, **30**: 831–838.

NORRIS, F. W., and S. B. SCHRYVER, 1925: The pectic substances of plants. III, The nature of pectinogen and its relation to pectic acid, *Biochem. J.*, **19**: 676–694.

——— and I. A. PREECE, 1930: Studies on hemicelluloses. I, The hemicelluloses of wheat bran, *Biochem. J.*, **24**: 59–66.

——— and C. E. RESCH, 1935: The analysis of carbohydrates of the cell wall of plants. I, The relation between uronic anhydride content and furfuraldehyde yield, *Biochem. J.*, **29**: 1590–1596.

O'DWYER, M. H., 1923: The hemicelluloses. III, The hemicellulose of American white oak, *Biochem. J.*, **17**: 501–509.

———, 1925: A note on the occurrence of a pectic substance in beech wood, *Biochem. J.*, **19**: 694–696.

———, 1928: Preliminary investigations on the constitution of the hemicelluloses of timber, *Biochem. J.*, **22**: 381–390.

OHGA, I., 1926: On changes of osmotic concentration in certain plants, *Bot. Mag. Tokyo*, **60**: 587–590.

OSTERHOUT, W. J. V., 1906: The resistance of certain marine algae to changes in osmotic pressure and temperature, *Univ. Calif. Pub.* 2(8).

———, 1917: Tolerance of fresh water by marine plants and its relation to adaptation, *Bot. Gaz.*, **63**: 146–149.

PEACOCK, W. M., 1931: The healing of potatoes skinned during harvest as affected by temperature, humidity, and solar and sky radiation, *Proc. Am. Soc. Hort. Sci.*, **28**: 266–269.

PFEIFFER, M., 1933: Der osmotische Wert im Baum, *Planta*, **19**: 272–278; *Biol. Abst.*, **9**: 42.

PHILLIPS, M., 1931: The isolation of normal propyl guaiacol as a degradation product of lignin, *Sci.*, **73**: 568–570.

———, 1934: Report on lignin, *J. Assoc. Off. Agr. Chem.*, **17**: 277–282.

———, 1934: The chemistry of lignin, *Chem. Rev.*, **14**: 103–170.

——— and M. J. GOSS, 1935: The composition of the leaves and stalks of barley at successive stages of growth with special reference to the formation of lignin, *J. Agr. Res.*, **51**: 301–320.

POIRAULT, G., 1893: Recherches anatomiques sur les cryptogames vasculaires, *Ann. Sci. Nat. Bot.*, 7 ser., **18**: 113–256.

POLITIS, J., 1921: Du rôle du chondriome dans la formation des essences dans les plantes, *C. R. Acad. Sci. Paris*, **173**: 98–100.

POWELL, D., 1925: The development and distribution of chlorophyll in roots of flowering plants grown in the light, *Ann. Bot.*, **39**: 503–515.

PREECE, I. A., 1930: Studies on hemicelluloses. II, The hemicelluloses of maize cobs, *Biochem. J.*, **24**: 973–979.

———, 1931: Studies on hemicelluloses. IV, The proximate analysis of boxwood and the nature of its furfuraldehyde-yielding constituents, *Biochem. J.*, **25**: 1304–1319.

PRESTON, R. D., 1934: The organization of the cell wall of the conifer tracheid, *Phil. Trans. Roy. Soc. London*, **224**: 131–174.

PRIESTLEY, J. H., and A. IRVING, 1907: The structure of the chloroplast considered in relation to its function, *Ann. Bot.*, **21**: 407–413.

———, 1921: Suberin and cutin, *New Phytol.*, **20**: 17–29.

——— and L. M. WOFFENDEN, 1923: The healing of wounds in potato tubers and their propagation by cut sets, *Ann. App. Biol.*, **10**: 96–115.

——— and A. WORMALL, 1925: On the solutes exuded by root pressure from vines, *New Phytol.*, **24**: 24–38.

———, 1929: The biology of the living chloroplast. A critical abstract of Professor Lubimenko's review of recent Russian work, *New Phytol.*, **28**: 197–217.

PRINGSHEIM, H., and A. M. VON MERKALZ, 1919: Fermentversuche an Zellulose-abbauprodukten, *Z. physiol. Chem.*, **105**: 173–178.

——— and H. MAGNUS, 1919: Über den Acetylgehalt des Lignins, *Z. physiol. Chem.*, **105**: 179–186.

RANDOLPH, L. F., 1922: Cytology of chlorophyll types of maize, *Bot. Gaz.*, **73**: 337–376.

REA, M. W., and J. SMALL, 1927: The hydrion concentration of plant tissues. V, The tissues of *Vicia faba*, *Protoplasma*, **2**: 45–58.

——— and ———, 1927: The hydrion concentration of plant tissues. VI, Stem tissue reactions throughout the year, *Protoplasma*, **2**: 428–459.

REED, H. S., 1921: Growth and sap concentration, *J. Agr. Res.*, **21**: 81–98.

——— and F. F. HALMA, 1926: Further observations on the relations between growth and sap concentration in citrus trees, *J. Agr. Res.*, **32**: 1177–1195.

REINHARD, H., 1933: Ueber die Teilung der Chlorplasten, *Protoplasma*, **19**: 541–564.

RHODES, E., 1925: The chemical nature of the membrane of potato cork, *Biochem. J.*, **19**: 454–463.

RITTER, G. J., and L. C. FLECK, 1917: Chemistry of wood. V, Result of the analysis of some American woods, *J. Ind. Eng. Chem.*, **14**: 1050–1055.

———, 1925: The distribution of lignin in wood, *Ind. and Eng. Chem.*, **17**: 1194–1197.

ROBERTSON, I. M., and A. M. SMITH, 1931: A study of the hydrogen-ion concentration of the potato tuber, *Biochem. J.*, **25**: 763–769.

RUDOLPH, K., 1912: Chondriosomen und Chromatophoren, *Ber. bot. Ges.*, **30**; 605–627.

RUHLAND, W., and K. WETZEL, 1927, 1929: Zur Physiologie der organischen Säuren in grünen Pflanzen. III, V, *Planta*, **3**: 765–769; **7**: 503*ff*. Quoted by Bennet-Clark and Woodruff, 1935.

SANDERS, J. P., and F. K. CAMERON, 1933: Unit cell of cellulose in cotton stalks and cusps, *Ind. and Eng. Chem.*, **25**: 1371–1373.

SANDS, L., and W. Y. GARY, 1933: The hemicelluloses of mesquite wood, *J. Biol. Chem.*, **101**: 573–581.

——— and P. NUTTER, 1935: The hemicelluloses extracted from mesquite wood after chlorination, *J. Biol. Chem.*, **110**: 17–22.

SAYRE, J. D., and V. H. MORRIS, 1931: Use of expressed sap in physiologic studies of corn, *Plant Physiol.*, **6**: 137–148.

SCARTH, G. W., 1927: The structural organization of plant protoplasm, *Protoplasma*, **2**: 189–205.

———, R. D. GIBBS, and J. D. SPIER, 1929: Studies of the cell walls in wood. I, The structure of the cell wall and the local distribution of the chemical constituents, *Trans. Roy. Soc. Canada*, **23**: 269–279.

SCHAFER, E. R., and F. A. SIMMONDS, 1929: Physical and chemical characteristics of hemp stalks and of seed flax straw, *Ind. and Eng. Chem.*, **21**: 1241–1244.

Schellenberg, H. C., 1896: Beiträge zur Kenntniss der verholzten Zellmembran, *Jahrb. wiss. Bot.*, **29**: 237–266.

Schimper, A. F. W., 1885: Untersuchungen über die Chlorophyllkörper und die ihnen homologen Gebilde, *Jahrb. wiss. Bot.*, **16**: 1–246.

Schmidt, E. W., 1912: Pflanzliche Mitochondrien, *Progr. rei. Bot.*, **4**: 163–181.

Schorger, A. W., 1917: The chemistry of wood. I, Methods and results of analysis of some American species, *J. Ind. Eng. Chem.*, **9**: 556–561.

———, 1917: The chemistry of wood. II, Discussion of methods and results, *J. Ind. Eng. Chem.*, **9**: 561–566.

———, 1917: The chemistry of wood. III, Mannan content of the gymnosperms, *J. Ind. Eng. Chem.*, **9**: 748–750.

———, 1924: The constitution of cellulose, *Ind. and Eng. Chem.*, **16**: 1274–1275.

Schryver, S. B., and D. **Haynes**, 1916: The pectic substances of plants, *Biochem. J.*, **10**: 539–547.

——— and E. M. **Thomas**, 1923: The hemicelluloses. II, The hemicellulose content of starches, *Biochem. J.*, **17**: 497–500.

Schulze, E., 1892: Zur Chemie der pflanzlichen Zellmembranen. II, *Z. physiol. Chem.*, **16**: 387–438.

———, 1893: Zur Chemie der pflanzlichen Zellmembranen. III, *Z. physiol. Chem.*, **19**: 38–69.

——— and N. **Castoro**, 1902: Beiträge zur Kenntniss der Hemicellulosen. *Z. physiol. Chem.*, **37**: 40–53.

——— and ———, 1903: Beiträge zur Kenntniss der Hemicellulosen. II, *Z. physiol. Chem.*, **39**: 318–329.

——— and U. **Pfenniger**, 1910: Ueber das Vorkommen von Hemicellulosen in den Samenhülsen von *Pisum sativum* und von *Phaseolus vulgaris*, *Z. physiol. Chem.*, **68**: 93–108.

Schürhoff, P. N., 1924: "Die Plastiden," 225 pp., Figs. 57, Berlin.

Seifriz, W., 1920: Viscosity values as determined by micro-dissection, *Bot. Gaz.*, **70**: 360–387.

———, 1930: The alveolar structure of protoplasm, *Protoplasma*, **9**: 177–208.

———, 1931: The Spierer lens and what it reveals in cellulose and protoplasm, *J. Phys. Chem.*, **35**: 118–129.

———, 1931: The structure of protoplasm, *Sci.*, **73**: 648–649.

———, 1934: The origin, composition and structure of cellulose in the living plant, *Protoplasma*, **21**: 129–159.

———, 1935: The effects of acids and alkalies on the viscous and structural properties of protoplasm, *Physics*, **6**: 159–161.

———, 1935: The structure of protoplasm, *Bot. Rev.*, **1**: 18–36.

———, 1936: "Protoplasm," 584 pp., McGraw-Hill Book Company, Inc., New York.

———, 1936: Spierer lens and colloidal structure, *Ind. and Eng. Chem.*, **28**: 136–140.

Sharp, L. W., 1921: "Introduction to Cytology," 582 pp., McGraw-Hill Book Company, Inc., New York.

Shedd, O. M., 1915: Variations in mineral composition of sap, leaves, and stems of the wild grape vine and sugar-maple tree, *J. Agr. Res.*, **5**: 529–543.

Singh, L., 1922: A study of the relation of pectin and acidity in jelly making, *J. Ind. Eng. Chem.*, **14**: 710–712.

Skinner, H. J., 1932: Cellulose in industry, *Ind. and Eng. Chem.*, **24**: 694–704.

Small, J., 1929: "Hydrogen-ion Concentration in Plant Cell and Tissues," 421 pp., Protoplasma Monographien, II.

Smith, E. F. and A. J. **Quirk**, 1926: A begonia immune to crown gall with observations on immune or semi-immune plants, *Phytopath.*, **16**: 491–509.

Smith, O., 1933: The studies of potato storage, *N. Y. Sta.* (*Cornell*) *Bull.* 553.

Smith, W. W., 1935: The course of stone cell formation in pear fruits, *Plant Physiol.*, **10**: 589–612.

Sponsler, O. L., 1926: Molecular structure of plant fibers determined by X-rays, *J. Gen. Physiol.*, **9**: 677–695.

———, 1926: The structure of ramie cellulose as derived from X-ray data, *Colloid. Sym. Mono.*, **4**:174–202.

———, 1928: The molecular structure of the cell wall of fibers. A summary of X-ray investigations, *Am. J. Bot.*, **15**: 525–537.

———, 1929: Mechanism of cell wall formation, *Plant Physiol.*, **4**: 329–337.

———, 1930: Orientation of cellulose space lattice in the cell wall, *Protoplasma*, **12**: 241–254.

———, 1931: Orientation of cellulose space lattice in the cell wall. Additional X-ray data from *Valonia* cell wall, *Protoplasma*, **12**: 241–254.

———, 1933: The molecule in biological structures as determined by X-ray methods, *Quart. Rev. Biol.*, **8**: 1–30.

———, 1934: Rate of carbohydrate condensation on a cell wall, *J. Am. Chem. Soc.*, **56**: 1599–1601.

——— and W. H. **Dore**, 1936: X-ray studies on the structure of compounds of biochemical interest, *Ann. Rev. Biochem.*, **5**: 63–80.

Staehelin, M., 1919: Die Rolle der Oxalsäure in der Pflanze, *Biochem. Z.*, **96**: 1–49.

Stanfield, J. S., 1932: Osmotic pressure of leaves of *Pinus scopulorum* and certain environmental factors, *Bot. Gaz.*, **93**: 453–465.

Stange, B., 1892: Beziehungen zwischen Substratkonzentration, Turgor, und Wachsthum bei einigen phanerogamen Pflanzen, *Bot. Ztg.*, **50**: 253–259, 273–285.

Stern, A. L., 1904: The so-called hydro-cellulose, *J. Chem. Soc.* (London), **85**: 336–340.

Stevens, N. E., 1932: Thickness of cuticle in cranberry fruits, *Am. J. Bot.*, **19**: 432–435.

Stoddart, L. A., 1935, 1936: Osmotic pressure and water content of prairie plants, *Plant Physiol.*, **10**: 661–680; thesis, University of Nebraska, January, 1936.

Stone, W. E., 1932: The origin, development, and increase of chloroplasts in the potato, *J. Agr. Res.*, **45**: 421–435.

Stout, A. B., 1915: The establishment of varieties in Coleus by the selection of somatic variations, *Carnegie Inst. Wash. Pub.* **218**: 1–80.

Stratton, F. C., 1931: Changes in osmotic pressure of bananas during ripening, *Plant Physiol.*, **6**: 361–365.

Thatcher, R. W., 1922: The role of celluloses in plant life, *Sci.*, **55**: 80.

Thielman, M., and L. **Berzin**, 1927: Über den osmotischen Wert kultivierter Pflanzenzellen, *Arch. Exp. Zellforsch.*, **4**: 273–327; *Biol. Abst.*, **4**: 685.

Thiessen, R., 1932: Physical structure of coal, cellulose fiber, and wood as shown by Spierer lens, *Ind. and Eng. Chem.*, **24**: 1032–1041.

Thoday, D., 1933: Some physiological aspects of differentiation, *New Phytol.*, **32**: 274–287.

Tottingham, W. E., R. H. **Roberts**, and S. **Lepkovsky**, 1921: Hemicellulose of apple wood, *J. Biol. Chem.*, **45**: 407–414.

———, 1933: Character of hemicellulose in certain fruit trees, *Plant Physiol.*, **8**: 559–561.

TRUOG, E., and M. R. MEACHAM, 1919: Soil acidity. II, Its relation to the acidity of the plant juice, *Soil Sci.*, **7**: 469–475.

TUCKER, L. R., 1934: Soluble solids in the water melon, *Plant Physiol.*, **9**: 181–182.

TUPPER-CAREY, R. M., and J. H. PRIESTLEY, 1923: The composition of the cell wall of the apical meristem of stem and root, *Proc. Roy. Soc.*, B, **95**: 109–131.

——— and ———, 1924: The cell wall in the radicle of *Vicia faba* and the shape of the meristematic cells, *New Phytol.*, **23**: 156–159.

TUTIN, F., 1921: The behavior of pectin towards alkalis and pectase, *Biochem. J.*, **15**: 494–498.

———, 1923: A note on the hydrolysis of pectin, *Biochem. J.*, **17**: 83–84.

———, 1923: Pectin and its hypothetical precursor, protopectin, *Biochem. J.*, **17**: 510–514.

TWISS, W. C., 1919: A study of plastids and mitochondria in *Preissia* and corn, *Am. J. Bot.*, **6**: 217–235.

UBER, F. M., and T. H. GOODSPEED, 1935: Microincineration studies. I, Localization of inorganic elements in plant cell walls, *Proc. Nat. Acad. Sci.*, **21**: 428–433.

VAN WISSELINGH, C., 1888: Sur la pario des cellules subéreuses, *Arch. Néerl. des Sc. Exp. Nat.*, **22**: 253.

———, 1892: Sur la lamelle subéreuse et la subérine, *Arch. Néerl. des Sc. Exp. Nat.*, **26**: 304–353.

———, 1894: Sur la cuticularisation et la cutine, *Arch. Néerl. des Sc. Exp. Nat.*, **28**: 373–410.

VON FELLENBERG, T., 1918: Über die Konstitution der Pektinkörper, *Biochem. Z.*, **85**: 118–161.

VON SCHMIDT, M., 1903: Zur Kenntniss der Korksubstanz, *Sitzungsber. Akad. Wiss. Wien*, 112 abt., 11 b.; *Bot. Cent.*, **96**: 318.

———, 1910: Zur Kenntniss der Korksubstanz, *Monatshefte Chem.*, **31**: 347.

———, 1910: Zur Kenntniss der Korksubstanz. III, *Sitzungsber. Akad. Wiss. Wien*, **119**: 233–242.

WAKSMAN, S. A., and R. A. DIEHM, 1931: On the decomposition of hemicelluloses by microorganisms. I, Nature, occurrence, preparation, and decomposition of hemicelluloses, *Soil Sci.*, **32**: 73–95.

WARBURG, O., 1885: Ueber die Stellung der organischen Säuren im Stoffwechsel der Pflanzen, *Ber. bot. Ges.*, **3**: 280–289.

———, 1886: Über die Bedeutung der organischen Säuren für das Lebensprocess der Pflanzen, *Unters. bot. Inst. Tübingen*, **2**: 57–150.

WEBBER, H. A., 1929: Cellulose from corn stalks, *Ind. and Eng. Chem.*, **21**: 270–275.

WEIER, T. E., 1930: Notes on the plastid and other cytoplasmic bodies during sporogenesis and spermatogenesis in *Polytrichum commune*, *Proc. Nat. Acad. Sci.*, **16**: 536–543.

———, 1932: A comparison of the plastid with the Golgi zone, *Biol. Bull.*, **62**: 126–139.

———, 1933: A critique of the vacuome hypothesis, *Protoplasma*, **19**: 589–601.

———, 1933: On the structure of the *Anthoceros* plastid in reflected light, *Sci.*, **78**: 264–265.

WEIMER, J. L., and L. L. HARTER, 1921: Wound cork formation in the sweet potato, *J. Agr. Res.*, **21**: 637–647.

WELLS, S. D., 1929: Cellulose from cereal straws, *Ind. and Eng. Chem.*, **21**: 275–278.

WHITFIELD, C. J., 1932: Osmotic concentrations of chaparral, coastal sage brush and dune species of southern California, *Ecology*, **13**: 279–285.

WIDDOWSON, E. M., 1932: Chemical studies in the physiology of apples. XIII, The starch and hemicellulose content of developing apples, *Ann. Bot.*, **46**: 597–632.

WIESNER, J., 1877: Ueber das Vorkommen und die Entstehung von Etiolin und Chlorophyll in der Kartoffel, *Osterr. Bot. Z.*, **27**: 7–11.

WOOD, F. M., 1924: Contributions to an investigation of the chemical nature of the cellulose membrane, *Ann. Bot.*, **38**: 273–298.

———, 1926: Further investigations of the chemical nature of the cell wall membrane, *Ann. Bot.*, **40**: 547–571.

———, 1931: Observations on the action of bromine on plant tissues, *Ann. Bot.*, **45**: 421–424.

WOODHEAD, N., 1934: Studies in growth and differentiation. V, Histological and metabolic changes during wound healing in *Kleinia articulata* Haw., *Ann. Bot.*, **48**: 467–480.

WORMALL, A., 1924: The constituents of the sap of the vine, *Biochem. J.*, **18**: 1187–1202.

WURDACK, M. E., 1923: Chemical composition of the walls of certain algae, *Ohio J. Sci.*, **23**: 181–192.

ZIRKLE, C., 1926: The structure of the chloroplast in certain higher plants. I, II, *Am. J. Bot.*, **13**: 301–342.

———, 1927: The growth and development of plastids in *Lunularia vulgaris*, *Elodea canadensis* and *Zea mays*, *Am. J. Bot.*, **14**: 429–446.

———: 1929: Development of normal and divergent plastid types in *Zea mays*, *Bot. Gaz.*, **88**: 186–203.

———, 1937: The plant vacuole, *Bot. Rev.*, **3**: 1–30.

CHAPTER II

SOLUTIONS AND MEMBRANES IN RELATION TO THE PLANT CELL

In considering the relation of solutions and membranes to the physiological functions of the plant, the student should keep the two following salient facts clearly in mind as a background for the discussion: (1) Owing to the structural nature of the plant cell, all materials that enter or leave a plant or that move from one cell to another must pass through membranes. (2) Through the cell walls, protoplasm, and vacuoles of the living cells that compose a green plant, water is continuous from the exterior walls of the root hairs which come in contact with the soil particles to the exterior surface of the outer cells of the leaves which are exposed to the air. On this account and because membranes must be penetrated, all materials that enter a plant, whether from the soil or from the air, or that migrate from one part of the plant to another must do so dissolved in water. The entrance of water and dissolved substances into a plant and their translocation after they have once entered depend thus upon certain behaviors and relationships of membranes and solutions. In order to understand how materials enter a plant or how they migrate in it, one must have a considerable knowledge of the behavior of membranes and solutions, and for that reason it is the intention here to discuss in a brief and elementary manner some of the more important facts in regard to them.

I. SOLUTIONS

A. Nature

A solution may be defined as a homogeneous mixture of two or more substances having the same physical and chemical properties throughout, the proportion of whose components may be altered continuously within certain limits without producing any abrupt change in any property of the mixture. The term "true solution" is now frequently used to designate a mixture of the nature above described to distinguish it from a colloidal solution, which is a heterogeneous mixture occupying an intermediate position between a true solution on the one hand and the more pronounced heterogeneous mixtures called "suspensoids" on the other. The most important difference between a true solution and a colloidal solution is the difference in the degree of subdivision or dispersion of the dissolved substances. In a true solution the dissolved substance or substances are generally in the ionic or molecular condition and the magnitude of the particles is 1mμ[1] or less in diameter, while in a colloidal solution the degree of dispersion is not so great and the particles vary from 1mμ to 0.1μ in diameter. True solutions may be of three general types, gaseous, liquid, or solid; but in plant physiology we are concerned wholly with liquid solutions, and it is to this type that our attention will be devoted.

In a solution, that component which is present in the larger proportion is called the "solvent," while the component or components that are present in smaller amounts are termed the "solute" or "solutes." It is generally stated that the solute is dissolved in the solvent, but it should be remembered that this method of designation is purely arbitrary and does not necessarily imply that one component possesses greater solvent powers than the other. The case of a solution of two liquids that are miscible in all proportions, as alcohol and water, is illuminating in discussing this relation of solvent and solute. If we are considering a 1 per cent solution of alcohol in water, we say that the alcohol is dissolved in water, although it would be just as correct to say that the water is dissolved in the alcohol. If the concentration of alcohol is 51 per cent, it is spoken of as the solvent and the water is termed the solute. If there is a 50 per cent solution of alcohol in water, however, either liquid may be called the solvent and either the solute. Water is practically the only solvent that functions in plants, but the solutes concerned are numerous and include all the soluble inorganic salts that may be present in the soil, as well as oxygen, carbon dioxide, sugars, organic acids, and various other soluble organic compounds.

B. Concentration

There are numerous ways in which the concentration and composition of a solution may be expressed, but the four following are the ones most generally used.

1. In Terms of Percentage.—This method does not show the relative number of molecules of the solute that are present in the solution. Since many of the physical reactions of solutions are resultants of the relative number of molecules present in the solvent, the percentage method of stating the concentration of solutions is not a very desirable one.

2. In Terms of Gram-molecular Weight or Formula Weight per Liter of Solution.—A solution made up after this manner is termed a "molar," "formal," "molecular," "gram-molecular," or "molecular normal" solution. It is made by dissolving as many grams of a substance as there are units in its molecular weight so as to make 1 l. of solution.

[1] $1\mu = \frac{1}{1{,}000}$ mm. $1\text{m}\mu = \frac{1}{1{,}000{,}000}$ mm., or $\frac{1}{1{,}000}\mu$.

Thus, in the case of sulphuric acid (H_2SO_4), the molecular weight of which is 98.08, a molecular solution would contain 98.08 g. of sulphuric acid in 1 l. of solution. In the case of hydrochloric acid, the molecular weight of which is 36.46, a molecular solution would contain 36.46 g. of hydrochloric acid in 1 l. of solution.

3. In Terms of Equivalent Weight per Liter of Solution.—An equivalent solution is one that contains 1 g. atomic weight of hydrogen or its equivalent per liter. A solution made up after this manner is also termed an "equivalent normal solution." In the case of monovalent compounds, a molecular solution and equivalent solution are one and the same. In the case of bivalent compounds, however, an equivalent solution contains only one-half the molecular weight of the substance in grams dissolved in water so as to make 1 l. of solution. An equivalent solution of sulphuric acid would thus contain 49.04 g. of this acid in 1 l. of solution. All equivalent solutions, regardless of the substances that compose them, thus contain the same number of hydrogen ions or their equivalent.

4. In Terms of Osmotic Pressure.—In biological work the concentration of a solution is frequently stated in terms of the osmotic pressure that would be produced if the solution were separated from pure water by a perfectly semipermeable membrane. The value of this osmotic pressure is commonly expressed in units of atmospheric pressure.

C. Diffusion

One of the distinct characteristics of a solution is the property of diffusion or the ability of the molecules or ions of the solvent and solute to move more or less freely throughout it. For a clear understanding of the entrance and translocation of materials in plants, the following simple but important facts in regard to the behavior of solvent and solutes in a solution must be remembered: (1) The molecules or ions of the solutes and solvent move or diffuse constantly throughout the solution. (2) The molecules or ions of the solute and solvent move in the solution independently of each other. (3) When more than one solute is present in the solution, the molecules or ions of each solute move independently of the other solutes.[1] (4) More molecules or ions of a given solute move in a unit of time from the higher toward the lower concentration of the solute than in the opposite direction. (5) More molecules of the solvent move in a unit of time from the higher toward the lower concentration of the solvent than in the opposite direction. These properties of diffusion thus soon bring about a uniform distribution of the solute or solutes and solvent in any given volume of solution. These facts in regard to the

[1] This statement is not strictly correct. See Oswald (1891), Osborne and Jackson (1914), and Bowen (1931).

diffusion of the solute and the solvent can best be understood by considering a hypothetical case as diagramed in Fig. 4. Let us assume that we have two solutions A and B in which water is the solvent and sodium chloride and potassium nitrate the solutes. Solution A has a concentration 0.1 N sodium chloride and 0.3 N potassium nitrate, which makes a total concentration of four-tenths normal. Solution B has a concentration of 0.1 N potassium nitrate and 0.2 N sodium chloride, making a

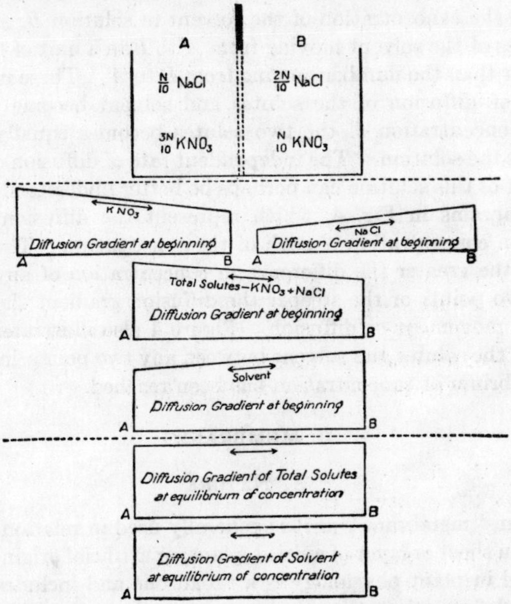

FIG. 4.—Diagrams showing the directions and relative rates of the diffusion of solvent and solutes. Description in the text.

total concentration of three-tenths normal. Now let solution A be placed in compartment A of a vessel and solution B in compartment B of the same vessel and the partition separating the two solutions be removed without disturbing them, thus allowing them to come in direct contact with each other. Let us now consider the behavior of the solutes and solvent when the two solutions thus come in contact.[1]

Let us consider first the behavior of the solutes. The number of molecules of sodium chloride moving from A to B in a unit of time is smaller than the number moving from B to A, since the concentration is greater in B. For the same reason the number of molecules of potassium

[1] For simplicity the two salts are considered here only in regard to their molecular concentration, and the fact of their ionization is not considered.

nitrate moving from A to B in a unit of time is greater than the number moving from B to A. The total number of molecules moving from A to B is greater in a unit of time than the number moving from B to A, since the total concentration of solutes in A (four-tenths normal) is higher than the total concentration of solutes in B (three-tenths normal). Let us now consider the behavior of the solvent. Since the total concentration of the solutes in solution A is equal to four-tenths normal and that in solution B to three-tenths normal, the concentration of the solvent in solution A is lower than the concentration of the solvent in solution B. The number of molecules of the solvent moving from A to B in a unit of time is therefore smaller than the number moving from B to A. These inequalities in the rates of diffusion of the solutes and solvent become less and less until the concentration of the two solutes becomes equally distributed throughout the solution. The independent rate of diffusion of the solutes and solvent of this solution can perhaps be better understood by considering the diagrams in Fig. 4, which represent the diffusion gradient or difference in concentration of each of the constituents. These diagrams show that the greater the difference in concentration of any component between two points or the steeper the diffusion gradient the more rapid the rate of movement or diffusion. Figure 4 also illustrates the rate of diffusion of the solutes and solvent between any two points in the solution after equilibrium of concentration has been reached.

II. MEMBRANES

A. Nature

The term "membrane" as it is generally used in relation to solutions means a thin sheet or layer of animal, plant, or artificial origin. The term as it is used in plant physiology is a broad one and includes on the one hand such definite tissues as the coats of seeds, scales of bulbs, or the epidermis of stems and leaves and, on the other hand, only the parts of the cell as the cell wall, the cytoplasmic layer, or even only the surface films of the latter bordering on the vacuole or on the cell wall. The consistency of these membranes thus varies from a semiliquid to a solid and ranges in thickness from a very delicate film that can be observed only with the high power of the microscope to a tissue whose thickness can easily be measured with the naked eye. In the performance of experiments to illustrate certain biological behaviors, artificial membranes of collodion, copper ferrocyanide, or parchment paper are frequently used, since it has been observed that artificial and natural membranes show in many cases striking similarities of behavior toward solutions.

Membranes differ greatly in their capacity for allowing substances to pass through them. A membrane is said to be permeable to a substance if it allows this substance to pass through and impermeable if it will not allow it to pass through. A membrane which will allow the solvent to pass through but not the solute is called a "semipermeable membrane." The capacity of a membrane for allowing substances to pass through it is termed its "permeability." Membranes differ not only in their permeability to a given substance but also in that their permeability is influenced by the general environmental factors, so that a membrane may have as many degrees of permeability toward a given substance as there are environmental conditions under which it may be placed. In speaking of the permeability of a membrane, therefore, it is necessary to define the system with which it is in contact.

None of the artificial membranes that have been made is perfectly semipermeable, although some of them prevent all but a trace of the solute from passing through. The nearest approach to a perfect semipermeable membrane is one of copper ferrocyanide in its relation to a solution of cane sugar in water. This membrane allows the water to diffuse through it readily but prevents almost completely the passage of the sugar. The cytoplasmic layer in plant cells behaves frequently as a more or less perfectly semipermeable membrane to various water solutions, although it is now considered that it is not nearly so semipermeable as formerly was supposed. There are some cases also where the cell walls behave as semipermeable membranes, as will be discussed later.

B. Imbibition of Water

The term "imbibition" as here used means the attraction of substances for water, a phenomenon that is especially characteristic of colloidal material in the gel condition. This process is best illustrated by the intake of water by dry wood, gelatin, starch grains, and dry seeds. Each particle of such substances has an attraction or affinity for water and will take it up from the surrounding medium against enormous pressures, which vary with the nature of the imbibing material and its water content (Shull and Shull, 1932). The water so attracted arranges itself around the particles in the form of films so that as the amount of water imbibed increases, the thickness of the watery films increases and thus the particles that compose the substance are crowded farther and farther apart. The limit to the distance that these particles may be separated from each other by the water thus taken up will depend upon the cohesive force of the particles of that substance. Thus in the case of wood the limit of separation is soon reached, while in the case of gelatin and similar substances the particles may go into colloidal solution. This phenomenon of imbibition of water is closely related to the behavior of membranes to watery solutions, since before there can be any passage of water, or of the solutes dissolved in it, through a membrane, the membrane must imbibe at least a certain amount of water. It is due to this characteristic imbibition of water by the protoplasm and the cell wall that water is continuous from the vacuole of one living cell to that of another. From the observed facts of the structure of the cell wall and the protoplasm and of the intake of water and of the migration of substances through them, we can crudely consider that these membranes are composed of a mixture

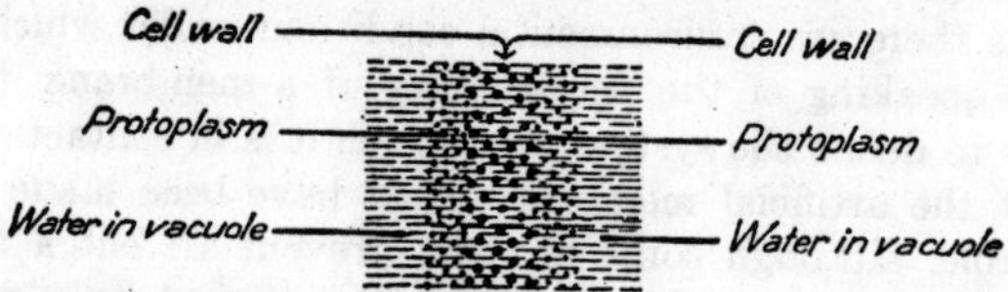

FIG. 5.—Diagram to illustrate that water is continuous from the vacuole of one cell through the protoplasm and cell wall to the vacuole of the adjoining cell. Description in the text.

of water and numerous and varied other substances. We can thus conceive that continuous waterways are provided along which the water-soluble substances may travel from cell to cell (Fig. 5).

C. DIFFUSION

When solutions are separated from each other or from the pure solvent by a membrane that is permeable to both the solvent and the solute or only to the solvent, certain phenomena occur which are not observable in the process of diffusion in solution when the membrane is not present. A discussion of this topic is necessary if we are to understand, in part at least, the intake, outgo, and translocation of materials in plants, since

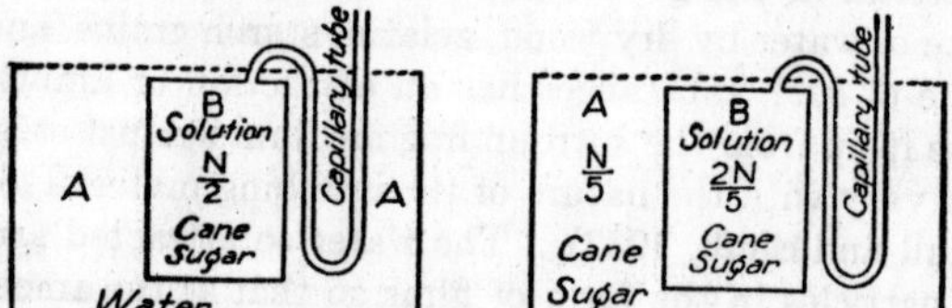

FIG. 6.—Diagram to aid in the discussion of osmosis and osmotic pressure. Description in the text.

these processes are apparently concerned primarily with the diffusion through membranes. It is the intention here to discuss the process of diffusion through membranes only in a very elementary and brief manner, since a more elaborate and theoretical discussion belongs to the province of physical chemistry and not to plant physiology. Let us consider first the behavior of the solvent and solute where the solution is separated from the pure solvent by a semipermeable membrane. In Fig. 6 let us consider that a half-normal solution of cane sugar completely fills a container *B*, the wall of which, or some portion of it, behaves as a semipermeable membrane to the sugar solution (in case only a portion of the wall is semipermeable, it is here assumed that the remaining portion is impermeable to both the solvent and the solute). Let us further suppose that a fine capillary tube be connected with the container in such a manner as to function as a manometer and that the vessel containing this sugar solution is immersed in distilled water in a larger container *A*. The solute will tend to diffuse very rapidly from *B* to *A* just as if no membrane intervened, for the diffusion gradient is steep, but the egress of the sugar from the vessel is prohibited, since the membrane is impermeable to it and equilibrium of concentration of the solute on both sides of the membrane cannot occur. Since the membrane is permeable to the solvent, water will readily move through it, diffusing more rapidly from *A* to *B* than it will from *B* to *A*, since the concentration of the solvent in *A* is higher than it is in *B*. The behavior of the solvent is thus similar to that which is attained when two solutions of unequal concentration are brought together without the intervention of a membrane, but with this difference, that the solute cannot diffuse into the pure solvent exterior to the membrane, so that the concentration of the solvent in *A* will always remain higher than it is in *B*.

Since water enters *B* more rapidly than it leaves, a hydrostatic pressure will be set up in *B* as indicated by the rise of the column of mercury in the manometer. Water will continue to move from *A* into *B* with a corresponding increase in the hydrostatic pressure until this pressure equals the force with which water tends to enter *B*. In the case of the half-normal solution of cane sugar, the hydrostatic pressure thus set up in *B* at 0°C. and 760 mm. of mercury will be approximately 11.2 atmospheres if the conditions are ideal and the membranes and walls of *B* are capable of sustaining such a pressure.

This discussion now brings us to the consideration of osmosis and osmotic pressure. The term "osmosis" has been used with such a variety of meanings by workers in biology that one is in doubt many times as to what it really signifies (Stark, 1928). The physicist and the physical chemist always consider an ideal condition for the illustration of this process similar to that just mentioned in the preceding discussion in which the solution is separated from the pure solvent by a perfectly semipermeable membrane. They would thus define osmosis as the passage of a solvent from one side of a semipermeable membrane to the other when the escaping tendency (or free energy) of the solvent on the two sides is unequal. The osmotic pressure would then be defined as the pressure difference which must be established upon the solution and pure solvent, respectively, in order to make the escaping tendency of the solvent the same for both of them. Osmotic pressure can be defined also as the equivalent of the hydrostatic pressure produced when solution and solvent are separated by a perfectly semipermeable membrane. It may likewise be expressed as the equivalent of the excess pressure that must be imposed on a solution to prevent the passage into it of the solvent through a perfectly semipermeable membrane (Findlay, 1913). Beck (1928) defines osmotic pressure as the difference of pressure on solution and solvent which produces a condition of equilibrium such that there is no tendency of the solvent to flow in either direction.

Let us now consider a case of diffusion through a membrane in which such ideal conditions do not prevail, since in plants a solution may be separated not only from the pure solvent but also from another solution not by a perfectly semipermeable membrane but by one that is permeable to both the solvent and solutes. This condition prevails in the cells of plants to as great extent as or greater than does semipermeability of membranes. For an illustration of the phenomena that occur under these conditions, let us assume (Fig. 6) that an apparatus *B* similar to that just described, but with a membrane permeable to both the solute and solvent, is filled with a two-fifths normal solution of cane sugar in water and immersed in a larger vessel *A* containing a fifth-normal solution of cane sugar in water. The solute under this condition will diffuse more rapidly from *B* to *A* than it will from *A* to *B* and thus tend to equalize the concentration of the solute on each side of the membrane just as it would act if the two solutions were in contact without the intervention of the membrane. The water diffuses more rapidly from *A* to *B* than it does from *B* to *A*, since the concentration of the solvent in *B* is lower than in *A*. The behavior of the solvent is thus similar to that which would occur if the two solutions were placed in contact without the intervening membranes, but with this difference that water as a general rule diffuses through a membrane at a more rapid rate than the solutes so that a hydrostatic pressure will be set up in *B* as indicated by the rise of the mercury in the manometer. This hydrostatic pressure, however, is only temporary, for as the concentration of the solute in *A* and *B* becomes more nearly equal, the solvent on both sides of the membrane more nearly reaches equilibrium and the hydrostatic pressure in *B* is diminished accordingly. When the concentration of the solute and solvent have become, respectively, the same in *A* and *B*, the rate of diffusion of the solvent from *A* to *B* is the same as it is from *B* to *A*, and as a result the hydrostatic pressure on the inside of *B* is reduced to zero.

It is evident from the above example that the phenomenon of osmosis and osmotic pressure may be exhibited when the membrane is not semipermeable. It would thus appear that osmosis should not be defined only in terms of semipermeable membranes but in terms of permeable membranes as well. Osmosis could thus be defined as the passage of the solvent from one side of a membrane to another where the escaping tendency of the solvent on the two sides is unequal. Because the diffusion of a salt solution through a membrane is similar to the diffusion of the water in which it is dissolved, there are those who believe that osmosis should be defined in terms of both the solvent and solute. Thus Stark (1928) proposed that osmosis be defined as diffusion through a membrane, the direction of the major movement being from a region of high concentration to a region of low concentration of the thing diffusing.

If the fifth-normal solution of cane sugar were placed in *B* and the two-fifths normal solution in *A*, the rate of water movement from *B* to *A* would be greater than from *A* to *B*, and as a result water would be withdrawn from *B*. Under such a condition the walls of *B* would contract or collapse unless they were strong enough to sustain the negative pressure developed therein. This phenomenon frequently prevails in plant cells under certain conditions, and their behavior under these conditions will be discussed in detail when the subject of plasmolysis is considered. The term "endosmosis" is used to designate the flow of either solute or solvent through the membranes of the plant cell into the vacuole, while the term "exosmosis" is used to designate the flow of solvent or solute outward from the cell vacuole. The connection of the phenomenon of osmosis with the water relations of the plant cell will be discussed in detail under the heading, Osmotic Relationship of the Plant Cell.

D. Membranes of the Plant Cell

With but few exceptions there are in each fully developed living plant cell two well-defined membranes through which materials must pass in moving from the vacuole or vacuoles of one cell to those of another or in entering the vacuoles of the external cells of the plant body from the medium in which the plant is growing. These two membranes are the cell wall and the cytoplasmic layer.

1. The Cell-wall or Nonliving Membrane.—With the exception of certain zoospores, sperms, unfertilized eggs, and in a few other cases, the protoplasm of plant cells is always enclosed by a definite compact solid membrane called the "cell wall." This membrane is frequently termed the "nonliving" membrane of the plant cell. The constituents that may enter into the composition of the cell wall have been named and described in considerable detail in Chap. I, so the intention here is to discuss only the general relationship of the cell wall to watery solutions.

a. Permeability to Water.—The cell wall is, for the most part, composed of an aggregation of colloidal particles in the gel condition, and like other gels it exhibits the power of imbibition and absorbs water when placed in contact with it. The amount of water thus absorbed before imbibitional saturation is reached varies greatly according to the constituents of the cell wall. Thus in the case of cutin and suberin the amount of water imbibed is very small, while in the case of the pectic substances the amount absorbed may be so great as practically to throw these substances into colloidal solution. Since all cell walls exhibit the power of imbibing water, it follows that none of them is absolutely impermeable to water. The suberized and cutinized cell walls are the least permeable to water, while the cell walls of embryonic tissue, root hairs, and parenchyma cells apparently offer the minimum retardation to water of any of the cell walls. The cell walls that are the most permeable to water, however, offer some resistance to water flow, and water never diffuses so readily through a cell-wall membrane as it would were there no membrane present. The principal factors that influence the rate of water movement through cell-wall membranes are the same as those that influence the behavior of practically all membranes. These main factors are the constitution of the membrane, the direction of flow, the concentration of the solutions in contact with it, and the temperature. The differences in the degree of permeability of the cell walls of different species of plants to water are shown by the work of Denny (1917) with seed coats and with the outer bulb scale of the onion. These membranes consist entirely of dead cells, so that with the possible exception of some nonliving material within the cavities of the cells, the behavior of the membranes toward the diffusion of water through them is due entirely to the nature of the walls of the cells composing them. The amount of water passing through equal areas (19.63 sq. mm.) of these membranes in a unit of time, when the external solution consisted of a saturated solution of sodium chloride with an osmotic pressure of approximately 375 atmospheres, is shown in the following table:

Membrane	Number of observations	Average amount of water passing per hour, milligram
Seed coat of grapefruit	3	0
Seed coat of *Cucurbita pepo*	4	4.0
Seed coat of *Cucurbita maxima*	5	9.3
Seed coat of cocklebur	4	18.2
Outer bulb scale of onion	6	24.0
Seed coat of almond	7	88.5
Seed coat of peanut	7	565.1

Denny investigated the nature of the constituents of these cell walls in order to determine if possible the cause of their differential permeability to water. His method consisted in determining the rate of flow of water through a given membrane and then after treating it with a given solvent to measure again the rate of water flow through it. The heating of the membranes in boiling water for 5 min. increased the permeability of the peanut and almond membranes but not that of the seed coats of the grapefruit or squash. The increase in permeability of the peanut coat averaged approximately 168 per cent, while the increase in permeability of the almond amounted to as much as 500 per cent. The treatment of these membranes by hot alcohol, hot acetone, and hot ether, which are solvents of fatty substances, increased greatly the permeability of the peanut, almond, squash, and cocklebur coats but not the grapefruit seed coats. The maximum increase in permeability of these membranes was 871, 242, 452, and 138 per cent, respectively, for the squash, peanut, almond, and cocklebur. After some of the membranes had been extracted with the fat solvents, they were immersed in the extract and exposed to evaporation so that they would again contain the materials that had been removed from them. Although the permeability of these membranes was reduced by this infiltration, it did not reach the low point exhibited by the membranes before their extraction. This fact indicates that the extracted materials which influenced the permeability of these membranes to water are distributed in an organized manner throughout the cell walls. This assumption is also strengthened by the fact that the permeability of some of these membranes is increased by simply heating them. Denny considered that the substances in the cell walls of the membranes mentioned above which are factors in determining their permeability to water are fatty substances and tannins. There was no evidence from his experiments that suberin plays any part in determining the permeability of these membranes. The squash seed coat, for example, is very resistant to the passage of water yet it has no distinctly suberized layers. Denny considered that the resistance of the seed coats of grapefruit to the passage of water is due to the presence in certain portions of the membranes of a large amount of fatty substances surrounded by thick pectinized walls. Denny made certain observations which showed that some of these membranes under consideration are more permeable to water in one direction than in another. Thus, for example, when the membrane of the almond was placed in an osmometer so that the water passed through the membrane, as it naturally would in entering the seed from the outside, the rate of flow was 48.5 mg. per hour. If the position of the membrane was reversed so that the water entered, as it would in passing from the seed to the external medium, the rate of flow was only 42.6 mg. per hour. In the case of the peanut membrane, the maximum percentage decrease of water flow from the "in" to the "out" amounted to 56 per cent. No difference in the rate of penetration of water in opposite directions could be observed in the case of the outer bulb scale of the onion and the seed coat of *Dioon edule*. This difference in the behavior of membranes is apparently correlated with their structure. The peanut and almond coats are composed of distinct layers and have surfaces of different physical and chemical natures, while the onion scale and *Cycad* membranes are apparently uniform in structure. A difference in the rate of water flow through membranes is characteristic of many living and nonliving animal membranes as well as many artificial ones.

According to Brauner (1930) the seed coat of *Aesculus hippocastanum* is 60 per cent more permeable when water passes from the outside to the inside than if in the opposite direction. The epidermis is the positive pole and a potential difference of several millivolts can be observed between the inner and the outer surface of the swollen membranes. The testa of this seed can thus be regarded as a "double membrane" containing electrolytes. This membrane consists of two layers differing in permeability and therefore the two sides are of different diffusion potentials. Because of this difference of potential, water flow across the membrane is promoted or hindered as the hydrodynamic potential varies with or against the water flow. The behavior of "double" membranes may have an important influence on the movement of water into and out of the plant cell and Denny asks the important question: "If this difference in the rate of diffusion is due to the presence of double membranes of different nature, or to the difference in surface on opposite sides, may not the plant cell itself show a difference in permeability in opposite directions, since such a system of double membranes is represented by the cell wall and ectoplast?"

Denny further made the very important observation in regard to the membranes with which he worked that when solutions of varying concentrations are placed on opposite sides of a membrane the relation between the rate of water movement and concentration difference is complex and that, in general, equal osmotic differences do not necessarily produce equal rates so that no mathematical relation can be noted between the concentration on opposite sides and the rate of water movement through the membranes.

b. Permeability to Solutes.—It has generally been supposed that if the cell wall were permeable to water, it would also be permeable to any solute dissolved in water. It has been shown, however, that there are numerous cell walls that are impermeable to various solutes, and this fact must be taken into consideration in studying the permeability of the plant cell. Brown (1907, 1909, 1915) studied especially the semipermeability of the covering of the barley grain. His first observation in this regard was that the blue pigment in the cells of the aleurone layer did not change in color even after the grains were soaked for several days in 1 per cent sulphuric acid, but, if the covering of the grain were punctured or otherwise injured, the pigment turned red at once, showing that the sulphuric acid had passed through the membrane. He further found that the seeds which had been soaked in the 1 per cent sulphuric acid absorbed water, swelled, and sprouted just as they did in distilled water, although as soon as the sprout ruptured the seed integument, the acid entered and turned the pigment red. It was further found that, when these seeds were placed in 4.9, 7.6, 9, 18, or even 36 per cent sulphuric acid, no acid entered after 44 hr., and these same seeds upon being washed and planted grew vigorously. The permeability of this membrane was studied in regard to copper sulphate, ferrous sulphate, potassium chromate, silver nitrate, and potassium ferrocyanide in a 5 per cent water solution. The covering was found to be semipermeable to all these, taking up water but as far as could be observed preventing the entrance of these solutes into the seed. The membrane was found to be permeable to mercuric chloride, acetic, formic, propionic, and butyric acids, ammonia, ethyl alcohol, and ethyl acetate among others. The degree of permeability to these substances, however, varied greatly. Ethyl alcohol, for example, entered at the same rate approximately as water, while acetic acid entered more rapidly, and ethyl acetate entered the most rapidly of the three. The seeds were subjected to boiling water for a 1-hr. period, but the results observed after this process were practically identical with those mentioned above, a fact which shows that this property of semipermeability lies in the cell walls of the covering of the grain and not in the protoplasm, since the latter was undoubtedly killed by this treatment. Brown also studied the rate of movement of various phenols through this seed membrane and concluded that when the osmotic pressures, vapor pressures, and viscosities of this series of solutions of soluble solutes are equal, their rate of diffusion through this membrane are inversely proportional to their surface tensions. The earlier work of Brown was followed by that of Schröder (1911), who found that the wheat grain possesses the same type of selective permeable coat as the barley grain, and by Tjebbes (1912), who observed that a part of the inner coat of the seed of the sugar beet shows the power of selective permeability. Becquerel (1907) showed that the dried seed coats of numerous plants are impermeable to various gases and to such penetrating substances as chloroform, absolute alcohol, and ether.

Shull (1913) extended the field of observation in regard to the permeability of seed coats and found that those of cocklebur, numerous grains, peach, apple, bean, and sunflower act as semipermeable membranes. His observations, however, were confined chiefly to the behavior of the seed coat of the cocklebur. This coat, he found, is composed of three rather definite layers, the outer of which does not function as a semipermeable membrane, the middle of which is several cells thick, and the inner of which is composed of a single row of cells. Both the middle and inner layers have osmotic properties, but neither functions so well separately as when they are together. The property of semipermeability in the case of the cocklebur coat is not dependent upon the protoplasmic membrane of the cells which compose it, since the same results are obtained after boiling the seeds or the membrane after it has been separated therefrom.

Shull tried the effect of soaking the dry cocklebur seeds in ether, chloroform, acetone, and absolute alcohol and found that no detectable amount of these substances penetrated through the seed coat and that the viability of the seed was not injured by this treatment. Thus, for example, seeds soaked in ether for 2 to 36 days and then dried and soaked in water and placed under germinative conditions after the testas were removed showed 87 per cent germination and the plants resulting were as vigorous as the controls. Shull investigated the permeability of this seed coat to approximately 40 different salts and found that sodium chloride, copper sulphate, potassium chromate, sodium thiosulphate, glycerol, hydrochloric acid, sugars, and tartaric acid were excluded while all the other compounds including silver nitrate, sodium nitrate, potassium nitrate, potassium chloride, ferrous sulphate, potassium hydroxide, sodium hydroxide, sulphuric, nitric, acetic, lactic, and citric acids were allowed to pass through, although the degree of permeability to these substances varied with the substance.

Experiments with a large number of seed types in 1.5 per cent solutions of potassium nitrate and manganese sulphate showed that when the seed coverings were intact any salts absorbed by the seeds were held superficially so that heavy losses in the amount of these salts resulted when the seeds were washed for 1 min. in water. When the seed coats were injured, a large amount of salts was absorbed but only a slight amount could be recovered by washing. Whitten (1914) found that the membranes of the corn seed are very impermeable to kerosene oils. Thus the grain may be immersed in kerosene for periods of 10 to 20 days without injury to the embryo.

The results obtained by Brown, Shull, Schröder, and others explain those reported by Giglioli (1895), who obtained a 20 per cent germination of alfalfa seed that had been in a saturated solution of mercuric chloride in absolute alcohol for 16 years.

Considerable work has been done to determine the structure, composition, and location of the portions of the seed coverings that prevent or retard the entrance of water and solutes into the seed. Collins (1918) considered that only a small part of the water absorbed by the grain of barley enters by the general surface of the grain, which he considered to be enveloped by cells with cuticularized cell walls. He believed that most of the water enters the micropyle or germinal region of the grain. It is here that the structure which is the seat of permeability must be sought, since he found that the solutes enter the grain as does the water through this restricted region. According to Brown (1931) the first intake of water by the seeds of *Lolium perenne* is through the micropyle. This diffuses upward and causes the endosperm to swell, which in turn induces greater permeability of the cuticular membranes of the seed coat. Although these membranes retard the absorption of water, their permeability is increased by the swelling and stretching of the seed.

In a study of the development of the testa of the wheat grain, Pugh, Johann, and Dickson (1932) noted the presence of suberin- or cutinlike compounds in the cell walls on the outer surface of one of the layers composing the testa. According to Hamly (1932) the impermeable region of the seed coat of *Melilotus alba* is formed by a layer of tightly appressed suberin layers.

The semipermeability of the grain of wheat is identified with the single layer of testa, and the cuticlelike membranes on its inner and outer surfaces (Brown, 1932). Microchemical tests showed that the testa is mucilaginous and that these cuticlelike membranes contain more true fat than does the cuticle of the normal foliage leaf.

In the seed the factors regulating the absorption of the solute would seem to be (1) electrical absorption in the cuticle-like membrane, (2) mechanical absorption in the testa and the tissues of the endosperm and embryo, (3) imbibitional pressure developed in the endosperm, and (4) the size of the intermolecular spaces of the semipermeable membranes. The solute will react to each of the above factors according to its own particular properties. It is thus extremely difficult to formulate general laws to cover the absorption of all chemical compounds.

Lavison (1910, 1911) studied the entrance of salts through the cell walls of the endodermis of the root. He found that the young cellulose periclinal walls of the endodermis are impermeable to the salts of iron, lead, copper, nickel, and mercury among others and that they behave in this regard like the protoplasm itself. We thus have a case in one of the most active translocation regions of the plant where the cell walls act in a semipermeable manner. Brooks (1917) found that the exterior cell wall of the epidermis of the inner surface of the bulb scale of the onion behaves as a semipermeable membrane toward a watery solution of calcium chloride, sodium chloride, aluminum chloride, sodium hydroxide, Bordeaux red, and eosin but that it allows a slight diffusion of hydrochloric acid through it. It is a striking fact that semipermeability to these substances is confined only to the exterior wall of the epidermal cell, while the other walls of the cell are permeable to them. The work that has thus far been done in studying the permeability of the cell-wall membrane makes it very evident that it is necessary to consider its influence in the interpretation of the permeability of plant tissues.

2. The Protoplasmic, Cytoplasmic, or Living Membrane.—One of the outstanding differences between the plant cell and the animal cell is the amount and distribution of the protoplasm within these units. An animal cell is composed almost entirely of protoplasm, and the growth of the cell simply means an increase in the amount of the living matter, which thus increases concurrently with the size of the cell. In the embryonic plant cell, the protoplasm occupies the whole space enclosed by the cell wall, but as the cell enlarges the protoplasm does not increase in amount proportionately but spreads out in a thin film around the inside of the cell wall and the outside of the vacuole which is developed under these conditions. A plant cell may increase in volume as much as 500 times or more and yet contain little more protoplasm than it did before it began to enlarge. This thin film of protoplasm thus forms a membrane through which materials entering or leaving the vacuole must pass and has been termed the "living membrane," the "protoplasmic membrane," or the "cytoplasmic membrane," the last term being used because the cytoplasm composes the greater portion of it.

a. Nature.—The protoplasmic membrane has more of the characteristics of a liquid than of a solid and evidently consists of a colloidal suspension and a true solution of numerous substances. It possesses considerable power of cohesion, is more or less elastic, and is stretched or held in its position by the pressure of the cell sap, as is evidenced by its rapid contraction inwardly as soon as the pressure within the vacuole is sufficiently reduced. Since the protoplasmic membrane is more or less liquid in its nature, it has been assumed that it exhibits certain surface phenomena which must here be discussed.

The surface of a liquid differs in physical properties from its general mass due to the fact that the molecules at the surface are not attracted equally on all sides. Any molecule throughout the general mass of the liquid is attracted equally in all directions by the molecules distributed around it and this mutual attraction results in keeping the pressure on any particular molecule uniform in every direction. A molecule at the surface of a liquid, however, is subjected to the attraction of the molecules in the interior and those adjoining it on the surface, but there is no attractive force at the exterior to balance this. As a result of this the surface layers of molecules are subjected to an inwardly directed attraction or pull at right angles to the surface. This phenomenon is called "surface tension" and because of it the surface of a liquid differs from its general mass in its behavior as a thin film or skin. Since the protoplasmic membrane is more or less of a liquid, it is generally assumed that its surfaces bordering on the cell wall and vacuole, respectively, exhibit characteristic surface-tension phenomena and also a difference in the degree of permeability as compared with the general mass of protoplasm.

The existence and nature of the surface layers of protoplasm have evinced much discussion, and probably no other topic in biophysics has been the subject of so much controversy. Some workers question its very existence, while others positively assert that an actual morphological and physiological definite surface layer can be clearly demonstrated. De Vries (1884, 1885) claimed that it has a morphological entity, Kite (1915) calls it a "hypothetical structure," while Fischer (1915) named it a "figment of the imagination," but Seifriz (1921) considered that the direct evidence from microdissection indicates a definite membrane on the surface of all protoplasm and considered that its thickness is about 0.1μ. The names used to designate this portion of the protoplasm are no less varied and numerous than the controversies concerning its existence. It has been called "surface layer," "surface film," "ectoplast," "phase boundary," "ectosarc," "hyaloplasm," and "plasma membrane." The term "plasma membrane" has been the one most commonly used. It was first used by Pfeffer (1877, 1890) to denote a surface layer or film of protoplasm that he considered to behave differently in regard to permeability from the general mass of the protoplasmic membrane. The term as it has been used by many authors and investigators is very loosely defined. In some cases it is employed only in the sense of a surface-tension film, in others it means a differentiated surface layer of the protoplasm of considerable depth, and by some it has been used to include the entire protoplasmic membrane. De Vries (1884) used the term "ectoplast" to denote the surface film of cytoplasm in contact with the cell wall and the term "tonoplast" to denote the surface bordering on the vacuole. Stiles (1924) suggested that the limiting surface of the protoplasm next to the cell wall be called the "external plasma membrane" and the surface bounding the vacuole the "internal plasma membrane."

Plowe (1931) in a study of plasmolyzed and unplasmolyzed cells concluded: (1) there is an external layer of the cytoplasm, the "plasma lemma" which, while fluid, is more elastic and more extensile than the general mass; (2) a similar but less elastic layer, the tonoplast, surrounds the vacuole; (3) the "mesoplasm" is much less extensile than the two membranes between which it lies; (4) the differential permeability of the protoplasm to water and dyes is a property of the surface layers rather than the mesoplasm; (5) the presence of differentially permeable membranes at the surfaces of the readily permeable mesoplasm makes the vacuolated plant cell a complex system, the behavior of which rests on osmotic as well as colloidal phenomena.

We must conclude therefore from Seifriz (1921), Stiles (1924), and Plowe (1931) that the surface of the protoplasmic membrane is not an invariable structure but will vary with the change in composition of the protoplasm and the external medium. The limiting surface of the protoplasmic membrane thus cannot be separated from the rest of the protoplasm, for it depends for its existence on and is closely related to the media that it separates, and it passes over into the phases on either side of it, sometimes abruptly and sometimes gradually, depending upon the environmental conditions and the composition of the mass of the cytoplasm and the surrounding medium.

The differences between the outer and inner surfaces of the protoplasmic membrane have been studied in two general ways:

1. *A Measurement of the Potential Difference between the Outer and Inner Layers.*—The works of Osterhout, Damon, and Jacques (1927), Osterhout and Harris (1928), Brooks and Gelfan (1928), Blinks (1929, 1932, 1935), Damon (1932), and Osterhout (1936) on the cells of *Valonia*, *Nitella*, and *Halicystis* show that when both surfaces of the protoplasmic membrane are in contact with the same solution there is a potential difference between them so that a current tends to flow through the electrometer from the positive to the negative surface. This potential difference ranges from a few to as many as 80 millivolts. The value of this potential difference, as well as the location of the negative and positive surfaces, is determined by the type of plant, the medium in which it is placed, and by the metabolism. Some of the substances or factors that may alter or even reverse the potential difference of the protoplasmic surfaces are distilled water, anesthetics, alcohol, other organic compounds, *e.g.*, guanidine and adrenaline, and temperature.

2. *A Comparison of the Effects of Chemicals upon the Two Surfaces.*—This comparison has been made by immersing plant tissues into chemicals, and by the injection of these same chemicals into the cells of another lot of tissue and noting any differences in effect upon the protoplasm. Thus Jacques and Osterhout (1928) found in *Nitella* and *Valonia* that the inner layer of the protoplasm was more sensitive to the action of chloroform than the outer layer. When manganese chloride was injected into the cells of *Valonia macrophysa*, they lived approximately one-half as long as when these cells were immersed in the same concentration of solution. This difference in longevity was not due to the injury by the needle used in injection since the immersed cells were stabbed in the same manner as the injected ones.

Skeen (1930) found that when certain compounds were injected into the root hairs of *Trianea bogotensis* they perished, but when the hairs were immersed in the same solution of these compounds they were uninjured for a considerable period of time. Osterhout and Harris (1928) found evidence in *Nitella* that the protoplasm in this case is made of layers differing considerably in their properties, each having a death curve of simple and regular form. These data indicate that the exterior portion or boundary of the protoplasm of these cells, under the conditions of this experiment, was of a different structure and composition from that portion bordering on the cell sap.

Whether or not the surface layers of the protoplasmic membrane differ in permeability from the greater portion of its bulk is of comparatively little moment. Many of the phenomena of diffusion that occur in plants, owing to the separation of solutions by the protoplasmic membrane, can be explained, partially at least, by assuming that as a unit it behaves as a permeable or impermeable membrane, as the case may be, without considering whether its qualities of permeability reside in its surface layers or in its general mass.

b. Permeability.—From our knowledge of the nature of protoplasm and its remarkable capacity to respond and adjust itself to its environment, we should suspect that as a membrane it would exhibit marked variations of permeability. It does, and so responsive is it to environmental conditions (only a few of which we at present comprehend) that the permeability of this membrane to a given substance may be strikingly different not only in adjoining cells but also in different parts of the same cell. Any statements, therefore, regarding the permeability of the protoplasmic membrane must be only relative to the given cell or portion of it, and on that account any discussion of its permeability to water or its solutes is inseparable from a discussion of the factors that influence permeability and the method of determining them.

The protoplasmic membrane imbibes a large amount of water (Seifriz, 1921), and as a general statement it can be said that it is readily permeable to water, but under certain conditions it is more so than under others. The greatest variations in permeability, however, prevail in the behavior of the protoplasmic membrane toward solutes, the gradations ranging from those cases where the protoplasm offers but little retardation to their passage on the one hand to practically absolute impermeability on the other. The degree of permeability depends primarily upon the environmental conditions, and these will be considered in detail when the factors influencing permeability are considered.

1. *Osmotic Relationships of the Plant Cell.*—In order to understand something of the factors that enter into the alteration of the permeability of the protoplasmic membrane as well as to comprehend the methods used in detecting these changes in permeability, it is necessary to have clearly in mind certain phenomena that occur in connection with the water relations of the plant cell. In a consideration of the osmotic relationships of a typically vacuolated plant cell, with an elastic cell wall, to the adjoining cells, or to the medium exterior to the plant, an understanding of the following factors, quantities, or forces is important:

(*a*) *Plasmolysis.*—In a mature living plant cell, the cell sap exerts a hydrostatic pressure against the protoplasm and the cell wall. As a consequence the protoplasm and cell wall are in close contact, and the cell is more or less expanded. If water is removed from the cell sap, the hydrostatic pressure against the protoplasm and the cell wall is released, and since they are elastic they contract. The cell wall soon reaches its limit of contraction, but the protoplasmic membrane which is more elastic will continue to contract as water is withdrawn and thus will move inwardly from its position along the cell wall.

The contraction of the protoplasmic membrane from the cell wall, owing to the release of the hydrostatic pressure of the cell sap against it, is called "evident plasmolysis." The term "plasmolysis" literally means "the loosening of the protoplasm." That condition in a plant cell when the cell wall has reached its limit of contraction but when the protoplasm has not yet receded from the wall is termed "incipient plasmolysis." The deficiency of water in the vacuole which causes plasmolysis in plant cells is generally due either to the difference in concentration of the solution external to the cell and that of the cell sap, or to the inability of the cell to replace the water that is being lost from it through evaporation.

(b) *Osmotic Value of the Cell Sap.*—By the osmotic value of the cell sap is meant the concentration of the osmotically active substances therein. This value may be expressed in terms of molar concentration, or in terms of the osmotic pressure, expressed in atmospheres, that the cell sap would be capable of producing if it were separated from pure water by a perfectly semipermeable membrane. It might also be appropriately named the "potential" or "latent" osmotic pressure of the contents of the vacuole, since it does not express the value of the osmotic pressure being exerted within the vacuole but rather that which such a solution would be capable of developing if it were placed under such ideal conditions.

It is evident from this definition that the osmotic value of the cell sap will be decreased as water enters and the cell expands, thus decreasing the concentration of the osmotically active substances. When water is withdrawn from the cell sap, the vacuole decreases in volume and the concentration of the osmotically active substances is increased, thus increasing the osmotic value of the cell sap. A plant cell is considered to function naturally as long as the protoplasm remains in contact with the cell wall. When evident plasmolysis occurs, the cell is under abnormal conditions which, if continued, will cause death. The smallest volume that the cell sap can attain and keep the protoplasm in contact with the cell wall is at the point of "incipient plasmolysis." The greatest volume that the cell sap can attain is when it is at equilibrium in distilled water. The osmotic value of the cell sap is thus the highest at incipient plasmolysis and the lowest when the cell is in equilibrium with distilled water.

A solution that has the same osmotic value as the cell sap is termed an "isotonic" solution, and if a plant cell is placed in such a solution it will not change in volume. A solution that has an osmotic value higher than that of the cell sap is called a "hypertonic" solution. A plant cell placed in such a solution will soon show evident plasmolysis. The term "hypotonic" is applied to a solution whose osmotic value is lower than that of the cell sap.

(*c*) *Osmotic Concentration of the External Solution.*—The term "external solution" as here used may refer to the cell sap of the cells adjoining the one under consideration or that which may obtain in the medium exterior to the plant. This external solution exerts an osmotic pressure that counterbalances to a greater or lesser degree that of the solution in the vacuole of the cell in question. If the osmotic concentration of the cell sap is higher than that of the external solution, water will be drawn into the cell with a force equal to the difference between the osmotic pressure of the cell sap and that of the external solution, provided that the cell wall and protoplasmic membrane exert no inward pressure upon the solution in the vacuole. If the osmotic value of the external solution is higher than that of the vacuole of the cell being considered, water will be withdrawn from the cell with a force equal to the difference between the osmotic pressure of the vacuole and the external medium, if the external medium is not subjected to any inwardly directed pressure.

This latter behavior can be observed in the laboratory by placing plant tissue in a solution, the concentration of which is higher than that of the cell sap of the cells which compose the tissue. Let us consider the phenomena that would prevail under these conditions if the protoplasmic membrane were (1) impermeable to the solutes of the exterior solution and those of the cell sap and (2) permeable to the solutes of the exterior solution but impermeable to those of the cell sap. It is assumed in both cases that the concentration of the solutes in the external solution is not of such a nature or of sufficient strength to be directly injurious to the protoplasm.

Let us now assume that a plant cell is placed in a solution whose osmotic value is higher than that of the cell sap and that the protoplasmic membrane is impermeable to the solutes of the exterior solution and of the cell sap (Fig. 7*A*). Under these conditions no diffusion of solutes through this membrane tending toward an equilibrium in their concentration could occur. Since the concentration of solutes in the exterior solution is higher than that in the vacuole of the cell, it follows that the concentration of solvent is lower in the former than it is in the latter and that as a result water will diffuse outwardly from the vacuole into the exterior solution more rapidly than it will move in the opposite direction. Hence water will be drawn from the vacuole of the cell and the hydrostatic pressure that is normally exerted against the protoplasm and the cell wall is lowered, and as a result these membranes contract inwardly. Since the cell wall is only slightly elastic, the limit of its contraction is soon reached, but this is not true of the protoplasmic membrane and it will continue further to contract inwardly and thus separate itself from the cell wall (Fig. 7*B*). This contraction of the protoplasm will continue until the osmotic concentration of the cell sap, owing to the unequal diffusion of water outwardly and also perhaps to the formation of new osmotically active materials by the protoplasm, has become equal to that in the external solution. When this state of equilibrium prevails, further contraction of the protoplasm will not occur, and, unless the protoplasm is capable of manufacturing osmotically active substances to increase further the osmotic concentration of the vacuole, the protoplasmic membrane will never regain its normal position, and death will shortly ensue.

Let us now consider the phenomena that occur if the protoplasm is permeable to the solutes of the medium in which the cell is placed but impermeable to the solutes within the vacuole. Since the solutes in the external solution are higher in concentration than those in the cell sap,

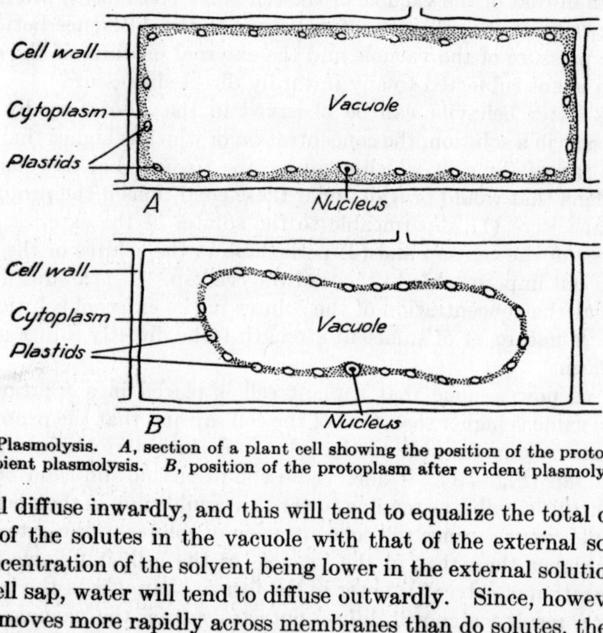

Fig. 7.—Plasmolysis. *A*, **section of a plant cell showing the position of the protoplasm at incipient plasmolysis.** *B*, **position of the protoplasm after evident plasmolysis.**

they will diffuse inwardly, and this will tend to equalize the total concentration of the solutes in the vacuole with that of the external solution. The concentration of the solvent being lower in the external solution than in the cell sap, water will tend to diffuse outwardly. Since, however, the solvent moves more rapidly across membranes than do solutes, the water of the cell sap will diffuse outwardly more rapidly than the solutes move inwardly. As a result, the hydrostatic pressure against the cell wall and the protoplasm will be reduced and these membranes will contract in the same manner as described in the previous example. The protoplasm will recede from the cell wall and continue to contract until, due to the diffusion outwardly of water and the diffusion inwardly of the solutes, the cell sap is in osmotic equilibrium with the external solution.

Since, however, the protoplasmic membrane is permeable to the solutes of the external solution, they will diffuse inwardly until the concentration of each solute is the same in the cell sap as it is in the external solution. As a result of this, the osmotic value of the cell sap will increase and water will move inwardly so that a hydrostatic pressure will be gradually set up which will force the contracted protoplasmic membrane against the wall into its normal position. Thus eventually the hydrostatic pressure against the protoplasmic membrane and the cell wall will be the same as that which prevailed before the cell was immersed in the solution.

According to King (1932) some cells die quickly when they are plasmolyzed, while others may remain alive for a few days to several weeks. The injury and recovery from plasmolysis will vary with the nature of the salt used, with the duration of plasmolysis, and with the kind of plant.

(*d*) *The Turgor Pressure and the Wall Pressure of the Cell.*—A vacuolated living plant cell under the ordinary conditions of plant life absorbs a sufficient amount of water to exert a hydrostatic pressure against the cell wall. Because the wall is elastic, it is stretched by the hydrostatic pressure, but its limit of extensibility is soon reached, and at that point the wall exerts a backward pressure against any further increase in the volume of the cell. When the limit of extensibility of the cell wall has been reached, its resilience or inward pressure is equal to the force with which water tends to enter the cell, and hence no more water will enter. When this process of expansion is under way or has reached its limit, the cell is rigid and is said to be "turgid." This condition or rigidity is spoken of as the "turgidity" of the cell.

The inwardly directed pressure of the cell wall against the protoplasm and the cell sap is termed the "wall pressure" of the cell. The hydrostatic pressure of the cell sap exerted against the cell wall is called the "turgor pressure" of the cell. This hydrostatic pressure is the equivalent of the osmotic pressure of the cell sap, but is designated as the turgor pressure only when it is considered in its relation to its pressure against the cell wall. When the cell has reached its limit of expansion, the wall pressure is at its highest possible value and is equal to the osmotic pressure, hydrostatic pressure, or turgor pressure at that time. At incipient plasmolysis the wall pressure and turgor pressure are zero. The cell-wall pressure is calculated at any point between incipient plasmolysis and the limit of cell expansion by assuming that any change in the pressure of the wall is proportional to the change in the volume of the cell. It appears, however, that such a relationship holds only approximately.

(*e*) *The Absorbing Power of the Cell.*—This term is used to designate the force with which water enters the plant cell under any given condition. In this relationship the terms suction tension, suction pressure, suction force, or suction of the cell have also been used. The use of these terms, however, has been adversely criticized by Shull (1927, 1932), and others, because water apparently moves into a plant cell and is in no sense sucked or pulled into it as these terms imply. The location and exact nature of the forces that cause water to enter the plant cell are not definitely known. These forces are in some manner the resultant of the reaction between the cell sap, the external solution, and the protoplasmic membrane separating these two solutions. In order to express the collective action of the forces involved in the intake of water by the cell, no term has been suggested or devised which is above criticism. The term "absorbing power" of the cell seems to be the least objectionable and is herein used.

The absorbing power of the plant cell, at any time, is equal to the osmotic value of the cell sap minus the osmotic value of the external solution minus the wall pressure.

Absorbing power = osmotic value of cell sap − (osmotic value of external solution + wall pressure)

When a plant is in equilibrium with pure water, the absorbing power of the cell is zero, while at the point of incipient plasmolysis it reaches its maximum value for any condition in which the protoplasm is in its normal position along the cell wall. The absorbing power of plant cells, under the conditions that generally prevail in the plant, is somewhere intermediate between the two extremes mentioned above.

The osmotic value of the cell sap at incipient plasmolysis has been wrongly termed by many investigators as the absorbing power of the cell under the conditions generally prevailing in the plant. These investigators considered the concentration of the cell sap at incipient plasmolysis to be the same as that which prevails in the cells of the plant under any condition. They thus did not take into consideration the shrinkage in the volume of the cell before incipient plasmolysis and the attending increase in the concentration of the cell sap. Ursprung (1916 to 1935), Ursprung and Blum (1916 to 1930), Molz (1926), Dumont (1926), Beck (1928 to 1935), Dixon (1930), Ernest (1931, 1934), Herrick (1933), and Lyon (1936) have studied the absorbing power of cells in different tissues under various conditions, and they have described and criticized the different methods that have been used in its determination.

(1) *Methods of Determining.*—Most of the information concerning the various osmotic forces of the plant cell has been obtained by using microscopic sections and noting the increase or decrease of the volume of the cell when solutions of various concentrations were applied. Ernest (1931) pointed out that the methods of determining the absorbing power of cells by means of sections are in error. In sections, the intact cells are in contact not only with the solution being used in the experiment but also with the cell sap of those cells which have been cut or ruptured in the sectioning of the material under observation. The absorbing power of the cells is thus not necessarily balanced by the solution that has been applied. The fact that the intact cells

expand abnormally because the adjoining cells are ruptured and their inwardly directed pressure is released also introduces another source of error in the use of sections.

The more simple method reported by Lyon (1936) seems to eliminate to a considerable extent the objections to the earlier methods. Blocks or strips of a tissue are immersed for some time in various concentrations of a sugar solution. The strips are measured microscopically for length before and after immersion. The concentration of the solution in which the strip neither lengthens nor shortens is a measure of the mean osmotic value of the cells, under the conditions prevailing in the plant. The first solution that gives the maximum shrinkage would be isotonic with the cell sap at incipient plasmolysis and would be a measure of the osmotic value and the absorbing power of the cell in that condition.

The solution that causes the first contraction of the tissue after it is in equilibrium with distilled water would indicate the osmotic value of the cell sap under that condition and would be the equivalent of the wall pressure of the cell at that time. The wall pressure is calculated at any point between these two extremes by assuming that the change in wall pressure is proportional to the change in volume of the cell.

(2) *Observations.*—As an example of the different quantities concerned in the osmotic relationship of the plant cell, those determined for the cells of the root of *Vicia faba* are given (Molz, 1926).

Osmotic Value, Absorbing Power, and Turgor Pressure as Determined in the Root Cells of *Vicia faba*

Tissue examined	Distance from growing point, millimeters	Osmotic[1] value of cell sap, atmospheres	Absorbing power of cells, atmospheres	Turgor pressure, atmospheres
Epidermis of cells of outer cortex......	1.5	14.3	7.4	3.9
	3.0	14.0	7.7	4.0
	5.0	13.3	10.0	2.7
	8.0	12.7	0.4	9.4

[1] At incipient plasmolysis.

These investigators found in the absorption zone of the root of *Vicia faba* an increase in the absorbing power of the cells from the epidermis toward the endodermis and then a marked decrease in the endodermis and wood parenchyma. It was also observed that the absorbing power of the cells of the upper portion of the stem and its leaves was higher than the cells of the same parts lower down. An increase in absorbing power from the base to the apex was found in the mid- and lateral veins of the ivy leaf. The observation of the absorbing power of a row of palisade cells that extended from the midvein into a portion of the leaf free from larger veins was especially noteworthy. From a minimum of 12 atmospheres near the midvein, the absorbing power rose regularly to a maximum of 32.6 atmospheres, which was attained in the two hundred and tenth palisade cell, from which point it decreased. In contrast to this regular rise in absorbing power, the osmotic value in the same row of palisade cells showed no regularity.

Buchinger (1927) observed that the absorbing power in 43 varieties of barley, varied from 23.4 to 35.2 atmospheres and that these differences were indicative of the geographical region to which the varieties are adapted. Herrick (1933) observed in *Ambrosia trifida* that if the competition for water was not severe neither the osmotic value nor the absorbing power of the cells reached a maximum in the upper part of the plant. Under these conditions there was no consistent gradient of these values throughout the plant. Increasing the internal competition for water, whether owing to seasonal or diurnal causes, resulted in an increase in the magnitude of the osmotic values throughout the plant and in the establishment of consistently increasing gradients of these values and absorbing power from the lower to the higher portions of the plant. Under conditions of severe competition for water, the osmotic value in any one leaf was essentially equal to the absorbing power of the same leaf. The osmotic value and absorbing power tended daily to reach maximum values between 1 and 3 P.M.

Ernest (1934) found that the absorbing power of the superficial mesophyll cells of the leaves of *Iris* maintain a remarkably constant value changing but little during a 24-hr. period. He observed a slight gradient of absorbing power from the water-conducting tissue to the surface tissue. He also observed that the mesophyll cells removed from a leaf that had been floating for some hours in water showed a considerable absorbing power which might amount to as much as 8 atmospheres. This absorbing power was due to the expansion of individual cells, which under natural conditions was resisted by the mutual pressure of the surrounding cells.

2. *Permeability and Absorption.*—The diffusion of any solute into a plant cell will depend not only upon the degree of permeability of the cell membranes to the substance in question but also on how much the equilibrium of the solute on the outside and inside of the cell is disturbed. On that account, therefore, the rate at which a substance enters a cell cannot be taken as a measure of the permeability of the cell membranes to that substance. Let us consider a case in which the protoplasmic membrane is only slightly permeable to a given solute, but as soon as this solute enters the cell it is chemically combined, precipitated, or adsorbed and thus rendered osmotically inactive. In that case, the concentration of this solute on the inside of the cell will always be lower than that on the outside and it will continue to enter, so that after a period of time a comparatively large amount of the substance has entered the cell even though the protoplasmic membrane is only slightly permeable to it. On the other hand, the protoplasmic membrane may be very permeable to a solute, but after its entrance into the cell, this solute may not be combined or adsorbed to any extent. In that case, equilibrium of concentration of this solute on the outside and inside of the cell will soon be reached and no more will enter. Thus after a period of time only a comparatively small amount of the solute has entered the cell, although the protoplasmic membrane is readily permeable to it. In discussing the methods used in determining the changes in the permeability of the protoplasmic membrane, the distinction between the degree of permeability of the membranes and the rate of absorption of a given solute must be kept clearly in mind. It should be remembered that these quantities have little to do with each other and that the rate of absorption is no criterion of permeability, although it has been so used by many investigators.

3. *Methods of Measuring the Permeability of the Protoplasmic Membrane to Solutes.*—The investigations of the changes in permeability of the protoplasmic membrane to solutes have shown marked differences in evidence secured by different methods and in the conclusions based thereon. These differences are in a large measure due to a lack of understanding of the limitations of the methods used and to assumptions which are unwarranted by the data obtained. A brief survey should thus be made of the methods that have been employed and their advantages and disadvantages mentioned so that the student may be able to judge better the value of the work that has been reported in this regard. The principal methods that have been used in the study of the permeability of the protoplasm are (*a*) chemical analysis of the tissue extracts or of the solutions external to the tissues; (*b*) observation of visible changes within the cells; (*c*) methods involving plasmolysis; and (*d*) electrical conductivity of tissues or masses of cells. These have been termed "indirect methods," since the cell contents are examined only in an indirect manner or not at all. A direct method has lately been devised that is applicable in certain cases, whereby the cell sap may be withdrawn directly from the cells and a quantity of material obtained sufficient to make a qualitative and in some cases a quantitative chemical determination.

(*a*) *Indirect Methods.* (1) *By Chemical Analysis.*—This method has been used primarily to determine what substances may enter or leave the plant cell through the protoplasmic membrane rather than a measure of the changes in its permeability. The method is more applicable in considering the amount of materials absorbed by a considerable portion of tissue or by the entire plant than to a consideration of a single cell or of a small amount of tissue. Two lines of procedure have been followed. One is to analyze the tissue or the tissue extract, and the other is to analyze the solution bathing the tissues. The principal objection to analysis of tissue extracts is that it is extremely difficult to obtain a sample that is representative of the cell sap by any of the present means of extraction. The analysis of the tissue or tissue extract also does not take into consideration the presence of salts in the intercellular spaces or in the cell wall where they may be held in solution or by adsorption. The analysis of the solution in which the plant is grown in order to determine the intake or outgo of solutes has many sources of error. The results in many cases can be observed only after several hours and, in many cases that have been reported, only after several days. Under such conditions some of the external cells are undoubtedly killed or are separated from the roots, especially the root tip, and thus give off their cell sap into the solution. Any increase in the diffusion outward of the solutes may be due not to increased permeability but to increased production within the cells under the conditions of the experiment. Brooks (1917) criticized the method, since it in no way distinguishes between the rate of absorption and the degree of permeability of the protoplasmic membrane. Steward (1928) concluded from a consideration of the effects of various treatments upon exosmosis that protracted leaching does not necessarily result in loss of semipermeability. It is evident thus that under certain conditions the chemical method may be an indicator at least of the degree of permeability of the protoplasmic membrane but that its limitations must be kept in mind.

(2) *By Visible Changes within the Cell.*—This method of studying permeability is of importance in the investigation of the diffusion of substances into the cell which will form therein precipitates, stains, crystals, or other compounds that may be observed. These substances include certain alkaloids that form intracellular precipitates; acids and alkalies that cause color changes of pigments within the cell; dyes that form precipitates; stains; salts that form crystals after their entrance; and sugar and glycerol whose penetration into the cells is made manifest by the formation of starch. This method of determining permeability is necessarily limited in its application, and it is also open to the objection that it is more a measure of the rate at which the substances are combined in the cell than a measure of the degree of permeability of the protoplasmic membrane to these substances.

(3) *By Plasmolytic Methods.*—Plasmolysis of a cell is caused by the inability of the solutes on the exterior of the cell to diffuse rapidly enough through the protoplasmic membrane to equalize the concentration of the cell sap with that of the exterior medium. This inequality of concentration may be due to the fact that the solutes are combined or rendered osmotically inactive after they have once entered the cell sap or that the protoplasmic membrane is not permeable to a sufficient degree to allow the solutes to enter rapidly enough. It is on the latter assumption that the plasmolytic methods for the studying of changes in permeability are based. Two methods have been the most generally used.

a) The Determination of Concentrations Necessary to Produce Plasmolysis.—If the protoplasmic membrane is only slightly permeable to the solute in the external solution, the attaining of equilibrium between the cell sap and a hypertonic solution takes place slowly, so that an exterior solution, the concentration of which is only slightly above that of the cell sap, will almost immediately cause plasmolysis. If, on the other hand, the protoplasmic membrane is readily permeable to the solute under consideration, the solute will enter the cell and equilibrium between it and the cell sap will be quickly restored and plasmolysis will not result. The external concentration of the solute must therefore be increased to such an extent that, even if the protoplasmic membrane is readily permeable to it, equilibrium will not be restored quickly enough to prevent plasmolysis. Thus if a cell or a portion of tissue under condition *A* is plasmolyzed, for example, by a 0.1 *N* sodium chloride solution and when placed under condition *B* it requires a fifth-normal solution of the same salt to produce plasmolysis, one may assume that the permeability of the protoplasmic membrane to sodium chloride is greater under condition *B* than it is under condition *A*.

b) The Determination of the Time Required to Recover from Plasmolysis.—If the formation of new osmotically active substances in the vacuole is not considered, the recovery of a cell from plasmolysis is due primarily to the diffusion of the solute of the exterior medium into the cell in sufficient amount to equalize the concentration of the cell sap and the exterior solution. Recovery from plasmolysis, therefore, will be more rapid if the protoplasmic membrane is readily permeable to the solute of the exterior solution than if it is only slightly permeable to it. Thus if a cell or the cells of a tissue are plasmolyzed by a hypertonic solution of sodium chloride and under condition *A* recovery from plasmolysis occurs in 30 min., while under condition *B* only 10 min. are required for recovery, it is considered that the permeability of the protoplasmic membrane to sodium chloride has increased under condition *B* as compared with what it was under condition *A*.

The main objection to the plasmolytic methods is that they do not take into consideration the possibility of the formation by the protoplasm of osmotically active substances in the cell sap, which if they were formed would have the same effect as solutes taken in from the external solution. The effect of ethyl alcohol on *Spirogyra* is of interest in this regard. Lepeschkin (1928) observed that the turgor pressure of this plant was decreased by a weak concentration of alcohol while a strong concentration of alcohol increased it to such an extent that the cells were torn to pieces. It is considered that the tannin in the cell sap is a colloidal solution in water, but in alcohol it becomes a molecular solution and thus increases the concentration of the cell sap. In a dilute solution of alcohol the permeability of the protoplasm is increased and the turgor pressure thus lessened.

(4) *By Electrical Conductivity of Tissues or Masses of Cells.*—"The conduction of an electrical current by a solution involves the passage through the solution of electrically charged ions of some substance. The rate at which the current will be conducted by the ions of a given salt will depend upon two factors, the potential gradient and the frictional or other resistance to the migration of the ions. If the potential gradient be kept constant we may follow fluctuations in the last factor by measurement of the current or by direct measurement of the electrical resistance. By the use of alternating currents of rather high frequencies, large effects due to the accumulation of ions at the surfaces impermeable to them are avoided" (Brooks, 1917). The electrical-conductivity method of measuring the permeability of the protoplasmic membrane is based on the assumption that the electrical conductivity of a tissue is to be regarded as a measure of the permeability of the protoplasmic membrane. If the current is forced through the living protoplasmic membrane in a solution, the resistance offered by this membrane to the passage of ions is considered to vary inversely as the permeability. This electrical method was first used by Bugarsky and Tangl (1897) and Stewart (1897) but has been modified, elaborated, and extensively used by Osterhout (1912, 1918) and his pupils. Much of the work of Osterhout in measuring changes in permeability by the electrical-conductivity method has been with the thallus of *Laminaria* after the following manner. One hundred to two hundred disks are cut from the fronds and packed together like a stack of coins. These form a cylinder 5 to 10 cm. high and about 1.3 cm. in diameter and are kept in position by means of glass rods. At each end of the cylinder and separated from it by a small amount of solution is a platinum electrode coated with platinum black and connected with an instrument for measuring the electrical conductivity. The general procedure is to immerse the stack of disks in a given solution and after drawing off the superflouus liquid to place the stack under the conditions mentioned above and measure the electrical conductivity. There is sufficient solution between each of the disks to make the circuit, and experiments have shown that they are not injured even by prolonged treatment.

The manner in which the electrical-conductivity method is used in determining changes in permeability can best be illustrated by a theoretical example. The resistance of a given column of disks after being immersed in a given solution *A* and then drained off, as has been previously mentioned, is 850 ohms. The same column after being immersed in solution *B*, which has the same conductivity as *A* but is of a different composition, and treated in the same manner has a resistance of 980 ohms. The assumption from the data is that the effect of solution *B* is to decrease the permeability of the protoplasmic membranes of the tissue under what it was in solution *A*. The question in dispute in regard to the reliability of this method is how far the electrical conductivity of tissue is a measure of the permeability of the protoplasmic membranes of the cells. Stiles and Jorgensen (1918) and Stiles (1924) have pointed out, however, that an increase in the conductivity of a tissue may be brought about in other ways than by an increase in the permeability of the protoplasmic membrane. They considered that changes in the content and nature of the electrolytes in the cell wall, protoplasm, and cell sap can have a marked influence on the alteration of the conductivity. Any factor such as diffusion, which would bring about an increase in the amount of free electrolytes in the cell parts or the breaking down of large complex molecules to smaller ones, which are more mobile and undergo more ionization, would increase the conductivity of the tissues. These changes might increase the permeability of the protoplasmic membrane, but the increase in conductivity could be due, in part at least, to other factors.

(*b*) *Direct Methods.*—The reliability of the various indirect methods that have been used in the study of permeability is open to question until they can be compared with results obtained by direct methods. In order to make direct determinations, it is necessary to remove the cell sap and then compare its composition with that of the solution exterior to the cell. The errors that are encountered in trying to analyze the extracts of cells and tissues have been enumerated above. The only known direct method is to use individual cells and to extract their cell sap without any contamination or change. Only a few cells, however, are large enough to give any satisfactory results by this method.

Wodehouse (1917) experimented with the marine alga *Valonia*, which has individual cells so large that the cell sap yields from 1 to 5 cc. of liquid. The cells were removed from the sea water, dried with filter paper, and then punctured with a needle. On pressing the cell wall, the cell sap exuded and was collected. Osterhout (1921) used the large multinucleate cells of a species of *Nitella* some of which reach a length of 6 in. and a diameter of 1⁄32 in. Two methods were used for obtaining the cell sap. By one method the cell was placed on glue or filter paper and pierced with the point of a capillary tube which had been drawn out to a fine tip. The cell sap was by this means drawn up into the tube by capillary action and was not contaminated by the chloroplasts or protoplasm. The other method was to suspend the cell by the upper end, cut off the lower end, and bring it in contact with a glass slide and then to apply a slight pressure to the cell wall, beginning at the top. In this manner the cell sap flowed out on the slide, and by uniting the drops from several cells a sufficient quantity of material was obtained for examination. By this method Osterhout (1922, 1927) was able to compare direct observations of permeability behavior with that indicated by the electrical-conductivity method and concluded that the conductivity method gives reliable information concerning permeability. The direct method of studying permeability has also been used by Irwin (1922) and Hoagland and Davis (1923). The results obtained by these various investigators will be discussed in full in Chap. V when the intake of solutes by the plant will be considered.

4. *Methods of Measuring the Permeability of the Protoplasmic Membrane to Water.* The number of methods used in determining the permeability of the protoplasmic membrane to water are very few. Delf (1916) used the hollow leaves of the onion and the hollow scape of the dandelion for investigation. They lend themselves to this experiment because they present on their inside a natural surface of uncutinized thin-walled cells. An apparatus was employed by which a current of liquid could flow continuously through a cylinder 30 mm. in length of the tissue under investigation. It was so arranged that a solution of cane sugar could be immediately substituted for a water current, and vice versa, without altering the temperature of the tissue. The data on the passage of water through the protoplasmic membrane were obtained by observing the rate of shrinkage of the tissue under consideration when in contact with solutions that tended to plasmolyze them. An optical lever in connection with the tissue enabled alterations in length of 0.003 mm. or 1 part in 10,000 to be detected. An experiment of this nature cannot be carried out satisfactorily with strong plasmolyzing solutions because of the rapid shrinkage of the tissue. The solution must be sufficiently weak so that contraction will be comparatively slow and thus confine the experimentation to the strictly normal state of falling turgor without the actual separation of the protoplasmic membrane from the cell wall. The rate of shrinkage of the tissue under these conditions is the resultant of two factors. One of these is the

compressing force of the cell-wall action on the protoplasmic membrane, which causes exudation of water, and the other is the permeability of the protoplasmic membrane for water. The influence of the permeability factor can be obtained only by comparing the tissues under consideration when the compressing force is equal. In the experiments of Delf, the points of 30, 50, and 70 per cent of the total contraction were used, but the most acceptable series for the two kinds of tissue under consideration was the 50 per cent series.

The change-in-weight-and-volume method has been used by Stiles and Jorgensen (1917) to determine the rate of water intake by the tissue of the potato tuber and the carrot root. The intake of water was measured by immersing the tissue of known weight and volume in water and weighing the tissue and measuring the increase in volume at certain intervals. The change in weight under these conditions is attributed to the amount of water taken up, since the weight of solutes that might diffuse outward during the experiment is considered negligible in comparison to the weight of water involved. The tissue in these experiments was cut in cylinders 1.8 cm. in diameter by means of a cork borer and then into disks 0.2 cm. in thickness by means of a hand microtone. Each experiment was run in triplicate or quadruplicate with a set of 20 disks.

Huber and Höfler (1930) determined the permeability of the protoplasmic membrane to water by noting the rate of change in volume of the cells during plasmolysis and subsequent recovery in water. Baptiste (1935) immersed uniform disks of the potato tuber and carrot in water and followed its uptake by weighing these disks periodically at frequent intervals.

5. *Factors That May Influence the Permeability of the Protoplasmic Membrane.* (*a*) *Temperature.* (1) *Influence on the Permeability of the Protoplasmic Membrane to Water.*—Rysselberghe (1901) studied the changes in the permeability by the plasmolytic methods and also by the rate of shrinkage of tissue in a 0.73 *N* solution of cane sugar. He observed especially the pith of the cells of the elder, *Spirogyra*, and the epidermal cells of *Tradescantia.* He found that the effect of temperature is significant from 0 to 20°C. but that from 20 to 30°C. the permeability of the protoplasmic membrane to water remains almost constant, as shown in the following table:

Temperature, °C.	Rate	Temperature, °C.	Rate
0	1.0	20	7.1
6	1.9	25	7.5
12	4.5	30	8.0
16	6.3		

Lepeschkin (1906) found that the volume of liquid extruded by hairs on the leaves of *Phaseolus multiflorus* increased from 0 to 20°C. and decreased from 20 to 35°C.

Delf (1916) found that the permeability of the protoplasmic membrane to water as measured by the rate of tissue shrinkage of the onion leaf and dandelion scape in a dilute sugar solution is increased continuously by temperature up to 42°C.—the highest temperature investigated. If the rate of permeability at 25°C. is taken as unity, the following relative values were established:

Temperature, degrees	Permeability	
	Onion	Dandelion
5	0.36	
10	0.44	0.22
15	0.50	0.30
20	0.66	0.50
25	1.0	1.0
30	1.7	1.9
35	2.9	3.0
40	5.0	5.0

Delf criticized the results of Rysselberghe in regard to the permeability remaining constant above 20°C. She considered that the sugar solution used by Rysselberghe was too concentrated for the best results, since a solution of that strength causes such a rapid contraction of the protoplasmic membrane that the cell wall cannot follow it. From her observations Delf considered that there is an upper limit to the rate at which the mechanical tissue system of the cell walls can collapse and shrink. If this assumption is correct, then no matter how rapidly the protoplasm contracts under the influence of the external solution, the cell wall cannot follow at a rate above a certain limit. In Rysselberghe's experiments that limit seemed to have been reached at 20°C. with a 0.73 *M* cane-sugar solution. Stiles and Jorgensen (1917) found that the temperature coefficients for the rate of swelling of the tissues of the potato tuber and root of carrot in distilled water were as follows:

Temperature, degrees centigrade	Temperature coefficients	
	Potato	Carrot
10 to 20	3.0	1.3
20 to 30	2.7	1.6

The rate of swelling, however, is dependent not only on the permeability of the protoplasmic membrane to water but on the elasticity of the cell wall and the permeability of the protoplasm to the dissolved materials in the cell sap. The previous history of the tissue under consideration also has a great influence on the results obtained.

The experimental evidence seems to indicate that the permeability of the protoplasmic membrane to water increases continuously from the lower to the higher temperatures so that if the other factors are the same the intake or outgo of water from the cells increases with a rise in temperature.

(2) *Influence on the Permeability of the Protoplasmic Membrane to Solutes.*—Rysselberghe (1901) found that the permeability of the protoplasmic membrane of the epidermal cells of *Tradescantia discolor* to glycerin, potassium nitrate, and urea increases from 0 to 30°C. Lepeschkin (1905) observed that the protoplasm of the leaves of *Phaseolus multiflorus* increased in permeability to the solutes of the cell sap from 0 to 20°C. and decreased from 20 to 35°C. Eckerson (1914) investigated the influence of temperature on the permeability of the protoplasmic membranes of the cells of the roots of numerous plants. The results were obtained on the assumption that an increase in the concentration which is necessary to cause plasmolysis corresponds to an increase in the permeability of the protoplasm. Potassium nitrate was used, for the most part, as the plasmolyzing agent, and the tests for permeability were carried on under controlled temperatures. The permeability of the protoplasm of the cells of the radish root to potassium nitrate was found to increase from 10 to 18°C. but did not change from 18 to 24°C., increased from 24 to 40°C., and then decreased. The permeability of these cells to glucose was observed to increase gradually from 10 to 40°C. and then decrease. The permeability to sucrose remained constant until temperatures above 42°C. were reached. *Helianthus annuus* showed no increase in permeability with increase in temperature but a decrease from 20 to 40°C. The primary roots of *Phaseolus multiflorus* exhibited no increase in permeability from 13 to 20°C., while the secondary ones increased in permeability from 6 to 25°C. Both types of roots decreased in permeability from 25 to 35°C.

The table on page 104 is a summary of the work of Eckerson on the relation of temperature to the changes in permeability to potassium nitrate.

The work of Eckerson indicates that the changes in permeability due to changes in temperature vary with the species, with different temperatures, and with the plasmolyzing agent used.

Sen (1928) found that the permeability of the plasma membrane to ions increased with rising temperature from 20 to 30°C. in *Aloe* leaves and in *Bassela* stems. From 30 to 35°C. there was little or no increase, while beyond 35 or 40°C. and up to the lethal temperature, the permeability decreased but is reversible up to 40 or 45°C. At the lethal temperature, the plasma membrane becomes highly permeable, a change which is irreversible (see also Dixon, 1924).

Permeability Variations of the Protoplasmic Membrane to Potassium Nitrate at Different Temperatures[1]

Cells of the root of	Degrees centigrade			
	+	0	+	−
Raphanus sativus	10 to 14	18 to 24	24 to 40	40 to 50
Pisum sativum	6 to 15	15 to 35	......	35 to 45
Sinapis alba	10 to 30	30 to 45		
Helianthus annuus	......	12 to 20	......	20 to 40
Phaseolus multiflorus:				
Primary roots	......	13 to 20	......	25 to 40
Secondary roots	6 to 25	......	......	25 to 40

[1] In the table a plus sign indicates increasing permeability; zero indicates no changes in permeability, while a minus sign indicates a decrease in permeability.

(*b*) *Light.*—Lepeschkin (1908) observed that light increases and darkness decreases the permeability of the protoplasm of the pulvinal cells of the legumes. The increase in the permeability of the protoplasmic membrane of these cells results in a decrease of their volume, while a decrease in permeability brings about an increased turgor pressure and an increase in volume. Tröndle (1909, 1910, 1918) found that the protoplasm of the cells of the leaves of *Tilia europaea*, *Buxus sempervirens*, *Salix babylonica*, and *Acer platanoides* are more permeable to sodium chloride in light than in darkness. The permeability of the cells of the leaves of *Buxus* and *Tilia* was found to be, respectively, 17.6 and 33 per cent greater in sunlight than in cloudy weather. If the leaves were held at the same temperature, increasing illumination increased the rate at which sodium chloride passed into the cell up to a certain maximum of light intensity that varies with the kind of plant. Lepeschkin (1930) studied the rate of accumulation of dye in the cell sap of the leaves of *Elodea* in light and in darkness. He concluded that the permeability of the protoplasm is increased in the light, and that not the mode of reception but the amount of light energy is the important factor. He noted that adsorption by the cell wall and protoplasm was not altered by light or darkness, and that the cells affected by light did not impart an increase in permeability to neighboring cells. The greater the change of the illumination, the greater was the change of permeability, but no proportionality between them was observed. The maximum change in permeability was observed when direct sunlight was diminished to 10 per cent of its full intensity. The rays most active in increasing permeability are those with a wave length of 320 to 420mμ, while the red rays are least active in this regard. Offord and d'Urbal (1931) found that sodium chlorate apparently penetrated *Nitella* more slowly when the light intensity was reduced. On the contrary, however, Ruhland (1911) found that light was without any measurable influence on the permeability of the protoplasmic membrane of the cells of the sugar beet to sucrose, glucose, and fructose.

According to Lepeschkin (1932), Zycha (1928) could detect no changes in permeability due to light; likewise, Jacques and Osterhout (1935) reported that the rate of entrance of potassium into *Nitella* cells was independent of illumination.

Hoagland and Davis (1923) observed that the alga *Nitella* absorbed more solutes from the surrounding medium in light than in darkness. They considered that light in this case is of primary importance as a source of energy in the process of absorption.

This assumption will be discussed in detail in Chap. V, when the absorption of solutes is considered.

(c) *The Nature of the Solutes.*—The influence of the various dissolved substances on the permeability of the protoplasm can best be understood by a summary of the results obtained by the various workers in that field. Osterhout (1911) found that a 0.2 *M* solution of calcium chloride just plasmolyzed the cells of *Spirogyra*, and a 0.38 *M* sodium chloride solution did the same. The cells, however, were not plasmolyzed by a 0.195 *M* calcium chloride solution or a 0.375 *M* sodium chloride solution. If, however, 10 cc. of a 0.195 *M* solution of calcium chloride is mixed with 1,000 cc. of the 0.375 *M* sodium chloride solution and Spirogyra cells placed therein, prompt plasmolysis occurs. The addition of the calcium chloride lowers the osmotic pressure of the solution, but in spite of that it plasmolyzes the cells. Thus by mixing two solutions, neither of which plasmolyzes, one is obtained that plasmolyzes very strongly. Osterhout (1915) used the electrical-conductivity method to determine the effects of various salts on permeability. Tissue of *Laminaria saccharina* in sea water had a net resistance of 770 ohms or a net conductance of 0.00130 reciprocal ohm. When the tissue was transferred to a 0.52 *M* solution of sodium chloride having the same conductivity as sea water, the resistance dropped to 580 ohms in 5 min. or to a net conductance of 0.001720 reciprocal ohm. The increase in permeability of the protoplasmic membrane thus amounted to 0.00042 reciprocal ohm or 32.3 per cent. This increase in conductance is not due to the increase in the number of ions of sodium chloride to which the protoplasmic membrane might be considered to be normally more permeable than those of ordinary sea water. For example, when the tissue was transferred from sea water to 0.52 *M* sodium chloride the conductance increased 25.2 per cent. The increase in sodium ions amounted to only 2 per cent, while the content of chlorine ions remained unchanged. Osterhout considered that this is evidence that there was a great increase in permeability of the protoplasm, which was induced by the sodium chloride. The effect of calcium chloride on the permeability of the protoplasm of *Laminaria* was found by the electrical-conductivity method as follows: The net resistance of a portion of tissue in sea water was 770 ohms. After transferring it to a 0.278 *M* solution of calcium chloride the net resistance rose within 10 min. to 1,260 ohms. This represents a decrease in permeability of 39.2 per cent. Osterhout found that potassium chloride, lithium chloride, ammonium chloride, sodium bromide, sodium iodide, sodium nitrate, potassium nitrate, and sodium sulphate all act, in general, like sodium chloride in increasing the permeability. All the bivalent cations investigated, *viz.*, magnesium, calcium, barium, strontium, manganese, cobalt, iron, zinc, cadmium, and tin, decreased permeability, and all the trivalent cations considered, *viz.*, lanthanum, iron, aluminum, cerium, and yttrium, did the same. Sodium hydroxide increased the permeability, while hydrochloric acid produced a rapid decrease followed at once by a rapid increase that continued until the death point was reached. Hydrogen is the only monovalent cation that was found able to decrease the permeability of the protoplasm.

These observations have been substantiated more recently by other investigators. Thus Ingold (1931) noted in the tissues of carrot, beet, and potato that the outward diffusion of solutes was increased by an immersion in a solution of potassium chloride, while treatments with the chlorides of calcium and lanthanum decreased the loss of solutes. Similar observations were made by Asprey (1935). Baptiste (1935) soaked disks of the potato and carrot in hypotonic solutions of the chlorides of potassium, sodium, ammonia, magnesium, and calcium, and then raised the absorbing power to a known, uniform level by controlled evaporation. The disks were then immersed in distilled water, and their rate of water intake determined. The effects of these cations on permeability to water were significant and followed the series,

$$K > NH_4 > Na > \text{Control} > Mg > Ca.$$

Kerr (1930, 1933) by microinjection into the root hairs of *Trianea bogotensis* and *Limnobium spongia* observed that sodium chloride and potassium chloride increased the fluidity of the protoplasm, while calcium chloride and magnesium chloride decreased it. Seifriz and Plowe (1930) noted that the chlorides of sodium, potassium, and lithium lowered while the chlorides of calcium and barium raised the elastic limit or extensibility of the protoplasm. DeHaan (1933) believed that the specific action of ions on the protoplasm is due to variations in its rate of swelling.

Raber (1920) investigated the effect of a series of anions on the permeability of *Laminaria agardhu* by means of the electrical-conductivity method. He used sodium sulphocyanide, iodide, nitrate, bromide, chloride, acetate, sulphate, tartrate, phosphate, and citrate in solutions having the same conductivity as a 0.51 *M* solution of sodium iodide. Measurements of the conductivity were taken, for the most part, at the end of 5, 25, 45, 85, and 105 min. and the effect of these anions upon the increase in permeability at the end of the experiment stood in the following order: $I > Br > SCN > Cl > NO_3 > \text{acetate} > SO_4 > \text{tartrate} > PO_4 > \text{citrate}$. From later work (1921) Raber considers that his results indicate that the effects upon permeability depend upon the valence of the anions regardless of whether the salts are organic or inorganic. He further found that bivalent and trivalent cations in combination with monovalent anions increase the electrical resistance of the protoplasm of *Laminaria*, but when combined with bivalent or trivalent anions the increase is less and may be entirely lacking. Chromium has a different initial effect upon the permeability of *Laminaria*, depending upon whether it occurs as the cation or the anion of the salt. If it is an anion, the first effect is a decrease in resistance, but if it is a cation, it increases the resistance. Osterhout (1913, 1916) observed that anesthetics, as ether and chloroform, increase or decrease permeability, depending upon the concentration used. Very low concentrations were found to produce little or no effect, but a point in concentration was soon reached in which a decrease in permeability could be maintained for several hours. This change is reversible and produces no injury. A higher concentration causes for a very short time a decrease and then an increase in permeability which is not reversible and continues until death ensues.

Stiles and Jorgensen (1917) studied the action of various organic substances on the permeability of plant cells by estimating the exosmosis of electrolytes from potato tissue especially immersed in watery solution of the substances. The exosmosis was estimated by the change in conductivity of the solution. They found that equimolecular solutions of different substances do not bring about the same exosmosis. In the homologous series of the monohydric alcohols, the more complex the molecule the greater the exosmosis produced by solutions of equimolecular strength. They were unable to find a critical concentration of solution below which exosmosis of electrolytes will not take place. True and Bartlett (1915) found that, when a plant like white lupine was grown in distilled water, leachings took place, and this exosmosis was increased in dilute solutions of sodium chloride and potassium chloride. In dilute solutions of calcium chloride, however, leaching stopped and absorption took place. The inference is that the permeability of the protoplasmic membrane of the roots of the plant was increased by the sodium chloride and potassium chloride and decreased by the calcium chloride.

Osterhout (1915) in experiments with *Laminaria saccharina* by the electrical-conductivity method showed that it is possible to bring about extreme changes in the permeability of the protoplasmic membrane without injury. Experiments were conducted over a period of 15 days in which the tissue each day was transferred from sea water to 0.52 *M* sodium chloride for 5 to 10 min. and then transferred again to sea water where it always regained its original permeability. The increase in permeability in the sodium chloride solution over that in sea water amounted to 20.3 per cent. The tissue that underwent this daily alteration in permeability retained its normal resistance in sea water until the tenth day, when it began to fall off. The control, however, showed the same results, so that the fall in resistance was due to other causes than the effects of the sodium chloride. The permeability was decreased 23 per cent by transference from normal sea water to sea water containing 0.01 *M* lanthanum nitrate for 5 min. After being again placed in normal sea water it regained its original permeability. The procedure was followed each day for 5 days, being repeated three or four times daily. At the end of the fifth day this tissue was from all appearances in as good condition as the control and had a slightly higher resistance. In another series of experiments covering a period of 3 days, the tissue was transferred from sea water to a 0.278 *M* solution of calcium chloride, which has the same conductivity as sea water, and then to a 0.52 *M* sodium chloride, and then again to sea water. The resistance of the tissue in sea water was 1,010 ohms; after 10 min. in the calcium chloride solution, it had risen to 1,500 ohms; and after being removed from the calcium chloride and transferred to the sodium chloride solution, it fell to 880 ohms. After 80 min. the tissue was again placed in sea water where it soon gained its normal resistance of 1,010 ohms. The permeability of the protoplasmic membrane in the calcium chloride solution decreased 39.2 per cent below the normal permeability in sea water, while in the sodium chloride solution it increased 20 per cent above normal. Regardless of these wide alterations of permeability, the tissue at the end of the experiments was to all appearances absolutely normal.

E. Theories of Membrane Action

Numerous theories have been put forward to explain the differential permeability of membranes but most of them can be grouped under one of the following four headings.

1. Sieve or Ultrafiltration Theory.—This theory assumes that the protoplasm acts as a sieve or ultra screen with pores that change in diameter according to surrounding conditions (Seifriz, 1936). The sieve thus is not a fixed one but is capable of constant change and adjustment. A change in the size of the pores can be brought about by aggregation or coagulation, by peptization or dispersion, by a change in sign and magnitude of electric charge, and by the reorientation of surface molecules. When small particles aggregate into large ones, the space between the particles will become larger, and the sieve will therefore become coarser. The reverse change takes place by peptization. A positively charged membrane would repel positive particles and thus keep them out, while a negatively charged membrane would permit them to enter. This theory was first put forward by Traube (1879) in order to explain the behavior of his precipitation membranes. It has received considerable support from the works of Küster (1911, 1918) and Ruhland (1912, 1913, 1914) on the intake of dyes by certain plant cells. These investigators studied the behavior of the cells of *Vicia faba, Allium cepa, Spirogyra, Helianthus annuus*, and *Coleus hybridus* with respect to over 80 different acidic and basic dyes.

Ruhland concluded that his work indicated that the entrance or nonentrance of a substance into the cell is related to the magnitude of its molecules or molecular aggregates and that the plasma membrane acts as an ultra filter. The behavior of the plasma membrane in the plant cells and of artificial parchment membranes toward dyes offers the strongest evidence for this theory, but there are numerous other cases where the observed facts are not in accord with the theory. Thus it has been shown that certain membranes are permeable to the large molecules of the alkaloids while they are impermeable to the relatively small molecules of the amino acids.

Thus in the case of glycerol and monoacetin, the molecules of which are very nearly equal in size, the latter enters the cell at a rate fifteen times more rapidly than the former.

Brooks (1930) considered that the behavior of the protoplasmic membrane of *Valonia macrophysa* to sea water indicated that the plasma membrane might consist of a mosaic of anion-permeable and cation-permeable areas that are of the nature of charged porous films. Such films exaggerate differences between the diffusion velocities of the ions to which they are permeable.

2. The Solution Theory.—This theory assumes that a membrane has definite solvent properties and that only those substances which are soluble in the membrane are capable of passing through it. The chief advocate of this theory in regard to the behavior of the protoplasmic membrane has been Overton (1895, 1896, 1899, 1900), who proposed the lipoid theory. This investigator from his observations on the permeability of the protoplasm to certain dyes and other organic substances concluded that the penetration of different substances into the protoplasm depends upon the solubility of these substances in the lipoids which with the proteins form the plasma membrane. This theory has only a very limited application, and the chief objection to it arises from the behavior of the numerous inorganic salts, which for the most part are insoluble in lipoid substances but which are known to enter the cells freely and in large amounts.

Apparently in many cases the question of the solubility of the substance overshadows other factors. Thus, as mentioned in the previous topic, the molecules of glycerin and monoacetin are very nearly of equal size, but the latter enters the cell at a rate fifteen times greater than does the former. Glycerin is very poorly soluble in lipoid material, while monoacetin is a good lipoid solvent. In this case solubility rather than size of molecule is apparently the determining factor in permeability. MacDougal (1928) in experiments with an artificial cell showed that the presence of a phospholipide such as lecithin alone lessens permeability and allows the attainment of a high degree of turgidity. Artificial cells, including both cholesterol and lecithin, showed certain behaviors in balanced solutions of sodium and calcium concurrent with those of living cells. It was found that ions penetrate a lipoidal cell more rapidly from solutions at pH 4.5 in contrast with the maximum for a lipoid-free cell at pH 5.4 of the solution. A lecithin-cholesterol suspension also stabilized the hydrogen-ion concentration of the cell content. This work of MacDougal indicates that certain lipoids in the protoplasm do influence the diffusion of organic and inorganic ions through it, but the exact manner in which they do so is not at present understood.

3. The Chemical Theory.—According to this theory the membrane is supposed to combine chemically with the substance to which it is permeable. This reaction is considered to be a reversible one, so that the diffusing substance is freed on the far side from its combination with the membrane. The chemical theory has received considerable support from numerous biological workers.

The temperature coefficient of permeability has been studied by numerous investigators in order to ascertain the nature of the absorption process through membranes. By the temperature coefficient of permeability is meant the increase in the permeability of the membrane for each 10°C. increase in temperature. The temperature coefficient for any reaction is commonly expressed by the symbol Q_{10}. That of chemical reactions is relatively high (2 to 3), and that of physical reactions is relatively low (1 to 1.5). Pfeffer (1877) measured the rate of water movement through the copper ferrocyanide membranes at different temperatures and found that from 7.1 to 17.6°C., $Q_{10} = 1.56$, and from 17.6 to 32.5°C., $Q_{10} = 1.27$. Krabbe (1896) observed that the temperature coefficient of permeability of the living pith cells of *Helianthus annuus* and of the roots of *Vicia faba,* from 0 to 20°C., was 2 to 2.5. He believed that this high coefficient indicated that purely physical forces were not operative but that the changes were due to the peculiar properties of the protoplasm. Rysselberghe (1901) studied the effect of temperature upon the permeability of the living protoplasm of the pith cells of *Sambucus nigra* and the epidermal cells of *Tradescantia* and found that from 0 to 30°C., Q_{10} averaged 2.0. Brown and Worley (1912) studied the rate of intake of water by the barley grains at different temperatures and obtained a temperature coefficient of 1.8 to 1.9. They considered that this indicated that chemical processes were involved in the penetration of water through the semipermeable seed coat of the barley grain. Shull (1920), however, found that the mean value of Q_{10} in *Xanthium* seeds was in one instance 1.55, in another 1.83, and in the case of split peas 1.6. He considered that his data do not support the theory of Brown and Worley but that the velocity of intake at any given moment in the seeds that were studied is approximately an inverse exponential function of the amount of water previously absorbed. He considered that his values do not indicate the absorption is conditioned by some single chemical change like simplifications of water to hydrone as the temperature rises but that absorption at different temperatures involves both physical and chemical changes.

Shull and Shull (1924), studying the absorption of water by seeds of corn, found that over the range from 5 to 55°C. the temperature coefficient averaged 1.54, somewhat above the coefficient for purely physical processes. The absorption at 55°C. was somewhat more than eight times as fast as at 5°C. whereas the chemical theory of absorption would call for a rate thirty-two times as great. Apparently many substances have a specific absorption behavior, and a rate law with wide applicability is not to be expected.

Osterhout (1935) found indications in *Valonia* that ammonia or ammonium hydroxide enters the cells by combining with a constituent of the protoplasm, and he considered that other electrolytes might also enter the cell by combining with one or more constituents of the protoplasm. His observations thus tend to substantiate the chemical nature of absorption.

Stiles and Jorgensen (1915) used 0.0011 *M* hydrochloric acid in which they placed plant tissue to observe the penetration of the hydrogen ions at various temperatures. They found that $Q_{10} = 2.22$ from 0 to 10°C. and 2.17 from 10 to 20°C. They believed that the high temperature coefficient thus obtained indicated that the process of absorption is a chemical and not a physical reaction. Osterhout (1917) criticized their conclusions on the ground that they, as many others, have not distinguished between permeability and absorption. The chemical process thus thought to have been observed might have been the union of the hydrogen ions with substances within the cell, and the permeability of the membranes thus might not have been considered. His experiments with the permeability of *Laminaria* gave a temperature coefficient of 1.3, which he considered suggested a physical reaction, although he was not positive. Denny (1917) measured the flow of water through the seed coat of *Arachis hypogaea* at temperatures ranging from 3.6 to 45°C. and found that Q_{10} ranged in value from 1.6 to 1.3, the coefficient being higher at the lower temperatures and lower at the higher temperatures, a fact that has been observed for a great many physical and chemical processes. The results obtained by Delf (1916), however, in measuring the temperature coefficient for increase of permeability for the onion leaf and the scape of the dandelion are in contrast to these. She found, for example, that the temperature coefficient between 10 and 20°C. was 1.5 and 2.3, respectively, for the onion leaf and dandelion scape, while from 20 to 30°C. it was 2.6 for the onion leaf and 3.8 for the dandelion scape. Denny (1917) makes the very important statement to the effect that the temperature effects on permeability thus observed may not be merely upon the membranes under consideration but also upon the cell contents. This latter effect may contribute to the total results so that the chemical reaction indicated by the relatively high temperature coefficient may have taken place in that part of the cell which is internal to the membrane under consideration. There seems thus to be no evidence that either physical or chemical processes are exclusively involved in the passage of materials through the membranes of the plant cell.

4. Colloidal Theories.—Since the protoplasmic membrane has been regarded as an emulsoid colloidal system, numerous hypotheses have been proposed to explain its changes in permeability upon the basis of changes in viscosity, phase inversion, or electrical reactions, all of which are common phenomena in colloidal suspensions or solutions.

The changes in viscosity, as suggested by Spaeth (1916), in the size of the colloidal particles, as offered by Lloyd (1915) and Free (1918), and the action of neutral salts on the colloidal membrane, as mentioned by Kaho (1921), without doubt have their effects upon the permeability of the protoplasm. These phenomena, however, need not be classified under a separate heading but may be considered simply as corollaries to the sieve theory.

The suggestion of Clowes (1916) that changes in permeability are brought about by the continuous and discontinuous phases changing places has received considerable attention. To support his theory he used data obtained in experiments on the effect of sodium hydroxide and calcium chloride in producing phase inversion in emulsion systems of olive oil and water.

However, there is no direct evidence whatsoever of phase reversal in the protoplasmic membrane. It is very unlikely that protoplasm could exist as a living substance if oil were the continuous phase, because so far as is known metabolic reactions take place only in an aqueous medium, and the amount of oil in the dispersed globules of an ultramicroscopic emulsion like the protoplasm is probably insufficient to enclose the aqueous medium.

No theory of membrane action has been proposed that will explain satisfactorily the behavior of the protoplasmic membrane. The passage of materials through it may occur to a certain extent after the manner proposed in the various theories that have herein been mentioned (Gurewitsch, 1929). When one considers, however, the highly complex nature of protoplasm, its rapid adjustment to its environmental surroundings, its highly specialized nature for different individuals, tissues, and even for the different parts of the same cell, it would seem most unlikely that any one theory could be proposed that would, under all conditions, cover its behavior toward the various substances that diffuse through it.

Bibliography

Asprey, G. F., 1935: On the relationship between exosmosis and salt absorption by potato tuber tissue previously treated with various salt solutions, *Protoplasma*, **23**: 497–504.

Bank, O., 1935: Zur Tonoplasten-frage. I, Die Natur der Vakuolenhaut. II, Die Reversibilität der Vakuolenform in Embryonalen Zellen, *Protoplasma*, **23**: 239–249.

Baptiste, E. C. D., 1935: The effect of some cations on the permeability of cells to water, *Ann. Bot.*, **49**: 345–366.

Beck, W. A., 1928: Osmotic pressure, osmotic value and suction tension, *Plant Physiol.*, **3**: 413–441.

———, 1929: Determining the osmotic value at incipient plasmolysis, *Trans. Am. Microsc.*, **48**: 204–208.

———, 1931: Variations in the O_g of plant tissues, *Plant Physiol.*, **6**: 315–323.

———, 1935: The effect of light on the O_g of plant tissues, I *Protoplasma*, **23**: 203–209.

Becquerel, P., 1907: Recherche sur la vie latente des graines, *Ann. Sci. Nat. Bot.*, IX, **5**: 193–320.

Blinks, L. R., 1929: Protoplasmic potentials in *Halicystis*, *J. Gen. Physiol.*, **13**: 223–229.

———, 1929: The direct-current resistance of *Valonia*, *J. Gen. Physiol.*, **13**: 361–378.

———, 1930: The variation of electrical resistance with applied potential. I, Intact *Valonia ventricosa*, *J. Gen. Physiol.*, **13**: 793–806.

———, 1932: Protoplasmic potentials in *Halicystis*. II, The effects of potassium on two species with different saps, *J. Gen. Physiol.*, **16**: 147–156.

———, 1935: Protoplasmic potentials in *Halicystis*. IV, Vacuolar perfusion with artificial sap and sea water, *J. Gen. Physiol.*, **18**: 409–420.

———, R. D. Rhodes, and G. A. McCallum, 1935: Protoplasmic potentials in *Halicystis*. V, The reversal of potential by unbalanced sodium chloride, *Proc. Nat. Acad. Sci.*, **21**: 123–125.

Blum, G., 1926: Untersuchungen über die Saugkraft einiger Alpenpflanzen, *Beih. bot. Cent.*, I., **43**: 1–100.

Bowen, J. L., 1931: Private communication.

Brauner, L., 1930: Über polare Permeabilität, *Ber. bot. Ges.*, **48**: 109–118.

Brooks, S. C., 1917: Methods of studying permeability of protoplasm to salts, *Bot. Gaz.*, **64**: 230–249.

———, 1917: A new method of studying permeability, *Bot. Gaz.*, **64**: 306–317.

———, 1917: Permeability of the cell walls of *Allium*, *Bot. Gaz.*, **64**: 509–512.

——— and S. Gelfan, 1928: Bioelectric potentials in *Nitella*, *Protoplasma*, **5**: 84–96.

———, 1930: The accumulation of ions in living cells—a non-equilibrium condition, *Protoplasma*, **8**: 389–412.

Brown, A. J., 1907: On the existence of a semipermeable membrane enclosing the seeds of the *Gramineae*, *Ann. Bot.*, **21**: 79–87.

———, 1909: The selective permeability of the coverings of the seeds of *Hordeum vulgare*, *Proc. Roy. Soc. London*, B, **81**: 82–93.

——— and F. P. Worley, 1912: The influence of temperature on the absorption of water by seeds of *Hordeum vulgare* in relation to the temperature coefficient of chemical change, *Proc. Roy. Soc. London*, B., **85**: 546–553.

——— and F. Tinker, 1916: The rate of absorption of various phenolic solutions by seeds of *Hordeum vulgare* and the factors governing the rate of diffusion of aqueous solutions across semipermeable membranes, *Proc. Roy. Soc. London*, B, **89**: 119–135.

——— and ———, 1916: Selective permeability: the absorption of phenol and other solutions by the seeds of *Hordeum vulgare*, *Proc. Roy. Soc. London*, B, **89**: 373–379.

Brown, R., 1931: The absorption of water by seeds of *Lolium perenne* L. and certain other *Gramineae*, *Ann. App. Biol.*, **18**: 559–573.

———, 1932: The absorption of the solute from aqueous solutions by the grain of wheat, *Ann. Bot.*, **46**: 571–581.

Buchinger, A., 1927: Saugkraftmessungen ("Osmotisches Verhalten") verschiedener Gerstensorten, *Fortschr. Landwirts.*, **2**: 344–350; *Biol. Abst.*, **3**: 468.

Bugarsky, S., and F. Tangl, 1897: Eine Methode zur Bestimmung des relativen Volums der Blutkörperchen und des Plasmas, *Zent. Physiol.*, **11**: 297.

Clowes, G. H. A., 1916: Protoplasmic equilibrium. I, Action of antagonistic electrolytes on emulsions and living cells, *J. Phys. Chem.*, **20**: 407–451.

Collins, E. J., 1918: The structure of the integumentary system of the barley grain in relation to localized water absorption and semipermeability, *Ann. Bot.*, **32**: 381–414.

Crozier, W. J., 919: Intracellular acidity in *Valonia*, *J. Gen. Physiol.*, **1**: 581–583.

Damon, E. B., 1929: Dissimilarity of inner and outer protoplasmic surfaces in *Valonia*. II, *J. Gen. Physiol.*, **13**: 207–221.

——— and W. J. V. Osterhout, 1930: The concentration effect with *Valonia:* potential differences with concentrated and diluted sea water, *J. Gen. Physiol.*, **13**: 445–457.

———, 1932: Dissimilarity of inner and outer protoplasmic surfaces in *Valonia*. III, *J. Gen. Physiol.*, **15**: 525–535.

DeHaan, I., 1933: Protoplasmaquellung und Wasserpermeabilität, *Rec. trav. bot. Néerl.*, **30**: 236–329. Quoted by Baptiste, 1935.

Delf, E. M., 1916: Studies of protoplasmic permeability by measurement of rate of shrinkage of turgid tissues. I, Influence of temperature on the permeability of protoplasm to water, *Ann. Bot.*, **30**: 283–310.

Denny, F. E., 1917: Permeability of certain plant membranes to water, *Bot. Gaz.*, **63**: 373–397.

———, 1917: Permeability of membranes as related to their composition, *Bot. Gaz.*,

De Vries, H., 1884: Eine Methode zur Analyse der Turgorkraft, *Jahrb. wiss. Bot.*, **14**: 427–601.

———, 1884: Zur plasmolytischen Methodik, *Bot. Ztg.*, **42**: 289–298.

———, 1885: Plasmolytische Studien über die Wand der Vacuolen, *Jahrb. wiss. Bot.*, **16**: 465–598.

Dixon, H., 1924: Variations in the permeability of leaf cells, *Sci. Proc. Roy. Dublin Soc.*, **17**: 349–356.

———, 1930: Über die Saugkraft, *Ber. bot. Ges.*, **48**: 428–432.

Dumont, A., 1926: Étude bibliographique et critique de la mesure de la concentration des liquides cellulaires végétaux, *Ann. Physiol. and Physico-chem. Biol.*, **2**: 215–256.

Eckerson, S., 1914: Thermotropism of roots, *Bot. Gaz.*, **58**: 254–263.

Eisenmenger, W. S., 1930: Toxicity of some aliphatic alcohols, *Plant Physiol.*, **5**: 131–156.

Ernest, E. C. M., 1931: Suction-pressure gradients and the measurement of suction pressure, *Ann. Bot.*, **45**: 717–731.

———, 1934: Studies in the suction pressure of plant cells. II, *Ann. Bot.*, **48**: 293–305.

———, 1934: The effect of intercellular pressure on the suction pressure of cells, *Ann. Bot.*, **48**: 915–918.

———, 1935: Factors rendering the plasmolytic method inapplicable in the estimation of osmotic values of plant cells, *Plant Physiol.*, **10**: 553–558.

Findlay, A., 1913: "Osmotic Pressure," 82 pp., Longmans, Green & Company, New York.

Fischer, M. H., 1915: "Oedema and Nephritis: A Critical, Experimental and Clinical Study of the Physiology and Pathology of Water Absorption in the Living Organism," 2d ed., John Wiley & Sons, Inc., New York.

Free, E. E., 1918: A colloidal hypothesis of protoplasmic permeability, *Plant World*, **21**: 141–150.

Fujita, A., 1925: Untersuchungen über elektrische Erscheinungen und Ionendurchlässigkeit von Membranen. I, Die Potentialdifferenz an der Apfelschale, *Biochem. Z.*, **158**: 11–27.

Giglioli, I., 1895: Latent vitality in seeds, *Nature*, **52**: 544–545.

Goldsmith, G. W., and J. H. C. Smith, 1926: Some physico-chemical properties of spruce sap and their seasonal and altitudinal variation, *Colorado Col. Pub. Sci. Ser.*, **13**: 13–71.

Gurewitsch, A., 1929: Untersuchungen über die Permeabilität der Hülle des Weizenkorns, *Jahrb. wiss. Bot.*, **70**: 657–706.

Hamly, D. H., 1932: Softening of the seeds of *Melilotus alba*, *Bot. Gaz.*, **93**: 345–375.

Hawthorne, W. C., 1930: Diffusion, osmosis, and osmotic pressure, *Sci. Month.*, **31**: 535–542.

Herrick, E. M., 1933: Seasonal and diurnal variations in the osmotic values and suction tension values in the aerial portions of *Ambrosia trifida*, *Am. J. Bot.*, **20**: 18–34.

Hoagland, D. R. and A. R. Davis, 1923: The composition of the cell sap of the plant in relation to the absorption of ions, *J. Gen. Physiol.*, **5**: 629–647.

——— and ———, 1923: Further experiments on the absorption of ions by plants, including observations on the effect of light, *J. Gen. Physiol.*, **6**: 47–62.

Hottes, C. F., and W. A. Huelsen, 1927: The determination of quality in sweet corn seed by means of the optical measurement of leached materials, *J. Agr. Res.*, **35**: 147–166.

HUBER, B., and K. HÖFLER, 1930: Die Wasserpermeabilität des Protoplasmas, *Jahr. wiss. Bot.*, **73**: 351–511.

INAMDAR, R. S., and K. V. VARADPANDE, 1929: The permeability of plant-cell membrane to sugar, *Nature*, **124**: 875–876.

INGOLD, C. T., 1931: On the effect of previous treatment with salt solutions on the subsequent outward diffusion of electrolytes from plant tissue, *Ann. Bot.*, **45**: 709–716.

IRWIN, M., 1922: Penetration of dyes into living cells, *Am. J. Physiol.*, **59**: *Proc.* 458.

———, 1923: The permeability of living cells to dyes as affected by hydrogen-ion concentration, *J. Gen. Physiol.*, **5**: 223–224.

———, 1923: The penetration of dyes as influenced by hydrogen-ion concentration, *J. Gen. Physiol.*, **5**: 727–740.

JACQUES, A. G. and W. J. V. OSTERHOUT, 1935: The kinetics of penetration. XI, Entrance of potassium into *Nitella*, *J. Gen. Physiol.*, **18**: 967–985.

KAHO, H., 1921: Ein Beitrag zur Permeabilität des Pflanzenplasmas für die Neutralsalze. IV, *Biochem. Z.*, **123**: 284–303; *Rev. Bot. Gaz.*, **81**: 350–353.

KERR, T., 1930: The effect of essential metals on injection into plant protoplasm and sap, *Am. J. Bot.*, **17**: 1043.

———, 1933: The injection of certain salts into the protoplasm and vacuoles of the root hairs of *Limnobium spongia*, *Protoplasma*, **18**: 420–440.

KING, R., 1932: Plasmolysis, *Proc. Ind. Acad. Sci.*, **41**: 135–136.

KITE, G. L., 1915: Studies on the permeability of the internal cytoplasm of animal and plant cells, *Am. J. Physiol.*, **37**: 282–299.

KOTOWSKI, F., 1927: Semipermeability of seed coverings and stimulation of seeds, *Plant Physiol.*, **2**: 177–186.

KRABBE, G., 1896: Über den Einfluss der Temperatur auf die osmotischen Processe lebender Zellen, *Jahrb. wiss. Bot.*, **29**: 441–498.

KÜSTER, E., 1911: Ueber die Aufnahme von Anilinfarben in lebende Pflanzenzellen, *Jahrb. wiss. Bot.*, **50**: 261–288.

———, 1918: Ueber Vitalfärbung der Pflanzenzellen, *Z. wiss. Mikros.* **35**: 95–100.

LAVISON, J., 1910: Du mode de la pénétration de quelques sels dans la plante vivante; rôle de l'endoderme, *Rev. Gén. Bot.*, **22**: 225–241.

———, 1911: Recherches sur la pénétration des sels dans le protoplasm et sur la nature de leur action toxique, *Ann. Sci. Nat. Bot.*, IX, **14**: 97–139.

———, 1911: Essai sur une théorie de la nutrition minérale des plantes vasculaires, *Rev. Gén. Bot.*, **23**: 177–211.

LEPESCHKIN, W. W., 1906: Zur Kenntnis des Mechanismus der aktiven Wasserausscheidung der Pflanzen, *Beih. bot. Cent.*, I, **19**: 409–452.

———, 1908: Über die osmotischen Eigenschaften und den Turgordruck der Blattgelenkzellen der Leguminosen, *Ber. bot. Ges.*, **26**: 231–237.

———, 1908: Zur Kenntnis des Mechanismus der Variationsbewegungen, *Ber. bot. Ges.*, **26**: 724–735.

———, 1909: Zur Kenntnis des Mechanismus des photonastischen Variationsbewegungen und der Einwirkung des Beleuchtungswechsels auf die Plasmamembran, *Beih. bot. Cent.*, I, **24**: 308–356.

———, 1928: The effect of ethyl alcohol on the turgor pressure of *Spirogyra*, *Am. J. Bot.*, **15**: 422–424.

———, 1929: The protective action of some substances on protoplasm, *Am. J. Bot.*, **16**: 207–218.

———, 1930: Light and the permeability of protoplasm, *Am. J. Bot.*, **17**: 953–971.

———, 1932: The influence of narcotics, mechanical agents, and light upon the permeability of protoplasm, *Am. J. Bot.*, **19**: 568–580.

LIVINGSTON, B. E., 1913: Osmotic pressure and related forces as environmental factors, *Plant World*, **16**: 165–176.

LLOYD, F. E., 1915: The behavior of protoplasm as a colloidal complex, *Carnegie Inst. Wash. Yearbook*, **14**: 66–69.

LOOMIS, W. E., 1931: A quantitative demonstration of osmotic equilibrium, *Plant Physiol.*, **6**: 365–367.

LYON, C. J., 1936: Analysis of osmotic relations by extending the simplified method, *Plant Physiol.*, **11**: 167–172.

MACDOUGAL, D. T., 1923: Permeability and the increase of volume contents of living and artificial cells, *Proc. Am. Phil. Soc.*, **62**: 1–26.

———, 1924: The arrangement and action of material in the plasmatic layers and cell walls of plants, *Proc. Am. Phil. Soc.*, **63**: 76–93.

——— and V. MORAVEK, 1927: The activities of a constructed colloidal cell, *Protoplasma*, **2**: 161–188.

———, 1928: Substances regulating the passage of material into and out of plant cells: the lipoids, *Proc. Am. Phil. Soc.*, **67**: 33–45.

MICHAELIS, L., and A. FUJITA, 1925: Untersuchungen über elektrische Erscheinungen und Ionendurchlässigkeit von Membranen. II, Die Permeabilität von Apfelschale, *Biochem. Z.*, **158**: 28–37.

MOLZ, F. J., 1926: A study of suction force by the simplified method. I, Effect of external factors, *Am. J. Bot.*, **13**: 433–463.

———, 1926: A study of suction force by the simplified method. II, Periodic variations and the influence of habitat, *Am. J. Bot.*, **13**: 465–502.

OFFORD, H. R., and R. P. D'URBAL, 1931: Toxic action of aqueous sodium chlorate on *Nitella*, *J. Agr. Res.*, **43**: 791–811.

OPPENHEIMER, H. R., 1930: Kritische Betrachtungen zu den Saugkraftmessungen von Ursprung und Blum, *Ber. bot. Ges.*, **48**: 131–140.

OSBORNE, W. A., and L. C. JACKSON, 1914: Counter diffusion in aqueous solution, *Biochem. J.*, **8**: 246–249.

OSTERHOUT, W. J. V., 1911: The permeability of living cells to salts in pure and balanced solutions, *Sci.*, **34**: 187–189.

———, 1912: Permeability of protoplasm to ions and the theory of antagonism, *Sci.*, **35**: 112–115.

———, 1912: Reversible changes in permeability produced by electrolytes, *Sci.*, **36**: 350–352.

———, 1913: Some quantitative researches on the permeability of the plant cell, *Plant World*, **16**: 129–144.

———, 1913: The effect of anesthetics upon permeability, *Sci.*, **37**: 111–112.

———, 1913: The organization of the cell with respect to permeability, *Sci.*, **38**: 408–409.

———, 1914: The effect of alkali on permeability, *J. Biol. Chem.*, **19**: 335–343.

———, 1914: The effect of acid on permeability, *J. Biol. Chem.*, **19**: 493–501.

———, 1915: Extreme alteration of permeability without injury, *Bot. Gaz.*, **59**: 242–253.

———, 1915: On the decrease of permeability due to certain bivalent kations, *Bot. Gaz.*, **59**: 317–330.

———, 1915: The effect of some trivalent and tetravalent kations on permeability, *Bot. Gaz.*, **59**: 464–473.

———, 1915: Normal and abnormal permeability, *Am. J. Bot.*, **2**: 93–94.

———, 1916: The decrease of permeability produced by anesthetics, *Bot. Gaz.*, **61**: 148–158.

———, 1917: Similarity in the effects of potassium cyanide and of ether, *Bot. Gaz.*, **63**: 77–80.
———, 1917: Does the temperature coefficient of permeability indicate that it is chemical in nature? *Bot. Gaz.*, **63**: 317–320.
———, 1918: Decrease of permeability and antagonistic effect caused by bile salts, *J. Gen. Physiol.*, **1**: 405–409.
———, 1918: A method of measuring the electrical conductivity of living tissues, *J. Biol. Chem.*, **36**: 557–568.
———, 1919: A comparative study of permeability in plants, *J. Gen. Physiol.*, **1**: 299–305.
———, 1919: A comparison of permeability in plant and animal cells, *J. Gen. Physiol.*, **1**: 409–415.
———, 1921: Conductivity and permeability, *J. Gen. Physiol.*, **4**: 1–9.
———, 1921: Direct and indirect determinations of permeability, *J. Gen. Physiol.*, **4**: 275–283.
———, 1922: Some aspects of selective absorption, *J. Gen. Physiol.*, **5**: 225–231.
———, 1922: "Injury, Recovery, and Death in Relation to Conductivity and Permeability," 259 pp., J. B. Lippincott Company, Philadelphia and London.
———, 1927: Some aspects on bioelectrical phenomena, *J. Gen. Physiol.*, **11**: 83–99.
———, E. B. Damon, and A. G. Jacques, 1927: Dissimilarity of inner and outer protoplasmic surfaces in *Valonia*, *J. Gen. Physiol.*, **11**: 193–205.
——— and E. S. Harris, 1928: Protoplasmic asymmetry in *Nitella* as shown by bioelectric measurements, *J. Gen. Physiol.*, **11**: 391–406.
——— and ———, 1928: The death wave in *Nitella*. I, Application of like solutions, *J. Gen. Physiol.*, **12**: 167–186.
——— and ———, 1928: The death wave in *Nitella*. II, Application of unlike solutions, *J. Gen. Physiol.*, **12**: 355–361.
——— and ———, 1929: Note on the nature of the current of injury in tissues, *J. Gen. Physiol.*, **13**: 47–56.
——— and E. S. Hill, 1930: Negative variations in *Nitella* produced by chloroform and by potassium chloride, *J. Gen. Physiol.*, **13**: 459–467.
———, 1930: Calculation of bioelectric potentials. I, Effects of KCl and NaCl on *Nitella*, *J. Gen. Physiol.*, **13**: 715–732.
———, 1930: The accumulation of electrolytes, *J. Gen. Physiol.*, **14**: 285–300.
———, 1935: Mechanism of salt absorption by plant cells, *Nature*, **136**: 1034–1035.
———, 1935: How do electrolytes enter the cell? *Proc. Nat. Acad. Sci.*, **21**: 125–132.
———, 1936: Electrical phenomena in large plant cells, *Physiol. Rev.*, **16**: 216–237.
Oswald, W., 1891: "Solutions," viii + 316 pp., Transl. by H. M. P. Muir. Longmans, Green & Company, New York.
Overton, E., 1895: Über die osmotischen Eigenschaften der lebenden Pflanzen- und Tierzelle, *Vierteljahrschr. Nat. Ges. Zürich*, **40**: 159–201.
———, 1897: Über die osmotischen Eigenschaften der Zelle in ihrer Bedeutung für die Toxikologie und Pharmakologie, *Vierteljahrschr. Nat. Ges. Zürich*, **41**: 383–406.
———, 1899: Ueber die allgemeinen osmotischen Eigenschaften der Zelle, ihre vermuthlichen Ursachen und ihre Bedeutung für die Physiologie, *Vierteljahrschr. Nat. Ges. Zürich*, **44**: 83–135.
———, 1900: Studien über die Aufnahme der Anilinfarben durch die lebende Zelle, *Jahrb. wiss. Bot.*, **34**: 669–701.
Pfeffer, W., 1877: "Osmotische Untersuchungen: Studien zur Zell-mechanik," Leipzig.

———, 1890: Zur Kenntnis der Plasmahaut und der Vacuolen, *Abh. Sächs. Ges. Wiss.*, **16**: 187–343.

PLOWE, J. Q., 1931: Membranes in the plant cell. I, Morphological membranes at protoplasmic surfaces, *Protoplasma*, **12**: 196–220.

———, 1931: Membranes in the plant cell. II, Localization of differential permeability in the plant protoplast, *Protoplasma*, **12**: 221–240.

PUGH, G. W., H. JOHANN, and J. G. DICKSON, 1932: Relation of the semipermeable membranes of the wheat kernel to infection by *Gibberella saubinetii*, *J. Agr. Res.*, **45**: 609–626.

PUMPHREY, R. J., 1932: Electrical properties of the inner epidermis of the onion scale, *Proc. Roy. Soc.*, B, **109**: 434–442.

RABER, O. L., 1920: A quantitative study of the effect of anions on the permeability of plant cells, *J. Gen. Physiol.*, **2**: 535–539.

———, 1920: The antagonistic action of anions, *J. Gen. Physiol.*, **2**: 541–544.

———, 1921: A quantitative study of the effect of anions on the permeability of plant cells. II, *Am. J. Bot.*, **8**: 366–369.

———, 1921: The effect of permeability on polyvalent cations in combination with polyvalent anions, *Am. J. Bot.*, **8**: 382–386.

———, 1921: The effect upon the permeability of (I) the same substance as a cation and anion and (II) changing the valency of the same ion, *Am. J. Bot.*, **8**: 464–470.

———, 1923: Permeability of the cell to electrolytes, *Bot. Gaz.*, **75**: 298–308.

RUBINSTEIN, D. L., and V. USPENSKAJA, 1934: Ueber den isoelektrischen Punkt der pflanzlichen Plasmahaut, *Protoplasma*, **21**: 191–225.

RUHLAND, W., 1911: Untersuchungen über den Kohlenhydratstoffwechsel von *Beta vulgaris*, *Jahrb. wiss. Bot.*, **50**: 200–257.

———, 1912: Studien über die Aufnahme von Kolloiden durch die pflanzliche Plasmahaut, *Jahrb. wiss. Bot.*, **51**: 376–431.

———, 1912: Die Plasmahaut als Ultrafilter bei der Kolloidaufnahme, *Ber. bot. Ges.*, **34**: 139–141.

———, 1913: Zur Kritik der Lipoid- und der Ultrafiltertheorie der Plasmahaut nebst Beobachtungen über die Bedeutung der elektrischen Ladung der Kolloide für ihre Vitalaufnahme, *Biochem. Z.*, **54**: 59–77.

———, 1913: Zur Kenntnis der Rolle des elektrischen Ladungssinnes bei der Kolloidaufnahme durch die Plasmahaut, *Ber. bot. Ges.*, **31**: 304–310.

———, 1913: Zur chemischen Organisation der Zelle, *Biol. Cent.*, **33**: 337–351.

———, 1914: Weitere Beiträge zur Kolloidchemie und physikalischen Chemie der Zelle, *Jahrb. wiss. Bot.*, **54**: 391–447.

RYSSELBERGHE, FR. VAN, 1901: Influence de la température sur la perméabilité du protoplasme vivant pour l'eau et des substances dissoutes, *Bull. Acad. Roy. Belg. Cl. Sci.*, 173–221; *Rec. Inst. Bot. Univ. Bruxelles*, **5**: 209–248 (1902).

SCHRÖDER, H., 1911: Über die selektiv permeabele Hülle des Weizenkornes, *Flora*, **102**: 186–208.

SEIFRIZ, W., 1921: Observations on some physical properties of protoplasm by aid of micro-dissection, *Ann. Bot.*, **35**: 269–296.

——— and J. PLOWE, 1930: The effects of salts on the extensibility of protoplasm, *Am. J. Bot.*, **17**: 1046.

———, 1936: "Protoplasm," 584 pp., McGraw-Hill Book Company, Inc., New York.

SEN, B., 1928: The effect of temperature on the permeability of the protoplasmic membrane, *Proc. Roy. Soc.*, B, **103**: 272–288.

———, 1931: A method for measuring the change of permeability to ions of single cells under electric stimulation, *Ann. Bot.*, **45**: 527–531.

SHOWALTER, A. M., 1929: The evolution of "plasmolysis," *Am. J. Bot.*, **16**: 523–524.

SHULL, C. A., 1913: Semipermeability of seed coats, *Bot. Gaz.*, **56**: 169–199.

———, 1920: Temperature and the rate of moisture intake in seeds, *Bot. Gaz.*, **69**: 261–390.

——— and S. P. SHULL, 1924: Temperature coefficient of absorption in seeds of corn, *Bot. Gaz.*, **77**: 262–279.

———, 1927: Suction forces in plant cells. A review, *Bot. Gaz.*, **83**: 213–214.

———, 1932: Textbook of plant physiology. A review, *Bot. Gaz.*, **93**: 108–109.

——— and S. P. SHULL, 1932: Irregularity in the rate of absorption by dry plant tissues, *Bot. Gaz.*, **93**: 376–399.

SKEEN, J. R., 1929: The tolerance limit of seedlings for aluminum and iron, and the antagonism of calcium, *Soil Sci.*, **27**: 69–80.

———, 1930: Experiments with *Trianea* on antagonism and absorption, *Plant Physiol.*, **5**: 105–118.

SPAETH, R. A., 1916 The vital equilibrium, *Sci.*, **43**: 502–509.

STARK, O. K., 1928: What is osmosis? *Sci.*, **67**: 556–557.

STEWARD, F. C., 1928: The maintenance of semipermeability in the plant cell during leaching experiments, *Proc. Leeds Phil. Soc.*, **1**: 258–270.

STEWART, G. N., 1897: Elektrische Leitfähigkeit thierischer Flüssigkeiten, *Zent. Physiol.*, **11**: 332–335.

STILES, W., and I. JORGENSEN, 1915: Studies in permeability. II, The effect of temperature on the permeability of plant cells to the hydrogen ion, *Ann. Bot.*, **29**: 611–618.

——— and ———, 1917: Studies in permeability. IV, The action of various organic substances on the permeability of the plant cell and its bearing on the Czapek's theory of plasma membrane, *Ann. Bot.*, **31**: 47–76.

——— and ———, 1917: Studies in permeability. V, The swelling of plant tissue in water and its relation to the temperature and various dissolved substances, *Ann. Bot.*, **31**: 415–434.

——— and ———, 1918: Quantitative measurements of permeability, *Bot. Gaz.*, **65**: 526–534.

———, 1922: The suction pressure of the plant cell. A note on nomenclature, *Biochem. J.*, **16**: 727–729.

———, 1924: "Permeability," 292 pp., Wheldon & Wesley, London.

TJEBBES, K., 1912: Keimproeven met suikerbietinzaad, *Inaug. Diss. Amsterdam.*

TRÖNDLE, A., 1909: Permeabilitätsänderung und osmotischer Druck in den assimilierenden Zellen des Laubblattes, *Ber. bot. Ges.*, **27**: 71–78.

———, 1910: Der Einfluss des Lichtes auf die Permeabilität der Plasmahaut, *Jahrb. wiss. Bot.*, **48**: 171–282.

———, 1918: Der Einfluss des Lichtes auf die Permeabilität der Plasmahaut und die Methode der Permeabilitäts-Koeffizienten, *Vierteljahrschr. Nat. Ges. Zürich*, **63**: 187–213.

URSPRUNG, A., and G. BLUM, 1916: Über die Verteilung des osmotischen Wertes in der Pflanze, *Ber. bot. Ges.*, **34**: 88–104.

——— and ———, 1916: Über die periodischen Schwankungen des osmotischen Wertes, *Ber. bot. Ges.*, **34**: 105–123.

——— and ———, 1916: Über den Einfluss der Aussenbedingungen auf den osmotischen Wert, *Ber. bot. Ges.*, **34**: 123–142.

——— and ———, 1916: Zur Methode der Saugkraftmessung, *Ber. bot. Ges.*, **34**: 525–539.

——— and ———, 1916, 1918: Zur Kenntnis der Saugkraft. I, II, *Ber. bot. Ges.*, **34**: 539–554; **36**: 577–599.

——— and ———, 1919: Zur Kenntnis der Saugkraft. III, *Ber. bot. Ges.*, **37**: 453–462.

——— and ———, 1920: Dürfen wir die Ausdrücke osmotischer Wert, osmotischer Druck, Turgordruck, Saugkraft synonym gebrauchen? *Biol. Zent.*, **40**: 193–216.

——— and ———, 1921: Zur Kenntnis der Saugkraft. IV, Die Absorptionszone der Wurzel. Der Endodermissprung, *Ber. bot. Ges.*, **39**: 70–79.

——— and ———, 1921: Zur Kenntnis der Saugkraft. V, Eine Methode zur Bestimmung des Widerstandes, den der Boden der Wasserabsorption durch die Wurzel entgegensetzt, *Ber. bot. Ges.*, **39**: 139–148.

———, 1923: Zur Kenntnis der Saugkraft. VII, Eine neue vereinfachte Methode zur Messung der Saugkraft, *Ber. bot. Ges.*, **41**: 338–343.

——— and G. Blum, 1924: Eine Methode zur Messung des Wand- und Turgordruckes der Zelle, nebst Anwendungen, *Jahrb. wiss. Bot.*, **63**: 1–110.

———, 1925: Einige Resultate der neuesten Saugkraftstudien, *Flora*, **118-119**: 566–599.

———, 1926: Über die gegenseitigen Beziehungen der osmotischen Zustandsgrössen, *Planta*, **2**: 640–660; *Biol. Abst.*, **3**: 1078.

——— and G. Blum, 1927: Eine Methode zur Messung der Saugkraft von Hartlaub, *Jahr. wiss. Bot.*, **67**: 334–348.

———, 1929: The osmotic quantities of the plant cell, *Proc. Int. Cong. Plant Sci.*, Ithaca, 1926, **2**: 1081–1094.

——— and G. Blum, 1930: Zwei neue Saugkraft-Messmethoden, *Jahrb. wiss. Bot.*, **72**: 254–334; *Biol. Abst.*, **6**: 2531.

———, 1935: Osmotic quantities of plant cells in given phases, *Plant Physiol.*, **10**: 115–133.

Walter, H., 1930: Saugkraft oder osmotischer Wert, *Z. Bot.*, **23**: 74–93.

Weber, F., 1929: Plasmolyse-Zeit-Method, *Protoplasma*, **5**: 622–624.

———, 1929: Plasmolysezeit und Lichtwirkung, *Protoplasma*, **7**: 256–258.

Whitten, J. H., 1914: The effect of kerosene and other petroleum oils on the viability of growth of *Zea mais*, *Bull. Ill. State Lab. Nat. Hist.*, **10**: 245–273.

Wodehouse, R. P., 1917: Direct determinations of permeability, *J. Biol. Chem.*, **29**: 453–458.

Zycha, H., 1928: Einfluss des Lichts auf die Permeabilität von Blattzellen für Salze, *Jahrb. wiss. Bot.*, **86**: 499–548.

CHAPTER III

THE ROOTS OF PLANTS

The root system of a plant includes all its roots considered collectively. The root systems are specific, *i.e.*, the root system of a given species, in its general appearance, can be distinguished from that of another type of plant. Weaver (1920) stated that the general characters of the root systems of species are often as marked and distinctive as are those of the aerial vegetative parts, and although these may be greatly modified when subjected to different environmental conditions, they still retain the characteristic impress of the species in its usual habitat. Kiesselbach and Weihing (1935) considered that root development in corn in subsequent hybrid generations exhibits heterosis as does stalk development. The root system of a given plant may, however, vary in its structure, extent, weight, number, and direction of its roots according to the conditions under which it is grown. The general behavior of roots in the soil is thus the resultant of the influence of many factors, the most important of which are moisture, nutrients, oxygen supply, temperature, physical texture of the soil, light, and gravity.

I. INFLUENCE OF ENVIRONMENT ON ROOTS

A. Influence of Soil Moisture on Root Development

The influence of varying amounts of moisture in the soil upon the roots has been studied along three general lines: (1) direction of root growth, (2) relative weight of roots and tops, and (3) lateral extent and depth of penetration into the soil. It is the purpose here to consider only the first two topics and to reserve the third for a later discussion under a separate heading.

1. Direction of Root Growth.—It is common knowledge that roots respond to the changes of moisture in the soil and that they are positive in their reaction as they turn from a lower to a higher water content. Some of the factors entering into this reaction were investigated by Hooker (1915) in working with the roots of *Lupinus albus*. He found that roots are positively hydrotropic in moist air and that they react only when the moisture of the air is between 80 and 100 per cent of saturation. He observed that the minimum, optimum, and maximum differences in moisture to which these roots reacted at 20°C. was a fall of 0.2, 0.4, and 0.5 per cent, respectively, per centimeter along a line at right angles to them. The best response of root reaction toward changes in moisture was thus obtained under conditions that would make a difference of only 0.04 per cent moisture between the opposite sides of a root 1 mm. thick. Hooker also found that under optimum conditions 6 hr. elapsed before the hydrotropic reaction was visible in the roots, and the sensitiveness of roots to moisture was exhibited chiefly at the root tip and to a lesser extent by the cells behind the root tip. From these facts it is evident that roots will turn toward or follow moisture in the soil only when they are in direct contact with it or in very close proximity to it.

FIG. 7A.—Root system of a corn plant grown under conditions of shallow penetration of moisture.

Loomis and Ewan (1936) grew over 7,000 seedlings, representing 29 genera and 14 families, to determine the hydrotropic response of primary roots of seedlings when grown on a steep gradient of soil moisture. This gradient was obtained by placing a layer of soil having a moisture percentage below the hygroscopic coefficient in contact with a layer having a moisture percentage slightly below the moisture equivalent. The data obtained indicated that hydrotropism in roots is by no means comparable in occurrence and intensity of reaction with geotropism or phototropism. In some cases the results suggested that hydrotropic responses may depend upon a genetic factor present in some plants and absent from closely related species or varieties.

In this connection it should be mentioned that the roots of the vast majority of plants cannot penetrate a layer of dry soil in order to reach a supply of moisture that might be beneath, since, in a dry soil, water would be extracted from the roots, and they would perish as a result.

Considerable experimental work has been done to determine the ability of roots to penetrate dry soil. Magistad and Breazeale (1929) noted that in a soil at the wilting coefficient the roots of some plants were covered with a moist adhesive sheath. They believed that the moisture content of this sheath was probably maintained under such conditions by an exudation of water by the roots. This water, according to Breazeale (1930), is furnished in some cases by other roots that have penetrated moisture regions of soil. In experiments with wheat and barley, he obtained evidence which indicated that the roots of these plants could absorb water at lower depths and transmit it to other roots that were in soil with a moisture content at or about the wilting coefficient. It was assumed that when roots were supplied with water after this fashion they might penetrate soils from which at that time they could not obtain water. In this regard Shantz (1927) stated that there are certain trees of the African grasslands that have the ability to extend their roots into dry soil and in this manner prepare to absorb moisture rapidly when it comes.

Shantz, however, stated that the roots of ordinary farm crops do not possess this ability, and that they will penetrate only about the length of the root cap in a soil reduced to the wilting coefficient. Hendrickson and Veihmeyer (1931), working with sunflower, and Robertson, Kezer, Sjogren, and Koonce (1934), working with wheat, found that the roots of these plants do not penetrate into a soil that has been depleted of its moisture to or below the wilting coefficient (see Fig. 7*A*).

2. Relative Weight of Roots and Tops.—Tucker and Von Seelhorst (1898) grew oats in soil containing 42, 45, and 49 per cent of its moisture-holding capacity and found that the ratio of the dry weight of the tops to the dry weight of the roots was 6.6, 11.5, and 13.2, respectively, for each of these soils. They found (*a*) that the weight of the aerial parts increased in both grain and straw with the rising water content of the soil and (*b*) that with but one exception the greatest weight of roots was produced in the soil with the smallest amount of water.

Kiesselbach (1910) at the Nebraska Experiment Station grew corn plants in soil with a moisture content of 98, 80, 60, 40, and 20 per cent of saturation. The dry weight of the tops of these plants was, respectively, 8.6, 6.7, 7.2, 6.1, and 5.2 times the weight of the roots. He concluded that root development varies inversely as the soil-water content, and that plants which have their early growth in a relatively dry soil may be able to withstand drought better on account of having a greater surface exposed to the soil particles. The observations of the above investigators have further been confirmed by Polle (1910) working with wheat and barley, by Harris (1914) and Miller and Duley (1925) working with corn, and by Gordienko (1930).

Weaver (1925) grew Nebraska White Prize corn for 5 weeks in fertile loess soil at water contents of 9 and 19 per cent, respectively, above the hygroscopic coefficient. In the wet soil the area of the tops (including the stem and counting both surfaces of the leaves) was 82 per cent of that of the roots, but in the drier soil the tops had only 46 per cent as great an area as the roots. The absorbing area of the roots (exclusive of root hairs) was 1.2 times as great as the area of tops in the wet soil and 2.2 times as great as the area of the tops of plants grown in the drier soil.

Jones (1931), and Jones and Haskins (1935) found that the water content of soil in porous pots was lower than that in nonporous ones. A comparison of the root systems of plants grown in porous and nonporous pots invariably showed a larger root system in the porous pot. The ratio of tops to roots in the former case was 3.3, while in the latter it was 4.0.

B. Soil Nutrients and Root Development

The effect of the kind and amount of nutrients in the soil, or other medium in which the plant is grown, upon root development has been studied along the three following lines: the influence (1) on the direction taken by the roots, (2) on the number of roots, and (3) on the relative weight of roots and tops.

1. Direction of Root Growth.—The experimental evidence in regard to the response in direction of roots to nutrients is very limited. It is also difficult to interpret the data that have been obtained in this regard, since there is no way of knowing whether the observed curvatures of the roots are due to the attraction or repulsion of the nutrients in question or to the injury of portions of the root by them. The main work in this regard has been done by Newcombe and Rhodes (1904), Lilienfeldt (1905), and Porodko (1911 to 1914). Newcombe and Rhodes (1904) worked especially with the primary root of the seedlings of *Lupinus albus* and *Cucurbita pepo*. Blocks of gelatin, in which were dissolved the nutrients or chemicals to be investigated, were pressed gently, but closely, against the root on two sides, and after a given period observations were made to determine whether the reaction of the root to the dissolved substances had been positive, negative, or neutral. In one experiment with a block on one side containing only distilled water, and the block on the other containing a 1.5 per cent solution of disodium phosphate, all the 10 roots of *Lupinus albus* under observation after 24 hr. had grown into the block containing the solution of phosphate, where they quickly perished. In another case with distilled water on one side and a 0.01 per cent solution of copper acetate on the other, 13 of the 16 roots of *Lupinus albus* of the experiment had grown into the distilled water after a period of 24 hr. In the same manner the behavior of the roots was tested out when two unlike salts of equal osmotic pressure were brought at the same time against the opposite sides of the root. The chemicals used were calcium nitrate, potassium nitrate, and magnesium sulphate, and the following table shows the observed direction of growth for the roots of *Lupinus albus:*

	Neutral	
$KNO_3 \leftarrow 9$	2	$0 \rightarrow Ca(NO_3)_2$
$KNO_3 \leftarrow 10$	0	$1 \rightarrow MgSO_4$
$Ca(NO_3)_2 \leftarrow 0$	4	$10 \rightarrow MgSO_4$

The roots of *Cucurbita pepo* under the same conditions as those of *Lupinus albus* did not behave in the same manner, which indicates that the response of roots to a given chemical will depend to some extent upon the nature of the plant. Lilienfeldt (1905) repeated the work of Newcombe and Rhodes, using their method and one of his own after the following manner: Small containers 12 cm. high were filled with 3 per cent of gelatin in distilled water. As soon as it was sufficiently hardened a hole of about 20 cc. capacity was dug out of the center and filled with a solution of the chemical to be tested. The seedlings of *Lupinus albus* were carefully planted without injury by pushing the roots into the soft gelatin and so placing them that they were at distances varying from 5 to 50 mm. from the cavity. These were left in the dark for varying periods of time and the effect of the chemical diffusing through the gelatin on their direction of growth was observed. He obtained results identical with those of Newcombe and Rhodes when he used their method and with his own method the results are the same, with the exception of the salts of copper, zinc, mercury, and lead. By the method of Newcombe and Rhodes the roots reacted in a positive manner to these chemicals, but by the method of Lilienfeldt they responded negatively to them. The difference in results is evidently due to the amount of the salt that reached the root. In the method of Lilienfeldt, the amount of the poisonous salts that reach the roots by diffusing through the gelatin is much smaller than that which comes in contact with them when the salts are directly dissolved in the gelatin. A small amount of the poisonous salts stimulates growth on the contact side of the root and thus brings about a negative reaction in its behavior. A larger amount of the chemical would injure the cells of the root on the contact side and thus bring about a positive reaction in growth.

Kisser and Beer (1934) observed a similar behavior of seedling stems after the application of a salt solution to an incision on one side of the stem. High concentrations caused a positive curvature in all cases as a result of mechanical and osmotic effects. With low concentrations the negative curvatures were due to unequal growth following chemotropic response. The latter was most marked with lithium chloride, ammonium chloride, and manganese chloride, less with sodium chloride and potassium chloride, and did not occur with calcium chloride, strontium chloride, barium chloride, or magnesium chloride. Porodko (1911, 1914) worked with *Lupinus albus* and *Helianthus annuus* grown in the presence of various salts in agar as a diffusion medium and with solutions varying from one-thousandth normal to tenth normal. He found in the reaction to chemicals that the upper and lower portions of the root remained straight and that the intermediate portion was concerned in the curvature. He considered that the region in which chemotropic sensitivity of the root may be exhibited is limited to about 1 to 2 mm. of the root tip. He found that the intensity of the curvature depended upon the intensity of the diffusing forces or stimuli and that this, in turn, varies with the nature of the salt or diffusion substance used, with the concentration, with the thickness of the agar, and with the length of time the diffusion current acts. In general, the reaction as to the direction of the curvature manifested itself by a check in the growth and a positive curvature of the roots. This effect was observed in all the substances tested, whether they were electrolytes or nonelectrolytes.

Watenpaugh (1936) found that the development of alfalfa roots was definitely correlated with the pH reaction of the soil. Thus a pH of 4.8 definitely stopped the growth of roots, while they grew well in a medium with a pH of 5.0 or above.

Although the experimental evidence is rather contradictory in regard to the response in direction of growth to chemicals in the soil, two facts seem to be definitely established. In the first place, positive and negative reactions occur only within limited concentrations of the substances that have been used. In the second place, the roots must come in contact with the chemicals before they can in any way react to them. Roots cannot detect a substance needed for their growth and development and grow toward it, so that any nutrients that are applied to a soil must be so placed that they will come in contact with the roots in a solution of water if they are to be effective for the needs of the plant.

A practical application of this fact is observed in the application of fertilizer to corn. In a study of the development of the roots of corn seedlings, Millar (1930) noted that these roots tended to develop horizontally and to remain at comparatively shallow depths for a considerable time. Thus when the roots had reached a length of 9 in. they were only 5 in. from the surface in a sandy soil and about 3 in. from the surface in muck soil. Therefore, fertilizers placed on either side of the seed and slightly below it at the time of planting will be in the direct line of growth of the roots and should thus be used earlier by the plant than fertilizers placed directly above or below the seed.

Sayre (1934) considered that fertilizer should be placed in the soil at such a distance from the roots that they will escape for a few days the excessive soluble salt concentration in and immediately adjoining the fertilizer band. These salts diffuse rapidly from the fertilizer band into the soil solution and the excessive concentration of soluble salts therein is thus reduced so that the enriched region then becomes an excellent medium for root development.

2. Number of Roots.—One of the classic experiments used to demonstrate the effects of fertilizers upon the development of roots was that performed by Nobbe (1862) with corn plants growing in an infertile clay soil contained in glass cylinders. The experiment was so arranged that the nutrient salts ammonium sulphate, calcium nitrate, and dipotassium phosphate were added only to definite regions of the soil contained in each of six cylinders. After a growing period of 4 months, the plants were harvested and the root systems carefully isolated by a system of washing. The general form of the root systems remained normal and practically the same in all the experiments. The only difference that appeared was the local variation in the number of root branches. In the unfertilized portion of the soil the number of branches per unit of length of both primary and secondary roots remained small, while in the fertilized portions of the soil the number of branches of both primary and secondary roots was strikingly large. In the cylinder where the fertilizer was equally mixed throughout the soil mass the number of branches per unit length of both primary and secondary roots was the same in all parts of the vessel. Häveler (1892) grew corn and bean plants in vessels that contained alternate layers of sand and fertile soil rich in humus. He found that whenever the roots passed through the fertile soil a profuse branching of the several root orders occurred, but where the roots passed through the sand very little or no branching took place.

Frank (1893) grew corn and peas in such a manner that half of the roots of each plant grew in separate vessels. Each vessel contained the same amount of sand and was watered with the same nutrient solution, with the exception that to one vessel a portion of calcium nitrate was added, while the other received no nitrogen compounds. In the vessel containing the nitrogen-free medium, only a few small branch roots were formed, while the vessel containing the nitrogen compound was filled with a thrifty root growth. This same effect of nitrogen compounds on the secondary root formation of vetch, clover, corn, beans, and alfalfa was observed by Müller-Thurgau (1894).

Sayre (1934) found that roots of beans, cabbage, and tomatoes, coming in contact with fertilizers, developed an unusual number of fine fibrous roots and that this development was accompanied by stimulated plant growth. Rogers (1933) observed that the application of manure to gooseberries, currants, and pear trees increased both the root and top growth and the ratio of the weight of stem to roots.

Von Seelhorst (1902) carried on experiments to determine whether or not the use of fertilizers affected the number of roots of spring wheat, rye, peas, flax, beans, potatoes, beets, and barley. These plants were grown separately on two series of plots. One of the series had been fertilized annually for 25 years with fertilizers containing nitrogen, phosphorus, and potassium, although the amounts stated are not given, while the other series received no fertilizers during that period. The following table is a summary of the average number of roots found in an area of 314 sq. cm. at the various depths stated:

Soil	Number of roots at a depth in centimeters of				
	25	50	75	100	125
Fertilized	500	307	197	109	52
Unfertilized	459	206	158	92	43

Weaver, Jean, and Crist (1922) investigating the absorption of nitrates by roots found that where roots came in contact with a fertilized layer of soil they not only developed much more abundantly and branched more profusely, but also such a fertilized layer apparently retarded normal penetration into the soil below.

In a study of the effects of different fertilizer constituents on root development and anchorage of corn plants, Pettinger (1933) found: (*a*) the application of nitrogenous fertilizers to a soil deficient in nitrogen slightly increased the anchorage of corn plants when the supply of available potassium was ample but decreased this anchorage when the supply of potassium was limited; (*b*) the application of phosphorus or potassium to a soil deficient in these elements greatly improved the anchorage of the plants; (*c*) under the conditions of the experiments, both potassium and phosphorus increased root anchorage to a greater degree than did nitrogen, while potassium acted to a greater degree than did phosphorus; and (*d*) fertile soils were more productive of root growth than unfertile soils.

3. Relative Weight of Roots and Tops.—In general, it may be stated that the application of fertilizers increases the weight of the roots but that the weight of the tops generally increases more in proportion under such conditions than the roots so that the ratio of the dry weight of tops to the dry weight of the roots is generally higher in a fertile than in an infertile soil. Nitrogenous fertilizers especially increase the growth of tops as compared to the roots, while phosphorus as a rule increases the weight of the roots in a greater proportion than the tops. Harris (1914) found that the ratio of the weight of tops to the weight of roots of wheat seedlings was always greater with plants grown in concentrated than in dilute soil solutions. The concentrated soil solutions always produced more tops than the dilute, but the root growth was always greater in the dilute solution. The following table from the work of Harris (1914) shows that the proportion of roots and tops is affected more by a change in water content than by a change in the amount of nutrients:

DRY MATTER OF WHEAT TOPS AND ROOTS AT VARIOUS STAGES WHEN GROWN IN SOIL WITH DIFFERENT AMOUNTS OF FERTILIZER AND MOISTURE

Fertilizer	Per cent soil moisture	Boot stage, grams			Bloom stage, grams			Maturity, grams		
		Tops	Roots	*T/R*	Tops	Roots	*T/R*	Tops	Roots	*T/R*
None..........	30	35.3	15.8	2.2	50.5	10.7	4.7	74.9	9.3	8.1
None..........	15	15.2	10.0	1.5	19.3	8.8	2.2	29.2	9.5	3.1
Complete......	30	45.4	22.8	2.0	76.2	14.3	5.3	102.0	11.7	8.7
Complete......	15	27.8	20.7	1.3	39.9	12.4	3.2	52.3	13.2	4.0

Turner (1922) found that barley and corn showed significant increases in the ratio of tops to roots as the nitrate concentration of the solution was increased. The effect of the nitrates was shown to be independent of the concentration of the solution and the hydrogen-ion concentration. Flax plants did not show this increased ratio even after a period of 40 days or more. He concluded that the increased ratio of tops to roots which results from increasing the amount of nitrate in the solution may be explained on the basis of the increased use of carbohydrates in the tops, due to their combination with the nitrogen. This results in a decrease in the supply of the carbohydrates for the roots, which may bring about an absolute or relative reduction of root growth. It has been found that barley grown in the shade has a greater ratio of top weight to root weight than that grown in the unobstructed sunlight. Turner grew roots in pure cultures with a supply of sugar and found that nitrates did not check root growth directly but that they increased the growth of roots in all cases. Willard and McClure (1932) found with bluegrass that heavy fertilization with nitrogen greatly decreased the relative amount of underground parts.

It was found by Johnston and Miller (1934) that heavy leaf-rust infection on a susceptible variety of wheat resulted in a rapid and severe deterioration of the roots as indicated by discoloration, a decrease in the number of fibrous roots, and a marked loss in weight. Murphy (1935) found in oats that the ratio of roots to tops was greatly decreased by crown-rust infection, the decrease being in proportion to the duration and severity of infection.

Biswell (1935) showed that half shade favored the growth of seedlings of black walnut, buckeye, red oak, shellbark hickory, and hard maple but retarded the growth of honey locust, box elder, and sycamore. However, the root systems were deeper and more branched in all cases when the seedlings grew in full light.

Reid (1933) found that the ratio of the green weight of tops to the wet weight of roots in velvet bent grass under different degrees of sunlight was 1.1 for plants that were in the sun all day, 3.0 for plants that were in the sun all forenoon only, 3.3 for plants in the sun in the afternoon only, and 5.9 for those plants in the shade all day. For metropolitan bent grass the ratio was 3.3 in full sunlight and 8.3 in partial shade.

C. Oxygen Supply and Root Development

1. Aeration of the Soil. *a. Methods.*—Numerous benefits are generally derived by crop plants through the cultivation of the soil previous to their planting and at intervals during their growing season. The main benefits thus derived are (1) the conservation of moisture by increasing its reception and retention, (2) the eradication of weeds, (3) the correction of the physical texture of the soil, and (4) the adequate aeration of the soil. The aeration of the soil is beneficial to the plants in an indirect way, since an adequate supply of oxygen is necessary for the proper functioning of the bacteria that elaborate the various materials of the soil into a form in which they can be absorbed and utilized by the plants. In a direct way the aeration of the soil benefits the plant by furnishing a supply of oxygen, which is essential for the proper functioning of the protoplasm of the cells of the root. The general texture of the soil will determine to a great extent the benefits to be derived by increasing the oxygen supply by cultivation. Many soils are so light and porous that a deficiency of oxygen in them is not a limiting factor for growth, and it is a well-known fact that in many regions cultivation during the growing season is unnecessary for crop production provided the surface of the soil is kept free of weeds and is not too compact to absorb the rainfall. In very compact and in waterlogged soils, special methods in addition to cultivation must be applied in order to insure the proper aeration for normal plant development.

Howard (1915, 1916, 1917, 1918) has reported that special methods are necessary to bring about the proper aeration of the soil in the Bihar region of India. The soils in this region are composed of fine particles, become very compact upon wetting, and do not crack when the surface dries, thus preventing natural aeration. This condition is corrected by manuring with straw, by subsoiling, by the application of fragments of tile and brick, and by the proper sequence of crops. *Cajanus indicus*, the pigeon pea, confers a great benefit on a crop following it, especially tobacco. The root system of the pigeon pea grows to a great depth and is extensive laterally. White ants eat out the roots and leave air passages throughout the soil. In converting rough grassland into cultivated fields, sweet potatoes are planted as the first crop. They seem to thrive even if aeration is poor, and the swelling of the roots bursts open the soil and admits air, so that, when the roots are removed, the numerous cavities furnish aeration for other crops that would not otherwise grow.

Bergman (1920) found by growing plants in soil in pots that were submerged in water that *Impatiens*, *Pelargonium*, and *Coleus* were sensitive to a scarcity of oxygen. If, however, the water surrounding these pots was aerated the plants would grow normally. The following experiment is worthy of note: Two plants of *Impatiens* were potted in soil and the pots submerged in water. Several plants of *Philotria*, an aquatic plant, were placed in the water surrounding one of the pots but not in that surrounding the other. The evolution of oxygen from the photosynthetic activity of these aquatic plants aerated the surrounding water, and thus the water of the soil in the submerged pots, sufficiently to provide for the oxygen needs of the roots of the plants growing therein. As a result this plant was in good vegetative condition after two months, while the other plant whose roots were not thus aerated wilted after three days and shortly thereafter perished.

A sensitive and practicable method for comparing the oxygen-supplying powers of the soil atmosphere under different environmental conditions has been devised by Hutchins (1926). This method is based essentially on a colorimetric determination of the time required to produce a standard color change in an alkaline aqueous solution of pyrogallol by the absorption of oxygen. This oxygen is brought to the indicator in a continuous stream, having been absorbed from the environment in question by a very thin-walled, water-impregnated, porous porcelain conical-shaped absorber. The absorber has the capacity for absorbing oxygen and delivering it at least as rapidly as that element is supplied to the absorbing surface of the medium in which it is placed. The numerical results are termed "indexes" of the environmental oxygen-supplying power and are expressed as milligrams of oxygen supplied per hour through a square meter of absorbing area exposed to the environment under consideration.

By this method it was found that the oxygen-supplying power of the soil is higher the less its depth, the dryer, and the more loosely it is packed. With a very wet loamy garden soil the index value for a depth of 40 cm. is only about 0.7 per cent as great as that for a depth of 15 cm The oxygen-supplying power varies also with the kind of soil and with the prevalence of aerobic organisms and readily oxidizable materials in the soil surrounding the absorber. Soil temperature and barometric pressure may also have an influence.

b. Effects on Plant Growth.—Two important observations have been made on the effect of soil aeration in the production of rice. As high as 20 per cent loss of the expected crop of rice has occurred due to a disease known as "straight head." In this trouble the flowers remain sterile and the panicles do not droop. Tisdale and Jenkins (1921) found that this condition of the plants was brought about mainly by the lack of proper soil aeration. Normal plants have few large roots and many branches and root hairs, while the plants affected with the disease show numerous large roots and very few branches and root hairs. For the control of this disease they recommended the irrigation of the plants ten days after they emerge or when they are 6 to 8 in. tall. The water should be allowed to drain off gradually, and if the soil is not free from water in 5 to 6 weeks after irrigation, it should be drained off and the land allowed to remain dry for 2 to 3 weeks. If the soil becomes dry enough to crack and the plants turn yellow, the results will be better. Water should then be turned on and allowed to remain until the end of the season. In the growing of swamp rice at Coimbatore, India, Harrison (1913) presented evidence to show that the surface film of algae and other organisms on these soils plays the leading part in the formation of the oxygen required by the roots of the rice. He believed that the aeration of these soils by the atmospheric oxygen is not so effective in promoting root aeration as is the aeration secured by the water that percolates through this surface film of algae. When certain rains fail to fall in the Bihar region, even if the soil is moist the rice crop is poor. This result is considered to be due to the failure of water to filter through the film algae and aerate the roots.

Albert and Armstrong (1931) found that a larger percentage of fruit buds were shed from cotton plants that were grown under conditions of high soil moisture than from plants grown under more nearly optimum conditions of soil moisture. The percentage of oxygen was lower and the percentage of carbon dioxide was higher in the soil with the higher content of moisture. They considered, therefore, that a high percentage of carbon dioxide and a low percentage of oxygen in the soil are associated with an increase in the percentage of shedding of the fruit buds of cotton.

c. Response of Roots.—Arker (1901) and Hall, Brenchley, and Underwood (1914) have found that the rate of root growth and top growth in both soil and water cultures may be increased by pulling a stream of air through the medium. The following table is a brief summary of the results obtained by Hall, Brenchley, and Underwood with barley and lupine plants grown in water cultures:

Plant	Treatment	Average dry weight, grams
Barley	Not aerated	1.31
	Aerated continually	2.12
Lupine	Not aerated	0.83
	Aerated continually	1.53

Bouygues (1928) noted that the roots of *Salix purpurea* stopped development after a short time when grown in an aqueous solution from which the air was excluded by an oily covering, but whenever the plants were removed to a solution in contact with air, the growth of roots was resumed.

Weaver and Himmel (1930) grew plants of the great bulrush (*Scirpus validus*), cattail (*Typha latifolia*), reed (*Phragmites communis*), and tall marsh grass (*Spartina michauxiana*) in dry soil, moist soil, alternately saturated and drained soil, and waterlogged soil. With the exception of *Scirpus*, the ratio of the dry weight of tops to roots was less in the drained than in the waterlogged soil. A response to poor aeration was shown in all experiments by the development of a shallow root system with fine, much-branched roots having a large surface in proportion to volume. Root growth increased with decreasing water content until moisture became a limiting factor. Similar observations were made by Dean (1933) on *Hibiscus*, *Acorus*, *Sagittaria*, and *Typha*, for in unaerated cultures the plants established shallow root systems, while in aerated cultures the roots ramified throughout the entire soil. Tincker (1935) noted with potted lilies that adequate drainage and aeration not only influenced the vigor of the leaves but also the size and color of the flowers.

Loehwing (1931) grew sunflower, flax, wheat, and soybeans in soil that was aerated for ½ to 5 hours daily. Two months after planting, the roots of the plants grown in aerated soil were more numerous, longer, and more fibrous than those grown in nonaerated soil. The total root surface of the aerated plants was at least twice that of the control plants. The root systems of aquatic plants, such as *Typha*, *Sagittaria*, and *Hibiscus*, also greatly increased in size in aerated media of sand, clay, or muck soils.

Heinicke (1932) considered that a low supply of oxygen is the primary cause of injury to apple trees when their roots are submerged.

Interference with the process of respiration is apparently the cause of the injury and eventual death of the roots due to the exclusion or depletion of the oxygen supply of the soil. The first visible effects of the lack of oxygen to the roots are the wilting of the plant and a lowering of the rate of transpiration. These effects are apparently due to the fact that the protoplasm does not function properly as an absorbing membrane. Hunter and Rich (1925) found that the rate of transpiration and the intensity of respiratory activity of the shoot of *Impatiens balsamina* are increased by the aeration of the soil about the root system.

It is suggested that these phenomena are due to the removal of carbon dioxide that has resulted from the respiration of roots and soil organisms. The concentration of this gas may become great enough to hinder or prevent absorption by the roots. The rate of elongation of the stem is governed by the degree of turgor pressure in the individual cells. It is thus conceivable that increased root activity in absorption, due to a lowering of the concentration of carbon dioxide by soil aeration, would tend to develop a higher degree of turgor pressure in the living cells of the shoot. This increased turgor would thus influence the rates of transpiration, growth, and respiration in that portion of the plant.

It has been shown by Cannon (1915 to 1925), Free and Livingston (1917), and Bergman (1920) that different species of plants vary widely in their response to the variations in the amount of oxygen in the soil, and these differences in responses are quite as significant as the responses to temperature and to available water. Cannon found that *Prosopis*—the mesquite—and *Opuntia*—a cactus—have roots that are unlike in their response to the oxygen supply in the soil. His experiments showed that if pure carbon dioxide filled the air space of the soil, growth ceased in the roots of both plants. The reduction of the oxygen to 12 to 15 per cent did not stop growth in the roots of *Prosopis* but did so in the roots of *Opuntia*. The roots of *Prosopis* continued to grow when only 2 per cent of oxygen was contained in the soil air. Free and Livingston (1917) found that the roots of the willow (*Salix nigra*) could endure successfully the almost complete exclusion of oxygen from the soil. Coleus and heliotrope were injured in a short time, so that within 12 to 24 hr. the absorption of water by the roots ceased.

2. Relation of Temperature to Oxygen Needs.—Cannon (1923 to 1925) has observed that the minimum oxygen supply is not a fixed amount but that it varies directly with temperature change. If the roots of *Potentilla anserina*, which inhabits swampy ground, are placed in a soil whose atmosphere consists of 1.2 per cent oxygen and the balance nitrogen gas, growth of the roots does not take place at 27 and 30°C., but at 18°C. it is about one-fourth normal for that temperature. In soils whose atmosphere contains 2 per cent oxygen, the growth rate is about one-fourth normal at 30°C., one-third normal at 27°C., and normal at 18°C. In the case of corn with the roots in soils containing 3 per cent oxygen, the rate of growth at 30°C. is about one-sixteenth normal, at 20°C. about one-fifth normal, and at 18°C. about one-third normal. When the percentage of oxygen is increased to 3.6, the rate of growth at 30°C. is about one-third the expected rate under normal conditions of aeration, while at 18°C., it is about two-thirds normal. These examples show that

> . . . there comes a point in the diminution of the oxygen content of the soil atmosphere when the growth of the root ceases because there is no longer a sufficient amount of this gas to supply the demands for energy correlated with physiological activities of higher temperatures.

On this account it can be inferred, for instance in the case of corn, that to attain a fair rate of growth at a time of high soil temperatures the aeration of the soil must be good, otherwise the rate of growth would be considerably cut down.

Collison (1935) noted that root elongation, and therefore absorption, assimilation, and respiration, took place in apple trees when the air temperature was below zero and apparently at a soil temperature not far above the freezing point.

3. The Sources of Oxygen.—The oxygen used by plants may have multiple origins. It may enter the plant directly from the atmosphere, it may enter the roots dissolved in water, or it may be liberated within the plant as a product of photosynthesis. The oxygen from the substratum may conceivably aerate the roots and perhaps the stem. That which is absorbed by the stem and leaves from the air may aerate the stem and perhaps the roots, while the oxygen that is liberated in photosynthesis may aerate both the stem and roots. The oxygen derived from photosynthesis has been designated "internal oxygen" by Cannon (1932). He noted in his experiments with sunflower and willow that in the majority of cases the roots showed a slower rate of oxygen consumption when the shoots were in sunlight than when they were in the dark. He considered that this indicated that the roots obtained a portion of their oxygen from the aerial parts exposed to sunlight. This indication was further substantiated by the work of Cannon, Demaree, and Purer (1933) who found in the willow that regardless of whether the temperature was high or low the expected drop in the rate of oxygen consumption by the roots occurred when the aerial portion of the plant was exposed to light.

4. General Observations.—Cannon (1917) has observed that an increased air supply to the roots of certain desert plants favored root branching. Miller (1916) and Weaver (1919) have noticed in isolating root systems in the field that a root upon passing through a cavity in the soil branches profusely so that in many cases it fills the entire cavity. This behavior of the roots is apparently due to the combined effects of moisture and the oxygen supply. The clogging of tile drains by roots is probably due to the same causes. The superiority of the roots of soybeans grown in aerated nutrient solutions over those in unaerated solutions was found by Allison and Shive (1923) to be due not to the higher dry weight of the substance produced but to the development of a more efficient absorbing system. The main roots in the aerated cultures were long and slender and thickly beset with well-developed laterals almost to the very tips.

The evidence in regard to the growth of roots toward an oxygen supply is mostly indirect. Many of the main roots of corn and sorghum plants growing in the field under ordinary agricultural conditions extend almost horizontally if there is sufficient moisture in the uppermost portion of the soil. The laterals arising from these horizontal roots have been observed by Miller (1916) to extend vertically upward to within ½ in. of the soil surface. This behavior is considered to be due to the influence of the supply of oxygen in the upper part of the soil. The same explanation is given to the behavior of roots that after being forced to grow downward to a considerable depth through a vertical tile have turned directly upward upon emerging from it. Ewart (1894) found that the roots of peas, hemp, and wheat grown over water deprived of oxygen either turned slightly upward or grew along the water surface, while if the water were aerated they grew directly downward into it.

Clements (1921) found in experiments with sunflowers growing in sealed pots which were glazed, with the exception of certain small, square surfaces, that in all cases the roots showed right-angled curvatures toward the unglazed squares.

Bennett (1904) worked with corn, squash, radish, peas, and lupine to determine whether any response could be obtained to oxygen and carbon dioxide from the roots of these plants. In general, the experiment consisted in allowing the gases to diffuse directly upon the roots growing in soil or water as a medium. Three hundred and sixty-two reactions were observed upon the roots of corn, radish, and squash growing in water upon which either carbon dioxide or oxygen was allowed to diffuse from one side only. Of this number 84 were positive, 72 were negative, and 206 were neutral. In an experiment in which carbon dioxide diffused against one side of the root and oxygen against the other, 166 observations were made of which 37 were positive in their reaction to oxygen, 22 were positive to carbon dioxide, and 107 were neutral to both gases. In the experiment in which plants growing in the soil were so arranged that oxygen could diffuse on one side of the primary root and carbon dioxide upon the other, the results showed no more than those mentioned above, since, of the 221 observations made, 45 reacted in a positive manner toward the oxygen, 48 in a negative way, and 128 in a neutral manner. The evidence in these three experiments refutes the idea that roots exhibit aerotropism.

D. Soil Temperatures and the Response of Roots

It is the intention here to discuss more particularly the response of roots as shown by curvatures induced by the changes in the temperature of the medium in which they are growing and to reserve for a subsequent chapter the effect of temperature changes upon the rate of root growth. Until 1914 our knowledge of the thermotropic curvature of roots was very confusing. The experimental results demonstrated that different plants exhibit little uniformity in behavior, since some roots were positive in their reaction at a given temperature, while others at the same temperature gave a negative or neutral reaction.

Eckerson (1914) investigated the response of roots when exposed to unequal temperatures on their two sides and arrived at results that cleared up the apparent discrepancies and contradictions in the results obtained by earlier investigators. In these experiments, Eckerson grew or placed the seedlings in sphagnum moss in a Ganong differential thermostat that furnished a gradation in temperatures from 5 to 58°C. If the roots curved toward a region of higher temperature, they were considered to have responded positively, while a curvature away from higher-temperature areas was listed as a negative response. After the roots of the radish had been in the thermostat 2 hr., they showed the following reactions in curvatures: 7 to 15° positive, 16 to 23° no curvatures, 24 to 36° positive, and 38 to 51° negative (see also Treitel, 1924). A study of the permeability of the cells of the roots of these plants at different temperatures by means of the plasmolytic method with potassium nitrate solutions showed that it increased from 10 to 18°, did not change from 18 to 24°, increased from 24 to 40°, and then decreased. A comparison of the changes in permeability and the response in curvatures at the different temperatures shows that in every case where the temperature induced an increased permeability there was a positive curvature, where no change in permeability occurred no curvatures resulted, while at temperatures causing a decrease in permeability, negative curvatures resulted. These observations show that with unequal temperatures on opposite sides of a root a curvature is produced only when the cells of the root are more permeable at one of the temperatures than the other. Those cells subjected to a temperature at which they are more permeable to dissolved substances are consequently less turgid, and this results in a shrinking of the tissue on that side of the root and a consequent mechanical curvature, and in all cases the more permeable side of the root becomes concave. Eckerson also studied the changes in the permeability of the cells of the roots of the various species of plants that had been investigated by Wortmann (1885) and by Klercker (1891) and found that the increase or decrease in the permeability at the various temperatures paralleled the positive or negative curvatures that had been observed by these investigators. She summed up her observations in the following statement: (1) Thermctropic curvatures of roots vary with the temperature and with the species. (2) The permeability of the cells of the roots to dissolved substances varies with the same factors, and wherever the thermotropic reaction of the root changes the permeability also changes. (3) Heat does not act as a stimulus but by affecting permeability is a direct factor in producing curvature, so that thermotropism is not to be considered a tropism but a turgor movement.

E. Soil Texture and Root Development

A hard, compact soil limits the extent of root systems to a marked degree, and Weaver (1919, 1920) observed that in a compact soil the roots are more or less contorted or kinked, while the branching is decidedly less than in a soil of loose texture.

Carlson (1925) in studying the effect of soil structure on the character of alfalfa root systems found that in compact soils all varieties and strains develop branch roots, while in open soil the taproots predominate. He also found that the different strains of alfalfa have inherent root characteristics which as a rule do not develop until after 3 or 4 months of normal growth. McCool (1928) observed the development of the roots of alfalfa, sweet clover, and red clover in three different types of soil. The results obtained showed that there were marked differences in the nature and extent of the roots of these plants. Thus the average length of alfalfa roots in one soil was 40 in. and in another soil only 8 in. Sweet-clover roots attained a length of 43 in. in one soil, 13 in. in another, and but 10 in. in a third soil.

Weaver and Crist (1922) made some observations on the relation of hardpan to root penetration. This soil layer, in general, delimits the depth of penetration of the roots of crop plants, but under exceptional moisture conditions they may penetrate the hardpan and extend 2 to 3 ft. beneath it. No modifications in root habits were apparent in the hardpan layer. Since roots will not grow into dry soil, the conclusion is that water penetrated the hardpan and the soil beneath.

Singh (1922) noted that the roots of wheat plants grew best in pure sand provided it was supplied with the proper amount of nutrients and water. At Rothamsted roots grew better in a mixture of sand and soil than in soil alone. Jensen (1931) showed that the type of soil has an important relation to the correct tillage methods of sugar cane.

It was noted by Taubenhaus, Ezekiel, and Rea (1931) that root strangulation of cotton occurs when the upper portions of the taproot and laterals of young seedlings are caught early in the season in a subsurface layer of hard, dry clay which prevents further development. Affected plants die when the constricted areas are killed by the hot dry soil or when the moisture supply that can be transported through the constricted area becomes inadequate for the needs of the plant.

In prairie soil in Nebraska, Weaver, Hougen, and Weldon (1935) studied the relation of root distribution to organic matter, in the case of *Andropogan scoparius* and *Andropogan furcatus*.

Except in the upper 6 in. of soil, there is an approximate linear relation between the amount of root material and the amount of organic matter in the various soil horizons.

F. The Response of Roots to Light

Some roots show a marked reaction in response to light. This is especially true of some members of the mustard family. The roots of these plants are those that are so frequently pictured in textbooks as examples of the negative action of roots to the stimulus of light. Many roots, however, show little or no response to light.

II. THE EXTENT OF ROOT SYSTEMS

This general subject can be conveniently discussed under two general headings: (*a*) the total number and combined length of the roots of a given plant, and (*b*) the extent of roots in the soil.

A. Total Number and Combined Length of Roots

Nobbe (1869) measured and counted the roots of winter wheat and rye and found that the roots of the first, second, third, and fourth order of 55-day-old plants growing in the soil numbered 16,000 for rye and 10,700 for wheat. The combined length of the roots of the rye plant measured 492 ft.; and that of the wheat plant, 283 ft. A mature wheat plant grown in the soil was found to possess 67,200 roots of the first four orders and to have a combined length of 1,730 ft.

Clark (1874) examined the roots of a squash plant growing in the greenhouse at the Massachusetts Agricultural College and estimated the total length of these roots at 15 miles. This estimate included not only the root system proper but also the roots with all their branches originating from the 70 nodes of the plant. In 1912 Gibson repeated this work with a closely related variety and found the combined length of both the main and nodal root systems to measure only 281 ft. at the time of full vegetative development. Hellriegel (1883) estimated the total length of the root systems of mature barley and oat plants grown in a fertile soil to be 134 and 168 ft., respectively. Weaver (1925) found that the root system of a single corn plant in the eighth leaf stage has from 8,000 to 10,000 laterals arising from 15 to 23 main roots. The area of the root system of winter wheat measured at various intervals from Sept. 30 to Nov. 29 exceeded the area of the aerial parts by 10 to 35 per cent, as shown by the following table:

Date	Total area of roots exclusive of root hairs, sq. cm.	Total area of tops, sq. cm.
Sept. 30................	9.6	7.6
Oct. 10................	31.4	22.5
Oct. 20................	61.5	50.5
Oct. 30................	105.1	89.0
Nov. 14................	237.5	215.3
Nov. 29................	309.7	280.0

Pavlychenko (1937) determined that the total extent of the roots of the first, second, and tertiary orders of mature wild oats, Marquis wheat, and prolific spring rye was 54, 44, and 49 miles, respectively, when the plants were spaced 10 feet from each other. However, when these plants were grown in 6-in. rows, with 18 to 20 plants per foot, the root systems were from 88 to 99 times smaller than in the former case.

Dittmer (1937) grew a plant of winter rye, *Secale cereale*, in soil, within a wooden container 12 in. square and 22 in. deep, for 4 weeks until the plant had reached a height of 20 in. The roots were then washed out and the number, area, and length of those of the first, second, third. and fourth orders determined. The number of root hairs, together with their area and extent, was also calculated. He found that the total length of the roots of the four orders was 387 miles with an area of 2,554 sq. ft. The total number of root hairs on the roots of this plant was calculated at about 14.5 billions, with a total length of 6,603 miles and an area of 4,321 sq. ft. The surface of the aerial parts of this plant totalled 51.5 sq. ft., so that the area of its subterranean parts was 130 times that of the tops.

B. Extent of Roots in the Soil

From 1880 to the present time, work has been done by Goff and Beckwith (1883 to 1888), Hays (1888 to 1894), King (1894 to 1893), Headden (1896), Georgeson and Payne (1897), Ten Eyck (1899, 1900, 1904), Cottrell (1902), Shepperd (1905), Schulze (1906), Rotmistrov (1907 to 1909), Miller (1916), Weaver (1915 to 1924), Collings and Warner (1927), Lee (1926), Arny and Johnson (1928), Gursky (1928 to 1929), Krassovsky (1929), Roxas and Villano (1930), Laird (1930), Farris (1934), West (1934), Sperry (1935), and Maximow and Kruzilin (1936) in studying the general distribution of the roots of cultivated plants in the soil and in perfecting methods for their isolation under field conditions. The general observations of these investigators and a brief summary of the conditions under which they worked are recorded in the table on page 139.

The roots of woody plants have been studied by Peren (1923), Allan (1926), Laitakari (1927), Gemmer (1928), Rogers and Vyvyan (1928 to

EXTENT OF ROOT SYSTEMS OF CULTIVATED PLANTS

Investigator, location, and date	Plants	Extent, inches		Remarks
		Vertical	Lateral	
Goff and Beckwith, New York (1883–1888)	18 garden plants	10 to 34	10 to 120	Clay loam 6 to 10 in. deep. Tenacious subsoil
Hays, Minnesota (1888)	Red clover	68	20	5-month-old plants. Soil fertile, 6 ft. deep
	Alfalfa	80	12	
	Mammoth clover	57	12	
King, Wisconsin (1892–1893)	Clover, oats, barley	48	....	Plants all mature. Clay-loam soil. Stiff subsoil
	Winter wheat, corn	48	24	
Headden, Colorado (1896)	Alfalfa	150	6	Plants 6 years old. Stiff clay soil
Georgeson and Payne, Kansas (1897)	White milo	....	72	In hills 8 ft. apart. Black loam soil 8 in. deep. Stiff clay subsoil
	White kafir	48	48 to 70	
	Corn	....	48	
Ten Eyck, North Dakota (1899–1900)	Spring wheat, oats	48	9	Soil a fertile loam 4 ft. deep. Oats and wheat in rows 18 in. apart
	Corn	42	42	
	Potatoes	18	24	
	Sugar beets	40	24	
Cottrell, Kansas (1902)	Alfalfa	120	....	Stiff subsoil. Did not find end of root
Ten Eyck, Kansas (1904)	Corn	48 to 60	....	Heavy clay-loam soil. Stiff clay subsoil
	Kafir	42	36 to 42	
	Winter wheat	48		
	Oats and barley	52		
Shepperd, North Dakota (1905)	Rye	36	....	Red clover 2 years old. Alfalfa 3 years old
	Spring wheat	48		
	Barley and oats	36		
	Red clover	48		
	Alfalfa	60		
Schulze, Germany (1906)	Rye	70		
	Wheat	74		
	Barley	70		
	Oats	90		
	Vetch	86		
	Potato	97		
Rotmistrov, Russia (1907–1909)	Corn	45	53	
	Potato	24	40	
	Yellow sorghum	42	44	
	Cucumber	42	33	
	Cotton	38	41	
	Rye	47	24	
	Wheat	41	37	
	Beet	58	44	
Miller, Kansas (1916)	Corn, kafir, milo	72	36 to 48	Grown on fall-irrigated, sandy loam 10 ft. deep
Weaver, Kansas, Colorado, South Dakota, Nebraska (1920–1921)	Rye	41	12	Short-grass plains
	Oats	41	9	
	Wheat	32	4 to 8	
	Rye	58	12	Mixed prairie
	Oats	50	9	
	Wheat	53	4 to 8	
	Rye	64	12	True prairie
	Oats	53	9	
	Wheat	65	4 to 8	
	Alfalfa	60	12 to 18	Grown on flood plain. Plants 116 days old. Eastern Nebraska
	Red clover	52	12	
	Sweet clover	64	12	
	Wheat, Durum	88	14	Upland silt-loam subsoil loose loess. Eastern Nebraska
	Wheat, Marquis	76	15	
	Barley	75	15	
	Corn	98	48	
	Potato	46	26	
Roxas and Villano, Philippines (1930)	Sugar cane	12 to 18	5	Aeration an important factor in determining depth
Laird, Florida (1930)	Bahia, Bermuda, St. Augustine, blue couch grasses	72 to 96		
Farris, New Jersey (1934)	Corn	24	24	Roots much less extensive than in semiarid regions
	Red clover (mature)	36		
	White potato	12		
West, Australia (1934)	Alfalfa	204	....	50 per cent of roots in upper 3.5 ft.
	Wheat	108	....	78 per cent of roots in upper 2.8 ft.
	Beans	36		
	Rice	36	....	82 per cent of roots in upper 4 in.
Sperry, Illinois (1935)	Big bluestem	50		
Maximov and Kruzilin, Russia (1936)	Spring wheat	....	....	80 per cent of roots in upper 40 cm. of soil, irrigated

EXTENT OF ROOT SYSTEMS OF TREES AND SHRUBS

Investigator, location, and date	Plants	Extent, ft. Vertical	Extent, ft. Lateral	Remarks
Peren, England (1923)	Cherry tree	....	30	15 years old
	Apple tree		24	15 years old
Allan, New Zealand (1926)............	*Podocarpus spicata*	....	65	Tree, 52 ft. high
Laitakari, Finland (1927)	*Pinus sylvestris*	....	60 to 90	On sandy heaths
Gemmer, S.E. United States (1928).......	*Pinus palustris*	....	30	
Rogers and Vyvyan, England (1928)	Lane's Prince Albert apple	....	....	10 years old; 10 per cent of fibrous roots within 1 yd. from tree; 40 per cent more than 2.5 yd. from trunk
Traub and Stansel, Texas (1930)	Fig tree	....	50	5 to 6 years old
Holch, Nebraska (1931)............	Bur oak	5.7	Exceeded spread of tops	First-year seedlings
	Shellbark hickory	2.5		
	Red oak	2.4		
	Linden	1.2		
	Walnut	4.5		
Beckenbach and Gourley, Ohio (1932)	Apple tree	5	....	Roots under sod, continuous mulch, and cover crop of rye. First foot, 411 roots; second foot, 225 roots; third foot, 98 roots; fourth foot, 61 roots; fifth foot, 44 roots
Oskamp, New York (1932)	Cherry trees	6	....	20 years old; 38 to 73 per cent of roots in surface foot; 17 to 30 per cent in second foot; very few in fifth and sixth foot
Rogers, England (1932)	Bramley seedling apple	3	20	Wet clay soil, grassland, water table at 3.5 ft.; tree, 26 years old
Weaver and Kramer, Nebraska (1932)	Bur oak	14	20 to 60	Mature trees, 50 to 60 years old, 35 to 45 ft. tall, 1 to 1.5 ft. in diameter, and 40 ft. apart
Heyward, Florida (1933)	Long-leaved pine	14	75	90 per cent of all laterals occurred in surface foot
Nutman, East Africa (1933, 1934)........	*Coffea arabica*	6	....	Mature plants
Rogers, England (1933)............	Gooseberry	8	....	Water table at 11 ft.
	Black currant	7	....	
	5-year pear trees	11	....	
Harmon and Snyder, California (1934)	20 species of grapes	1 to 5	....	21 per cent of roots in first foot; 48 per cent in second foot; 25 per cent in third foot; and 6 per cent in fourth foot
Marth, Maryland (1934)	Stayman apple trees	....	12	70 per cent more roots in first 6 in. than in second 6 in.
Rogers and Vyvyan, England (1934)	Lane's Prince Albert apple trees	9.5	2 to 3 times extent of branches	Trees 10 to 11 years old; only 10 per cent of fibrous roots in square meter surrounding the trunk
Duncan, Indiana (1935)	*Rhus*, *Rosa*, *Rubus*, *Sassafras*, and others	1 to 3.5		
McQuilkin, New Jersey (1935)............	*Pinus rigida*	....	25 to 35	Trees, 12 to 30 years old
Rogers, British Columbia (1935)	McIntosh apple tree	5	28	Trees 19 ft. high, on irrigated soil
Yeager, North Dakota (1935)	31 species of trees and shrubs	4	0.4 to 2.1 times the height of the tree	When some water supplied, roots spread less and penetrate deeper. 97 per cent of roots in first 4 feet

1935), Traub and Stansel (1930), Holch (1931), Weaver and Kramer (1932), Oskamp (1932), Rogers (1932, 1933, 1935), Beckenbach and Gourley (1932), Nutman (1933, 1934), Heyward (1933), Marth (1934), Harmon and Snyder (1934), Duncan (1935), and others. A summary of their observations and of the conditions under which they were made is shown in the table on page 140.

The investigations of the distribution of the roots in the soil have been undertaken for two main purposes: (1) to determine the effects of cultivation upon roots and (2) to obtain information concerning the relations of the roots to the water supply of the soil. Most of the investigations have been made with the latter point in view, and a study of the tables shows that most of the work with cultivated plants has been conducted in more or less semiarid regions where rainfall is the limiting factor in crop production.

Gericke (1923) noted in wheat that plants with roots 50 to 60 cm. in length had 20 tillers. If these roots were cut back to 30 cm., only 17 tillers developed, and, if pruned to 4 in. (10 cm.), only four tillers developed. He thus considered that the extent of tillering is dependent upon the size of the root system.

Some of the most recent work relative to the effects of cultivation upon the roots is that of Kiesselbach, Anderson, and Lyness (1935). They found in corn that a considerable range in cultivation may be practiced without materially affecting the yield. Thus, shallow, medium, deep, and close cultivation in one direction yielded respectively 32.9, 34.0, 35.2, and 34.5 bu. per acre during an 11-year period. Shallow cultivation was 4 in. deep and 9 in. from the plant, deep cultivation was 6 in. in depth and 9 in. from the plant, and close cultivation was 5 in. deep and 7 in. from the plant. The grain yield was lowered 12 per cent by pruning the roots 6 in. deep and 7 in. from the plant, and pruning the roots at a distance of 4 in. from the plant and 6 in. deep lowered the yield 20 per cent.

Nedrow (1937) pruned the roots of Sudan grass at a depth of 5 or 10 in. every 10 days. The pruning of 5-in. depths reduced the weight of tops 56 per cent and the weight of roots 49 per cent. The three successive prunings of the roots at the 10-in. depth decreased the weight of tops and roots 34 and 37 per cent, respectively. Deuber (1936) pruned the roots of small oak trees that had been injured by gas but found that after 2 years the tops of the pruned trees showed no measurable improvement over those of the controls.

1. Vertical Penetration.—Cannon (1911) made a study of the root systems of over 50 annual and perennial plants growing in the desert region of Tucson, Ariz. He found that the root systems of the annuals rarely penetrated the soil more than 8 in., and that the greatest root development took place in the upper 2 to 3 in. of soil. He observed that the depth of root penetration depended upon the depth to which the rains moistened the soils. The cacti have two specialized root systems—a vertical one serving as an anchorage for the plant and a horizontal one that functions mainly in absorption. Most of the horizontal roots of cacti were found in the upper 4 in. of soil. Cannon (1913) also found a similar distribution of the roots of plants growing in the Algerian Sahara. Weaver (1919) found that the roots of *Yucca glauca* or Spanish bayonet, a plant very common in the semiarid regions of the Western plains of the United States, penetrated the soil to a depth of 18 in. only, while the horizontal roots extended to a distance of 30 ft. or more.

Weaver (1920) made an extensive study of the distribution in the soil of the roots of both cultivated and wild plants growing in the grassland formation that extends from the Missouri River to the foothills of the Rocky Mountains. Due primarily to variations in rainfall and evaporation, three more or less distinct plant associations are to be recognized in this area and have been classified by Clements (1920) as (*a*) the true prairie, (*b*) the short-grass plains, and (*c*) the mixed prairie. The true prairies occur in regions of sufficiently favorable soil- and air-moisture conditions to permit the growth of large numbers of dominant tall grasses with which are intermixed a great number of other herbs, the whole forming a luxuriant vegetation. The plants in this region grow under a mean annual precipitation of 28 in. or over, of which 20 in. comes during the growing season. The short-grass plains are characterized by a few short, sod-forming grasses, especially *Bulbilis dactyloides* (buffalo grass) and *Bouteloua gracilis* (grama grass). This short-grass association extends from southwestern Nebraska and the western half of Kansas through eastern Colorado into northwestern Texas, northern New Mexico, and Arizona. In these regions the evaporating power of the air is high and the rainfall is limited, averaging usually 20 in. or less. The mixed prairie occupies a region of greater precipitation and more favorable conditions for plant growth than the short-grass plains, and one of the distinctive features of this association is the intimate mixing of the tall with the shorter grasses.

From his 1,500 observations of the root systems of 180 species of plants growing under varied conditions of this grassland formation in 25 stations in Kansas, Colorado, South Dakota, and Nebraska, Weaver concluded that the longest and most deeply penetrating roots are found where there is considerable rainfall, where much of this rainfall penetrates into the soil, and where the water table is relatively low. He emphasizes the fact that the maximum depth to which a few roots will penetrate is not of so much importance in the water relations of the plant as is the working or average depth to which the majority of the roots penetrate. It is also not so much a question of the depth to which roots penetrate as what proportion of the root system is functioning in absorption. Weaver found that 65 per cent of the species in the true prairie reached depths greater than 5 ft.; 42 per cent in the mixed prairie and only 13 per cent of the species examined in the short-grass plains reached beyond that depth.

Most of the species of the plains area were characterized by a fine system of wide-spreading laterals. This type of growth is evidently a response to the moisture in the upper portions of the soil resulting from summer showers. Practically all the deep-rooted plants in the short-grass plains area are well adapted to absorb moisture from the upper portion of the soil. Of the 45 species examined by Weaver (1920) in this region the lateral extent of the roots ranged from ½ to 3½ ft. This

HEIGHT OF TOPS, WORKING DEPTH, AND MAXIMUM DEPTH OF ROOT PENETRATION OF RYE, OATS, AND WHEAT IN THE THREE REGIONS OF THE GRASSLAND FORMATION

Station and crop	Height of tops, feet	Working depth of roots, feet	Maximum depth of roots, feet	Remarks
Winter rye:				
Short-grass plains: 5 stations, Colorado and Kansas......	2.7	2.8	3.4	Silt-loam soil
Mixed prairie: 3 stations, Colorado and Kansas..........	3.3	3.7	4.8	Sandy loam and silt loam
True prairie: 3 stations, eastern Nebraska............	4.9	4.0	5.3	Mostly silt-loam soil
Oats:				
Short-grass plains: 3 stations, Colorado and Kansas........	2.4	2.7	3.4	Silt-loam soil
Mixed prairie: 4 stations, Colorado, Kansas, and S. Dakota	2.7	3.4	4.2	Silt and sandy loam
True prairie: 5 stations, eastern Nebraska.............	2.8	3.4	4.4	Silt and clay loam
Wheat (turkey red):				
Short-grass plains: 7 stations, Colorado and Kansas....	2.1	2.3	2.7	Silt loam
Mixed prairie: 4 stations, Colorado, Kansas, and S. Dakota	2.8	3.6	4.4	Silt and sandy loam
True prairie: 6 stations, eastern Nebraska...........	3.3	3.8	5.4	Silt loam

contrasts markedly with plants growing in the true prairie region where the lateral spread of roots ranges from 1 to 4 in. The working depth of the roots of the majority of plants in the short-grass plains region was found to be from 1 to 5 ft. Weaver concluded that the root habit of a species is usually altered greatly when plants are grown under distinctly different conditions and that, in general, the distribution of roots conforms strikingly with that of the soil moisture, although a few species are found that show little or no variation to a change in habitat. The effect of the moisture supply on the depth of root penetration can be illustrated by quoting the data obtained by Weaver (1920) in his examination of the root systems of oats, winter wheat, and winter rye growing in the three regions of the grassland formation (see table above).

The mean annual precipitation was 16.7 in. for the short-grass plains, 18.4 in. for the mixed prairie, and 28.8 in. for the true prairie.

The relative development of the roots and tops of plants growing in the short-grass plains and mixed prairie as compared to those growing in the true prairie region is shown in the following table:

Relative Development of Cereal Crop Plants and Relative Precipitation in the Grassland Area

Plant	Short-grass plain	Mixed prairie	True prairie
Rye:			
Height of tops	56	66	100
Working depth of roots	69	92	100
Maximum depth of roots	65	90	100
Oats:			
Height of tops	85	94	100
Working depth of roots	79	95	100
Maximum depth of roots	77	94	100
Wheat:			
Height of tops	64	85	100
Working depth of roots	61	93	100
Maximum depth of roots	51	80	100
Mean annual precipitation	58	64	100
Precipitation July, 1918 to July, 1919	73	77	100
Precipitation January, 1919 to July, 1919	50	53	100

If the proper conditions of soil texture, moisture, oxygen, and temperature prevail, there is apparently nothing to limit the growth of plant roots downward except the time limit in the length of the life of the plant. The roots of cultivated perennial plants as a rule extend more deeply than those of annuals. The maximum depth of penetration of the roots of corn, sorghums, and the small grains that have been investigated is between 5 and 8 ft. The roots of alfalfa have been observed to grow downward to depths of from 6 to 14 ft. depending upon the age of the plant and the nature of the soil in which they were grown. Cottrell (1902) stated that the roots of alfalfa in a good soil grow to a depth of 30 ft., but he gives no experimental evidence to support his statement. Kiesselbach, Russel, and Anderson (1929) found under the conditions existing in eastern Nebraska that 6-year-old alfalfa drew upon the moisture of the subsoil to a depth of 33 ft., indicating that some of the roots had penetrated to that depth. Duley (1929) at the Kansas Agricultural Experiment Station observed that alfalfa used the moisture of the subsoil to a depth of from 10 to 25 ft., depending upon the age of the plants. Coburn (1907) in his book on alfalfa mentions an instance in which the roots of alfalfa grew downward through the soil into a mining tunnel 129 ft. beneath. This great length was possible in this case, since the roots followed the crevices that were present in the soil and rock overlying the tunnel. The major portion of the root systems of the cultivated plants, however, is found in the upper 4 ft. of soil, and in many cases most of the roots are in the uppermost 2 to 3 ft. of soil. In the case of alfalfa, where the maximum depth of root penetration has been observed to be 16 ft. or more, the major portion of the root system was in the upper 6 ft. of soil. In observations on the roots of sugar cane growing in the Hawaiian Islands, Lee (1926) found that more than 58 per cent of the roots of the plants growing in furrows were above the 8-in. level of the soil and that more than 85 per cent of the roots were in the uppermost 2 ft. of soil. On the other hand, the few roots that extend downward into the maximum depth undoubtedly do absorb considerable moisture and in periods of severe drought may be a very important factor in furnishing sufficient moisture to prevent the death of the plant.

The vertical penetration of the roots of the woody plants that have been studied is relatively shallow. A survey of the table on page 140 shows that the maximum depth of penetration of bur oak, long-leaved pine, apple trees, cherry, grapes, gooseberry, and currant ranged from 1 to 14 ft. The plants, however, that have been studied were those growing for the most part in shallow soil or over a high water table, conditions that are conducive to a shallow development of roots.

2. Lateral Extent.—The lateral extent of the roots of plants will depend upon two main factors: (*a*) the amount of rainfall and (*b*) the thickness of the stand. The difference in the lateral extent of roots of plants growing in the more arid regions as compared with those growing where rainfall is more abundant has been discussed under the heading of the vertical depth of root penetration, but it should be repeated here that the lateral extent of roots is greater the less the rainfall (see Fig. 7*A*). The thickness of the stand is an important factor in determining the lateral extent of roots. A plant growing by itself in the open will extend its roots farther laterally than one that is closely surrounded by other plants. Miller (1916) found that the roots of corn and sorghum which were grown in rows 44 in. apart extended from one row to the other and then turned downward. The roots of a single plant of corn or sorghum under such conditions thus extend through a cylinder of soil 7 ft. in diameter and 6 ft. deep and are distributed through a volume of soil that totals 230 cu. ft. The longest roots observed in this experiment were 10 ft. in length. They were those that extended laterally until they reached the opposite row and then turned downward. From the table on the extent of roots it can be observed that the lateral extent of the roots of the smaller cereals is only 4 to 12 in., due to the fact that the plants are in very close rows. The long lateral extent of the roots of corn and sorghum that were investigated by Georgeson and Payne (1897) was due to the fact that these plants were grown in hills 8 ft. apart.

The lateral extent of the roots of woody plants that have been studied is shown in the table on page 140. These roots show a wide lateral extent which is apparently due to the fact that most of the plants were growing under conditions that prevented or hindered deep vertical penetration. The lateral roots of woody plants extend from two to seven times farther than the spread of their branches. The minimum lateral extent shown in the table is 12 ft., and the maximum is 90 ft.

3. The Depth from Which Roots Absorb Water and Nutrients.—The question concerning the depth from which the roots of plants absorb is a very important one from the standpoint of practical agriculture. The extensive investigations concerning the depth of root penetration have shown that a large proportion of the roots of the cultivated plants extend to 3 or 4 ft., a depth much greater than that from which absorption was supposed to occur. This had led certain investigators to doubt the accuracy of statements that are found in the literature on soils concerning the depth from which plants absorb. Russel (1917) stated that "only the upper 6 or 8 in. of the soil is suited to plant life and the subsoil plays only an indirect part in plant nutrition." There seems to be considerable indirect evidence to refute the above statement, especially from observations of plant life growing in the Great Plains area. In the first place, the number of roots found in any given area of soil is no indication of their activity. It is a common observation that the first roots formed in the upper portion of the soil become suberized or cutinized so that they are totally incapable of absorbing, even if that portion of the soil is moist. It has also been frequently observed that during periods of drought the upper foot of soil has been depleted of its moisture below the wilting coefficient so that little or no absorption occurs there, yet the plant remains turgid and functions normally if there is moisture at the lower depths.

Miller (1916) found, in growing corn and sorghums on fall-irrigated land under seasonal conditions in which the rainfall penetrated the surface mulch little or not at all, that the moisture content of the soil became depleted as the roots penetrated downward. When the corn was tasseling and the sorghums heading, the moisture content of the upper 2 ft. of soil was below the wilting coefficient, yet the plants remained perfectly normal and matured a crop of grain. Determinations at this

Soil-moisture Content and Depth of Root Penetration of Corn, Kafir, and Milo in 1914 at Garden City, Kan.

Date	Percentage of moisture at depth in feet of										Greatest depth of roots, feet		
	1	2	3	4	5	6	7	8	9	10	Corn	Kafir	Milo
June 5	22.9	22.5	22.1	22.7	23.5	21.2	16.1	17.0	16.4	14.5			
July 2	14.6	20.2	21.2	22.8	22.1	21.8	....	....	....	....	1	1.5	1.5
July 10	11.8	17.1	19.5	23.6	24.6	21.4							
July 21	10.6	13.3	14.5	19.4	22.8	21.7	....	....	....	....	3	2.75	3
July 29	8.7	13.1	13.5	16.8	21.4	21.0	....	....	....	....	4	4	4
Aug. 9	9.4	13.5	13.2	14.5	20.3	19.5							
Aug. 22	8.4	13.4	12.4	14.4	19.2	19.7	....	....	....	....	6	6	6
Sept. 6	7.7	12.2	11.4	12.9	15.6	19.1	....	....	....	....	6	6	6
Wilting coefficient	12.7	14.5	14.5	16.3	17.1	16.1	18.2	18.9	16.7	13.0			

time showed that the roots had drawn heavily on the moisture supply of the soil to a depth of 6 ft., as shown in the table on page 146.

Weaver, Jean, and Crist (1922) reported control experiments in which they determined the depth from which certain crop plants could absorb water and nitrates from the soil. They grew the plants in large containers. The soil was tamped firmly in layers separated from each other by wax seals 2 to 3 mm. in thickness, which were composed of a mixture of 85 per cent paraffin and 15 per cent petrolatum. These seals permitted the penetration of the roots as they pushed downward, but they prevented the upward ascent of water and nutrients from the lower portions of the soil to the depleted portions above. Each layer of soil could be analyzed for moisture and nutrients at the close of the experiment, and the depth from which absorption took place could thus be determined. These investigators found that oats absorbed water from a depth of 2.5 ft. and barley from a depth of 3.5 ft. before the blossoming stage. Corn absorbed large quantities of water from the third and fourth foot and some from the fifth foot. Potatoes absorbed water from a depth of 2.5 ft. Barley at the age of 54 days removed 168 to 145 p.p.m. of nitrates from 1- to 1.5- and 1.5- to 2-ft. levels, respectively. When in blossom at the age of 73 days, it had removed 168 p.p.m. from the 2- to 2.5-ft. level and at maturity 186 p.p.m. from the 2.5- to 3-ft. level. Corn removed 203, 140, and 118 p.p.m. at depths of 3, 4, and 5 ft., respectively.

Crist and Weaver (1924) found that barley plants under controlled conditions absorbed nutrients in large amounts at every level to 30 in. The absorption of nutrients at levels below the surface foot materially affects the quantity and quality of the yield, and the additive effect of a nutrient salt present in the lower levels is not lost, even when the surface foot is abundantly supplied with the same nutrients. The effects of the nutritive salts are most marked on the quantity and quality of yield at an early and late period in the development of the plant, *i.e.*, when absorption is effected largely from the first foot of soil as the crop is tillering and again when the younger portions of the longer roots are absorbing from the deeper levels at the time of heading. Millar (1925) studied the relative growth of oats and inoculated sweet clover on the different horizons of Fox sandy loam and Miami silt loam and found indications that different crops may have quite markedly different feeding powers on the various soil horizons.

He also concluded (1925, 1933) that in Michigan the corn plant draws only very sparingly on the soil horizons beneath the surface so that the subsoil contributes little nutrient to the growth of the plant. Aldrich, Work, and Lewis (1935) found that the pear tree extracted approximately 34 per cent of its moisture from the top foot of soil, 62 per cent from the upper 2 ft. and 84 per cent from the upper 3 ft. Approximately 35, 68, and 89 per cent of the total roots were in these respective regions. The most rapid extraction of moisture by the roots occurred between 2 and 8 ft. from the trunk where the greatest number of roots was located. According to Conrad and Veihmeyer (1929), in California, when the oil is not in contact with a free-water surface, the capillary movement of moisture from moist soil to drier soil is too limited in extent to be effective for use by the roots of plants. Unless the depleted moisture be returned to the dry soil by rain or irrigation, additional water can be obtained only by the elongation of the absorbing portion of roots into new moist soil.

This view is substantiated by the observations of Kiesselbach, Anderson, and Russel (1934), in Nebraska, and of Grandfield and Metzger (1936) in Kansas, on alfalfa. They found that within 2 years this plant depleted the soil moisture to a depth of 25 ft., and the restoration of this depleted water to the subsoil is practically impossible except under the most favorable conditions. After such depletion the alfalfa plants must depend on current rainfall for their water supply unless root penetration is sufficient to reach moisture at a lower depth.

4. Rate of Growth.—It was observed by Stevens (1931) in New Hampshire that the lateral roots of 4- to 6-year-old white pines showed no growth from Nov. 15 to Apr. 1 when grown in the open. When plants were grown in the greenhouse, however, the roots grew approximately as fast in winter as in summer. In sandy soil the growth of the roots of 6-year-old pine trees averages 18 to 20 in. annually, while in a clay soil the growth was only 5 to 10 in. There was no correlation between the amount of root growth and the amount of top growth. The average daily growth in the length of the roots from April to November varied from 0.0006 to 0.089 in.

Kinman (1932) found that the average maximum extension of the roots of the plum, peach, and apricot for a 2-week period was 3.2, 3.4, and 4.5 in., respectively. The greatest seasonal activity of roots preceded the critical period of need during fruit setting. It was noted by Heinicke (1935) that the amount of growth of roots of the apple is relatively small during the dormant period as compared to that made during the several weeks preceding leaf fall. He considered that root growth in the fall might considerably increase the absorption of nutrients for use the following spring. Such new roots produced late in the season might be killed by low temperatures during the winter or early spring.

III. THE WEIGHT OF ROOTS

The weight of the roots of some of the more common agricultural plants is here considered. It has been mentioned that the weight of the roots as compared to the weight of the tops will depend upon numerous factors, so that in discussing this subject any data given must be considered relative to the conditions under which the plants were grown. The data here presented were obtained by growing plants in large cans containing approximately 700 lb. of soil that was obtained from the surface foot of soil in the field. The soil was in good tilth and the water content was kept practically constant during the growing season. The quantity of soil was sufficient to produce plants comparable in every way to those growing in the field. The data were obtained at the Kansas Agricultural Experiment Station during the years 1923 to 1926.

Crop	Date	Number of plants per can	Number of observations	Dry weight of stem, leaves, and grain, grams	Dry weight of roots, grams	Weight of roots in comparison to tops, per cent
Corn, Kansas sunflower	1923	1	4	810.5	66.5	8.2
Kafir, Blackhull	1924	1	3	385.4	19.8	5.1
Oats, Kanota	1925	6	5	236.3	17.9	7.6
Wheat, Kanred	1926	6	5	358.5	12.9	3.5
Alfalfa, common (2 years old)	1926	3	3	242.0[1]	36.1	15.0

[1] Total for three cuttings during the season.

If the number of corn plants per acre is taken as 8,000, the dry matter left in the soil by the roots would approximate 1,125 lb. per acre. This together with the weight of the crowns from which the roots originate would make approximately ¾ ton of dry matter left in the soil.

Eaton (1931) found that root development of the cotton plant is largely influenced by the character of its growth and fruitfulness. Thus the weights of the roots of plants without bolls were approximately three times those of the control plants. Willard and McClure (1932) found that the dry weight of the aerial parts of bluegrass grown on unfertilized soil was 4,010 lb. per acre, while the underground parts weighed 2,060 lb. per acre. On fertilized soil the dry weight of the tops and roots was 8,470 and 2,120 lb., respectively, per acre.

Weihing (1935) noted that the secondary root system of corn tended to increase as the tops increased. The dry weight of the tops in the small, medium, and large varieties was respectively 6.4, 4.9, and 3.5 times that of the roots. It was estimated that 445, 1,080, and 1,420 lb. of dry roots per acre were left in the soil by the current crop of these respective plants.

From the experimental evidence at hand we must conclude that the deeper parts of the soil are not only suited to plant life but also play an exceedingly important part in it and deserve a careful consideration in the study of plant production.

IV. ROOT HAIRS

The root tissue that functions in the absorption of materials from the soil is limited to the single outer layer of epidermal cells. Absorption, however, is not carried on over the whole surface of the root system, since after varying periods of time the older portions of the roots become so modified by the deposition of cutin or the formation of cork cells that absorption through them can in no way take place. Usually, then, the absorption area of the root is confined to a definite zone that begins at or near the growing tip and extends backward for varying distances, depending upon the species of plant and the conditions under which it is growing. In most plants a varying number of the epidermal cells of the root undergo a modification that greatly increases their surface of absorption. This modification consists of a tubular outgrowth of the exterior wall of the epidermal cell and is termed a "root hair." The root hair is in no sense a root and it should not be confused with the term "hair roots," which is used to a considerable extent in agricultural literature to denote the small

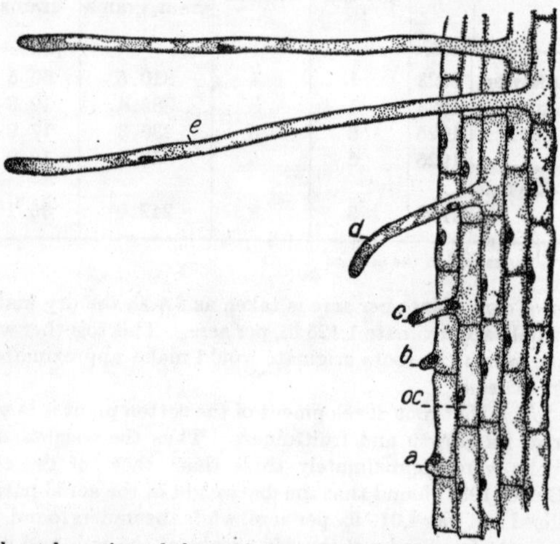

FIG. 8.—Enlarged portion of the epidermis of a root of lupine showing the origin and development of the root hairs. *a, b, c d, e,* successive stages in the development of root hairs from the younger portion of the root backward. *oc,* ordinary epidermal cell.

fibrous roots. The formation of root hairs usually begins at a distance of 1 to 3 mm. back of the growing tip, but in the case of the sedges, their formation begins at the tips of the roots. New root hairs arise in acropetal succession, and new ones are apparently only very rarely interpolated between the preexisting ones (Fig. 8).

Root hairs range in length from a fraction of a millimeter to 10 mm., but their length and general shape depend to a great extent upon the medium in which they are grown. In its passage through the soil the surface of the root hair retains the irregularity in the outlines of the objects to which it adheres. The changes in the shape of the root hair due to the various obstructive particles of the soil bring about a retardation of its general growth so that it often reaches only a small fraction of the length that it would attain if it could develop in moist air. The diameter of root hairs is about 10μ.

Root hairs increase the absorbing surface of the epidermal cells. Schwarz (1883) calculated that the absorbing surface of the roots was increased six times in corn and twelve times in peas by the root hairs that developed on these roots in moist air. He estimated the density of the root hairs on the roots of corn growing in moist air as 425 per sq. mm. while the number on the roots of the pea under the same conditions was 232 per sq. mm.

In most plants, each cell of the root epidermis is capable of forming a root hair, although under average conditions many do not so develop. It should be remembered, however, that even though an epidermal cell does not develop a root hair, it is still an effective organ for absorbing materials from the soil.

Popesco (1926) considered that the absorbing portion of the root lies between the part where elongation begins and the part where the walls of the endodermis are completely suberized. Hohn (1934) decided that the root hairs of *Vicia faba*, *Zea mays*, and *Tradescantia fluminensis* in water cultures are of no significance in the uptake of water under such conditions. Sierp and Brewig (1935) found in the roots of *Vicia faba* that the apical zone of approximately 5 mm. absorbed no water for transpiration and that the next zone without root hairs absorbed only an insignificant quantity. The zone of greatest water absorption was in the region 15 to 80 mm. from the apex. This is the region that ordinarily has the living root hairs.

A. Origin

Ordinarily it is not possible to predict which epidermal cells will develop root hairs. In some plants root hairs arise from a well-defined type of cell. It was first noted by Leavitt (1904) that certain higher cryptogams, monocotyledons, and water plants have short cells, interspersed among the longer epidermal cells of the root, which give rise to the root hairs. He named these short cells "trichoblasts." Bardell (1915) observed on the roots of corn, wheat, squash, beans, and peas, grown under various conditions, that the average length of the cells producing root hairs was less than the hairless epidermal cells. Bartoo (1929) also observed this peculiarity in the development of root hairs in *Schizaea rupestris*.

It was further observed by Cormack (1935) that the epidermis of the roots of certain species of *Brassica* was composed of long and short cells arranged in distinct rows, and that the root hairs were produced from the short cells. Wilson (1936) found that the root hairs of *Elodea canadensis*, *Potamogeton densus*, and *Alisma plantago* are produced from specialized cells. He considered that the two types of cells are differentiated early. The short cells are distinguished in their early stages by thicker walls, larger nuclei, and denser cytoplasm than the other epidermal cells. The differentiation is evident long before the root hairs are developed. The long cells are three or four times longer than the short ones. The proportion of these root-hair-forming cells laid down at the apex of the root is remarkably constant over a wide range of conditions. Cormack (1935) considered that the long cells have a metabolism that differs from that of the short cells. He based this assertion on the facts that their walls did not react with calcium, and that they had a considerably more acid reaction than the hair-forming cells.

B. Conditions Affecting the Formation of Root Hairs

The stimulus that initiates the production of root hairs is not definitely known. Some consider that the retardation of the longitudinal growth of the epidermal cells is necessary for the initiation of the root hairs. Others believe that their formation is an accompaniment of optimal conditions for growth. Both these theories may be correct in part, because the conditions that prevail when root-hair formation occurs may be opposite in different parts of the same cell.

The number of root hairs that will develop will thus depend primarily upon environmental factors. These factors include: the available water, temperature, obstructions, hydrogen-ion concentration, toxic substances, and certain elements.

The frequency of root hairs has generally been expressed as the number per unit area. Wilson (1936) considered that a better indication of the intensity of root-hair development would be an expression showing the proportion of the epidermal cells that produce hairs. Such an expression is analogous to the stomatal index, described in Chap. VII, and may be termed the "root-hair index." This may be expressed after the manner

$$\text{Root-hair index} = \frac{\text{Frequency of root-hair-forming cells}}{\text{Frequency of all epidermal cells}} \times 100$$

Snow (1904, 1905) found that a saturated soil tends to suppress root-hair production. Roots of corn and wheat grown in garden soil kept moist were found well covered with hairs after 7 days. Similar roots growing in a saturated soil for the same period showed few or no root hairs, but after this soil was allowed to dry, root hairs appeared as numerously as in the case of the moistened soil.

Spalding (1904) found better and more root-hair development in the creosote bush growing in a soil comparatively dry than when growing in a soil well supplied with water. Bardell (1915) grew corn, wheat, squash, beans, and peas in water and found that root hairs were produced under these conditions. She found that light conditions have no effect upon root-hair production.

In this connection it should be mentioned that the roots of *Lemna minor*, *Nymphaea alba*, *Hippurus vulgaris*, and others do not develop root hairs, while *Scirpus sylvaticus* and *Carex paludosa* growing under similar conditions produce them fairly abundantly. The roots of *Elodea canadensis*, *Acorus calamus*, *Nuphar luteum*, and others produce no root hairs so long as they are immersed in water but develop them abundantly upon entering the soil. Snow (1904, 1905) observed that high temperature with sufficient moisture tends to decrease the number of root hairs per unit of area by increasing the elongation of the epidermal cells. She also observed that the slower the growth the more root hairs were developed per unit area, and that not only did lower temperatures aid this, but also obstructions in the pathway of the root produced the same results.

Mer (1879, 1884) considered that kinking of the roots is due primarily to growth retardation that is accompanied by a good production of root hairs. Bardell (1915) noted that curves and swellings are favorable to root-hair formation, the hairs being most abundant on the concave side of the root.

The effect of bog water on the production of root hairs of the cuttings of *Tradescantia* was investigated by Rigg (1913). He found that practically no root hairs were produced normally by this plant when grown in bog water from Puget Sound. Of the specimens examined, 18 per cent produced no root hairs, 66 per cent produced hairs that were much stunted, and 16 per cent produced hairs only slightly stunted. He also found that effects on root-hair formation similar to those produced by bog water occurred in each of the following solutions: sea water diluted three times; carbolic acid, 0.001 per cent; formalin, 0.001 per cent; weak tea; and coffee. The root hairs developed normally in the filtered bog water and fairly well in the diluted bog water. Rigg believed that many plants are excluded from bogs on account of their inability to produce root hairs under those conditions, thus having a too limited absorbing surface. Addoms (1923) concluded that the abnormal root development and decreased growth of plants grown in nutrient solutions containing relatively large amounts of potassium dihydrogen phosphate are due to the coagulation of the protoplasm of the root hairs. This coagulation, which is accompanied with flocculation, was found to be induced by the hydrogen ions formed by dissociation of the phosphate. The hydrogen-ion concentration of the nutrient solutions used varied from pH 3.94 to 3.47.

Parker and Sampson (1930) found that when the plants of *Stipa pulchra* and *Bromus hordeaceus* were clipped frequently the roots were devoid of root hairs.

According to Cormack (1935) it has been shown by Magowan (1908), Hansteen-Cranner (1914), Kisser (1925), Farr (1925 to 1928), and Sorokin and Sommer (1929) that the presence of calcium is important for the development of root hairs. Mevius (1926), however, observed that in corn the development of root hairs is not related to the presence or absence of calcium. Pearsall and Wray (1927) found that the root hairs of *Eriophorum angustifolium* tend to be longer and more abundant in solutions of low calcium content.

Cormack (1935) found in studying the roots of Georgia collards that if the amount of calcium available for wall formation is reduced the root hairs frequently become inflated and branched, indicating that the cell walls were too soft to withstand the outward pressure. With further reduction in the amount of calcium there is no formation of root hairs, and the epidermal cells expand more or less symmetrically. The walls of the hairs in solutions minus calcium retain the pectic acid layer. Apparently the hydrogen-ion concentration plays a very important role in the development of root hairs, and this concentration is maintained at optimum by the presence of the proper amount of calcium. Pectic acid and calcium fail to unite if the medium is too acid, and the hydrogen-ion concentration must be such that the reaction between the calcium salt and the pectic acid of the cell takes place gradually. Cormack found that when the roots of Georgia collard were allowed to grow in a solution with sufficient calcium and with a pH of 7.2 to 8.2, the long cells produced typical hairs. When the medium was sufficiently alkaline, the walls hardened so rapidly that no hairs were produced. If the hardening of the walls of the epidermal cells is gradual, vertical elongation is arrested, and, owing to increasing pressure from within, the wall is pushed out at its softest point. In the case of Georgia collards this is at or near the distal end of the cell. The progressive hardening of the outer layer confines this soft spot to a narrow area at the growing tip of the hair where new wall is constantly being laid down. When the short cells begin to develop hairs they showed a pH value of 5.8 to 6.8, while the developing hairs varied from pH 6.6 to 7.2. The long cells in the same region showed a pH of 4.6 to 4.8.

C. Growth of Root Hairs

Haberlandt (1914) observed the manner of the longitudinal growth of the root hairs of *Cucurbita pepo*, *Pisum sativum*, and *Helianthus annuus* by dusting the hairs with rice powder. He found that the longitudinal growth is confined strictly to the dome-shaped tip of the hair. Haberlandt considered that this manner of growth greatly facilitates the power of the root hair to push itself between the solid particles of the soil.

Jeffs (1922) studied the rate of growth of root hairs in moist chambers in the dark at a temperature of approximately 22°C. He found that root hairs grow slowly at first, gradually increasing their rate of growth up to the time root elongation ceases, and that a definite relationship exists between the elongation of the root hair and the parent root. About the time root elongation ceases, the rate of growth of the root hairs in that region becomes stabilized, and there are only slight variations in their rate of growth until about 2 hr. before their growth ceases. The average rate of growth of root hairs of corn, each of which was on a different root, showed an initial rate of 6.5μ per 15-min. interval. This increased to 16.1μ per 15-min. interval at the end of the first hour and 33.6μ per 15-min. interval at the end of the third hour. After this there was no further root elongation in this region, and the root hairs grew at the average rate of 35μ per 15-min. period for the next 3 hr., after which they were studied no longer. It has been demonstrated that the rate of growth of a root hair diminishes very rapidly as it approaches maturity. The approximate length of five root hairs of the radish was 1,824μ at 2 hr. before growth ceased, when they were elongating at the rate of 44.5μ per 15-min. interval. At 1 hr. before growth ceased they were elongating at the rate of 36μ per interval, and at 15 min. before growth ceased their rate of elongation was only 15μ per ¼-hr. period. Growth ceased on an average of 2 hr. after the retardation in their growth rate could be observed.

Jeffs (1925) investigated the influence of light and temperature on the rate of elongation of root hairs and found that those of *Raphanus sativa* and *Sinapis alba* showed no noticeable growth reactions in the light intensities used. The root hairs of these two species grown under a constant temperature within the limits of 17 and 27°C. showed a decided reaction to temperature changes of from 1 to 6° produced within a period of from 5 to 15 min. A drop in temperature resulted in a decrease in the growth rate and sometimes even in a contraction of the root hair. A rise in temperature of 2.5 to 3° within the 5- to 15-min. period resulted in a temporarily increased growth rate, but this soon returned to that which characterized the root hair before the temperature was raised. Jeffs concluded that the effect of temperature upon the rate of growth of tissues is, in part at least, an effect upon the osmotic pressure and imbibition of the cell colloids, rather than simply an effect upon the rate of chemical reactions within the tissues.

Farr (1925) compared the rate of growth of the root hairs of *Tradescantia fluminensis* in tap water and in Knop's solution. The average growth per minute in tap water was 0.916μ, while that in Knop's solution was 0.437μ. The fastest growth recorded in tap water during any 10-min. period was 24μ as compared with 10.8μ in Knop's solution. The average growth per hour for any series in tap water was 91.8μ; while in Knop's solution it was 32.9μ. Farr's experiment shows that substances in solution may modify not only the form of root hairs but also their rate of elongation. Farr (1927 to 1928), in a series of the most extensive experiments yet undertaken on the growth of root hairs in solution, found that the root hairs of the Georgia collards (*Brassica oleracea*) do not grow at all during a period of 13 to 16 hr. after immersion in distilled water. When they do grow, however, they do so at a rate of 60μ per hour in tap water and about 50μ per hour in Knop's solution. They also do not grow at all during the period of 13 to 16 hr. after immersion in concentrations of calcium nitrate below 0.002 M. Their maximum growth rate of 61μ per hour in 0.012 M calcium nitrate is more or less closely approached when growing in concentrations between 0.007 and 0.012, but it drops abruptly at higher concentrations to 40μ at 0.013 M. Farr also observed that the graph representing the effects of varying concentrations of calcium chloride upon the growth rate of root hairs is almost identical with that for calcium nitrate in form and in the location of the optimum molar concentration. The importance of calcium in the root hair and in root growth is attributed largely to its being the only constituent of the cell wall that is of direct external origin, although it must be recognized that calcium also plays a part in antagonism, hydrogen-ion concentration, and permeability. In a study of the effect of the hydrogen-ion concentration in calcium solutions upon root-hair growth, it was found that the maximum elongation of the root hairs of the Georgia collards was 109.8μ per hour in 0.020 *M* calcium chloride at pH 6.9. The hydrogen-ion concentration range for root-hair elongation was from 3.4 to 11.9, varying with the concentration of the solution. It was also concluded that the entire root responds to changes in the hydrogen-ion concentration in much the same way as do the root hairs. Farr states that, in producing root hairs, roots exhibit the power of adaptation to a wider range of hydrogen-ion concentration than any other organ, organism, or physiological process that has thus far been studied, although enzymes have been found to tolerate a still wider range.

D. The Longevity of Root Hairs

The length of life of root hairs depends on the species of the plant and upon the moisture conditions of the soil. Root hairs of some species live for only a few days or weeks, while those of others exist for several years, although they do not apparently function in absorption for a period that long.

McDougall (1921) observed that the root hairs of *Gleditsia triacanthos* became thick walled and brown in color and that they persist as long as the root epidermis, which may be for several months. The older root hairs examined by him had a cell-wall thickness of 2μ, which is approximately four times the thickness of the cell wall of the root hairs of this plant when they are first formed. These hairs are only about 0.4 mm. in length and are stiff and rigid and do not shrivel upon exposure to the air even after several months. McDougall found that the root hairs of *Gymnocladus dioica* and *Cercis canadensis* sometimes become thick walled and brown, but that this is not nearly so characteristic of these as it is of *Gleditsia triacanthos.* Whitaker (1923) confirmed the work of McDougall and found that the root hairs in certain species of the *Compositae* persist for as long as 3 years. Weaver (1925) found in the case of wheat growing in the soil under field conditions that the root hairs did not begin to slough off from the oldest parts of the primary root until 7 weeks after planting and that at the end of 10 weeks more than 99 per cent of the root area still possessed functioning root hairs. In this case the soil was in excellent condition for their retention, but under conditions of a low water supply in the upper portion of the soil not only the older root hairs but also many of the more delicate branch roots would perish.

At 6-month intervals, Stoddart (1935) examined the roots of a group of prairie grasses 2 in. below the soil surface. During a 2-year period these roots were subjected to all variations of soil moisture from below the wilting coefficient to saturation, and to temperatures from 0° to 112°F. In all the plants observed the roots lived for at least a year and many in excess of 2 years, some new roots being produced each season.

E. The Cell Wall of Root Hairs

Schwarz (1883) found that the cell wall of the root hair of *Taxus baccata* or yew is of two parts—an inner layer that stains blue with chloriodide of zinc and an outer layer that stains yellow brown. He concluded that the contact between the root hairs and the soil particles is rendered more permanent and intimate by the gelatinization of the outermost layers of the root membrane. Hesse (1904) found that the thickness of the cell wall of the root hair of any plant varied with the medium in which it was grown.

The composition of the cell wall of the root hairs has been variously reported. Apparently it depends to a considerable degree upon the conditions prevailing at the time of their formation.

Ridgway (1913) observed the occurrence of callose in the cell wall of certain root hairs. Roberts (1916) found in a majority of the forms investigated that the walls of root hairs growing in moist air are made up of two parts—an inner membrane that gave the cellulose reaction with 75 per cent sulphuric acid and the potassium iodide iodine solution and an outer membrane that stained red with ruthenium red denoting pectic compounds. That this membrane is calcium pectate was shown by the fact that on the addition of ammonium oxalate the membrane broke down and calcium oxalate crystals were formed. Callose was observed at the tip of the root hairs in mustard, pea, and *Tradescantia* amongst others. The relative amount of the cellulose and pectin found in the cell walls of various root hairs is shown in the following table:

Plant	Nature of the cell wall of the root hair	
	Inner layer	Outer layer
Alfalfa	Thin cellulose	Thick pectin
Barley	Thin cellulose	Thick pectin
Corn	Thick cellulose	No pectin
Cabbage	Thin cellulose	Thin pectin
Pea	Thick cellulose	Thick pectin
Tobacco	Thin cellulose	Thin pectin
Mustard	Thin cellulose	Thin pectin

The calcium pectate membrane on the outer wall of the epidermal cell and the root hair is a continuation of the middle lamella of the other walls of the cell. There is thus a secretion of calcium pectate around the entire cell, but the layer on the external wall is thicker than that on the other walls of the cell. The layer of pectic material is thinner toward the tip than it is at the base of the root hair. The high efficiency of the root hair as an absorbing organ is thus due to the fact that the soil particles are held to it by this pectic substance.

Howe (1921) examined the root hairs of 20 economic plants growing in sand and in loam soil. Some of the plants examined were beans, cabbage, corn, cucumber, radish, peas, squash, tomatoes, and watermelons. She found no cellulose in the cell wall of the root hairs grown under these conditions. The wall of the root hair contained a layer of pectic material on the outside and on the inside a layer of callose that varied in thickness for different plants, but in all cases it was a little thicker at the tips than in the lateral walls. As the wall of the epidermal cell bulges to form the root hairs, the inner layer, which is composed of cellulose, apparently stretches to its capacity and breaks and no more cellulose is formed. The outer pectic layer of the wall, however, is not ruptured but is extended out to form the root hair. Howe found that the root hairs formed either in loam or in sand gave an acid reaction in all cases, but the root hairs produced in loam soil were usually somewhat the more acid. The pH values of the root hairs ranged from 6.0 to 6.8 in most cases. Whether the acidity of the root hair can be ascribed to the presence of pectic materials or to some other cause was not determined.

According to Cormack (1935) it was reported by Ziegenspek (1920) that the inner layer of cellulose in the cell wall of the root hair was replaced by amyloid at the tip, and he considered that this plastic amyloid made possible the growth of the hair. Hopmann (1930), however, reported the absence of amyloid from root hairs that had developed in water, but reported the presence of cellulose under all conditions over the whole inner surface of the wall of the hair. Cormack found in tomato, radish, endive, white mustard, corn, and Georgia collards that the cell wall of root hairs grown in moist air and in tap water had layers of cellulose and pectic materials continuous with corresponding layers in the epidermal cells. Tests were negative for all other wall constituents. On the dome-shaped tip of the root hair the pectic layer was composed either of pectic acid or of a softer layer of calcium pectate than the mature portion of the wall.

V. EXCRETIONS OF ROOTS

From the discussions in Chap. II concerning the factors affecting the permeability of the cell membranes, it can readily be conceived that numerous conditions, both internal and external, might arise under which the outward diffusion of the cell contents of the roots into the soil might occur. It is not the intention here to consider the possible loss of the inorganic salts from the roots but to discuss more particularly the excretion of those compounds that are waste products or synthetic substances of the protoplasm.

The two topics to be considered are (*a*) the excretion of acids, and (*b*) the excretion of enzymes.

A. Acids—Carbonic Acid

Liebig (1858) first observed the dissolving action of roots on limestone, and his observations were verified by Sachs in 1860. Sachs advanced two theories as to the cause of this dissolving action. He considered that it was brought about either by the liberation of carbon dioxide or by the formation of acid due to the decomposition of the cell walls of the cells of the root. Czapek (1896) studied the excretion of acids from roots by the use of plates made with mixtures of aluminum phosphate and plaster of Paris. Plates of this composition are soluble in hydrochloric acid, nitric acid, sulphuric acid, orthophosphoric acid, formic, oxalic, succinic, lactic, malic, citric, and tartaric acids but are insoluble in carbonic, acetic, propionic, and butyric acids. These plates showed no signs of etching when the roots of various plants came in contact with them, from which it was considered that none of these acids was excreted by the roots under the conditions of the experiment. Plants were grown in Congo red, which becomes brownish red when acted upon by carbon dioxide and bright blue through the action of acetic and propionic acids. The results showed that carbon dioxide alone was excreted by the roots. Kunze (1906) also observed that free mineral acids are not excreted from the roots of higher plants, and Haas (1916) showed that carbonic acid was the only acid excreted by the roots of sweet corn. Stocklasa and Ernest (1909) found that if the roots of plants have a normal oxygen supply no acids either organic or inorganic except carbonic acid are excreted. If, however, the oxygen supply is cut off or is limited, organic acids only are produced. They found that the roots of buckwheat and barley under limited supplies of oxygen produced formic and acetic acids, while the roots of oats and maize under like conditions excreted formic acid, and the roots of beets, oxalic acid. These organic acids arise from incomplete oxidation and under normal conditions of aeration would be oxidized to carbon dioxide and water.

1. Relative Amounts.—The evidence indicates that under normal conditions carbonic acid is the only acid liberated by the living cells of the roots. Coupin (1918) considered that this was excreted more particularly by the superficial cells of the rootlet between the root hairs and the root tip and by that portion from which the root hairs have disappeared.

Carbon dioxide is an end product of respiration, a process that is going on continuously in every living cell, so that the amount of carbon dioxide thus formed will depend, then, upon the environmental conditions and upon the nature of the plant. Since the intensity of respiration varies directly with the intensity of metabolic activity, it would also be expected that the amount of carbon dioxide excreted by the roots would vary during the growing season of that plant.

Kossowitch (1904) found that the roots of mustard in pots and water cultures produced increasing amounts of carbon dioxide up to the time of blooming, and Barakov (1910) showed that the carbon dioxide content of the soil was insignificant at the beginning of vegetative growth but increased rapidly and reached its maximum at the blooming period and declined rapidly to the minimum at the ripening period. He observed that the greater the plant growth the more carbon dioxide there was in the soil. Most of this came from the respiration within the roots and varied with different crops and different stages of growth. With lupines the maximum was reached at the blooming stage and with oats about 2 weeks before blooming. The maximum production of carbon dioxide by the roots of potatoes was reached after blooming, while its excretion by sugar beets reached its highest point at the period of maximum growth.

Lau (1906) observed that potatoes and legumes gave off more carbon dioxide than other crops, while Stocklasa and Ernest (1905) found that red clover, beet, and oat roots produced more carbon dioxide in the order named than the other plants considered.

Russell and Appleyard (1917) observed a considerable increase in the production of carbon dioxide in the soil during the period of rapid growth of the crop and another and larger increase at the time of ripening. Bizzell and Lyon (1918) found the greatest apparent production of carbon dioxide by oats to be at the time of blooming, after which a marked decrease occurred.

Stocklasa (1934) found that the evolution of carbon dioxide by the roots was highest with leguminous plants and lowest with wheat and barley. Turpin (1920) found that the oat crop increased the production of carbon dioxide in the soil. This increase became marked from the first month after seeding and increased to a maximum just previous to or after these plants had headed, following which decline took place. Millet produced about the same amount of carbon dioxide as oats. A cropped soil after harvest maintained a higher carbon dioxide content than bare soil, and it was considered that this was due to the decomposition of roots.

Smith and Brown (1931, 1932) considered that the production of carbon dioxide in the soil under crops is such a complex procedure that the amount present in the soil cannot serve as an index to the production of this gas by the roots.

Turpin considered that plants at their most active period of growth produce many times as much carbon dioxide as is produced by soil organisms, and that the excess of carbon dioxide in the soil of growing crops is due to the respiratory activity of the roots of the plants rather than to the decay of root particles from the crop. Newton (1924), however, considered that there is good evidence to prove that the quantity of carbon dioxide produced in soils by the decomposition of organic matter is greater than the quantity given off by the respiration of the roots of the growing crop.

2. Solvent Action.—The solvent power of the carbonic acid excreted by the roots upon the minerals of the soil is a matter of question. Hall (1906) considered that carbonic acid is the only acid excreted by the plant roots, and that it is capable of dissolving everything needed by the plant. Aberson (1910) considered that the mucilaginous covering of the root hairs contains a saturated solution of carbon dioxide sufficient to bring into solution the insoluble soil constituents with which it comes in contact, especially the phosphates. Pfeiffer and Blanck (1912), however, showed that in soils treated with phosphates the carbon dioxide given off by the roots of the plant does not form a solvent sufficiently powerful to account for all the mineral nutrients obtained by the plant from the soil. Fred and Haas (1919) studied the etching power of the roots of Canada field peas upon polished marble in sterile soil and in soil that had been inoculated with various bacterial cultures other than those of the nitrifying bacteria. It was found that in every case the marble slabs that had the greatest degree of etching were those in the soil that had been inoculated. They considered that the greater etching power of these roots in the presence of bacteria can be attributed to the normal carbon dioxide excretions of the living cells together with the carbon dioxide and acids set free from dead or dying root cells whose decomposition is accelerated by the presence of bacteria.

Parker (1924) is of the opinion that the carbon dioxide excreted by the roots is more effective in bringing plant food into solution within reach of the plant than a unit of carbon dioxide produced in the soil by the decomposition of organic matter. He, however, could find no relation between the carbon dioxide production of roots and the feeding power of the plants for calcium, magnesium, phosphorus, and potassium in a poor sandy soil. More carbon dioxide was excreted by the roots of cowpeas than by those of any other plant, while buckwheat roots gave off very little carbon dioxide but had the greatest feeding power of any of the plants used. For each gram of carbon dioxide excreted by the roots, buckwheat absorbed 41.5 mg. of calcium, while sorghum, cowpeas, and soybeans, respectively, absorbed 5.0, 12.7, and 21.2 mg. of this element for each gram of carbon dioxide excreted by the roots. It has long been recognized that species of plants differ widely in their ability to grow on infertile soils.

It is well known that beans, rye, and buckwheat can grow apparently to good advantage on poor land, while pennyroyal, sassafras, poverty grass, and others are indicators of impoverished soil upon which plants like corn, barley, wheat, blue grass, and others grow poorly or do not thrive at all. Dustman (1925) studied the rate of excretion of carbon dioxide from the roots of these two types of plants. He found in his experiments that the production of carbon dioxide whether estimated from the standpoint of dry matter of roots or unit area of root system was greater for buckwheat and rye than for corn and barley. He is of the opinion that the output of carbon dioxide in connection with the relative size and area of root systems is an important factor in aiding certain kinds of plants to secure the essential elements from the soil.

Stocklasa (1934) found that, when plants were grown in pulverized minerals, to which had been added a sufficient amount of nitrogen, the absorptions of phosphorus and potassium were in the order oats > rye > wheat > barley. The low absorption by barley was correlated with the smaller excretion of carbon dioxide by the roots of that plant. He believed that plants growing in radioactive soil can probably utilize insoluble minerals to a greater degree than similar plants in inactive soils.

Metzger (1928) also considered that the carbon dioxide excreted by roots exerts a measurable solvent action on soil nutrients. He used the concentration of bicarbonates as a means for determining the action of the carbon dioxide excreted by the roots. In 66 determinations with corn, kafir, wheat, and other plants, 53 showed higher concentration of bicarbonate in samples from around roots, 4 showed equal, and 9 lower concentrations than in samples taken at short distances from the root.

B. Enzymes

The question of the excretion by the roots of substances that are capable of oxidizing, reducing, or digesting materials in the soil has been investigated and is worthy of note. Molisch (1887) first demonstrated the oxidizing power of intact roots. He showed that there was considerable active secretion on the surface of grown roots and considered that this secretion had definite powers to effect changes in organic substances. Raciborski (1905) found that all the roots of the phanerogams examined by him had the property of extracellular oxidation. Schreiner and Reed (1909) considered that plant roots are able to carry on active extracellular oxidation chiefly by means of enzymes that they secrete. These investigators used wheat plants of the varieties Harvest Queen and Chul and found that their roots were able to oxidize organic substances such as chromogens at a fairly rapid rate. When the roots were grown in solutions containing α-naphthylamine, benzidine, vanillin, vanillic acid, or esculin, insoluble colored oxidation compounds were yielded which were deposited mainly upon the surface of the roots. Color changes that indicated oxidation were also observed when plants were grown in solutions containing phenolphthalein, aloin, and leuco-rosolic acid. The oxidizing power appeared to be most energetic in the region of the root where the root hairs are found and increasingly less active as that portion of the root becomes older. Schreiner and Reed found that the roots had greater oxidizing power in extracts of productive soil than in those of unproductive soil and that the presence of toxic organic substances in solution was extremely deleterious to the oxidizing power of the roots. Schreiner and Sullivan (1911) found that the roots carry on oxidation and reduction concurrently. The oxidation processes are weak in the young seedling but increase in strength as the seedling grows. The reduction processes are predominant in the early stages of the seedling growth but are less manifest at a later stage. Reduction is mostly intracellular, while oxidation, which seems to be the more prominent of these two properties of the roots, is mostly extracellular. The reducing power of the roots of grasses has also been observed by Sani (1919), and the oxidizing power of roots upon ammonium ferrosulphate has been observed by Borowski (1919). He also considered that this oxidizing ability is located in the region of the root hairs and at the growing tips. He found that the oxidizing ability of the roots of different plants varies. In his experiments the roots of *Sinapis* showed the weakest, while *Phaseolus* showed the greatest power of oxidation. Lyon and Wilson (1921) working with nutrient solutions in which corn, oats, peas, and vetch were growing could detect reducing substances by some tests but not by others. They were not able, however, to show to their satisfaction that oxidizing substances were present in these solutions. Neller (1922), on the other hand, found in all his experiments with soybeans, field peas, and buckwheat that these plants had a beneficial influence upon the oxidation activities in the substrata in which the plants were grown and suggested a symbiotic relationship between the soil oxidizing organisms and the roots of these growing plants.

The ability of the plant to absorb from the soil more or less complex organic compounds makes the excretion of digestive enzymes by the roots an interesting question. This subject has been investigated by Knudson and Smith (1919) and by Knudson (1920). They found that neither corn nor Canada field peas are capable of utilizing soluble starch directly or indirectly, nor is there any appreciable excretion of diastase by the roots of these plants under the conditions of the experiments. When these plants were grown in the presence of sucrose there was an increase in the amount of reducing sugars in the cultural solution, but incubation experiments gave negative results with respect to the presence of invertase. Knudson considered that the increase in reducing sugars in his experiments was due to their excretion from the roots into the solution rather than to the secretion of invertase with a consequent digestion of nonreducing sugars.

C. Organic Substances

The theory that plant roots excrete or eliminate into the soil certain waste products that are especially injurious to other plants is an old one. De Candolle (1832) advocated a theory of crop rotation on the basis of root excretions in which he claimed that plants excreted from their roots substances that are harmful to succeeding crops of closely related plants. His theory was based partly on his own observations and partly on the statement of earlier writers. Experiments carried out according to his directions tended rather to disprove than to support his hypothesis. The subject of root excretions was given very little attention until about 1900, when the matter was investigated by Pickering in England and by the

Bureau of Soils of the U. S. Department of Agriculture. The investigators in this department were at first inclined to favor the theory of root excretions to account for the deleterious substances that occur in soils. Later, however, it was considered that these detrimental substances might be caused by the decomposition of vegetable matter. In the case of growing plants, especially, it is considered that the decomposition of the contents of the disintegrating epidermal cells and the cells of the root cap, which are continually being lost into the soil, might produce the compounds that are toxic to other plants.

Fletcher (1912) reported that the roots of corn excreted a substance that was detrimental to the development of the plants of sesamum growing in adjoining rows. Shull (1932) grew plants of sesamum and sorghum together and noted that their roots intertwined without any injurious effects upon either plant. He concluded therefore that the roots of these plants did not secrete any substance that was detrimental to the growth of the other.

It is now rather well established that the roots of plants do not excrete substances that are beneficial or detrimental to either accompanying or succeeding crops. However, there are some cases in which the roots excrete substances that serve some definite purpose. Thus Saunders (1933) demonstrated that the host plants of the witchweed (*Striga lutea*) exude from their roots a substance that is necessary for the germination of the witchweed seeds. The precise nature of this substance is not known, but it is easily destroyed by high temperatures and loses its activating power on exposure to air. The liberation of this substance is not dependent upon the mineral nutrition of the host. Its origin is due apparently to the metabolic activity of the protoplasm of the host cells. Its action is upon the cell contents of the seed of the witchweed and not upon the seed coats.

VI. BENEFICIAL AND TOXIC EFFECTS OF ROOTS

It has been known from the earliest times that one species of plant may have a detrimental or beneficial influence upon other plants of the same or different species that are growing in close proximity to it or that follow it in succession. Let us consider first the beneficial influence of one plant upon another.

A. The Beneficial Influence of One Plant on Another

It has been known for several thousand years that grapes grow well in association with the elm tree and that certain grains produce better when grown in combination than when they are grown separately. Furthermore, mixed cropping apparently had a prominent place in primitive agriculture.

Dandeno (1908, 1909) grew squash and corn seedlings together and separately and found that the growth when grown together was greater than when grown separately. He also found that the growing of Canada thistle with oats, wheat, and barley was beneficial to them. Zavitz (1922) grew oats, barley, spring wheat, and peas separately and in various combinations for the production of grain for six years. Six mixtures having two classes of grain in each mixture, four having three classes of grain in each mixture, and one having all four classes of grain in combination were used each year and varieties were selected that matured uniformly. It was found with but one exception in the 11 mixtures that the grains grown in combination gave a greater yield per acre than the average production of the same grains when grown separately. The yield of straw was the highest in every instance in the case of the mixed grains. The different classes of grains exerted an influence on the productiveness of the mixtures in the following order: oats, barley, peas, and wheat—oats having the greatest effect and wheat least. It was found after 15 years of experimentation that the greatest yield of grain per acre in the case of barley and oats was produced by using a mixture of 48 lb. of barley and 34 lb. of oats per acre. The use of mixtures of different varieties of the same grain showed no beneficial results over the grains when grown by themselves.

The yields of grain and straw of the different mixtures and when grown separately are shown in the following table:

Varieties of grain grown in mixtures	Yield per acre			
	Straw, tons		Grain, pounds	
	Grown separately	Grown in mixture	Grown separately	Grown in mixture
Barley and oats	1.56	1.74	1,935	2,261
Barley, peas, and oats	1.47	1.67	1,489	2,101
Barley, wheat, and oats	1.47	1.72	1,683	2,067
Peas and oats	1.52	1.77	1,873	1,988
Barley, peas, wheat, and oats	1.43	1.71	1,682	1,955
Wheat and oats	1.52	1.68	1,624	1,921
Peas, wheat, and oats	1.44	1.73	1,642	1,860
Barley and peas	1.33	1.56	1,740	1,760
Barley, peas, and wheat	1.32	1.57	1,553	1,665
Wheat and barley	1.33	1.41	1,491	1,558
Peas and wheat	1.29	1.37	1,429	1,322

Lyon and Bizzell (1913) spaced plants of wheat, oats, and timothy so that they would not interfere with one another. After they had attained some size, seeds of different kinds of plants were placed between them. In the majority of the tests a larger yield of the primary crop resulted in the mixtures than when it was grown alone. These increases amounted to as little as two per cent and as much as 52 per cent. The plants that were particularly active in promoting better growth of the primary crop were mustard, buckwheat, maize, and red clover. In growing wheat and other plants in Knop's solution, Mukerji (1920) found that the solutions which had previously grown a set of plants supported better growth than the fresh solutions. It was noted by Peralta and Estioko (1923) that *Cyperus* and water lily (*Monochoria hastata*) are beneficial to the production of rice. Relevant to the possible production of beneficial substances by the roots of plants, it should be mentioned that Krogh and Lange (1930) showed that fresh-water algae lost into the surrounding water approximately 10 per cent of the organic matter synthesized, and that most of this organic matter came from dead and decomposed cells of these plants.

It was also noticed by Morrish (1934) that oats and barley produced a higher yield when they were grown together than they produced on an equal area, of which half was used for pure oats and the other half for pure barley.

Arny, Stoa, McKee, and Dillman (1929) under certain conditions and in some regions thought that mixtures of flax with wheat, oats, or barley, was beneficial in producing a greater yield but that under other conditions it was of no advantage.

It was reported by Thornton and Nicol (1934) that Italian rye grass grown in the presence of lucerne in sand for 18 weeks contained 2.25 times as much nitrogen as did the grass similarly grown except for the absence of lucerne. In another case, 0.33 gram of sodium nitrate was added per pot, and after a growth of 13 and 18 weeks the rye grass contained 2.5 and 5.5 times as much nitrogen, respectively, as was supplied by the added nitrate. They made no attempt to trace the course of the supposed transfer of nitrogen from this legume to the nonlegume. According to these authors, Virtanen (1933) found in pure sand, in which inoculated legumes were growing, the same percentage of combined nitrogen as occurred in the plants themselves.

According to Nicol (1935) it was observed by Kaserer (1911) that there was little or no interpenetration of the roots of plants of the same species when grown together, but that there was a marked interpenetration of roots when the plants of two different species were so grown. The maximum interpenetration was shown when a legume and nonlegume were grown together. Nicol found that the roots of alfalfa and of grass growing in sand were difficult to separate when these plants were grown together, but that no such difficulty was experienced when these plants were grown in single cultures.

B. The Deleterious Influence of One Plant on Another

It has long been a matter of common observation that the soil becomes more and more unproductive for a given crop that is grown continuously upon it, and also that the preceding crop may have a detrimental effect upon the succeeding one of a different species. And it has also been observed, especially in comparatively recent years, that plants growing in close relationships may have a detrimental influence on each other that cannot be wholly attributed to the competition of these plants for the raw materials of the soil and air or for the supply of sunlight.

Pickering (1903, 1917) and Bedford and Pickering (1914, 1919) reported their observations dating from 1898 upon the deleterious effects of one plant upon another. They considered that it is established with a reasonable amount of certainty that the detrimental effect of one growing plant upon another is of general occurrence. They developed a method for growing two plants in a single pot in order to study the influence of one upon the other. One of the plants was grown in a layer of soil contained in a tray that was superimposed on the soil in which the other plant was growing. The water was applied to the tray so that the lower soil received the drainage from the upper one. They found that pears, plums, apples, cherries, certain forest trees, mustard, tobacco, tomatoes, barley, clover, and some grasses are susceptible to this detrimental effect, which may be exerted by apple seedlings, mustard, tobacco, tomatoes, clover, and 16 varieties of grasses. This deleterious effect varies greatly but is especially manifest in the reduction of growth, which varies from 6 to 97 per cent. Fruit trees that are affected by the harmful products of certain grasses not only show reduced growth but also exhibit alterations in the coloring of the bark, leaves, and fruit that are unlike those attending other forms of ill-treatment. These investigators found the above results so manifest that expert fruit growers were unable to name correctly the varieties thus affected. Pickering (1917) considered that different individual plants of the same species when grown together exert a deleterious effect upon one another. For example, it has frequently been observed that when the seeds of a given plant are sown and then 3 or 4 days later more seeds of the same kind are sown among those already in the ground, the plants resulting from the seeds first sown outdistance those from the later-sown seeds in their general vigor and rate of growth out of all proportion to their respective ages. In the case of mustard, a difference of only 4 days in the date of planting showed 3 months later a difference of 60 to 70 per cent inferiority in growth of the plants from the seeds sown later. In one case, a difference in germination of 4 days of one-half of the mustard seeds in a given plot showed a reduction in yield by weight of 20 per cent.

Hedrick (1905) found that young peach trees were hindered in their development when grown in the same pot cultures with oats, potatoes, tomatoes, mustard, and rape even though the optimum conditions of moisture, aeration, and food supply were maintained. Dandeno (1908, 1909) observed that the growth of buckwheat was considerably deterred by Canada thistles when the two plants were grown together. He also observed that poplars and oaks were injurious to the growth of corn. The injurious effects extended to some distance from the trees, a fact which showed that shade and moisture were not the deciding factors. It has also been observed by Jones and Morse (1902), Cook (1921), Massey (1925), and Schneiderhan (1926) that walnut trees have a detrimental effect upon the growth and vigor of various plants. These trees prevent the growth of cinquefoil (*Potentilla fruticosa*) within a large area surrounding them and cause the wilting, stunting, and frequently the death of corn, tomatoes, potatoes, alfalfa, and apple trees when their roots come in contact or close proximity to those of the walnut.

It was found by Prianischnikov (1914) that when plants were grown in sand cultures there was a marked decrease in the amount of the harvest of the second and third crops. Perlata and Estioko (1923) found that zacate (*Leersia hexandria*), the most common forage plant in the Philippines, is detrimental to the rice that follows it. Rice is also detrimental to the rice crops that follow.

It is a common observation in the Great Plains area of the United States (Sewell, 1923) that wheat planted in the fall after the various types of sorghums (*Andropogon sorghum*) does not yield so well as wheat planted after corn. At the Kansas Agricultural Experiment Station, winter wheat grown after a kafir yielded three bushels less per acre per year over a period of six years than when grown after corn.

The probable causes for the detrimental influence of one plant or crop on another have been attributed to two general factors: (1) the depletion of the nutrients in the soil so that there is not a sufficient supply for the plants that follow, and (2) the production in the soil, by the decomposition of roots, stems, leaves, and the cells that are lost from the growing root, of compounds that are deleterious to the roots of the plants with which they come in contact. Let us now consider the evidence for the two general theories.

1. The Influence of the Depletion of the Nutrients in the Soil on Plant Growth.—It was noted by Livingston, Schreiner, Skinner, and Reed (1905 to 1912) that frequently there is no correlation between the mineral composition of soils and their crop-producing power, since the amount of nutrients removed from a given soil by a crop is not sufficient to account for the diminished yield of the following crop.

Hartwell and his workers (1918, 1927) have reported extensive work in regard to the influence of certain crops on those that follow. The general plan of their experiment was to grow 16 different crops on that number of parallel plots for two successive seasons and then every third year to grow one of these crops over the entire area. The plants that were grown were onion, potato, mangel beet, rutabaga, cabbage, buckwheat, corn, millet, oats, rye, carrot, redtop, timothy, alsike clover, and red clover. The crops that were grown uniformly over all the plots were onions in 1910, buckwheat in 1913, and alsike clover in 1916. The yield of onions following the various crops will serve as an example of the general facts observed. This was from 13 to 17 bu. per acre following cabbage, mangel beets, rutabaga, turnips, and buckwheat; 35 to 87 bu. following rye and potatoes; 131 to 178 bu. following corn, millet, onions, oats, and red clover; 240 to 314 bu. following squash, timothy, and alsike clover; and 406 to 412 bu. following mixed timothy and redtop and redtop alone. These field experiments were checked by pot experiments in the greenhouse, and the following conclusions drawn:

> The divergent effect of crops on those which follow seems not to be attributable at least principally to differences in the amount of nutrients removed by the crops grown previously. That is, the smallest yields may not occur after the crop which removed the largest amount of even the most needed nutrient.

They also found that the soil acidity was affected differently by the crops under consideration and that in the case of the onion, a plant that is sensitive to acid conditions, the best yields followed the crops giving rise to the least acidity. It was also found that the divergent effect of crops on those that follow was less if the soil acidity was corrected by liming. Garner (1925) and his workers from extensive studies on tobacco culture considered that the tobacco plant is particularly sensitive to the effects of preceding crops. It may be seriously retarded as a result of the effects of preceding crops of tobacco itself or of various other plants. Under intensive methods of production, when tobacco follows tobacco, reduced yields are not due, proportionately, to a reduction in the general productiveness of the soil.

Odland and Smith (1933) over a period of 23 years grew 10 uniform crops following each of 16 different crops. In some years very striking differences were obtained in the yields of the uniform crops following the various preceding ones, but at other times these differences were not significant. They considered that several factors were operative in causing these differences in growth. The variations in soil acidity caused by the growth of different crops showed sufficient correlation with the subsequent growth of other plants to prove that this is an important factor. In the case of corn, the relative removal of nitrogen by the preceding crops was correlated with yield but the results with other crops were not conclusive with respect to this factor.

It was observed by Wilkins and Hughes (1934) in Iowa that corn yielded 50.8, 49.0, and 47.8 bu. per acre after Sudan grass, soybeans, and oats, respectively. The Sudan grass produced over twice as much dry matter as soybeans and oats, removed over twice as much nitrogen as oats and considerably more than soybeans. Regardless of this, however, the Sudan grass plots after 13 years of cropping appeared to be just as productive as the plots of soybeans and oats. They believed that the same results would not have been obtained on infertile soils. Dodd and Pohlman (1935) in West Virginia found that the yield of buckwheat and potatoes was not affected differently by preceding crops of oats and soybean. Oats following oats, buckwheat, or corn were poorer than oats following wheat and potatoes. Oats, buckwheat, and potatoes were similar to wheat in their effects on the succeeding crop of soybeans. Russell and McDonald (1935) reported from England and Australia that continuous cropping of wheat is not detrimental to the soil nor is there any diminishing return, provided that fertilizer is supplied in sufficient quantities. However, on the incompletely fertilized plots there has been a falling-off in yield.

It is now generally accepted, at least frequently, that the detrimental effect of a crop on the succeeding one is not due to the exhaustion of the nutrients of the soil by the preceding crop.

2. The Influence of the Decomposition of Roots in the Soil on Plant Growth.—It is now universally accepted that the productivity of the soil may be decreased by reason of more or less complex organic substances that may be produced in it. Livingston, Britton, and Reid (1905) and Livingston (1909) found that the unproductive soils which they examined contained soluble organic substances that were toxic to wheat plants. They observed that these soils were greatly benefited without any alteration of the nutrient conditions by the use of practically insoluble bodies such as ferric hydrate and carbon black which apparently rendered the harmful substances inactive. Tannic acid and pyrogallol also benefited these soils. This beneficial action they considered due to the chemical combination of these compounds with the toxic substances, which so altered them that they became noninjurious to the plants. They considered that the beneficial effects derived from the use of stable manures on unproductive soils are due more to the addition of organic matter, which alters the toxic substances of the soil, than to the increased amount of salts added to the soil. Breazeale (1906) showed that the extracts of certain soils were injurious to wheat plants when grown in water cultures and that this toxicity could be overcome wholly or in part by the treatment of the extract with carbon black, calcium carbonate, ferric hydroxide, and other solids.

Schreiner, Skinner, Shorey, Reed, and their coworkers (1907 to 1912) in some of the outstanding work of the century in organic chemistry have studied the effects and nature of the organic compounds of the soil. They developed means of extraction and identification of many of these substances. The following are some of the compounds of the soil that have been isolated and identified: dihydroxystearic acid, picolinecarboxylic acid, agroceric acid, hydroxy-fatty acids, paraffinic acid, lignoceric acid, resin, glycerides of fatty acids, phytosterol, pentosan, pentose, histidine, arginine, pyrimidine, oxalic acid, succinic acid, purine nucleic acid, xanthine, choline, adenine, creatine, tyrosine, vanillin, cumarin, and others. Some of these compounds were found to be noninjurious and even beneficial to plant growth, while others were found to be injurious. Among those which were injurious were cumarin, vanillin, dihydroxystearic acid, alanine, glycocoll, neurine, quinone, pyridine compounds, and tyrosine, the degree of injury depending upon the nature and concentration of the substances under observation. These investigators found that when absorbing agents or certain fertilizers were used, the toxicity of the above-mentioned compounds was greatly reduced. The various fertilizer salts acted differently in overcoming the harmful effects of certain compounds. To neutralize cumarin, phosphate fertilizers were the most effective. In the case of vanillin, nitrogenous fertilizers were the most efficient antidoting agents, while potassium salts were most effective in overcoming the toxic action of quinone. The action of fertilizers on nonproductive soils is not definitely known, but the results observed add weight to the evidence against the assumption that their effect is due altogether to the increase in plant nutrients that they add to the soil.

A glance at the list of the organic compounds that have been isolated from the soil shows that many of these substances are those which commonly result from the decomposition of animal and vegetable matter. The sources of the organic matter in the soil are the roots and stubbles of the plants which have previously grown in it and the cells which may be lost from the root tips of the plants growing therein. The kind of products that may arise from the decomposition of the organic matter in the soil will thus depend upon many and varied factors. Among these are the kind of plant that produced the organic matter, since each species of plant is made up of more or less specific compounds, and the age of the plants (Waksman and Tenney, 1928). The products produced will depend upon the amount of moisture in the soil, the thoroughness of aeration, the character of the microorganisms in the soil, and the mutual relation of the organic compounds and the mineral constituents. The factors that have just been mentioned are in turn influenced by cultural practices, fertilizer, drainage, irrigation, and inoculation. Thus the resulting compounds from plant material in the process of decomposition may vary greatly so that the substances which may arise from a given type of plant material under one set of conditions may be very toxic to plants, while those which may arise under another set of conditions may not injure the following plants at all. The aeration of the soil is a very important factor in determining or overcoming the toxic properties of many organic compounds. Pickering (1917) found that if the soil leachings that were toxic to plants were exposed to the air for 24 hr. their toxic effect disappeared entirely, and these leachings afterward became very beneficial. Schreiner and his coworkers (1906 to 1912) found that the detrimental influence of many of the organic compounds of the soil could be overcome by the addition of certain substances to oxidize more thoroughly the compounds in question and suggested that the beneficial effect of fertilizers upon plants might be due in part to their capacity as oxidizers.

The idea that the stubble and roots of certain plants may give rise to decomposition products that will be injurious to the crops which follow has been strengthened by the work of Sewell (1923), Breazeale (1924), Collison (1924), and Hawkins (1925). As has previously been mentioned, wheat does not yield so well when grown after a crop of sorghum as it does when grown after corn. In pot cultures where wheat received its moisture solely from water percolating through perforated trays (after the method of Pickering) in which corn and kafir were growing, the growth of the kafir plants inhibited the growth of the wheat. When kafir and corn were grown in these percolating trays, harvested, and then wheat planted so that it would receive the leaching from the trays, the kafir soil leachings depressed the growth of the wheat. From this Sewell concluded that there are decomposition products from the crop residue of kafir which have a retarding effect on the growth of the wheat that follows. Breazeale (1924) by growing wheat in water cultures containing the chopped-up pieces of the stubble of kafir found that a toxic property was developed during the decomposition of the stubble that was injurious to the wheat plants. He observed that this toxic body or bodies were shortly either volatilized or decomposed into nontoxic substances. The decomposition of the stubble is hastened by an increase in temperature, and a good nutrient solution can be produced from finely chopped sorghum stubble in 3 or 4 weeks. From a mixture of water and stubble that was toxic to wheat plants, Breazeale distilled off a poisonous compound that would kill wheat plants within a few hours. Hawkins (1925) found that the stalk residues of the sorghums were more detrimental to plants that follow than those from the roots. He also observed that certain crop plants such as field peas do not use the upper stratum of soil as do the sorghum plants and are not depressed in growth following a crop of sorghum so much as other crops such as vetch, which have root systems more like sorghums. The detrimental effects of sorghum disappear in a few months after the crop has been harvested. It was found by Conrad (1927) that sorghum roots may contain as much as fifteen times the amount of sugar as the roots of corn. When this is liberated into the soil after harvesting, it stimulates the growth of certain microorganisms which thus compete with the following plants for nitrogen and probably for other elements. This released sugar may also retard the process of nitrification. He considered, then, that the immediate cause of the deleterious effects of sorghums on crops that follow may be due to one or more elements becoming a limiting factor. He believed that the result can be overcome to a great extent by a system of cultivation that would lead to the rapid decay of the roots. The work of Wilson and Wilson (1928) suggests that the injurious aftereffects of sorghum may be associated with the comparative ease with which the roots are oxidized in the soil. This process, which is accompanied by an increase in the number of soil organisms and an increase in the assimilation of nitrate nitrogen, would tend to deplete the soil of available nitrogen.

McKinley (1931) found indications that sorghum residues, under favorable conditions and not in excess of 2 tons per acre, increase the yield of crops over those of untreated soils. The increased yield from these residues may or may not be less than that from corn or other plants. He considered that part of the decreased yield of crops following sorghums may be partially explained on the basis of the readily decomposable materials in the residues.

Watery extracts of cereal straws, timothy residues, and alfalfa were found by Collison (1924) to be very toxic to plants grown in them, and he considered that there is no doubt but that these and other plant residues are important contributors to the soil of compounds toxic to plant growth. Whether the amounts of material thus contributed to the soil would ever become toxic to a crop in the field would depend upon the nature of the soil and the treatment that it receives. Collison further suggests that if these toxic compounds are present in the fresh sap they might serve as inhibiting agents against parasitic fungi.

Davis (1928) in a search for the substances that might cause the toxic action of walnut trees upon plants growing near them succeeded in extracting from the hulls and roots of the walnut a substance that is apparently identical with juglone. This extract when purified and crystallized proved to be an exceedingly powerful toxin when injected into the stems of tomato and alfalfa plants.

The oxidation reduction system of juglone to hydrojuglone was studied by Davis (1931). The determination was made on the root-hair cells of *Trianea bogotensis*, and the data indicated that juglone would be reduced if injected into these cells. It is thought that an explanation of the resistance to walnut toxicity might be associated with the reduction of the toxic juglone to hydrojuglone, its nontoxic reductant. Whether this is the substance that exerts the toxic action by the roots in nature, however, is not known.

The problem of the toxic or beneficial effect of one plant on another is a complicated one and the causes are yet in doubt. A summary of this subject can best be stated in the words of Thatcher (1923):

> There is yet no positive proof of the nature of the causative agent or agencies for either the beneficial or injurious effect of one plant on another. It may vary widely in different cases and may be chemical or bacterial in character. Definite proof that observed injurious effects on a second crop are due to toxic chemical substances in the soil produced by or in association with the first crop has not been established.

Bibliography

ABERSON, J. H., 1910: Ein Beitrag zur Kenntnis der Natur der Wurzelausscheidungen, *Jahrb. wiss. Bot.*, **47**: 41–56.

ADDOMS, R. M., 1923: The effect of the hydrogen ion on the protoplasm of the root hairs of wheat, *Am. J. Bot.*, **10**: 211–220.

———, 1927: Toxicity as evinced by changes in the protoplasmic structure of root hairs of wheat, *Am. J. Bot.*, **14**: 147–165.

ALBERT, W. B., and G. M. ARMSTRONG, 1931: Effects of high soil moisture and lack of soil aeration upon fruiting behavior of young cotton plants, *Plant Physiol.*, **6**: 585–591.

ALDRICH, W. W., R. A. WORK, and M. R. LEWIS, 1935: Pear root concentration in relation to soil moisture extraction in heavy clay soil, *J. Agr. Res.*, **50**: 975–988.

ALLAN, H. H., 1926: The surface roots of an individual matai, *N. Z. J. Sci. and Tech.*, **8**: 233–234. Quoted by Weaver, 1932.

ALLISON, R. V., and J. W. SHIVE, 1923: Studies on the relation of aeration and continuous renewal of nutrient solution on the growth of soybeans in artificial culture, *Am. J. Bot.*, **10**: 554–567.

ARKER, J., 1901: Die Beeinflussung des Wachsthums der Wurzeln durch das umgebende Medium, *Bot. Cent.*, **87**: 433–434.

ARNY, A. C. and I. J. JOHNSON, 1928: The roots of flax plants, *J. Am. Soc. Agron.*, **20**: 373–381.

———, T. E. STOA, C. MCKEE, and A. C. DILLMAN, 1929: Flax cropping in mixture with wheat, oats and barley, *U. S. Dept. Agr. Tech. Bull.* 133.

BARAKOV, P., 1910: The carbon dioxide content of soils during different stages of growth of plants, *Zhur. Opuitn. Agron.* (*Russ. J. Expt. Landw.*), **11**: 321–343 (Russian); *Exp. Sta. Rec.*, **23**: 523.

BARDELL, E. M., 1915: Production of root hairs in water, *Univ. Wash. Pub. Bot.*, **1**: 1–9.

BARTOO, D. R., 1929: Origin and development of tissues in root of *Schizaea rupestris*, *Bot. Gaz.*, **87**: 642–652.

Beckenbach, J., and J. H. Gourley, 1932: Some effects of different cultural methods upon root distribution of apple trees, *Proc. Am. Soc. Hort. Sci.*, **29**: 202–204.

Beckwith, M. H., 1885: Report of the assistant horticulturist, *N. Y. Sta. Rept.*, 261–267

Bennett, M. E., 1904: Are roots aerotropic? *Bot. Gaz.*, **37**: 241–259.

Bergman, H. F., 1920: The relation of aeration to the growth and activity of roots and its influence on the ecesis of plants in swamps, *Ann. Bot.*, **34**: 13–33.

Biswell, H. H., 1935: Effect of environment upon the root habits of certain deciduous forest trees, *Bot. Gaz.*, **96**: 676–708.

Bizzell, J. A., and T. L. Lyon, 1918: The effect of certain factors on the carbon dioxide content of soil air, *J. Am. Soc. Agron.*, **10**: 97–113.

Borowski, R., 1919: Beitrag zur Kenntnis des Oxydationsvermögens der Wurzeln der höheren Pflanzen, *Landw. Versuchs-Stat.*, **94**: 265–284.

Bouygues, H., 1928: À propos de la nécessité de l'oxygène de l'air sur l'accroissement des racines poussant dans l'eau, *Bull. Soc. Bot. France*, **75**: 469–472.

Breazeale, J. F., 1906: Effect of certain solids upon the growth of seedlings in water cultures, *Bot. Gaz.*, **41**: 54–63.

———, 1924: The injurious after effects of sorghum, *J. Am. Soc. Agron.*, **16**: 689–701.

———, 1930: Maintenance of moisture equilibrium and nutrition of plants at and below the wilting percentage, *Ariz. Sta. Tech. Bull.* 29.

Bryant, A. E., 1934: A demonstration of the connection of the protoplasts of the endodermal cells with the Casparian strips in the roots of barley, *New Phytol.*, **33**: 231.

Büsgen, M., and E. Münch, 1929: "The Structure and Life of Forest Trees," pp. 269–300. John Wiley & Sons, Inc., N. Y. Quoted by Weaver and Kramer, 1932.

Cannon, W. A., 1911: Root habits of desert plants, *Carnegie Inst. Wash. Pub.* 131.

———, 1912: Physical relations of roots to soil factors, *Carnegie Inst. Wash. Yearbook*, **11**: 61–62.

———, 1913: Botanical features of the Algerian Sahara, *Carnegie Inst. Wash. Pub.* 178.

———, 1915: On the relation of root growth and development to the temperature and aeration of the soil, *Am. J. Bot.*, **2**: 211–224.

———, 1916: Distribution of the cacti with especial reference to the role played by the root response to soil temperature and soil moisture, *Am. Nat.*, **50**: 435–442.

———, 1916: On the relation between the rate of root growth and the oxygen of the soil, *Carnegie Inst. Wash. Yearbook*, **15**: 74–75.

———, 1917: Root growth of *Prosopis velutina* and *Opuntia versicolor* under conditions of a small oxygen supply in the soil, *Carnegie Inst. Wash. Yearbook*, **16**: 82–83.

——— and E. E. Free, 1917: The ecological significance of soil aeration, *Sci.*, **45**: 178–180.

———, 1918: Root growth in desert plants and the oxygen supply of the soil, *Carnegie Inst. Wash. Yearbook*, **17**: 81–83.

———, 1923: The influence of the temperature of the soil on the relation of roots to oxygen, *Sci.*, **58**: 331–332.

———, 1924: A note on the relation of root growth in the soil to the oxygen supply: The growth ratio, *Ecology*, **5**: 319–322.

———, 1925: On the upper critical concentration of oxygen in root growth, *Sci.*, **61**: 118–120.

———, 1925: Physiological features of roots with especial reference to the relation of roots to the aeration of the soil, *Carnegie Inst. Wash. Pub.* 368.

———, 1932: Absorption of oxygen by roots when the shoot is in darkness or in light, *Plant Physiol.*, **7**: 673–684.

———, 1932: Experimental studies on the relation of roots to oxygen and on the aeration of plant tissues, *Carnegie Inst. Wash. Yearbook*, **31**: 294–296.

———, D. Demaree, and E. A. Purer, 1933: Evaporation, transpiration and oxygen consumption by roots, *Sci.*, **78**: 388–389.

Carlson, F. A., 1925: The effect of soil structure on the character of alfalfa root systems, *J. Am. Soc. Agron.*, **17**: 336–345.

Clark, W. S., 1874: Observations upon the phenomena of plant life, 22d *Ann. Rept. Mass. State Board Agr.*, 204–312.

Clements, F. E., 1920: Plant Indicators, *Carnegie Inst. Wash. Pub.* 290.

———, 1921: Aeration and air content. The role of oxygen in root activity, *Carnegie Inst. Wash. Pub.* 315.

Collings, G. H., and J. D. Warner, 1927: Root development of cotton on Cecil sandy loam during 1926, *J. Am. Soc. Agron.*, **19**: 839–842.

Collison, R. C., 1924: The presence of certain organic compounds in plants and their relation to the growth of other plants, *J. Am. Soc. Agron.*, **17**: 58–68.

———, 1935: Lysimeter investigations. IV, Water movement, soil temperatures, and root activity under apple trees, *N. Y. Sta. (Geneva) Tech. Bull.* 237.

Conrad, J. P., 1927: Some causes of the injurious after effects of sorghums and suggested remedies, *J. Am. Soc. Agron.*, **19**: 1091–1111.

——— and F. J. Veihmeyer, 1929: Root development and soil moisture, *Hilgardia*, **4**: 113–134.

Cook, M. T., 1921: Wilting caused by walnut trees, *Phytopath.*, **11**: 346.

Cormack, R. G. H., 1935: Investigations on the development of root hairs, *New Phytol.*, **34**: 30–54.

Cottrell, H. M., 1902: Growing alfalfa in Kansas, *Kans. Sta. Bull.*, **114**: 61–62.

Crider, F. J., 1933: Selective absorption of ions not confined to young rootlets, *Sci.*, **78**: 169.

Crist, J. W., and J. E. Weaver, 1924: Absorption of nutrients from the subsoil in relation to crop yield, *Bot. Gaz.*, **77**: 121–149.

——— and G. J. Stout, 1929: Relation between top and root size in herbaceous plants, *Plant Physiol.*, **4**: 63–85.

Curry, J., R. Platt, and S. F. Trelease, 1935: Does dilute heavy water influence biological processes? *Sci.*, **81**: 275–277.

Czapek, F., 1896. Zur Lehre von den Wurzelausscheidungen, *Jahrb. wiss. Bot.*, **29**: 321–390.

Dandeno, J. B., 1908, 1909: Mutual interaction of plant roots, *Rept. Mich. Acad. Sci.*, **10**: 32–36; **11**: 24–25.

Davis, E. F., 1928: The toxic principle of *Juglans nigra* as identified with synthetic juglone, and its toxic effects on tomato and alfalfa plants, *Am. J. Bot.*, **15**: 620.

———, 1931: Observations from the microinjection of acid-base and oxidation reduction indicators into the root-hair cells of *Trianea bogotensis*, *Am. J. Bot.*, **18**: 896.

Dean, B. E., 1933: Effect of soil type and aeration upon root systems of certain aquatic plants, *Plant Physiol.*, **8**: 203–222.

De Candolle, A. P., 1832: *Physiologie végétale*, **1**: 248–251; **3**: 1474–1475, 1493–1520, Paris.

Dittmer, H. J., 1937: A quantitative study of the roots and root hairs of a winter rye plant (*Secale cereale*), unpublished data, *Iowa State College.*

Dodd, D. R., and G. G. Pohlman, 1935: Some factors affecting the influence of soybeans, oats and other crops on the succeeding crop, *W. Va. Sta. Bull.* 265.

Duley, F. L., 1929: The effect of alfalfa on soil moisture, *J. Am. Soc. Agron.*, **21**: 224–231.

———, 1929: Unpublished data, *Kans. Sta.*

DUNCAN, W. H., 1935: Root systems of woody plants of old fields of Indiana, *Ecology*, **16**: 554–567.

DUSTMAN, R. B., 1925: Inherent factors related to absorption of mineral elements by plants, *Bot. Gaz.*, **79**: 233–264.

EATON, F. M., 1931: Root development as related to character of growth and fruitfulness of the cotton plant, *J. Agr. Res.*, **43**: 875–883.

ECKERSON, S., 1914: Thermotropism of roots, *Bot. Gaz.*, **58**: 254–263.

EWART, A. J., 1894: Oxytropic irritability of roots, *Trans. Liverpool Biol. Soc.*, **8**: 238; **10**: 190.

FARR, C. H., 1925: Root-hair elongation in Knop's solution and in tap water, *Am. J. Bot.*, **12**: 372–383.

———, 1925: The formation of root hairs in water, *Proc. Iowa Acad. Sci.*, **32**: 157–165.

———, 1927: Studies on the growth of root hairs in solutions. I, The Problem, previous work, and procedure, *Am. J. Bot.*, **14**: 446–457.

———, 1927: Studies on the growth of root hairs in solutions. II, The effects of the concentration of calcium nitrate, *Am. J. Bot.*, **14**: 497–515.

———, 1927: Studies on the growth of root hairs in solutions. III, The effect of concentrations of $CaCl_2$ and $Ca(OH)_2$, *Am. J. Bot.*, **14**: 553–565.

———, 1928: Studies on the growth of root hairs in solutions. IV, The pH-molar rate relations for collards in calcium chloride, *Am. J. Bot.*, **15**: 6–32.

———, 1928: Studies on the growth of root hairs in solutions. V, Root-hair elongation as an index of root development, *Am. J. Bot.*, **15**: 103–114.

———, 1928: Studies on the growth of root hairs in solutions. VI, Structural responses to toxic pH and molar concentrations of calcium chloride, *Am. J. Bot.*, **15**: 171–179.

———, 1928: Studies on the growth of root hairs in solutions. VII, Further investigations on collards in calcium hydroxide, *Bull. Torrey Bot. Club*, **55**: 223–247.

———, 1928: Studies on the growth of root hairs in solutions. VIII, Structural and intracellular features of collards in calcium nitrate, *Bull. Torrey Bot. Club*, **55**: 529–553.

FARRIS, N. F., 1934: Root habits of certain crop plants as observed in the humid soils of New Jersey, *Soil Sci.*, **38**: 87–112.

FLETCHER, F., 1912: Toxic excreta of plants, *J. Agr. Sci.*, **4**: 245–247.

FRANK, B., 1893: Die Assimilation des freien Stickstoffs durch die Pflanzenwelt, *Bot. Ztg.*, **51**: 139–156.

FRED, E. B., and A. R. C. HAAS, 1919: The etching of marble by roots in the presence and absence of bacteria, *J. Gen. Physiol.*, **1**: 631–639.

FREE, E. E., and B. E. LIVINGSTON, 1917: The effect of deficient soil oxygen on the roots of higher plants, *Johns Hopkins Univ. Circ.*, n. s., **3**: 182–185.

GARNER, W. W., W. M. LUNN and D. E. BROWN, 1925: Effects of crops on the yields of succeeding crops in the rotation with special reference to tobacco, *J. Agr. Res.*, **30**: 1095–1133.

GAUTHERET, R., 1933: Artificial cultures of corn root tip meristems, *C. R. Acad. Sci. Paris*, **197**: 85–87; *Exp. Sta. Rec.*, **73**: 23.

GEMMER, E. W., 1928: The root system of a long-leaf pine, *Sci. Month.*, **27**: 384.

GEORGESON, C. C., and J. E. PAYNE, 1897: Investigations of the root development of some forage plants, *Kans. Sta. Bull.* **75**: 212–222.

GERICKE, W. F., 1923: Further notes on effect of extent of root systems on tillering of wheat, *Bot. Gaz.*, **75**: 320–321.

GIBSON, R. G. H., 1912: The extent of the root system of *Cucumis sativus*, *Ann. Bot.*, **26**: 951–952.

GOFF, E. S., 1883–1888: Report of the horticulturist, *N. Y. Sta. 2d Ann. Rept.*, 219; *3d Ann. Rept.*, 305–315 (1884); *5th Ann. Rept.*, 159–168 (1886); *6th Ann. Rept.*, 90–96 (1887); *7th Ann. Rept.*, 171–178 (1888).

GORDIENKO, M., 1930: Über die Beziehungen zwischen Bodenbeschaffenheit und Wurzelgestaltung bei jungen Pflanzen, *Landw. Jahrb.*, **72**: 125–139; *Biol. Abst.*, **6**: 2510.

GRANDFIELD, C. O., and W. H. METZGER, 1936: Relation of fallow to restoration of subsoil moisture in an old alfalfa field and subsequent depletion after reseeding, *J. Am. Soc. Agron.*, **28**: 115–123.

GURSKY, A. V., 1928–1929: The root systems of *Fraxinus excelsior*, *Fraxinus Pennsylvanica* and *Acer Negundo* on the black soils of Kuban, *Bull. App. Genetics and Plant Breeding*, **21**: 145–181.

HAAS, A. R. C., 1916: The excretion of acids by roots, *Sci.*, **44**: 685.

———, 1917: The excretion of acids by roots, *Proc. Nat. Acad. Sci.*, **2**: 561–566.

HABERLANDT, G., 1914: "Physiological Plant Anatomy," Eng. transl., 222 pp., Macmillan & Company, Ltd., London.

HALL, A. D., 1906: The solvent action of roots upon the soil particles, *Sci. Prog. Twentieth Cent.*, **1**: 51–57.

——— W. E. BRENCHLEY, and L. M. UNDERWOOD, 1914: The soil solution and the mineral constituents of the soil, *J. Agr. Sci.*, **6**: 278–301.

HANSTEEN-CRANNER, B., 1914: Über das Verhalten der Kulturpflanzen zu den Bodensalzen. III, *Jahrb. wiss. Bot.*, **53**: 536–598. Quoted by Cormack, 1935.

HARMON, F. N., and E. SNYDER, 1934: Grape root distribution studies, *Proc. Am. Soc. Hort. Sci.*, **32**: 370–373.

HARRIS, F. S., 1914: The effect of soil moisture, plant food and age on the ratio of tops to roots in plants, *J. Am. Soc. Agron.*, **6**: 65–75.

HARRISON, W. H., and P. A. S. AIYER, 1913: The gases of swamp rice soils; their composition and their relation to the crop, *Mem. Dept. Agr. in India, Chem. Ser.*, **3**: 65–106.

HARTWELL, B. L., and S. C. DAMON, 1918: The influence of crop plants on those which follow. I, *R. I. Sta. Bull.* 175.

———, F. R. PEMBER, and G. E. MERKLE, 1919: The influence of crop plants on those which follow. II, *R. I. Sta. Bull.* 176.

———, J. B. SMITH, and S. C. DAMON, 1927: The influence of crop plants on those which follow. III, *R. I. Sta. Bull.* 210.

HÄVELER, W., 1892: Ueber die Verwerthung des Humus bei der Ernährung der chlorophyllführenden Pflanzen, *Jahrb. wiss. Bot.*, **24**: 283–316.

HAWKINS, R. S., 1925: The deleterious effect of sorghum on the soil and on the succeeding crops, *J. Am. Soc. Agron.*, **17**: 91–92.

HAYS, W. M., 1888: Roots of clovers, *5th Bien. Rept., Univ. Minn., Suppl.* I, *Dept. Agr. Exp. Sta.*, 188–198.

———, 1894: The roots of crops, *N. Dak. Sta. Bull.* **10**: 47–50.

HEADDEN, W. P., 1896: Alfalfa, *Colo. Sta. Bull.* **35**: 47–59.

HEDRICK, U. P., 1905: The relationship of plants in the orchard, *Proc. West N. Y. Hort. Soc.*, *50th ann. meet.*, 17–28.

HEINICKE, A. J., 1932: The effect of submerging the roots of apple trees at different seasons of the year, *Proc. Am. Soc. Hort. Sci.*, **29**: 203–207.

———, 1935: Root growth in young apple trees made shortly before and after defoliation, *Proc. Am. Soc. Hort. Sci.*, **33**: 164–165.

HELLRIEGEL, H., 1883: Wurzel und Bodenvolumen, *Beiträge zu den naturwiss. Grundlagen des Ackerbaus*, 119–280.

HENDRICKSON, A. H., and F. J. VEIHMEYER, 1931: Influence of dry soil on root extension, *Plant Physiol.*, **6**: 567–576.

HESSE, H., 1904: Beiträge zur Morphologie und Biologie der Wurzelnhaare, *Inaug. Diss.*, Jena.

HEYWARD, F., 1933: The root system of long-leaf pine on the deep sands of western Florida, *Ecology*, **14**: 136–148.

HÖHN, K., 1934: Die Bedeutung der Wurzelhaare für die Wasseraufnahme der Pflanzen, *Z. Bot.*, **27**: 529–564.

HOLCH, A. E., 1931: Development of roots and shoots of certain deciduous tree seedlings in different forest sites, *Ecology*, **12**: 259–298.

HOLE, R. S., 1918: Recent investigations on soil aeration, *Indian Forester*, **1918**: 202.

HOOKER, H. D., 1914: Thermotropism in roots, *Plant World*, **17**: 135–153.

———, 1915: Hydrotropism in roots of *Lupinus albus*, *Ann. Bot.*, **29**: 265–285.

HOPMANN, O., 1930: Über das Vorkommen und die Mikrochemie des Amyloids in pflanzlichen Zellhauten, *Inaug. Diss.*, Münster. Quoted by Cormack, 1935.

HOWARD, A. and G. L. C. HOWARD, 1915: Soil ventilation, *Agr. Res. Inst. Pusa Bull.* **52**: 1–35.

———, 1916: Soil aeration in agriculture, *Agr. Res. Inst. Pusa Bull.* **61**: 1–22.

——— and G. L. C. HOWARD, 1917: The economic significance of the root development of agricultural crops, *Agr. J. India, Special Indian Sci. Congo Number.*

———, 1918: Recent investigations on soil aeration, *Indian Forester*, **1918**: 187.

HOWE, C. G., 1921: Pectic materials in root hairs, *Bot. Gaz.*, **72**: 313–320.

HUNTER, C., 1912: Some observations on the effect of soil aeration on plant growth, *Proc. Phil. Soc. Univ. Durham*, **4**: 183.

———, and E. M. RICH, 1925: The effect of artificial aeration of the soil on *Impatiens balsamina*, *New Phytol.*, **24**: 257–271.

HUTCHINS, L. M., and B. E. LIVINGSTON, 1923: Oxygen-supplying power of the soil as indicated by color changes in alkaline pyrogallol solution, *J. Agr. Res.*, **25**: 133–140.

———, 1926: Studies on the oxygen-supplying power of the soil together with quantitative observations on the oxygen-supplying power requisite for seed germination, *Plant Physiol.*, **1**: 95–150.

JEAN, F. C., and J. E. WEAVER, 1924: Root behavior and crop yield under irrigation, *Carnegie Inst. Wash. Pub.* 357.

JEFFS, R. E., 1922: The grand period of growth of root hairs, *Proc. Okla. Acad. Sci.*, (Okla. Univ.) **2**: 53–56.

———, 1925: The elongation of root hairs as affected by light and temperature, *Am. J. Bot.*, **12**: 577–606.

JENSEN, J. H., 1931: Some studies on root habits of sugar cane in Cuba, *Sci. Contr. Tropical Plant Res. Found.* 21.

JOHNSTON, C. O., and E. C. MILLER, 1934: Relation of leaf rust infection to yield, growth, and water economy of two varieties of wheat, *J. Agr. Res.*, **49**: 955–981.

JONES, L. H., 1931: Flower pot composition and its effect on plant growth, *Mass. Sta. Bull.* **277**: 147–161.

——— and H. D. HASKINS, 1935: Distribution of roots in porous and nonporous plant containers, *Plant Physiol.*, **10**: 511–519.

JONES, L. R., and W. J. MORSE, 1902–1903: The shrubby cinquefoil as a weed, 16*th Ann. Rept. Vt. Sta.*, 188–190.

KASERER, H., 1911: Beobachtungen über die Bewurzelung der Kulturpflanzen bei Reinsaat und bei Mischsaat, *Z. landw. Versuchs. in Oesterreich*, **14**: 1022. Quoted by Nicol, 1935.

KERR, T., 1933: The injection of certain salts into the protoplasm and vacuoles of the root hairs of *Limnobium spongia*, *Protoplasma*, **18**: 420–440; *Biol. Abst.*, **9**: 750.

KIESSELBACH, T. A., 1910: Transpiration experiments with the corn plant, *Nebr. Sta. 23rd Ann. Rept.*, 125–139.

———, J. C. RUSSEL, and A. ANDERSON, 1929: The significance of subsoil moisture in alfalfa production, *J. Am. Soc. Agron.*, **21**: 241–268.

———, A. ANDERSON, and J. C. RUSSEL, 1934: Subsoil moisture and crop sequence in relation to alfalfa production, *J. Am. Soc. Agron.*, **26**: 422–442.

———, ———, and W. E. LYNESS, 1935: Cultural practices in corn production, *Neb. Sta. Bull.* 293.

——— and R. M. WEIHING, 1935: The comparative root development of selfed lines of corn and their F_1 and F_2 hybrids, *J. Am. Soc. Agron.*, **27**: 538–541.

KING, F. H., 1892–1893: Natural distribution of roots in field soils, *Wis. Sta. Rept.*, 112–121; 160–167 (1893).

KINMAN, C. F., 1932: A preliminary report on root growth studies with some orchard trees, *Proc. Am. Soc. Hort. Sci.*, **29**: 220–224.

KISSER, J., 1925: Über das Verhalten von Wurzeln in feuchter Luft, *Jahrb. wiss. Bot.*, **64**: 416–439. Quoted by Cormack, 1935.

——— and I. BEER, 1934: Chemotropic sensitivity of dicotyledonous seedlings, *Jahrb. wiss. Bot.*, **80**: 301–335; *Chem. Abst.*, **29**: 7384.

KLERCKER, J. A., 1891: Über calorotropische Erscheinungen bei einigen Keimwurzeln, *Oefvers. Vetensk. Akad. Förhandl. Stockholm*, **48**: 765–790.

KNIGHT, R. C., 1924: The response of plants in soil and in water culture to aeration of the roots, *Ann. Bot.*, **38**: 305–325.

KNUDSON, L., and R. S. SMITH, 1919: Secretion of amylase by plant roots, *Bot. Gaz.*, **68**: 460–466.

———, 1920: The secretion of invertase by plant roots, *Am. J. Bot.*, **7**: 371–379.

KOSSOWITCH, P., 1904: The quantitative determination of carbon dioxide produced by the roots of plants during the period of their development (Russian), *Zhur. Opuitn. Agron.*, **5**: 482–493 (quoted by Turpin, 1920).

KRASSOVSKY, I., 1929: The root system of different varieties of flax, *Bull. App. Bot. and Plant Breeding*, **22**: 43–100.

KROGH, A., and E. LANGE, 1930: On the organic matter given off by algae, *Biochem. J.*, **24**: 1666–1671.

KUNZE, G., 1906: Ueber Säureausscheidung bei Wurzeln und Pilzhyphen und ihre Bedeutung, *Jahrb. wiss. Bot.*, **42**: 357–393.

LAIRD, A. S., 1930: A study of the root systems of some important sod-forming grasses, *Fla. Sta. Bull.* 211.

LAITAKARI, E., 1927: Männyn juuristo, morfologinen tutkimus, *Acta Forestalia Fennica*, **33**: 1–380. Quoted by Weaver and Kramer, 1932.

LAU, E., 1906: Beiträge zur Kenntnis der Zusammensetzung der in Ackerboden befindlichen Luft, *Inaug. Diss.*, Rostock.

LEAVITT, R. G., 1904: Trichomes of the root in vascular cryptogams on angiosperms, *Proc. Boston Soc. Nat. Hist.*, **31**: 273–313. Quoted by Cormack, 1935.

LEE, H. A., 1926: The distribution of the roots of sugar cane in the soil in the Hawaiian Islands, *Plant Physiol.*, **1**: 363–378.

LIEBIG, J., 1858: Ueber einige Eigenschaften der Ackerkrume, *Ann. Chem. Pharm.*, **105**: 109–144.

LILIENFELD, M., 1905: Ueber den Chemotropismus der Wurzel, *Ber. bot. Ges.* **23**: 91–98.

LIVINGSTON, B. E., J. C. BRITTON, and F. R. REID, 1905: Studies on the properties of an unproductive soil, *U. S. Bur. Soils Bull.* **28**.

———, C. A. Jensen, J. F. Breazeale, F. R. Pember, and J. J. Skinner, 1907: Further studies on the properties of unproductive soils, *U. S. Bur. Soils Bull.* 36.

———, 1909: An advance in root physiology, *Plant World*, **12**: 127–130.

———, 1923: Some physiological aspects of soil toxicity, *J. Am. Soc. Agron.*, **15**: 313–323.

Loehwing, W F., 1931: Effect of soil aeration on growth and root development, *Proc. Iowa Acad. Sci.*, **38**: 71–72.

Loomis, W. E., and L. W. Ewan, 1936: Hydrotropic responses of roots in soil, *Bot. Gaz.*, **97**: 728–743.

Lyon, T. L., and J. A. Bizzell, 1913: Is there a mutual stimulation of plants through root influence? *J. Am. Soc. Agron.*, **5**: 38–44.

——— and J. K. Wilson, 1921: Liberation of organic matter by roots of growing plants, *N. Y. Sta. (Cornell) Mem.* 40.

Magistad, O. C., and J. F. Breazeale, 1929: Plant and soil relations at and below the wilting percentage, *Ariz. Sta. Tech. Bull.* 25.

Magowan, F. N., 1908: The toxic effect of certain common salts in the soil on plants, *Bot. Gaz.*, **45**: 45–49. Quoted by Cormack, 1935.

Marth, P. C., 1934: A study of root distribution of Stayman apple trees in Maryland, *Proc. Am. Soc. Hort. Sci.*, **32**: 334–337.

Martin, T. L., 1927: Decomposition studies of alfalfa and sweet clover roots and straw, *Soil Sci.*, **24**: 309–316.

Massey, A. B., 1925: Antagonism of the walnuts (*Juglans nigra* L. and *J. cinerea* L.) in certain plant associations, *Phytopath.*, **15**: 773–785.

Maximcw, N. A., and A. S. Kružilin, 1936: Influence of irrigation on the development of the root system of spring wheat plants, *Plant Physiol.*, **11**: 457–463.

McDougall, W. B., 1921: Thick-walled root hairs of *Gleditsia* and related genera, *Am. J. Bot.*, **8**: 171–176.

McKinley, A. D., 1931: Effects of sorghum residues on crop yields, *J. Am. Soc. Agron.*, **23**: 844–850.

McQuilkin, W. E., 1935: Root development of pitch pine with some comparative observations on shortleaf pine, *J. Agr. Res.*, **51**: 983–1016.

Mer, E., 1879: Recherches expérimentales sur les conditions de développement des poils radicaux, *C. R. Acad. Sci. Paris*, **88**: 665–668.

———, 1884: Nouvelles recherches sur les conditions du développement des poils radicaux, *C. R. Acad. Sci. Paris*, **98**: 583–586.

Metzger, W. H., 1928: The effect of growing plants on solubility of soil nutrients, *Soil Sci.*, **25**: 273–281.

Mevius, W., 1926: Kalzium-Ion und Wurzelwachstum, *Jahrb. wiss. Bot.*, **66**: 183–253. Quoted by Cormack, 1935.

Millar, C. E., 1925: The feeding power of plants in different soil horizons, *J. Am. Soc. Agron.*, **17**: 150–157.

———, 1925: Availability of nutrients in subsoils, *Soil Sci.*, **19**: 275–285.

———, 1930: Root systems of young corn plants in relation to fertilizer applications, *J. Am. Soc. Agron.*, **22**: 868–874.

———, 1933: Availability to corn of nutrients in the A_2 and B horizons of Hillsdale loam, *J. Am. Soc. Agron.*, **25**: 418–426.

Miller, E. C., 1916: Comparative study of the root systems and leaf areas of corn and the sorghums, *J. Agr. Res.*, **6**: 311–333.

———, 1916: The root systems of agricultural plants, *J. Am. Soc. Agron.*, **8**: 129–154.

Miller, M. F., and F. L. Duley, 1925: The effect of varying moisture supply upon the development and composition of the maize plant at different periods of growth, *Mo. Sta. Tech. Bull.* 76.

MININA, E. G., 1927: On the acidity of root secretions, *Bull. Inst. Rech. Biol. Univ. Perm.*, **5**: 233–258; *Biol. Abst.*, **4**: 62.

MOHAMMAD, A., and R. B. DESHPANDE, 1929: Studies in Indian chillies. II, The root systems, *Agr. J. India*, **24**: 251–258. Quoted by Weihing, 1935.

MOLISCH, H., 1885: Ueber die Ablenkung der Wurzeln von ihrer normalen Wachsthums-richtung durch Gase (Aerotropismus), *Sitzungsber. Akad. Wiss. Wien.*, **90**: 111.

———, 1887: Über Wurzelausscheidungen und deren Einwirkung auf organische Substanzen, *Sitzungsber. Akad. Wiss. Wien.*, **96**: 84–109.

MOLLIARD, M., 1913: Sur la sécrétion par les racines de substances toxiques pour la plante, *Soc. Bot. France Bull.* **60**: 442–446.

MORRISH, R. H., 1934: Crop mixture trials in Michigan, *Mich. Sta. Spec. Bull.* 256.

MUKERJI, J. N., 1920: The excretion of toxins from the roots of plants, *Agr. J. India*, **15**: 502–507.

MÜLLER-THURGAU, H., 1894: Einfluss des Stickstoffs auf die Wurzel-bildung, *Jahrb. Versuchs-Stat. Wädensweil*, **4**: 48–52.

MURPHY, H. C., 1935: Effect of crown rust infection on yield and water requirement of oats, *J. Agr. Res.*, **50**: 387–412.

NEDROW, W. W., 1937: Studies on the ecology of roots, *J. Ecology*, **18**: 27–51.

NELLER, J. R., 1922: The influence of growing plants upon oxidation processes in the soil, *Soil Sci.*, **13**: 139–160.

NEWCOMBE, F. C., and A. RHODES, 1904: Chemotropism of roots, *Bot. Gaz.*, **37**: 23–35.

NEWTON, J. D., 1924: Measurement of carbon dioxide evolved from the roots of various crop plants, *Sci. Agr.*, **4**: 268–274.

NICOL, H., 1935: Mixed cropping in primitive agriculture, *Empire J. Exp. Agr.*, **3**: 189–195.

NOBBE, F., 1862: Über die feinere Verästelung der Pflanzenwurzeln, *Landw. Versuchs-Stat.*, **4**: 212–224.

———, 1869: Die Pflanzencultur in Wasser, und ihre Bedeutung für die Landwirthschaft, *Landw. Versuchs-Stat.*, **11**: 106–120.

NOVEZ, A. E., 1930: On the distribution of tabular roots in *Ceiba* (*Bombocaceae*), *Proc. Nat. Acad. Sci.*, **16**: 339–344.

NUTMAN, F. J., 1933, 1934: The root system of *Coffea arabica*. I, II, III, *Empire J. Exp. Agr.*, **1**: 271–284; 285–296; **2**: 293–302.

ODLAND, T. E., and J. B. SMITH, 1933: The effect of certain crops on succeeding crops, *J. Am. Soc. Agron.*, **25**: 612–618.

OSKAMP, J., 1932: The rooting habit of deciduous fruits on different soils, *Proc. Am. Soc. Hort. Sci.*, **29**: 213–219.

PARKER, F. W., 1924: Carbon dioxide production of plant roots as a factor in the feeding power of plants, *Soil Sci.*, **17**: 229–249.

———, 1925: The carbon dioxide content of the soil air as a factor in the absorption of inorganic elements by plants, *Soil Sci.*, **20**: 39–45.

PARKER, K. W., and A. W. SAMPSON, 1930: Influence of leafage removal on anatomical structure of roots of *Stipa pulchra* and *Bromus hordeaceus*, *Plant Physiol.*, **5**: 545–553.

PAVLYCHENKO, T. K., 1937: Quantitative study of the entire root systems of weeds and crop plants under field conditions, *J. Ecology*, **18**: 62–79.

PEARSALL, W. H., and E. M. WRAY, 1927: The physiology and ecology of the calcifuge habit in *Eriophorum angustifolium*, *Ecology*, **25**: 1.

PERALTA, F. DE, and R. P. ESTIOKO, 1923: A tentative study of the effect of root excretion of common paddy weeds upon crop production of lowland rice, *Philippine Agr.*, **11**: 205–216.

Peren, G. S., 1923: Data on the lateral spread of the roots of fruit trees, *J. Pom. and Hort. Sci.*, **3**: 96–102.

Persecke, K., 1877: Über die Formveränderungen der Wurzel in Wasser und Erde. *Diss.*, Leipzig.

Pettinger, N. A., 1933: The effect of fertilizers, crop rotation, and weather conditions on the anchorage of corn plants, *Va. Polytech. Inst. Va. Sta. Tech. Bull.* 46.

Pfeiffer, F., and E. Blanck, 1912: Die Säureausscheidung der Wurzeln und die Löslichkeit der Bodennährstoffe in kohlensäurehaltigem Wasser, *Landw. Versuchs-Stat.*, **77**: 217–268.

Pickering, S. U., 1903: The effect of grass on apple trees, *J. Roy. Agr. Soc. England*, **64**: 365–376.

——— and Duke of Bedford, 1914: Effect of one crop upon another, *J. Agr. Sci.*, **6**: 136–151.

———, 1917: The effect of one plant on another, *Ann. Bot.*, **31**: 181–187.

——— and Duke of Bedford, 1919: "Science and Fruit Growing," Chap. 28, pp. 291–312.

Polle, R., 1910: Über den Einfluss verschiedener hohen Wassergehalts, verschiedener Düngung und Festigkeit des Bodens auf die Wurzelentwicklung des Weizens und der Gerste im ersten Vegetationsstadium, *J. Landw.*, **58**: 297–345.

Popesco, S., 1926: Recherches sur la région absorbante de la racine, *Buletinul Agriculturii (Rumania)*, IV, No. 10–12. 136 pp.

Porodko, T., 1911: Über den Chemotropismus der Pflanzenwurzeln, *Jahrb. wiss. Bot.*, **49**: 307–388.

———, 1912: Vergleichende Untersuchungen über die Tropismen. II, Mitt. Thermotropismus der Pflanzenwurzeln, *Ber. bot. Ges.*, **30**: 305–313.

———, 1914: Vergleichende Untersuchungen über die Tropismen. VI, Der relative chemotrope Wirkungswert von Alkali- und Erdalkalisalzen für Keimwurzeln von *Lupinus albus*, *Ber. bot. Ges.*, **32**: 25–35.

———, 1914: Vergleichende Untersuchungen über die Tropismen. VII, Der relative chemotrope Wirkungswert von Schwermetallsalzen für Keimwurzeln von *Lupinus albus*, *Ber. bot. Ges.*, **32**: 271–275.

———, 1914: Vergleichende Untersuchungen über die Tropismen. VIII. Bedingungen der chemotropen Krummungsreaktion bei den Pflanzenwurzeln, *Ber. bot. Ges.*, **32**: 275–282.

Pratt, I. W., 1930: Preliminary study of elongation of roots of Georgia collards as affected by sodium luminal, *Plant Physiol.*, **5**: 587–597.

Prianischnikov, D., 1914: Sur la question des excrétions nuisibles des racines. *Rev. Gén. Bot.*, **25**: 563–582.

Raciborski, M. M., 1905: Oxydierende und reduzierende Eigenschaften der lebenden Zelle, *Bull. Acad. Sci. Cracovie*, **1905**: 338, 668, 693.

Rastek, J. C., 1934: Evidence that some plant roots give off organic acids. *Proc. Am. Soc. Hort. Sci.*, **32**: 620–622.

Reid, M., 1930: The influence of nutritive conditions of seeds and cuttings upon the development of roots, *Rpt. and Proc. Ninth Int. Hort. Cong.*, Group A, 165–169.

———, 1933: Effects of shade on the growth of velvet bent and metropolitan creeping bent, *Bull. U. S. Golf Assoc., Green Sec.*, **13**: 131–135.

Ridgway, C. S., 1913: The occurrence of callose in root hairs, *Plant World*, **16**: 116–122.

Rigg, G. B., 1913: The effect of some Puget Sound bog waters on the root hairs of *Tradescantia*, *Bot. Gaz.*, **55**: 314–326.

——— and E. S. Harrar, 1931: The root systems of trees growing in sphagnum, *Am. J. Bot.*, **18**: 391–398.

Roberts, E. A., 1916: The epidermal cells of roots, *Bot. Gaz.*, **62**: 488–506.

Robertson, D. W., A. Kezer, J. Sjogren, and D. Koonce, 1934: Studies on the critical period for applying irrigation water to wheat, *Col. Sta. Tech. Bull.* 11.

Rogers, W. S., and M. C. Vyvyan, 1928: The root systems of some ten-year-old apple trees on two different root stocks and their relation to tree performance, *Ann. Rpt. East Malling Res. Sta.* **1926-1927**: 31–43.

———, 1932: Root studies. II, The root development of an apple tree in a wet clay soil, *J. Pom. and Hort. Sci.*, **10**: 219–227.

———, 1933: Root studies. III, Pear, gooseberry and black-currant root systems under different soil fertility conditions with some observations on root stock and scion effect in pears, *J. Pom. and Hort. Sci.*, **11**: 1–18.

——— and M. C. Vyvyan, 1934: Root studies. V, Root stock and soil effect on apple root systems, *J. Pom. and Hort. Sci.*, **12**: 110–150.

———, 1935: Root studies. VI, Apple roots under irrigated conditions, with notes on use of a soil-moisture meter, *J. Pom. and Hort. Sci.*, **13**: 190–201.

———, 1935: Soil factors in relation to root growth, *Trans. Third Int. Cong. Soil Sci.*, **1**: 249–253.

Rotmistrov, V., 1909: "Root Systems of Cultivated Plants of One Year's Growth," 57 pp., 22 pl., Odessa.

Roxas, M. L., and M. Villano, 1930: Studies on the development of the sugar-cane plant in the Philippines. Root and shoot development of M-1900, *Philippine J. Sci.*, **43**: 367–403.

Russell, E. J., 1917: "Soil Conditions and Plant Growth," 190 pp., Longmans, Green & Company, New York.

——— and A. Appleyard, 1917: The influence of soil conditions on the decomposition of organic matter in the soil, *J. Agr. Sci.*, **8**: 385–417.

Russell, J., and A. H. E. McDonald, 1935: Wheat after wheat. No inherent harm in continuous wheat cropping, *Agr. Gaz. N. S. Wales*, **45**: 183–184.

Sachs, J., 1860: Bericht über die physiologische Thätigkeit an der Versuchs station in Tharandt. II, Wurzel-Studien, *Landw. Versuchs-Stat.*, **2**: 1–31.

Saunders, A. R., 1933: Studies in phanerogamic parasitism with particular reference to *Striga lutea* Lour., *Dept. Agr. Union South Africa Sci. Bull.* 128.

Sayre, C. B., 1934: Root development of beans, cabbage, and tomatoes as affected by fertilizer placement, *Proc. Am. Soc. Hort. Sci.*, **32**: 564–571.

Schneiderhan, F. J., 1926: Apple disease studies in northern Virginia, *Va. Polytech. Inst. Va. Sta. Bull.* **245**: 29–30.

Schreiner, O., and H. S. Reed, 1907: Some factors influencing soil fertility, *U. S. Bur. Soils Bull.* 40.

———, ———, and J. J. Skinner, 1907: Certain organic constituents of soils in relation to soil fertility, *U. S. Bur. Soils Bull.* 47.

——— and ———, 1907: The production of deleterious excretions by roots, *Bull. Torrey Bot. Club*, **34**: 279–303.

——— and ———, 1908: The toxic action of certain organic plant constituents, *Bot. Gaz.*, **45**: 73–102.

——— and ———, 1909: The role of oxidation in soil fertility, *U. S. Bur. Soils Bull.* 56.

——— and ———, 1909: Studies on the oxidizing power of roots, *Bot. Gaz.*, **47**: 355–388.

——— and E. C. Shorey, 1909: The isolation of harmful organic substances from soils, *U. S. Bur. Soils Bull.* 53.

——— and ———, 1910: Chemical nature of soil organic matter, *U. S. Bur. Soils Bull.* 74.

——— and J. J. SKINNER, 1910: Some effects of a harmful organic soil constituent, *U. S. Bur. Soils Bull.* 70.

——— and M. X. SULLIVAN, 1911: Reduction by roots, *Bot. Gaz.*, **51**: 121–130.

——— and ———, 1911: Concurrent oxidation and reduction by roots, *Bot. Gaz.*, **51**: 273–285.

——— and J. J. SKINNER, 1912: The toxic action of organic compounds as modified by fertilizer salts, *Bot. Gaz.*, **54**: 31–48.

———, 1923: Toxic organic soil constituents and the influence of oxidation, *J. Am. Soc. Agron.*, **15**: 270–277.

SCHULZE, B., 1906: Studien über die Bewurzelung unserer Kulturpflanzen, *Festschrift*, 50 *Jubiläum Agr. Chem. Vers. und Kontroll Stat.*, Breslau, 67–95.

SCHWARZ, F., 1883: Die Wurzelnhaare der Pflanzen, *Untersuch. Bot. Inst. Tübingen*, **1**: 135–188.

SEWELL, M. C., 1923: Effect of *Andropogon sorghum* on succeeding crops of *Triticum sativum vulgare*, *Bot. Gaz.*, **75**: 1–26.

SHANTZ, H. L., 1927: Drought resistance and soil moisture, *Ecology*, **8**: 145–157.

SHEPPERD, J. H., 1905: Root systems of field crops, *N. Dak. Sta. Bull.* **64**: 529–535.

SHULL, C. A., 1932: Toxicity of root excretions, *Plant Physiol.*, **7**: 339–341.

SIEMENS, H. J., 1929: The development of secondary seminal roots in corn seedlings, *Sci. Agr.*, **9**: 747–759.

SIERP, H., and A. BREWIG, 1935: Quantitative Untersuchungen über die Wasserabsorptionszone der Wurzeln, *Jahrb. wiss. Bot.*, **82**: 99–122.

SINGH, K., 1922: Development of the root system of wheat in different kinds of soil and with different methods of watering, *Ann. Bot.*, **36**: 353–361.

SKINNER, J. J., 1913: Illustration of the effect of previous vegetation on a following crop: cabbage after sesame, *Plant World*, **16**: 342–346.

SMITH, F. B., and P. E. BROWN, 1931: Soil respiration, *J. Am. Soc. Agron.*, **23**: 909–916.

——— and ———, 1932: Further studies on soil respiration, *J. Am. Soc. Agron.*, **24**: 577–583.

SNOW, L. M., 1904: The effects of external agents on the production of root hairs, *Bot. Gaz.*, **37**: 143–145.

———, 1905: Development of root hairs, *Bot. Gaz.*, **40**: 12–48.

SOROKIN, H., and A. L. SOMMER, 1929: Changes in the cells and tissues of root tips induced by the absence of calcium, *Am. J. Bot.*, **16**: 23–39.

SPALDING, V. M., 1904: Biological relations of certain desert shrubs. I, The creosote bush (*Covillea tridentata*) in its relation to water supply, *Bot. Gaz.*, **38**: 122–138.

SPERRY, T. M., 1935: Root systems in Illinois prairie, *Ecology*, **16**: 178–202.

STEVENS, C. L., 1931: Root growth of white pine, *Yale Forestry Bull.* 32.

STEWART, G. R., and J. C. MARTIN, 1921: Effect of various crops upon the water extract of a typical silty clay loam soil, *J. Agr. Res.*, **20**: 663–668.

STIELER, G., 1903: Ueber das Verhalten der Wurzelhärchen gegen Lösungen, *Diss.*, Kiel (abs. *Bot. Cent.*, **96**: 541 (1904)).

STOCKLASA, J., and A. ERNEST, 1905: Ueber den Ursprung, die Menge, und die Bedeutung der Kohlendioxyds im Boden, *Cent. Bakt.*, **2**: 14: 723–736.

——— and ———, 1909: Beiträge zur Lösung der Frage der chemischen Natur des Wurzelsekretes, *Jahrb. wiss. Bot.*, **46**: 55–102.

———, 1934: Decomposition of raw phosphate by the root secretions of individual culture plants, *Phosphorsäure*, **4**: 129–148; *Chem. Abst.*, **28**: 4100–4101.

STODDART, L. A., 1935: How long do roots of grasses live? *Sci.*, **81**: 544.

TAUBENHAUS, J. J., W. N. EZEKIEL, and H. E. REA, 1931: Strangulation of cotton roots, *Plant Physiol.*, **6**: 161–166.

Ten Eyck, A. M., 1899: A study of the root systems of wheat, oats, flax, corn, potatoes, and sugar beets and of the soil in which they grew, *N. Dak. Sta. Bull.* **36**: 333–362.

———, 1900: A study of the root systems of cultivated plants grown as farm crops, *N. Dak. Sta. Bull.* **43**: 535–550.

———, 1904: The roots of plants, *Kans. Sta. Bull.* **127**: 199–252.

Thatcher, R. W., 1923: The effect of one crop on another, *J. Am. Soc. Agron.*, **15**: 331–338.

Thomas, W., 1930: The feeding power of plants *Plant Physiol.*, **5**: 443–491.

Thornton, H. G., and H. Nicol, 1934: Further evidence upon the nitrogen uptake of grass grown with lucerne. *J. Agr. Sci.* **24**: 540–543.

Tincker M. A. H., 1935: Experiments with lilies at Wisley. *R. H. S. Lily Year Book*, 68–75.

Tisdale, W. H., and J. M. Jenkins, 1921: Straighthead of rice and its control, *U. S. Farmers' Bull.* 1212.

Traub, H. P., and R. H. Stansel, 1930: The lateral root spread of the fig tree. *Proc. Am. Soc. Hort. Sci.*, **27**: 109–113.

Treitel, O., 1924: Thermotropismus bei Wurzeln, *Bot. Archiv.*, **7**: 375–388.

Tucker, M., and C. von Seelhorst, 1898: Der Einfluss, welchen der Wassergehalt und der Reichtum des Bodens auf die Ausbildung der Wurzeln und der oberirdischen Organe der Haferpflanze ausüben, *J. Landw.*, **46**: 52–63.

Turner, T. W., 1922: Studies of the mechanism of the physiological effects of certain minerals salts in altering the ratio of top growth to root growth in seed plants, *Am. J. Bot.*, **9**: 415–446.

Turpin, H. W., 1920: The carbon dioxide of the soil air, *N. Y. Sta (Cornell) Mem.* 32.

Venkatraman, R. S., and R. V. Vittal, 1928: Coimbatre seedling canes, *Agr. J India*, **23**: 37–38. Quoted by Weihing, 1935.

Virtanen, A. I., 1933: Nitrogen uptake of plants, *Nature*, **131**: 534 (1934); also, *Ann. Acad. Sci. Fenn. Ser A.* 1933, **36**, No. 12.

von Seelhorst, C., 1902: Beobachtungen über die Zahl und den Tiefgang der Wurzeln verschiedener Pflanzen bei verschiedener Düngung des Bodens, *J. Landw.*, **50**: 91–104.

Waksman, S. A., and F. G. Tenney, 1928: Composition of natural organic materials and their decomposition in the soil. III, The influence of nature of plant upon the rapidity of its decomposition. *Soil Sci.*, **26**: 155–171.

Weaver, J. E., 1915: A study of the root systems of prairie plants of southeastern Washington. I. *Plant World*, **18**: 227–248.

———, 1919: The ecological relations of roots, *Carnegie Inst. Wash. Pub.* 286.

———, 1920: Root development in the grassland formation. *Carnegie Inst. Wash. Pub.* 292.

——— and J. W. Crist, 1922: Relation of hardpan to root penetration in the Great Plains, *Ecology*, **3**: 237–250.

———, F. C. Jean, and J. W. Crist, 1922: Development and activities of roots of crop plant, *Carnegie Inst. Wash. Pub.* 316.

———, J. Kramer, and M. Read, 1924: Development of root and shoot of winter wheat under field environment, *Ecology*, **5**: 26–50.

———, 1925: Investigations on the root habits of plants, *Am. J. Bot.*, **12**: 502–510.

———, 1926: "Root Development of Field Crops," 290 pp., McGraw-Hill Book Company, Inc., New York.

——— and W. E. Bruner, 1927: "Root Development of Vegetable Crops," 351 pp. McGraw-Hill Book Company, Inc., New York.

——— and W. J. Himmel, 1930: Relation of increased water content and decreased aeration to root development in hydrophytes, *Plant Physiol.*, **5**: 69–92.

——— and J. Kramer, 1932: Root system of *Quercus macrocarpa* in relation to the invasion of prairie, *Bot. Gaz.*, **94**: 51–85.

——— and ———, 1935: Relative efficiency of roots and tops of plants in protecting the soil from erosion, *Sci.*, **82**: 354–355.

———, V. H. Hougen, and M. D. Weldon, 1935: Relation of root distribution to organic matter in prairie soil, *Bot. Gaz.*, **96**: 389–420.

Weihing, R. M., 1935: The comparative root development of regional types of corn, *J. Am. Soc. Agron.*, **27**: 526–537.

West, E. S., 1934: The root distribution of some agricultural plants, *J. Council Sci. and Ind. Res. Australia*, **7**: 87–93.

Whitaker, E. S., 1923: Root hairs and secondary thickening in the *Compositae*, *Bot. Gaz.*, **76**: 30–60.

Whitney, M., and F. K. Cameron, 1904: Investigations in soil fertility, *U. S. Bur. Soils Bull.* 23.

Wilkins, F. S., and H. D. Hughes, 1934: Effect of Sudan grass and of soybeans on yield of corn, *J. Am. Soc. Agron.*, **26**: 901–908.

Willard, C. J., and G. M. McClure, 1932: The quantitative development of tops and roots in bluegrass with an improved method of obtaining root yields, *J. Am. Soc. Agron.*, **24**: 509–514.

Wilson, B. D., and J. K. Wilson, 1928: Relation of sorghum roots to certain biological processes, *J. Am. Soc. Agron.*, **20**: 747–754.

Wilson, K., 1936: The production of root hairs in relation to the development of the piliferous layer, *Ann. Bot.*, **50**: 121–154.

Woodroof, J. G., and N. C. Woodroof, 1934: Pecan root growth and development, *J. Agr. Res.*, **49**: 511–530.

Work, R. A., and M. R. Lewis, 1936: The relation of soil moisture to pear tree wilting in heavy clay soil, *J. Am. Soc. Agron.*, **28**: 124–134.

Wortmann, J., 1885: Über den Thermotropismus der Wurzeln, *Bot. Ztg.*, **43**: 193–200, 209–216, 225–235.

Worzella, W. W., 1932: Root development in hardy and nonhardy winter wheat varieties, *J. Am. Soc. Agron.*, **24**: 626–636.

Yeager, A. F., 1935: Root systems of certain trees and shrubs grown on prairie soils, *J. Agr. Res.*, **51**: 1085–1092.

Zavitz, C. A., 1922: Grains grown in combination for grain production, *J. Am. Soc. Agron.*, **14**: 225–228.

Ziegenspek, H., 1920: Das amyloid jugendlicher Organe. Das Amyloid in den wachsenden Wurzelhaaren und seine Beziehungen zum Zellwachstum, *Ber. bot. Ges.*, **38**: 328–333.

CHAPTER IV

THE INTAKE OF WATER BY PLANTS

I. SOURCES OF THE PLANT'S WATER SUPPLY

It is an accepted fact that plants obtain through their roots from the soil the major portion of the water that is necessary for their needs. There is, however, considerable dispute in regard to the ability of plants to absorb water through their stem and leaves from the atmosphere, so the intention here is to review the experimental evidence.

Boussingault (1878) believed that the absorption of rain and dew by leaves is of common occurrence, while Burt (1893) and Dandeno (1901) demonstrated that certain leaves can absorb water when it is placed in direct contact with them. Ganong (1894) grew plants in soil in sealed containers and after allowing them to wilt badly placed them under condition favorable for the absorption of water by their leaves. In one experiment the wilted leaves were covered with moist filter paper, in another they were sprayed with water, in a third case the wilted plants were immersed entirely or partially in water and left for a considerable period of time, while in the fourth experiment they were placed in a saturated atmosphere. In all these experiments the wilted leaves showed no signs of regaining their turgidity. This indicated that the amounts of water absorbed by the leaves were of little significance in the welfare of the plant. Spalding (1906) immersed the cut branches of various species of desert plants in water for 3 hr. and found that some of these parts absorbed no water while others absorbed amounts ranging from 2 to 6 per cent of their original weight. The stems and leaves that did absorb water, however, lost it all in a few minutes when exposed to the air. The absorption of water by the stem and leaves was shown most strikingly by those plants which showed marked mesophytic tendencies. Spalding also investigated the ability of these plants to obtain moisture from the atmosphere. A review of the data obtained shows that of the 12 species of desert perennials subjected to the experiment, all exhibited some slight capacity for the absorption of water vapor from the atmosphere, but in general the amounts absorbed were inconsiderable in comparison with that given off in corresponding periods in dry air. Spalding considered that as yet there is no evidence to indicate that the small quantities of water absorbed are utilized in the body of the plant, since the promptness with which the water thus absorbed is returned to the air suggests a process purely physical in its nature and of no physiological significance. Spalding stated that the most important conclusion to be drawn from his experiments is the patent fact that the roots of desert plants constitute their only reliable active agent in providing a normal water supply no matter how far some of these plants have departed from the habits of the present-day mesophytes. In Australia all members of the genus *Atriplex* and to a less extent species of *Kochia baosia* and *Rhagodia* absorb an appreciable amount of water from saturated atmospheres (Wood, 1925). The much greater absorption in the species *Atriplex* appears to be correlated with the high salt content of the leaves of these plants, the percentage of salt ranging from 12 to 30 per cent as compared with 2 to 7 per cent for the other leaves in question. Henslow (1908) concluded that the stems and leaves of various horticultural plants immersed in water could absorb some water under these conditions, since the loss by transpiration subsequent to immersion was lower than it was previous to the immersion. His experiments on the absorption of dew by leaves is worthy of note. He used the leaves of 20 varieties of plants which were plucked in the evening between 4 and 5 P.M. and exposed to sunlight until nightfall, after which they were exposed to a heavy dew during the night. The specimens were dried with a cloth in the morning before the sunlight had had any effect and weighed. With but one exception these leaves showed a gain in weight of from 2 to 35 per cent over that of the previous evening. The conditions under which plants grow in the field would seem to eliminate, however, the absorption of any appreciable amount of moisture from the dew deposited on the surface of their leaves. Even when the moisture supply in the soil is extremely low and the plants are wilted badly during the day, the leaves rapidly regain their turgidity during the early evening so that loss of water from guttation frequently occurs during the night. It is thus difficult to conceive how the leaves could absorb any water of consequence through their surface under such conditions. In many semiarid regions where under extreme conditions the absorption of dew might be of some little benefit, the weather conditions are such that dew rarely if ever falls.

Various experiments have been conducted by Gates (1914), Grundmann (1923), Wetzel (1924), Nakajima (1928), Darrow and Sherwood (1931), Williams (1932), Kessler (1933), Brierley (1934), and Smyth (1934) to determine the amount of water absorbed by wilted leaves when submerged in water or when it is applied to their surfaces. Some of the plants that have been shown to absorb some water under such conditions include *Vitis*, *Lactuca*, *Solanum*, *Beta*, *Lycopersicum*, *Phaseolus*, *Trifolium*, and *Rubus*. In all, the leaves of more than 100 genera have been reported as capable of absorbing water. The uptake of water by leaves is influenced by (1) the permeability of the cuticle and epidermis, (2) the hairiness of the upper surface, (3) the ease of wetting the surface, and (4) the internal deficit of water in the parenchymatous cells bordering the epidermis.

There seems to be no doubt but that leaves in general are capable under certain conditions of absorbing moisture, but the conditions for such absorption seldom occur in nature. When such absorption of water does occur, the amount is so small compared to that lost by transpiration that it is generally of no physiological significance in the water economy of the plant.

The benefit to a plant of a light shower that does not penetrate to the roots or of a humid atmosphere is to reduce the evaporating power of the air so that the loss of water from the leaves is cut down to a rate that is lower than that of absorption by the roots, and as a result of this the cells of the leaves and stem become turgid and have the appearance of having absorbed water from the air. These statements refer more particularly to plants that are rooted in the soil. Epiphytic plants such as certain ferns, orchids, and hanging mosses, however, secure all their water from the atmosphere.

II. WILTING COEFFICIENT

A plant cannot remove all the moisture from the soil in which it grows, so that when a plant permanently wilts there will always be some moisture retained by the soil, regardless of how thoroughly roots may permeate it. A plant is said to be permanently wilted when its leaves have become so flaccid as a result of a deficiency in the soil moisture supply that they will not recover their normal turgidity when placed in an approximately saturated atmosphere without the addition of water to the soil. Briggs and Shantz (1912) first used the term "wilting coefficient" to denote the moisture content of the soil (expressed as percentage of the dry weight) at the time of the permanent wilting of the plants growing therein. These authors also found that the loss of moisture from the soil through the plant does not cease at permanent wilting but goes on to some extent even after the death of the plant—the plant during the drying stage acting simply as a medium for the transference of water from the soil to the air.

A. Methods of Determining the Wilting Coefficient

1. Direct Method. The Observed Wilting Coefficient.—The wilting coefficient is determined directly by growing plants in sealed containers until they are so wilted that they will not revive when placed in a saturated atmosphere. The soil is then sampled and the percentage of moisture on a dry basis determined. The wilting coefficient obtained after this manner is termed the "observed wilting coefficient." The direct method of determining the wilting coefficient has been used, for the most part, for seedling plants in small containers but has not been used successfully for plants at a more advanced vegetative state. In the case of more mature plants it is almost impossible to obtain a sharp end point for wilting, and in large containers the soil moisture is not evenly depleted throughout the soil mass.

Koketsu (1928) noted that the content of water in the leaves at the wilting point differed widely for different plants as did also the ratio of the critical water content of the leaves to the content at full turgidity. This ratio appeared to be a specific quality of a given plant and to indicate the power of resistance of the plant to wilting. He also noted that the water content of the leaves at various stages of wilting is affected by cultural methods. Breazeale (1930) stated that there is no direct method of determining the wilting coefficient which does not involve several errors. The wilting coefficient is largely a function of the layer of soil that lies in direct contact with the absorbing zone of the roots (Work and Lewis, 1936).

No accurate method of separating this layer of soil from the roots for a determination of its moisture content has yet been devised. On that account, if plants are grown in a certain volume of soil until they permanently wilt, and all the soil is then used in the moisture determination, the results obtained will represent the average percentage of soil moisture and not the percentage in the soil that has been in contact with the roots. As the amount of soil in contact with the roots may be relatively small in comparison with all the soil in the culture, the error may be highly significant.

Moinat (1932) considered that permanent wilting is not necessarily an accurate indicator of the wilting coefficient. Thus in the seedling of the bean a relatively large proportion of the water in the plant is to be found in the lower part of the hypocotyl. During wilting the transpiring leaves draw upon this reserve water in the stem when the supply from the roots is cut off so the critical point of soil moisture may be reached previous to permanent wilting.

2. Indirect Method. *a. The Calculated Wilting Coefficient.*—The determination of the wilting point, especially with plants growing in the field, is beset with many difficulties, among which are the direct evaporation from the soil, local variation in soil texture, lack of uniformity in root development, and the lack of conditions to produce wilting. Briggs and Shantz, therefore, upon the assumption that the wilting coefficient is practically the same for a given soil for any plant under any climatic condition, formulated several indirect methods of determining the moisture content of the soil at the wilting point, based upon the relationship of the wilting point to the moisture retentiveness of the soil as measured by certain physical methods. By these methods it is not necessary to work with the plants at all to determine the amount of moisture that would remain in the soil when they wilt permanently. By determining in the laboratory certain factors on the moisture retentiveness of a sample of soil in which the plants are growing the wilting coefficient can be calculated and is then termed the "calculated wilting coefficient." Of the methods thus formulated by Briggs and Shantz, three that have been used to a considerable extent in ecological and agronomic work are here considered.

1. *Relation to the Moisture Equivalent of the Soil.*—The moisture equivalent of the soil is the percentage of water that it can retain in opposition to a centrifugal force 1,000 times that of gravity. The ratio of the moisture equivalent to the wilting coefficient was found to be 1.84 ± 0.013, or, expressed in another way,

$$\text{Wilting coefficient} = \frac{\text{moisture equivalent}}{1.84 \pm 0.013}$$

Subsequent work by Veihmeyer and Hendrickson (1927, 1928), Veihmeyer, Osekowsky and Tester (1927), Hendrickson and Veihmeyer (1929), and Wadsworth and Das (1930) show that the wilting coefficient cannot be determined for all soils by using the factor 1.84. They concluded that, although there is a remarkable constancy of the residual moisture content for a given soil when permanent wilting is attained, a common factor to evaluate the amount of water that remains in all soils at permanent wilting cannot be used. These workers found the ratio of the moisture equivalent to the wilting coefficient to be as low as 1.3 for some soils and as high as 3.8 for others. Thus they concluded that the amount of water available for plant growth cannot be obtained from the moisture equivalent alone by using any single definite factor for all soils. The factor used must be determined for each soil under consideration because plants are apparently able to reduce the soil moisture content of different soils to different degrees of dryness, relative to the moisture equivalent before they become permanently wilted. Thus for accurate work the amount of moisture readily available to plants can be obtained only when the residual moisture at permanent wilting is known.

2. *Relation to the Hygroscopic Coefficient of the Soil.*—When a dry soil is placed in a saturated atmosphere it will absorb water vapor until a condition of approximate equilibrium is attained. The moisture content of a soil under such conditions is known as the "hygroscopic coefficient" of that soil. This term should not be confused with the term hygroscopic water content of the soil, which is used in some plant-physiological tests to signify the water content of "air-dry" soil. The mean ratio of the hygroscopic coefficient to the wilting coefficient was considered by Briggs

and Shantz to be 0.68 ± 0.012, or

$$\text{Wilting coefficient} = \frac{\text{hygroscopic coefficient}}{0.68 \pm 0.012}$$

3. *Relation to the Moisture-holding Capacity of the Soil.*—The moisture-holding capacity of a soil is the percentage of water that it can retain in opposition to the force of gravity when free drainage is provided. The experiments of Briggs and Shantz indicated that

$$\text{Wilting coefficient} = \frac{\text{moisture-holding capacity} - 21}{2.9 \pm 0.06}$$

b. The Index of Transpiring Power as an Indicator of Permanent Wilting in Plants.—It has been observed that the transpiration rate falls suddenly with wilting but rises thereafter to a considerable extent subsequent to a fall once more as desiccation occurs. This initial fall in the transpiration rate represents the increase in incipient drying within the leaves. During this initial decrease in transpiration the water columns in the conducting tissues remain intact. As the drying of the leaves continues, the strain on the columns of water increases so that the breaking point is eventually reached. When the water columns break, the tension is relieved and the transpiration rate is accelerated. When the water of the broken columns has been evaporated, the rate of transpiration will again fall. Bakke (1915) considered that the stage of permanent wilting represents the most intense wilting possible without the rupture of the water columns of the plant. He considered further (1918) that the transpiring power of plants as determined by standardized hygrometric paper gives an accurate knowledge of the internal water relations of the plant, and that the permanent wilting point of the plant corresponds to that point at which the transpiration rate indicates that the water columns of the stem have been ruptured.

B. Factors Influencing the Wilting Coefficient

The amount of water that remains in the soil at the time of permanent wilting will depend upon various factors, the chief of which are (1) the type of soil, and (2) the kind of plant.

1. Type of Soil.—Sachs (1859) was the first to observe that the type of soil determined the amount of water that remained when the plants wilted permanently. He grew tobacco plants in three different soils—a vegetable mold, a loam, and a sand—and found that they wilted permanently while 12.3, 8.0, and 1.5 per cent of water, respectively, remained in the soil. The most extensive work on the relation of the water remaining in the soil at the time of permanent wilting of the plant and the type of soil has been done by Briggs and Shantz (1912). These investigators determined the wilting coefficient for 20 different types of soil ranging from a coarse sand to a clay loam. The following table has been compiled by selecting a few examples at random from their 1,300 observations:

Plant	Wilting coefficient of				
	Coarse sand	Fine sand	Sandy loam	Loam	Clay loam
Moisture equivalent	1.55	5.5	12.0	18.9	27.4
Corn, Boone County, white	1.07	3.1	6.5	9.9	15.5
Sorghum, Dwarf milo	0.94	3.6	5.9	10.0	14.1
Wheat, Kubanka	0.88	3.3	6.3	10.3	14.5
Oats, Kherson	1.07	3.5	5.9	11.1	14.8
Pea, Canada field	1.02	3.3	6.9	12.4	16.6
Tomato, Golden Queen	1.11	3.3	6.9	11.7	15.3
Rice, Japan	0.96	2.7	5.6	10.1	13.0

These data show that the finer the soil particles the greater is the amount of water remaining in the soil when the plant dies from the lack of a moisture supply.

2. Kind of Plant.—It has generally been supposed that the various groups of plants differ in their ability to reduce the moisture content of the soil. It has also been assumed that the ability of certain plants to withstand drought is due to their power to obtain water by depleting the soil moisture to a lower point than could other plants which consequently perish from lack of moisture. These suppositions seem to be based, however, on little or no experimental evidence, and it is the intention here to consider the facts that have been observed in this regard. Briggs and Shantz (1912) found in their experiments that the plants with which they worked differ only slightly in their ability to deplete the moisture content of a given soil. If 100 is taken to represent the average wilting coefficient for all the plants considered, their results show that the value for corn would be 103, lègumes 101, wheat 99, oats 99, sorghum 98, millet 97, barley 97, grasses 97, and rice 94. According to the figures just quoted, in a soil whose mean wilting coefficient is 10 per cent, a corn plant would wilt with a moisture content of 10.3 per cent, while the sorghums would wilt with a moisture content of 9.8.

Results quite comparable to these were obtained by Capalungan and Murphy (1930) for a relatively large number of crop plants in Oklahoma.

The advantage to be gained by this difference in depletion of soil moisture is very slight, and Briggs and Shantz state "that if the roots were distributed to a depth of 4 ft. this would correspond to an additional amount of water for the sorghum plants in the 4 ft. of soil equivalent to a rain of $3/10$ in., all of which penetrated to the root zone." They considered that this additional amount of moisture would be of slight benefit to a plant and would scarcely suffice to ameliorate conditions during a drought to any greater extent than a cool, cloudy day. Alway (1913) grew Red Fife and Kubanka wheat, milo, corn, Mexican beans, and the desert legumes, *Prosopis velutina*, *Acacia Greggi*, and *Acacia constricta* in water-tight galvanized-iron cylinders. These cylinders were 6 ft. long and 6 in. in diameter and contained known amounts of soil and water. The plants were allowed to grow without the further addition of water until they died. The cylinders were then opened and the moisture content determined at different depths. Alway found that, in their ability to exhaust the moisture of the soil before dying, wheat, corn, milo, and Mexican beans showed little difference, but that, in their ability to continue alive after first showing injury from drought, they exhibited marked differences. The interval between wilting and death in the case of the beans amounted to only a few days, while in the case of wheat and milo it amounted frequently to several weeks. He found that if no unfavorable conditions occurred before the death of the plants, the moisture content could be reduced by any of them almost to the hygroscopic coefficient. The desert legumes mentioned above remained alive after the water content of the soil had been depleted slightly below the hygroscopic coefficient after the above-mentioned annual crops had died. Although this does not mean that these plants can utilize for growth the small amount of this water that they obtain from the soil of so low a moisture content, it does indicate that this portion of water has a very high value for the maintenance of life.

The experimental evidence at hand thus indicates that, in the cases observed, one kind of plant cannot deplete the soil moisture in a given soil appreciably more than can another, before wilting permanently. The experimental work, however, has been done for the most part with seedling plants or with plants in a comparatively early stage of development (Crump 1913) and is open to criticism in that regard. It has been found difficult to determine the wilting point in the case of fully developed plants, since as the water content in the soil becomes low the lower leaves of the plant droop, dry up, and drop off while the upper ones continue to be turgid and function normally. Powers (1922) from observations of plants growing in the field under arid conditions concluded that the wilting point varies more than has commonly been supposed when judged in the light of crop appearance, soil moisture, and the yield of dry matter. It appears to vary more for different crops on a soil that is rather heavy in texture than on a soil of narrow moisture limits.

It has been shown that the cells of the roots, particularly of crop plants, show little difference in the tractile force that they exert upon the water of the soil. There are, however, growth and morphological differences that will enable certain plants to reduce the moisture of the soil more than others before permanent wilting occurs.

Wolfe (1926) showed that the back pull of the soil for water is approximately 25 atmospheres at the hygroscopic coefficient. Thus if the plant is obtaining any moisture from the soil under this condition, its roots are exerting an osmotic pull somewhat greater than 25 atmospheres. The distribution of the root system has a marked influence upon the degree of moisture depletion of the soil. Briggs and Shantz (1912) found that rice, a plant that grows for the most part in a saturated soil, depleted the soil moisture to a lower degree than did plants that grow in the more arid regions. They attributed this to the fact that the former plants possessed abundant fine fibrous roots that permeated all parts of the soil. This is no doubt true also in the case of the sorghums as compared to corn. Miller (1916) found that the sorghum plants have almost twice as many fibrous roots as have the corn plants, so that the soil is more thoroughly penetrated by roots, and the water in the soil thus has a much shorter distance to move before it enters the roots of the sorghums than it does in the case of the corn, where the roots are more sparsely distributed. Under mild atmospheric conditions where evaporation is low, these characteristics would be of little importance, but under conditions of exceedingly high evaporation, as encountered in the field in actual agricultural practice, they become very significant. While Briggs and Shantz found that the sorghum seedlings showed but little more ability to reduce the moisture content of the soil than did corn under the mild conditions of the greenhouse, the fact that the sorghum plants in the field survive under severe climatic conditions while corn wilts and dries up gives indirect evidence that these sorghum plants under these conditions can deplete the moisture content of the soil lower than can corn before permanent wilting occurs. These plants survive, not necessarily because their individual root hairs exert a greater tractile force on the water in the soil than does corn, but simply because the water in the soil has a shorter distance to move in order to come in contact with the roots of the sorghums and thus keeps the water lost from the leaves replaced and the turgidity of the leaves prolonged for a longer period than in the case of corn.

Apparently, the roots in the upper region of the soil can deplete the soil moisture below the wilting coefficient provided that there is a sufficient water supply in the lower levels and that enough roots penetrate therein to furnish the bulk of the water necessary to replace that lost from the leaves. The roots in the upper portion of the soil thus continue to absorb a small amount of water and pass it on to the stem, whereas, if this small amount of water in the upper portion of soil were all that was available, the plant would die from lack of moisture, and the water in the soil at that level would not be depleted to so low a degree.

The following table shows the calculated wilting coefficient and percentage of moisture at different depths under corn and milo at Garden

Depth of sample, feet	Per cent of moisture	Calculated wilting coefficient
1	7.7	12.7
2	12.2	14.5
3	11.4	14.5
4	12.9	16.3
5	15.6	17.1
6	19.1	16.1

City, Kan., at the close of the growing season, which extended from June 5 to Sept. 6, 1914.

These wilting coefficients were obtained by determining the moisture equivalent of the soil samples at these depths and using 1.84 as the factor, as given by Briggs and Shantz (1912). Consequently, the figures are subject to the previously mentioned criticisms of Veihmeyer and Hendrickson (1927, 1928), and Wadsworth and Das (1930).

Magistad and Breazeale (1929), and Breazeale (1930) considered that roots remain alive for a considerable period in a soil that has been depleted in moisture to the wilting coefficient because under these conditions the roots secrete around themselves a sheath with adhesive properties. They assumed that this sheath is kept at a moisture content near the wilting coefficient by an exudation of water from the root. They considered that this water is obtained from other roots which have penetrated into moist soil or from water stored in the plant.

A small leaf surface is no doubt an important factor in delaying the wilting of plants when the moisture supply of the soil is low. In general, the smaller the leaf surface the smaller the amount of water lost in a given time so that even the low rate of absorption that prevails when the lower limits of the water supply in the soil have been reached would suffice to replace the water lost by evaporation from a small leaf surface. Thus the soil moisture could be reduced to a lower level than would be possible if the amount of water lost was large owing to a large leaf surface. The case of the desert legumes mentioned by Alway (1913) illustrates this point. As the water supply in the soil became low their leaf surface was greatly decreased by the shedding of the lower leaves, while the upper ones continued to live until the water content had been reduced even below the hygroscopic coefficient.

C. Climatic Conditions and Wilting

The extensive observations of Briggs and Shantz (1912) on the wilting of plants were made for the most part upon plants growing in the greenhouse where the temperature was about 70°F. and where the rate of loss of water from the plants was relatively low. These investigators, however, considered that their observations warranted the conclusion that the wilting point for a given soil would be a constant regardless of the nature of the climatic conditions prevailing at the time of wilting.

Subsequent to the work of Briggs and Shantz, numerous investigators have studied the relation of the wilting point of plants to the climatic conditions prevailing at the time of wilting and the relationship of the observed wilting point to the calculated one obtained from the formulas mentioned above. Brown (1912) placed potted plants of *Martynia* (Devil's Claw) in positions having varying degrees of evaporation and observed the moisture left in the soil when these plants wilted from lack of water. The following table gives an example of his results:

Position	Evaporation from atmometer. Rate per hour previous to wilting, cubic centimeters	Soil moisture at wilting, per cent
Open	7.2	15.4
Lath shelter	4.1	11.3
Room	2.8	6.8
Moist chamber	0.9	5.2

He showed thus that the rate of evaporation previous to the time of wilting determined the percentage of moisture that the plant cannot extract from the soil. Brown transferred plants that had wilted under a high rate of evaporation to a position where a lower rate of evaporation occurred. After this transfer the plants revived quickly without the addition of water to the soil and removed therefrom a greater quantity of water. The case of three plants that were placed in the open, where evaporation was relatively high, and allowed to wilt badly, after which they were transferred to a moist chamber, where the evaporation was very low, will serve as an illustration of the facts observed (see page 199).

Caldwell (1913) repeated some of the work of Brown and reached the following conclusions concerning the values of the observed and calculated wilting points: When wilting of plants occurs in a partial shelter or in the open with relatively high evaporating rates, the ratio of the

	Hourly rate of evaporation from atmometer, cubic centimeters	Moisture remaining in the soil at the time of wilting, per cent
Plant 1:		
A[1]	4.2	11.3
B[2]	0.3	5.9
Plant 2:		
A	7.5	17.5
B	1.6	6.6
Plant 3:		
A	3.2	10.5
B	0.9	5.2

[1] *A* represents the open.
[2] *B* represents the moist chamber.

observed to the calculated wilting point is always greater than unity. When wilting occurs in a nearly saturated atmosphere, the ratio of the soil-moisture content at permanent wilting as found by actual observations to that obtained by calculations from physical constants is practically unity. Caldwell makes the significant statement that "for a series of plants grown in any one soil and wilted under a number of aerial conditions, as many different soil-moisture contents are obtained as there are sets of conditions." He observed that the moisture remaining in the soil at the time of wilting at the maximum evaporation rates at Tucson, Ariz., varied from 30 to 40 per cent in excess of that taking place when plants were wilted in damp chambers. Shive and Livingston (1914) grew potted plants of corn, common beans, and Spanish pepper under different evaporating conditions and determined the ratio of the actual water content at the time of wilting to the calculated moisture content. The following table shows the result they obtained with corn 4 weeks of age growing in a soil with a calculated wilting coefficient of 3.80:

Position	Evaporation for 2-hr. period previous to the wilting, cubic centimeters	Actual moisture content of soil at wilting, per cent	Ratio of actual to calculated moisture at wilting
Glass box	0.3	4.30	1.13
Lath shelter	2.4	5.78	1.52
Cheesecloth shelter	3.0	6.04	1.58
Open	3.7	6.16	1.68

The great bulk of the experimental evidence thus indicates that the amount of water remaining in a given soil when a plant dies from lack of moisture will depend upon the atmospheric conditions prevailing just previous to the time of wilting. The wilting of a plant is due to the excess of the water loss from the stem and leaves over the amount absorbed by the roots. The rate of water loss from a plant is influenced primarily by the evaporating power of the air. If it is high, the rate of water loss from the plant is relatively high; and, if it is low, the rate of water loss from the plant is relatively low. At the time that the water available to the plant has been depleted nearly to the critical stage, the rate of absorption by the roots becomes low. If the rate of water loss from the plant under such conditions is relatively high, the plant will soon die, even though the roots are still absorbing considerable moisture from the soil. If the rate of water loss, however, is slow, owing to mild atmospheric conditions, the roots can continue for a considerable length of time to absorb sufficient water from the soil to supply the loss from the leaves and as a consequence deplete the moisture in the soil to a point lower than they could if the evaporating rate were high.

D. Water-supplying Power of the Soil at the Time of Wilting

The rate at which water is delivered to the roots of plants by the soil has been termed the "water-supplying power" of the soil. The intention here is to consider the causes for the failure of a plant to obtain its needed water supply at the time it begins to wilt. The experimental work in this regard has been conducted along two main lines—(1) a study of the back pull of the water of the soil at the time of wilting, and (2) a study of the speed at which water moves from the soil to the roots at the lower limits of the water supply of the plant.

1. The Back Pull of the Soil for Water.—Shull (1916) investigated the back pull of the water of the soil at various percentages of moisture content by means of the amount of moisture that air-dry seeds can extract from it. Seeds will imbibe water under a high pressure or against an enormous back pull. Shull placed air-dry seeds of the cocklebur (*Xanthium pennsylvanicum*) in salt solutions of various concentrations and determined the percentage of moisture that they could imbibe under these different conditions. The first table on page 201 shows the intake of water by these seeds in the different concentrations after 48 hr.

Since the percentage of water absorbed by these seeds is an indicator of the force or back pull exerted by the external medium to its entrance, they were used by Shull after the following manner to measure the back pull exerted by the water of the soil at different percentages of water content. Sixty grams of soil of the desired moisture content were placed with seeds in 200-cc. bottles, sealed, and kept agitated for 15 days. At the end of that period the bottles were opened and the moisture contents of the seeds and of the soil were determined. The moisture content of the seeds and soil at the end of the experiment shows the equilibrium

Solution	Osmotic pressure or back pull of the water of the solution, atmospheres	Intake of water by the seeds, per cent
H_2O	0.0	51.6
0.1 *M* NaCl	3.8	46.3
0.2 *M* NaCl	7.6	45.5
0.3 *M* NaCl	11.4	42.0
0.4 *M* NaCl	15.2	40.3
0.5 *M* NaCl	19.0	39.0
0.6 *M* NaCl	22.8	35.2
0.7 *M* NaCl	26.6	32.8
0.8 *M* NaCl	30.4	31.1
1.0 *M* NaCl	38.0	26.7
2.0 *M* NaCl	72.0	18.5
4.0 *M* NaCl	130.0	11.8
Saturated NaCl	375.0	6.4
Saturated LiCl	965.0	− 0.3

relation of that soil-moisture content. The following abbreviated table shows the relation of the soil moisture in the subsoil of Oswego silt loam to the water intake of *Xanthium* seeds. This soil had a moisture equivalent of 35.2 and a calculated wilting coefficient of 19.1.

Moisture in soil, per cent air dry	Intake of moisture by seeds, per cent	Back pull of water of the soil, atmospheres
5.83	0.0 sat. LiCl	965
9.36	6.47 sat. NaCl	375
11.79	11.94 4 *M* NaCl	130
13.16	21.36 2 *M* NaCl	72
14.88	28.61 1 *M* NaCl	38
17.10	37.70 0.5 *M* NaCl	19
17.93	43.20 0.3 *M* NaCl	11.4
18.07	45.15 0.2 *M* NaCl	7.6
18.87	47.26 0.1 *M* NaCl	3.8
19.34	49.31	
20.04	50.50 sat.	0.0

An examination of this table shows that in the case of the subsoils of the Oswego silt loam the back pull of the water in the soil when it is air dry reaches the enormous force of 965 atmospheres. As the moisture content of the soil increases, this back pull becomes less and less, so that by the time the former has reached about 19 per cent, the wilting coefficient of that soil, the back pull of the soil water is somewhat less than that of a $0.1M$ sodium chloride solution, which is equivalent to between 3 and 4 atmospheres. Numerous types of soils were used ranging from sands to clay loams, and it was found that the average intake of water by the seeds from the seven different types of soil used was approximately 49 per cent at the percentage of soil moisture that represents their calculated wilting coefficient. Shull concluded that these results probably mean that the wilting coefficient represents a fairly definite water-holding power for the soil particle regardless of its size. Thus the force with which the soil particles of any type of soil withhold water from the root hairs of plants at the wilting coefficient is no greater than 3 to 4 atmospheres. Let us now consider how the soil forces at the wilting coefficient compare with the osmotic pull of the root hair. In Chap. I it was stated that their osmotic pressure varied with the conditions under which they were grown and that the work of Roberts (1916) showed that the root hairs adjust their osmotic pressure as the medium changes, keeping it approximately 4 atmospheres above that of the surrounding medium.

Stocker (1930) noted in this regard that the absorbing power of the cells of the roots of plants growing on the steppes of Hungary was generally only 2 to 7 atmospheres higher than the back pull of the soil for water. The root hairs of these plants die when the pull of the soil approaches 30 atmospheres. It thus seems reasonable to assume that as the soil dries out even below the wilting coefficient the root hairs maintain an osmotic pressure a few atmospheres in excess of the soil forces until these forces become relatively very high. Shull concluded that the wilting of the plant at the wilting coefficient of the soil cannot be due to the lack of moisture in the soil nor the lack of a gradient of forces tending to move the water toward the plant, but the wilting of the plants at this critical soil-moisture content must be due to the increasing slowness of the water movement from soil particle to soil particle and from these to the root hairs, the rate of movement falling below that necessary to replace that lost from the aerial parts even under conditions of low water loss from the plant.

2. The Rate of Water Movement in the Soil. *a. Methods of Measuring.*—Methods for measuring the water-supplying power of the soil have been studied and devised by Livingston and Hawkins (1915), Pulling and Livingston (1915), Livingston and Koketsu (1920), Mason (1922), Livingston and Ohga (1926), Hardy (1923), Baldwin (1928), and Wilson (1927). The method devised by Livingston and Koketsu (1920) has proved very satisfactory and gives promise of furnishing a means of predicting the need and time of application of water to plants. This method consists of pushing into the soil porcelain cones that have a known area of unglazed surface. The cones are weighed before insertion into the soil and after remaining there for a given period are brushed free from soil particles and weighed. In this manner the amount of water that they absorb is determined. The amount of water absorbed by these cones is taken as an index of the water-supplying power of a given soil under a given condition. Two points must be considered in the use of an absorbing instrument to measure the water-supplying power of the soil: (1) whether the absorbing surface is in such capillary contact with the soil as to allow the water to move from the soil to the instrument as rapidly as it arrives at the soil surface adjacent to it and (2) whether the absorbing surface itself is able to absorb water as rapidly as it is delivered to it by the soil. Numerous investigations by the authors mentioned on page 202 indicate that the soil points of Livingston and Koketsu (1920) are reliable in these respects.

Livingston and Koketsu (1920) studied *the critical water-supplying power of the soil* or the rate at which the water moves toward the plant when the moisture content has been depleted to or near the wilting coefficient. Their method of procedure was somewhat as follows: Plants of *Coleus blumei* and *Triticum sativum* were grown in 4-in. flowerpots and allowed to grow until they had reduced the soil moisture to or near the wilting point. At this point the dry unglazed porcelain cones were pushed into the soil and allowed to remain approximately 2 hr. These were then removed and their intake of water determined. The amount of water received by these cones, the area of which approximated 10 sq. cm., in a given time was taken as an index to the water-supplying power of the soil at that moisture content. The investigators worked with 12 types and mixtures of soils and found that the water-supplying power of the soil at the wilting coefficient does not vary with the physical make-up of the soil and is thus the same for sand, loam, humus, or any mixture of these.

b. Factors Influencing the Rate.—The two main environmental conditions that influence the relation of plants to water are (1) the capacity of the soil to deliver water to the plant roots, and (2) the capacity of the aerial surroundings to cause water loss by transpiration. The water-supplying power of the soil is dependent upon the temperature, upon the gradient of the moisture content, and upon the level of this gradient. The loss of water in transpiration is influenced not only by the climatic conditions but by the nature of the plant under consideration.

Kramer (1934) by means of soil points showed that an increase in the soil temperature increased the water-supplying power of the soil as shown in the table at the top of page 204.

Magistad and Breazeale (1929) mentioned that the rate of water movement in the soil is dependent not only upon the gradient of the moisture content but also upon the

Series A		Series B		Series C	
Temperature, Centigrade	Water absorbed, mg.	Temperature, Centigrade	Water absorbed, mg.	Temperature, Centigrade	Water absorbed, mg.
0.0	5.8	0.0	57.2	0.0	27.9
8.5	8.5	8.2	96.6	9.5	38.3
23.5	12.2	24.0	132.2	23.8	49.0
35.0	15.1	34.8	171.8	35.0	59.8
				43.0	66.5

amount of moisture present. Thus the rate of water movement in a soil from a region of 4 per cent moisture to one of 3 per cent is much lower than that from a water content of 20 per cent to one of 19 per cent although the gradient is the same in either case. It was first noted by Livingston and Koketsu (1920) that the average water-supplying power of the soils with which they worked was, at the wilting coefficient, 85 mg. during each 2-hr. period for the type of soil points used. They found that the movement of water when the soil was drain-saturated was 6,000 times more rapid than when the moisture content was reduced to the wilting coefficient.

The term "soil-moisture index" was used by Wilson and Livingston (1932) to express the water-supplying power of the soil under different plants. This term indicates the number of milligrams of water absorbed per square centimeter of soil point during 1 hr. An index value of 100 indicates that the soil can supply water to the absorbing surface at the rate of 8 mg. per square centimeter for the first hour. This value was taken as a basic point because, in a majority of the cases studied, drought injury became apparent at or near the time that the water-supplying power of the soil was reduced to this rate.

When death from lack of water was first noted, the soil-moisture indexes at a depth of 6 cm. varied considerably for different plants as shown below:

Plant	Soil-moisture index	Plant	Soil-moisture index
Tall oat grass	52	Perennial rye grass	38
Velvet bent grass	44	Timothy and red top	37
Meadow fescue	43	Bermuda grass	34
Brome grass	42	Alsike clover	30
Kentucky bluegrass	41	Red fescue	29

The influence of the plant on the water-supplying power of the soil has been studied also by Welton and Wilson (1931). They determined the water-supplying power of the soil under Kentucky bluegrass, narrow-leaved fescue, and Washington bent grasses at different moisture contents of the soil. In a majority of the 40 observations the water-supplying power of the soil under the narrow-leaved fescue was 60 per cent greater than that under Kentucky bluegrass. These results suggest that the former plant places a smaller demand on the available soil moisture and that this characteristic is probably a significant factor in enabling this species to survive drought periods better than bluegrass. Nutman (1933) found that the maximum rate of the entry of water into the roots of *Coffea arabica* in East Africa was 25 cc. per square meter per hour.

If the absorbing surfaces of the root are fixed in position and extent, the maximum rate of absorption is limited to any instant by the rate at which the soil system can supply water to these surfaces (Wilson and Livingston, 1932). Conditions in the plant, however, will control the rate of absorption of water as long as it is slower than the highest possible rate of the external supply. When an absorbing surface of a root initially comes in contact with a new system of soil films by the penetration of roots into untouched soil, absorption may be relatively rapid and the water-supplying power of the soil system may be greater for a time than the absorbing power of the root surface. This, however, does not long continue. Under conditions of a diminishing soil water supply the roots must advance into unpenetrated moist soil at an adequate rate if the aerial parts are to be supplied with sufficient moisture for their needs. The process of root extension thus becomes of great importance to plants in times of drought.

Veihmeyer (1927) and Veihmeyer and Hendrickson (1927, 1934) noticed that under comparable atmospheric conditions the rate of extraction of moisture by prune and peach trees was the same regardless of the moisture content of the soil above the wilting point. Apparently the roots of these trees were able to obtain water as readily when the soil-moisture content had been reduced almost to the wilting coefficient as when the soil was filled with water to its maximum field capacity. Lewis, Work, and Aldrich (1935), however, in Oregon found that the growth of the fruit of pear trees growing on heavy soil was reduced whenever the soil moisture in the major portion of the root zone was reduced below 70 per cent of the available capacity.

The work of the various investigators shows that the rate of movement of water in the soil from particle to particle toward the roots is one of the most important factors that influence the intake of water by the plant, especially when the water supply becomes low. This substantiates the statement of Livingston and Hawkins (1915) that, whatever may be the nature of the forces producing the migration of water from soil to plant, it seems that osmotic pressure does not play a prime and direct role in this movement, but the power of the soil to deliver water to root surfaces is the principal condition that determines the moisture supply to plants in nonsaline and nontoxic soils.

III. ENTRANCE OF WATER INTO THE ROOT

A. The Structure of the Root

In order to discuss the forces that are considered to act in the intake of water by the root, its structure must be clearly understood. From Fig. 9*A* it is seen that water must transverse three definite tissues in moving through the root from the soil to the stele or fibrovascular bundle and the cells enclosing it. The three tissues are the epidermis, the cortex, and the endodermis. The general nature of the epidermal cells of the root has been discussed in considerable detail in Chap. III and need not be considered here. The cortical cells are of the typical thin-walled vacuolated parenchyma type that is very common in plant tissue. The endodermis is the most outstandingly differentiated of the three tissues. It is a cylindrical layer of cells one layer thick which surrounds the fibrovascular bundle and tissues adjoining it (Eames and MacDaniels, 1925). The cells of this layer are living and are closely packed together and devoid of any intercellular spaces. The differentiations peculiar to the endodermis lie principally in the structure of the cell walls that compose it. In the young root these cells are thin walled, but the end and radial walls are thickened by a band or strip that is known as the "Casparian strip" (Fig. 9*B* and *C*). The development of the endodermis has recently been studied by Priestley, Rhodes, and North (1922, 1926)

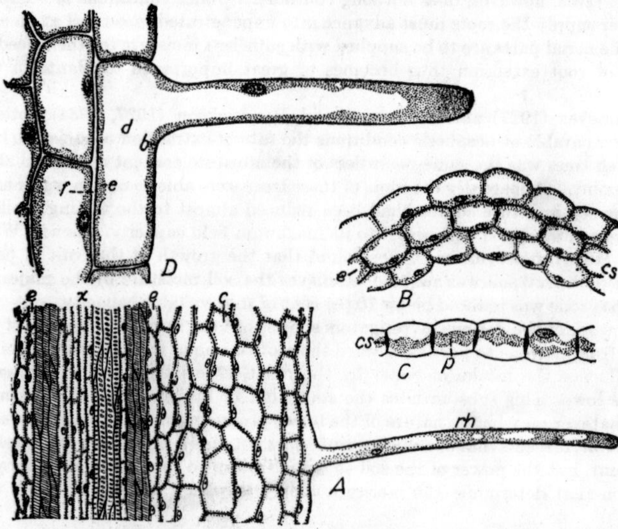

Fig. 9.—Drawings showing the cells of the root that are concerned in the absorption of water and its translocation to the central cylinder. *A*, longitudinal section of the absorbing portion of a root of the lupine; *rh*, root hair; *c*, cortex, *e*, endodermis; *x*, central cylinder. *B*, enlarged portion of the endodermis and adjoining cells; *e*, endodermis; *cs*, Casparian strip. *C*, the endodermis showing the Casparian strip and the attachment of the protoplasm to it at plasmolysis. *D*, diagram illustrating the manner in which water may be absorbed and translocated across the root. Description in the text.

especially in regard to the nature of the substances that compose the cell wall and the Casparian strip. They observed that the first noticeable differentiation of the endodermal cells from the apical meristem is the appearance of the Casparian strip, which is laid down even before cellulose can be detected in the walls of these cells. The wall of the cells of the endodermis with the exception of the Casparian strip develops into a thin cellulose wall that is soluble in sulphuric acid. The Casparian strip, however, does not so dissolve and remains intact when the other portions of the cell wall go into solution. It is more resistant to sulphuric acid and less resistant to oxidizing agents than the cellulose wall. The various reactions suggest the presence in the Casparian strip of basal substances that at a very early time in the development have been impregnated with complex substances containing derivatives of certain compounds that are apparently allied with suberogenic and cutinogenic acids and with substances giving reactions to lignins. The basal material of the Casparian strip is not definitely known. Priestley and Rhodes (1926) stated that upon treatment of this material with concentrated nitric acid it yielded a brown body, insoluble in water and dilute acids, and an ether-soluble body of a waxy appearance. Neither of these bodies, however, could be identified. The cell wall of the endodermal cells at a later stage of the development of the root becomes greatly thickened, lignified, and suberized, but from the standpoint of water absorption that later differentiation need not be considered.

The endodermal cylinder of the young root has thus in the Casparian strip an anatomical structure that renders difficult the radial passage of water through it by way of the radial walls. The passage of water through the endodermal cells of the young root must then take place through the protoplasm. The protoplasm of the endodermal cell is apparently closely attached to or interwoven with the Casparian strip, since it has been shown by De Rufz de Lavison that, when the cells are vigorously plasmolyzed, the protoplasm does not withdraw from the Casparian strip but remains suspended across the cell from strip to strip (Fig. 9*C*) after plasmolysis around the remainder of the cell wall has been complete (Bryant, 1934). The roots of certain plants have some cells in the endodermis that remain unsuberized and are termed "passage cells." It was observed by Scott (1928) that lead salts and dyes which cannot penetrate the suberized walls of the endodermis pass through these passage cells.

Priestley and Tupper-Carey (1922) obtained evidence to show that very little water can enter the root through the root apex. They stated that the meristematic tissue or cap closing the endodermal cylinder in the young root will prevent the passage of water under a pressure of 2 atmospheres. This tissue is also relatively impermeable to salts in solution. The stem apex, on the other hand, is very permeable to water and solutes. Its cell walls are composed of cellulose, while the cells that make up the apex of the root have walls of an unknown substance that apparently may be converted into cellulose by boiling with potassium hydroxide.

B. Modes of Water Entrance into the Roots

Two main forces are apparently concerned in the entrance of water into the root from the soil and its movement across it to the fibrovascular bundle. The forces are imbibitional forces and osmotic forces.

1. Imbibition.—The role of imbibition in the processes of plants has been shown by Shull (1924) and others to be of much import. The attraction or affinity of dry cell walls and protoplasm for water amounts in many cases to hundreds of and even 1,000 atmospheres (Shull, 1914 to 1916), and it has also been shown that very strong imbibitional forces may be developed within the plant even when the cell walls and protoplasm contain considerable water. If moisture is withdrawn from any region of an imbibing system, the surface tension of the watery films in that region is increased, and a corresponding force or pull is exerted upon the surrounding medium for the movement of water into the region of loss to restore the equilibrium of the surface tension of these films. This deficiency of water in any region of an imbibing system has been termed by Livingston and Brown (1912) "incipient drying" and by Renner (1912) "the saturation deficit."

All the external contacts of the plant that are concerned in the intake of water and also in water loss are made through the medium of imbibing surfaces. The cell wall of an epidermal cell of the young root is colloidal in nature and has a strong affinity for the water of the soil particles with which it is in close contact. The protoplasm that borders upon the cell wall likewise has a strong attraction for water. Let us now consider an epidermal cell of a young root in which the imbibitional and osmotic forces are in equilibrium with the soil and the surrounding cells. Let us now assume that water is withdrawn from the exterior surface of the cell wall (Fig. 9*Df*). A saturation deficit is set up at *f* and a pull of equal magnitude is exerted upon the water farther back in the cell wall. When the imbibing forces of the cell wall are thus set in action and cause a movement of water toward it, the supply of water must be secured from the protoplasm *e* with which it is in close contact. Holle (1915) found that during wilting the cell wall keeps in close contact with the protoplasm, thus permitting the continual withdrawal of water from it. The withdrawal of water from the protoplasm produces in it a saturation deficit that is satisfied by imbibing free water from the cell sap *d*. This action increases the absorbing power of the cell sap and causes the entrance of imbibed water from the protoplasm at *b*, which draws finally upon the exterior cell wall whose saturation deficit must be supplied by the water of the soil. The migration of water across the root to the endodermis could be explained, in part at least, under certain conditions after the same manner as has been described for the movement of water across the epidermal cell. Kunkel (1912) considered that imbibitional factors were the main ones in the intake of water by the root. It seems very evident, however, that imbibitional factors do not act independently but in connection with osmotic factors, since the imbibitional disturbances create osmotic disturbances and both together are equally important in the process of water intake and translocation across the root. Some water apparently may enter the cell wall of the epidermal cell and travel across the root through the cell walls of adjacent cells as far as the Casparian strip of the endodermis without the aid of the protoplasm or the cell sap. In such a case, imbibitional forces alone would be concerned. The intake of water by the young cells of the root where no vacuole is yet developed is apparently a purely imbibitional process. The amount of water, however, that may enter the plant under the direct influence of imbibitional forces alone is very small in amount compared to the intake through the vacuolated cells.

When transpiration is proceeding even at a low rate during the daylight hours the leaves tend to dry out on account of water being abstracted from them by evaporation, and a saturation deficit is produced. This results in the water of the vessels passing into a state of tension that is transmitted equally in all directions, so that under these conditions water is not being forced up through the fibrovascular bundles by the roots but is apparently being pulled up the stem under great stress by the forces operative in the leaves. Under these conditions when a stem is cut off water does not exude from the stump, and if the cut surface of the stump is immersed in water, water will be sucked back into it, showing that there is a negative pressure in the conducting vessels as far as the roots are concerned. Dixon (1910) and Renner (1912) considered that the water-absorbing power of the roots is directly related to the saturation deficit in the leaves and that the tensile stress developed in the cells of the leaves and conducting vessels on this account is transmitted to the root and has the ultimate effect of drying out the root surface. According to this theory of water intake in the absence of root pressure, the passage of water from the soil into the root is purely passive, and the root acts merely as a filter in the process. Of what importance the osmotic forces of the root are under the conditions of passive absorption is not known. Pulling and Livingston (1915) considered that the osmotic forces may influence the process of absorption in an indirect manner by maintaining the form of the roots and their contact with the water films of the surrounding soil. They also stated that it has been shown that roots do not take up water at rates proportional to the osmotic pressures of the cells, and that absorption may go on equally well with low or high turgidity of roots as long as the protoplasts of the cells involved are not deformed by plasmolysis.

2. Osmotic Forces.—The osmotic value of the cell sap of the epidermal cells of the roots is always higher than that of the soil solution, and apparently, as mentioned in Chap. I, this superiority is maintained as the concentration of the external medium is increased. The effective force, however, with which water will be drawn into the cell by osmotic action will be the difference between the osmotic value of the cell sap and the osmotic value of the external soil solution less the turgor pressure of the cell wall (Thoday, 1918). This absorbing power of the epidermal cell is equivalent to 1 atmosphere or more in the cases that have been observed. The conditions are fulfilled for the migration of water into the root, provided that there is an increasing gradient of absorbing power from cell to cell across the cortex to the fibrovascular bundle. The absorbing power of the cells of the root has been investigated by Ursprung and his workers, and they have determined that there is a gradient of absorbing power proceeding from the epidermal cell across the cortex to the endodermis and then a marked decrease in the endodermis and wood parenchyma. The following table is taken from Molz (1926):

Absorbing Power in the Absorbing Zone of the Root of *Vicia faba*

Position of cells	Absorbing power, atmospheres	Position of cells	Absorbing power, atmospheres
Epidermis	0.7	Fifth row	2.8
Cortex cells		Sixth row	3.0
First row	1.4	Endodermis	1.7
Third row	1.5	Wood parenchyma	0.9
Fourth row	2.1		

In the intake of water by the root and in its movement across the same, osmotic forces always act in connection with imbibitional forces. There is in this connection as well as in the general movement of water through plant cells a tendency for these two forces to proceed toward equilibrium (Shull, 1924). Wherever imbibitional forces exceed osmotic forces, water will move into the imbibing matter, or if the osmotic forces are the larger, water will move from the imbibed material into the cell vacuole.

Under conditions of low transpiration or in the spring before the leaves are unfolded, water is forced into the conducting vessels of the root and up through the stem under pressures varying from a fraction of an atmosphere to several atmospheres. This pressure is evidenced by the bleeding of cut vines and branches of certain species of plants and is apparently connected with the exudation of water from the leaves of plants, which occurs under certain conditions. This pressure, which is set up in the fibrovascular bundles of the stem and root owing to the water that is being forced in them, is known as "root pressure." The forces that are involved in the transference of water across the endodermis into the fibrovascular bundle are not known. The effects of the deprivation of oxygen and of changes in temperature suggest that root pressure cannot be explained by means of osmotic and imbibitional forces alone and indicate that the vital relations of the cell involving energy transference are concerned (Blackman, 1921). The nature of the Casparian strip, which prevents the transference of water across the radial walls of the endodermal cells, would indicate that the water transference across the endodermis is under control of the protoplasm of these cells (Priestley, 1920). The manner in which the protoplasm brings about this migration of water into the bundles has not been determined experimentally, and any suggestions that have been made are purely theoretical. Most of the theories proposed assume that the processes involved are of a secretory or excretory nature. Since little or nothing is known, however, concerning the nature of the factors involved in the processes of secretion or excretion in either plants or animals, to state that water transference into the bundle from the endodermis is a process of secretion does not illuminate the matter at all. We must, therefore, in the words of Blackman (1921), have a much greater knowledge of cell dynamics than we possess at present before we can explain the exudation of water and root pressure with any degree of satisfaction.

Under conditions of little or no transpiration and an abundant water supply, the intake of water and its movement across the root can be explained by purely osmotic and imbibitional forces in the root, provided that a saturation deficit is created in the endodermis. At present, however, there is no satisfactory way of explaining how this saturation deficit is brought about by the endodermal cells.

Kramer (1932, 1933) noted that positive pressures were manifested at the cut surfaces of stems only when the roots were alive, and that the immersion of the living root systems into sucrose solutions stopped the exudation of water from cut stems. He further observed that transpiring woody and herbaceous plants would remain alive and unwilted for several days after their roots had been killed. Under these conditions, however, considerable quantities of water were absorbed from the soil through the dead root systems. When suction was applied to cut stems, much more water was taken up after the roots were killed than before. He believed that absorption by osmosis is too slow to replace the water lost by transpiration and that under conditions of high evaporation the role of living cells in absorption is apparently a passive one. They are important in absorbing surfaces in preventing the entrance of air into the vessels. It was thought by Reed and Bartholomew (1927) that the pectic materials of the cell walls of immature fruits absorb water by imbibition since the majority of such cells are some distance from the ultimate division of the bundle. Henderson (1934), however, found a correlation between the volume of water absorbed and the rate of respiration in the roots. This indicates that these organs expend energy in the process of absorption.

It is evident that the intake of water by the root and its movement across it is a very complex process and that numerous factors and forces are involved. Imbibition, osmotic forces, and passive absorption are evidently concerned in the process and under certain conditions act separately and under others jointly. That these forces can supply the plant with sufficient water to replace that lost during vigorous transpiration seems doubtful, and, in the words of Overton (1921), little is actually known about the forces that are operative in the passage of water from the soil through the root into the vessels in sufficient quantity to supply the transpiration needs.

Bibliography

ALWAY, F. J., 1913: Studies of the relation of the non-available water of the soil to the hygroscopic coefficient, *Nebr. Sta. Res. Bull.* **3**: 5–122.

BAKKE, A. L., 1915: The index of foliar transpiring power as an indicator of permanent wilting in plants, *Bot. Gaz.*, **60**: 314–319.

———, 1918: Determination of wilting, *Bot. Gaz.*, **66**: 81–116.

BALDWIN, H. I., 1928: A comparison of the available moisture in sod and open soil by the soil point method, *Bull. Torrey Bot. Club*, **55**: 251–255.

BLACKMAN, V. H., 1921: Osmotic pressure, root pressure, and exudation, *New Phytol.*, **20**: 106–115.

BOUSSINGAULT, J., 1878: Étude sur les fonctions physiques des feuilles: transpiration, absorption de la vapeur aqueuse, de l'eau des matières salines, *Ann. Chim. et Phys.*, Ser. 5, **13**: 289–394. Quoted by Williams, 1932.

BOWEN, E. J., 1931: Water conduction in *Polytrichum commune*, *Ann. Bot.*, **45**: 175–200.

BREAZEALE, J. F., 1930: Maintenance of moisture equilibrium and nutrition of plants at and below the wilting percentage, *Ariz. Sta. Tech. Bull.* 29.

BRIERLEY, W. G., 1934: Absorption of water by the foliage of some common fruit species, *Proc. Am. Soc. Hort. Sci.*, **32**: 277–283.

BRIGGS, L. J., and H. L. SHANTZ, 1911: Application of wilting-coefficient determinations in agronomic investigations, *J. Am. Soc. Agron.*, **3**: 250–260.

——— and ———, 1912: The wilting coefficient for different plants and its indirect determination, *U. S. Bur. Plant Ind. Bull.* 230.

——— and ———, 1912: The wilting coefficient and its indirect determination, *Bot. Gaz.*, **53**: 20–37.

——— and ———, 1912: The relative wilting coefficient for different plants, *Bot. Gaz.*, **53**: 229–235.

——— and ———, 1913: Die relativen Welkungskoeffizienten verschiedener Pflanzen, *Flora*, **105**: 224–240.

BROWN, W. H., 1912: The relation of evaporation to the water content of the soil at the time of wilting, *Plant World*, **15**: 121–134.

BRYANT, A. E., 1934: A demonstration of the connection of the protoplasts of the endodermal cells with the Casparian strips in the roots of barley, *New Phytol.* **33**: 321.

BURT, E. A., 1893: Do the leaves of our ordinary land plants absorb water? *Sci.*, **22**: 51–52.

CALDWELL, J. S., 1913: The relation of environmental conditions to the phenomenon of permanent wilting in plants, *Physiol. Res.*, **1**: 1–56.

Capalungan, A. V., and H. F. Murphy, 1930: Wilting coefficient studies, *J. Am. Soc. Agron.*, **22**: 842–847.

Crump, W. B., 1913: Notes on the water content and the wilting point, *J. Ecol.*, **1**: 96–100.

Dandeno, J. B., 1901: An investigation into the effects of water and aqueous solutions of some of the common inorganic substances of foliage leaves, *Trans. Can. Inst.*, **7**: 238–350.

Darrow, G. M., and H. Sherwood, 1931: Transpiration studies on strawberries. *Proc. Am. Soc. Hort. Sci.*, **28**: 225–230.

Dixon, H. H., 1910: Transpiration and the ascent of sap, *Prog. rei Bot.*, **3**: 1–66.

Eames, A. J., and L. H. MacDaniels, 1925: "An Introduction to Plant Anatomy," pp. 100–103, McGraw-Hill Book Company, Inc., New York.

Ganong, W. F., 1894: On the absorption of water by the green parts of plants, *Bot. Gaz.*, **19**: 136–143.

Gates, F. C., 1914: Winter as a factor in the xerophily of certain evergreen ericads, *Bot. Gaz.*, **57**: 445–489.

Grundmann, W., 1923: Die Wasseraufnahme der Pflanzen durch die Blätter, *Z. Pflanzenernähr. Düng. u. Bodenk.*, **33**: 88–90.

Haines, F. M., 1932: A self-recording potometer with a note on rate of transpiration under pressure, *Ann. Bot.*, **46**: 1051–1060.

Halket, A. C., 1911: Some experiments on absorption by the aerial parts of certain salt plants, *New Phytol.*, **10**: 121–139.

Hardy, F., 1923: The soil-point method for directly estimating the water-supplying power of a soil in the field, *J. Agr. Sci.*, **13**: 355–360.

Henderson, L., 1934: Relation between root respiration and absorption, *Plant Physiol.*, **9**: 283–300.

Hendrickson, A. H., and F. J. Veihmeyer, 1929: Some facts concerning soil moisture of interest to horticulturists, *Proc. Am. Soc. Hort. Sci.*, **1929**: 105–108.

——— and ———, 1934: Irrigation experiments with prunes, *Calif. Sta. Bull.* 573.

Henslow, G., 1908: On the absorption of rain and dew by the green parts of plants, *J. Roy. Hort. Soc.*, **34**: 167–178.

Holle, H., 1915: Untersuchungen über Welken, Vertrocknen und Wiederstraffwerden, *Flora*, **108**: 73–126.

Kessler, O. W., 1933: Die Wasseraufnahme der Pflanzen durch die Blätter und ihre Bedeutung für die Feldberegnung, *Die Feldberegnung*, **38**: 111–118.

Koketsu, R., 1928: Variation of the water content of leaves as related to the wilting of plants, *J. Dept. Agr. Kyushu Imp. Univ.*, **2**: 93–116; *Exp. Sta. Rec.*, **61**: 62.

Kramer, P. J., 1932: The absorption of water by root systems of plants, *Am. J. Bot.*, **19**: 148–164.

———, 1933: The intake of water through dead root systems and its relation to the problem of absorption by transpiring plants, *Am. J. Bot.*, **20**: 481–492.

———, 1934: Effect of soil temperature on the absorption of water by plants, *Sci.*, **79**: 371–372.

Kunkel, L. O., 1912: A study of the problem of water absorption, *Mo. Bot. Gard. Ann. Rept.*, **23**: 26–40.

Lewis, M. R., R. A. Work, and W. W. Aldrich, 1935: Influence of different quantities of moisture in a heavy soil on rate of growth of pears, *Plant Physiol.*, **10**: 309–323.

Livingston, B. E., and W. H. Brown, 1912: Relation of the daily march of transpiration to variations in the water content of foliage leaves, *Bot. Gaz.*, **53**: 309–330.

——— and L. A. Hawkins, 1915: The water relation between plant and the soil, *Carnegie Inst. Wash. Pub.*, **204**: 1–48.

——— and R. KOKETSU, 1920: The water supplying power of the soil as related to the wilting of plants, *Soil Sci.*, **9**: 469–485.

——— and I. OHGA, 1926: The summer march of soil-moisture conditions as determined by porcelain soil points, *Ecology*, **7**: 427–439.

———, T. HEMMI, and J. D. WILSON, 1926: Growth of young wheat plants in autoirrigated soils, as related to the water supplying power of the soil and to the adjustment of the autoirrigator, *Plant Physiol.*, **1**: 387–395.

MACDOUGAL, D. T., 1925: Absorption and exudation pressures of sap in plants, *Am. Phil. Proc.*, **64**: 102–130.

MAGISTAD, O. C., and J. F. BREAZEALE, 1929: Plant and soil relations at and below the wilting percentage, *Ariz. Sta. Tech. Bull.* 25.

MASON, T. G., 1922: The physiological humidity of the soil and its direct determination, *W. Ind. Bull.*, **19**: 137–154.

MILLER, E. C., 1916: Comparative study of the root systems and leaf areas of corn and the sorghums, *J. Agr. Res.*, **6**: 311–331.

MOINAT, A. D., 1932: Available water and the wilting of plants, *Plant Physiol.*, **7**: 35–46.

NAKAJIMA, Y., 1928: Wirkung des der Luft entleerten Wassers auf die Wasseraufnahme verschiedener Körper, *Sci. Rept. Tôhoku Imp. Univ.*, Fourth Series, (Biol.), **3**: 279–298; *Biol. Abst.*, **4**: 686.

NUTMAN, F. J., 1933, 1934: The root system of *Coffea Arabica*. I, II, and III, *Empire J. Exp. Agr.*, **1**: 271–284, 285–296 (1933); **2**: 293–302 (1934).

OVERTON, J. B., 1921: The mechanism of root pressure and its relation to sap flow, *Am. J. Bot.*, **8**: 369–374.

POWERS, W. L., 1922: Field moisture capacity and wilting point of soils, *Soil Sci.*, **14**: 159–167.

PRIESTLEY, J. H., 1920: The mechanism of root pressure, *New Phytol.*, **19**: 189–200.

———, 1922: Further observations upon the mechanism of root pressure, *New Phytol.*, **21**: 41–48.

——— and E. E. NORTH, 1922: Physiological studies in plant anatomy. III, The structure of the endodermis in relation to its function, *New Phytol.*, **21**: 113–139.

——— and E. RHODES, 1926: On the macro-chemistry of the endodermis, *Proc. Roy. Soc.*, B, **100**: 119–129.

PULLING, H. E., and B. E. LIVINGSTON, 1915: The water supplying power of the soil as indicated by osmometers, *Carnegie Inst. Wash. Pub.*, **204**: 49–84.

———, 1917: The rate of water movement in aerated soils, *Soil Sci.*, **4**: 239–268.

REED, H. S., and E. T. BARTHOLOMEW, 1927: Water translocation in young fruits, *Sci.*, **66**: 382–385.

RENNER, O., 1912: Versuche zur Mechanik der Wasserversorgung. I, Der Druck in den Leitungsbahnen von Freilandpflanzen. II, Über Wurzeltätigkeit, *Ber. bot. Ges.*, **30**: 576–580, 642–648.

SACHS, J., 1859: Ueber den Einfluss der chemischen und physikalischen Beschaffenheit des Bodens auf die Transpiration der Pflanzen, *Landw. Versuchs-Stat.*, **1**: 203–241.

SCOTT, L. I., and J. H. PRIESTLEY, 1928: The root as an absorbing organ. I, A reconsideration of the entry of water and salts into the absorbing region, *New Phytol.*, **27**: 125–141.

———, 1928: The root as an absorbing organ. II, The delimitation of the absorbing zone, *New Phytol.*, **27**: 141–174.

SHIVE, J. W., and B. E. LIVINGSTON, 1914: The relation of atmospheric evaporating power to the soil moisture content at permanent wilting in plants, *Plant World*, **17**: 81–121.

Shull, C. A., 1913: Semipermeability of seed coats, *Bot. Gaz.*, **56**: 169–199.
———, 1914: Measurement of the internal forces of seeds, *Trans. Kans. Acad. Sci.*, **27**: 65–70.
———, 1916: Measurement of the surface forces in soils, *Bot. Gaz.*, **62**: 1–29; *Sci.*, **43**: 361.
———, 1924: Imbibition in relation to absorption and transportation of water in plants, *Ecology*, **5**: 230–241.
———, 1930: Absorption of water by plants and the forces involved, *J. Am. Soc. Agron.*, **22**: 459–471.
Smyth, E. S., 1934: A contribution to the physiology and ecology of *Peltigera cania* and *P. polydactyla*, *Ann. Bot.*, **48**: 781–818.
Spalding, V. M., 1906: Biological relation of desert shrubs. II, Absorption of water by leaves, *Bot. Gaz.*, **41**: 262–283.
Stocker, O., 1930: Über die Messung von Bodensaugkräften und ihrem Verhältnis zu den Wurzelsaugkräften, *Z. Bot.*, **23**: 27–56; *Biol. Abst.*, **6**: 678.
Thoday, D., 1918: On turgescence and the absorption of water by the cells of plants, *New Phytol.*, **17**: 108–113.
Veihmeyer, F. J., 1927: Some factors affecting the irrigation requirements of deciduous orchards, *Hilgardia*, **2**: 125–290.
——— and A. H. Hendrickson, 1928: Soil moisture at permanent wilting of plants, *Plant Physiol.*, **3**: 355–357.
Wadsworth, H. A., and U. K. Das, 1930: Some observations on the wilting coefficient of a selected Waipio soil, *Hawaiian Planters' Rec.*, **34**: 289–294; *Biol. Abst.*, **5**: 2659.
Welton, F. A., and J. D. Wilson, 1931: Water-supplying power of the soil under different species of grass and with different rates of water application, *Plant Physiol.*, **6**: 485–494.
Wetzel, K., 1924: Die Wasseraufnahme der höheren Pflanzen gemässigter Klimate durch oberirdische Organe, *Flora*, **117**: 221–269.
Williams, H. F., 1932: Absorption of water by the leaves of common mesophytes, *J. Elisha Mitchell Sci. Soc.*, **48**: 83–100; *Biol. Abst.*, **8**: 1653.
Wilson, J. D., 1927: The measurement and interpretation of the water supplying power of the soil with special reference to lawn grasses and some other plants, *Plant Physiol.*, **2**: 385–440.
———, and B. E. Livingston, 1932: Wilting and withering of grasses in greenhouse cultures as related to water supplying power of the soil, *Plant Physiol.*, **7**: 1–35.
Wolfe, H. S., 1926: Surface forces of soils within the range of hygroscopic moisture, *Bot. Gaz.*, **82**: 195–207.
Wood, J. G., 1925: The selective absorption of chlorine ions and the absorption of water by the leaves in the genus *Atriplex*, *Australian J. Exp. Biol. Med. Sci.*, **2**: 45–46; *Bot. Abst.*, **15**: 1164.
Work, R. A., and M. R. Lewis, 1936: The relation of soil moisture to pear tree wilting in heavy clay soil, *J. Am. Soc. Agron.*, **28**: 124–134.

CHAPTER V

THE INTAKE OF SOLUTES BY THE PLANT

Under this heading it is intended to discuss more particularly the entrance of the inorganic compounds into those cells of the plant which border upon the medium in which the plant is growing. The discussion will include not only the mode of entrance of the solutes but also the general influence of the environmental factors on the intake of materials by the plant as well as the nature and composition of the solutions that are commonly used in the study of plant nutrition.

I. MODE OF ENTRANCE OF SOLUTES

Numerous experiments have been performed to show that inorganic salts readily penetrate the protoplasmic membrane and enter the vacuole of the cell, but a simple experiment by Osterhout (1909) in this regard will suffice as an example. He noted that *Dianthus barbatus* could be grown with its root hairs entirely free from calcium oxalate if the seeds of this plant were simply placed on the surface of distilled water and allowed to grow. After the root hairs thus grown had reached the desired length they were transferred to a calcium solution and examined from time to time under the microscope for the appearance of calcium oxalate crystals. Controls in the distilled water were examined at the same time and treated in the same manner as the plants in the solution. In no case did the controls produce crystals, but calcium oxalate crystals were always produced in the root hairs when grown in the following solutions: $0.05M$ calcium sulphate, $0.005M$ calcium nitrate, $0.005M$ calcium chloride, and $0.001M$ calcium hydroxide, dilute sea water, and tap water. In each of these solutions the crystals of calcium oxalate made their appearance within the root hairs inside 4 hr. at 30°C. The shortest time required for the formation of crystals was $1\frac{1}{2}$ hr. in the case of calcium sulphate solution. It was also found that the root hairs in contact with particles of precipitated calcium carbonate and tricalcium phosphate produced calcium oxalate crystals, but only after several days. It has generally been considered that inorganic salts penetrate the protoplasm in the dissociated or ionic form. Recently, however, work by Osterhout (1925), Irwin (1926), Cooper, Dorcas, and Osterhout (1929), and Osterhout, Kamerling, and Stanley (1934) indicates that, in some cases at least, certain compounds enter the plant cell through the protoplasm only in the undissociated form.

Brooks (1931) adversely criticized the molecular hypothesis of diffusion.

A. Passive or Physical Absorption

Passive or physical absorption may be defined as the absorption of solutes by cells according to the ordinary laws of diffusion.

It has generally been observed that the protoplasmic membrane of the cells bordering upon the soil or the medium in which the plants are growing is permeable to practically all the inorganic solutes present in the soil solution. It has thus been considered that the entrance of any given solute into the vacuole of these cells is a process of diffusion from the higher concentration of this solute in the soil to a lower concentration in the cell sap. If no disposition were made of the solute which entered, equilibrium in concentration between the cell sap and the soil solution in regard to this solute would be reached and no more would enter. If, however, some or all of this solute were removed from the cell, combined, precipitated, adsorbed, or otherwise rendered osmotically inactive, the concentration of the solute would continue to be lower in the cell sap than in the surrounding medium, so that it would continue to enter the cell. This principle can be illustrated by an experiment, as shown in Fig. 10. A solution of tannin is separated from a solution of ferric chloride by a membrane of collodion or parchment paper. The tannin will not diffuse through the membrane, but the ferric chloride will. As soon as the iron salt enters *A*, it is precipitated as iron tannate so that the concentration of ferric chloride in *A* never becomes equal to that in *B* and more will continue to enter. All the ferric chloride in *B* will pass through into *A*, provided there is a sufficient quantity of tannin in *A* to combine with the iron salt. Frequently the analysis of certain tissues has indicated a higher concentration of certain elements or ions in the cell sap than in the surrounding medium, and this fact has been explained until recently by assuming that a portion of the ions in question was adsorbed, combined, or precipitated so that the concentration was never greater than that of the surrounding medium.

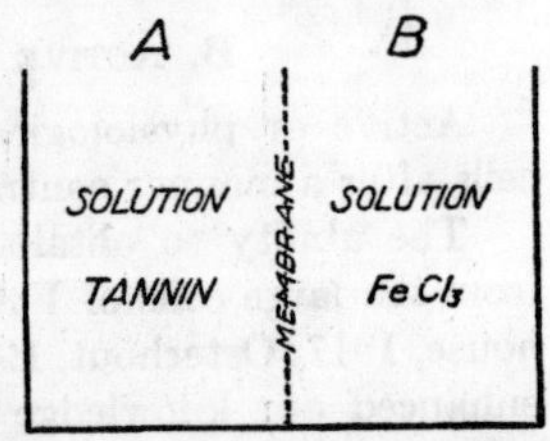

Fig. 10.—Diagram to illustrate the principle of physical absorption. Description in the text.

In a study of the relation between the composition of young orange trees and the concentration of nutrient solutions, Reed and Haas (1924) concluded, however, that there is apparently no evidence to support the assumption that the absorbed materials are converted into insoluble compounds as rapidly as they accumulate in the plant, or that absorption necessarily depends upon the precipitation of ions within the plant.

Gračanin (1932) observed in the plants of barley and corn that the amount of absorption of phosphorus ions by the roots was according to the ordinary physical laws of diffusion. Thus plants growing in solutions with concentrations of 0.001 and 0.05 per cent of phosphorus pentoxide, supplied as monocalcium phosphate, absorbed from equal concentrations of solutions the same quantity of phosphate ions regardless of the degree of illumination.

Examinations of the uncontaminated sap of individual cells show that, in some cases at least, the concentration of the osmotically active solutes is higher in the cell sap than in the surrounding medium. This indicates that the intake of solutes is not always so simple a process as has been mentioned in this topic.

B. Active or Physiological Absorption

Active or physiological absorption is the absorption of solutes by cells after a manner contrary to the ordinary laws of diffusion.

The ability to obtain a sufficient quantity of uncontaminated sap from the large cells of *Valonia* and *Nitella* for chemical analyses (Wodehouse, 1917; Osterhout, 1922; and Hoagland and Davis, 1923) has greatly enhanced our knowledge concerning the concentration of the cell sap as compared to the external medium. Osterhout (1922) made an analysis of the cell sap of *Valonia* and found the concentration of certain elements in it as compared to the sea water in which it grew to be as follows:

Element	Concentration in sea-water, p.p.t.	Concentration in cell-sap, p.p.t.
Chlorine	19.6	21.2
Sodium	10.9	2.1
Potassium	0.5	20.1
Calcium	0.45	0.7
Magnesium	1.31	Trace
SO_4	3.3	0.005

The data show that some factor or factors prevented the sodium, magnesium, and sulphate ions from reaching as high a concentration in the cell sap as in the medium in which the cells were growing. The chlorine content was practically the same inside the cell as in the sea water. The most striking fact of the experiment was the high concentration of potassium in the cell sap as compared to the external concentration. Its concentration in the sap in this case was approximately forty times as high as in the sea water. Practically all the potassium in the cell sap was in the form of chlorides, so that this accumulation could not be attributed to its combination with other compounds.

Blinks and Jacques (1930) noted that the ratio of potassium to sodium in the cell sap of *Halicystis* was relatively low, amounting to only 0.0011, and Jacques and Osterhout (1932) found in *Valonia macrophysa* that potassium was the only cation that accumulated in the cell, and it became approximately 40 times more concentrated inside than outside. Cooper and Osterhout (1930) found that when 0.005*M* ammonium chloride was added to sea water containing cells of *Valonia macrophysa* ammonia soon appeared in the sap and reached as high a concentration as the potassium mentioned above.

The concentration of ions in the cell sap of *Nitella clavata* compared to that of the surrounding medium was investigated by Hoagland and Davis (1923). A summary of their work is given in the following table:

Items considered	K	Na	Ca	Mg	Cl	SO_4	PO_4	NO_3
	p.p.m.							
Pond water in which plants grew	Trace	5	31	41	32	31	0.4	34
Sap from cells	2,120	230	410	430	3,220	800	350	0
Ratio of cell sap concentration to external concentration	2,000	46	13	10	100	26	870	

They considered that the major portion of the inorganic elements in the cell sap was in the ionic state since its conductivity was slightly greater than that of a 0.1 *N* potassium chloride solution. By a comparison of the positive and negative ion equivalents it was possible to balance 80 to 90 per cent of the positive ions by negative ions. The remaining ions were ascribed to organic acids and other organic compounds.

The analysis of the sap showed that the concentration of each of the ions with the exception of NO_3 was higher than in the water in which the cells were growing. The accumulation of potassium was the highest and that of magnesium the lowest. The insoluble or combined elements found in the cell wall and protoplasm included calcium, magnesium, sulphur, silicon, iron, and aluminum, the calcium being the most abundant. No potassium was present in the insoluble form. Hoagland (1919) and Hoagland and Davis (1925) observed that the expressed sap of the leaves of barley growing in an ordinary cultural solution had a conductivity four to fifty times greater than that of the solution from which the plants obtained their inorganic nutrients. This also indicates that the concentration of the ions within the cell is higher than that of the surrounding medium.

In a study of the composition of the cell sap of the plants of corn, sorghum, and Sudan grass, Pierre and Pohlman (1933) found that the total content of electrolytes, as measured by conductivity, was higher in some cases and lower in others than the displaced soil solution. The amount of phosphate ranged between 150 and 400 p.p.m., silica averaged approximately 250 p.p.m., calcium 80 p.p.m., chlorides 90 p.p.m., while nitrates varied from a trace to 344 p.p.m. Phosphorus and silica were found in much higher concentrations in the sap than in the displaced soil solution, the concentration factors being 552 to 4,967, and 15.2 to 348, respectively. The concentrations of chlorine and calcium in the cell were considerably lower than those of the soil solution.

Stiles (1924) in a study of the absorption of salts by storage tissues concluded that the absorption of salt by them is not a simple process of diffusion of the salt through a cell membrane. The vital force that overcomes the physical equilibrium and enables the cell to accumulate salts in concentration greater than the outside is termed "epictesis" by Lapique (1925).

There are evidently many and varied forces concerned in physiological absorption, since the simplest living plant cell is a very complex system that changes continually in response to its environment. The soil in which a plant grows is a heterogeneous medium, and the plant responds to such a medium in a heterogeneous manner by effecting a sort of physiological integration of the various factors concerned (Hoagland, 1930).

Some of the various forces that may be concerned in the accumulation of solutes in the cell sap are: (1) the attraction of colloids for ions, (2) the amphoteric nature of protoplasm, (3) Donnan equilibrium, (4) ionic exchange, and (5) metabolic activity.

1. The Attraction of Colloids for Ions.—In the opinion of Breazeale (1923), the demand for the various elements originates in the tissues of the plants and is carried to the absorbing surface of the root by means of unsaturated colloidal compounds bearing a plus or minus electric charge. For example, let us suppose that the protoplasm in a leaf cell removes an atom of potassium from a colloidal compound and uses it in building up a permanent compound that is to be one of the final constituents of its tissues. This leaves the colloidal compound or molecule out of equilibrium and with a minus charge. This charge is transmitted by replacement and not by bodily movement down through the cells to the epidermal cells of the root and there appears as a minus charge. If an ion of potassium were present in the soil solution, it would be attracted to the negative charge. The potassium could in this manner be transported from the absorbing surface of the root to any part of the plant without the bodily movement of a potassium compound through the sap. Breazeale further considered that if an ion of potassium bearing a plus charge is removed from the soil solution by the root, this ion can be filled by replacement and the charge carried along through the soil, where the source of supply of potassium exists. If this is true, the plant will not be dependent upon the soil grains that come in contact with the epidermal cells of the root but may actually obtain its nutrients at a distance from the roots. The objections to this theory will be brought out in the following discussions.

2. Amphoteric Nature of the Protoplasm.—The proteins and protoplasm of a plant cell are ampholytes. They are thus substances that can behave as either cations or anions depending upon the hydrogen-ion concentration of the medium in which they are placed. The isoelectric point of an ampholyte is the pH value of the solution in which it is placed at which the ampholyte acts as neither an anion nor a cation. The isoelectric point is sometimes called the "neutral point" of the ampholyte in question. Thus an ampholyte on the acid side of its isoelectric point behaves as a cation and combines with anions, while on the alkaline side of the isoelectric point it behaves as an anion and combines with cations. Since the absorption of ions is intimately connected with certain characteristics of the protoplasm and since many of these characteristics are bound up apparently with the peculiar properties of its constituent proteins, it would seem possible that the amphoteric nature of the plant proteins would exert an influence on the absorption of ions (Robbins, 1923).

Lapique (1925) considered that the outer surface of the protoplasm of a plant cell may be in contact with a medium that is either neutral or alkaline to it, while at the vacuole it may be in contact with a medium that is on the acid side of its isoelectric point. As the protoplasm rotates or circulates in the cell, every part of it is alternately in contact with the exterior medium; when it is on the alkaline side of its isoelectric point, it combines with cations and releases anions. As circulation continues it comes in contact with the acid cell sap, releases its cations, and combines with anions. This theory accounts in part for the "heaping up" or accumulation of cations in the cell sap.

Davidson (1933) assumed that there is a wide range in the isoelectric points of the plant ampholytes allowing the occurrence of both electropositive and electronegative ampholytes within certain limits of hydrogen-ion concentration. This makes possible the simultaneous accumulation of cations and anions. A change in the reaction of the medium may modify to some extent the reaction within the plant cells and thus shift some of the electropositive ampholytes to the negative side, and vice versa. This might cause an increased absorption of cations and anions.

3. Donnan Equilibrium.—According to Briggs and Petrie (1928), the conception of a simple Donnan equilibrium operating between two homogeneous phases is inadequate to explain the phenomenon of the ionic intake by plants, since experimental evidence indicates that the product of the apparent internal concentration of cation and anion, estimated on the basis of the total volume of the tissue, may be greater than that of the concentrations of the two ions in the external medium instead of equal to it, as the above conception would require. According to these authors, the interior of the plant tissues comprises a number of phases, each of which may be in Donnan equilibrium with the external solution. The product of the apparent internal ionic concentrations resulting from the total effect of all these internal phases is shown under such circumstances to be greater than the external product. Even in such a system as this, however, if the ratio of the apparent internal to the external concentration is greater than unity, the same ratio of the corresponding anion would be less than unity. Experimental evidence, however, indicates that both may be greater than unity. Briggs and Petrie suggest that this behavior may be accounted for by assuming that one internal phase contains indiffusible cations and that another contains indiffusible anions.

Thomas (1930) believed that for plants in the presence of colloidal silica the increased absorption of phosphorus pentoxide may be explained better on the basis of the Gibbs-Donnan distribution law than by any other theory.

Osterhout and Stanley (1932) demonstrated that the accumulation of electrolytes in *Valonia* does not depend on Donnan equilibrium or the presence within the cell of ions and molecules that cannot pass out through the protoplasm.

4. Ionic Exchange.—It was considered by Casale (1921) that the ectoplasm yields hydrogen ions to the soil solution and thereby establishes a difference of potential between the plant cell and the soil particles. In the process of equalization of those charges, the more distant particles yield cations to the nearer ones, and these in turn yield them to the ectoplasm. Thus fertilizers act, in part, by changing the difference of potential between cells of the plant root and the soil particles.

McGeorge (1932) noted that air-dried tissues of alfalfa showed a definite property of absorbing bases and hydrogen and exchanging them for bases in chemically equivalent proportions. He suggested that living tissues may exhibit this same property and that it may be related to the permeability and cationic concentration within the plant system. It was suggested by Briggs (1932) that, from weak solutions, salt absorption by mature plant cells might consist essentially of an exchange of anions between the solution and the cell sap, and an exchange of cations between the solution and the cytoplasm.

5. Metabolic Activity.—The accumulation or "heaping up" of a given salt in the cell sap must necessarily involve the expenditure of energy to produce and maintain a condition not in conformity with physico-chemical equilibrium. Steward (1932, 1933), Steward, Wright, and Berry (1932), Asprey (1933), and Steward and Berry (1934, 1936) have studied the accumulation of salts in the tissues of a variety of storage organs including carrot, parsnip, beet, mangold, turnip, artichoke, dahlia, potato, and pear. Petrie (1933), Lunderårdh and Burström (1933), Potapov and Stankov (1934), Hoagland and Broyer (1936), and Prevot and Steward (1936) studied the accumulation of salts by the roots of plants. All these investigators considered that salt accumulation in cells is dependent upon metabolic phenomena that presumably supply the necessary energy. In storage tissues, the accumulation of salts is apparently a characteristic of the cells that are starting on a new cycle of intense metabolism which may culminate in cell division. The evidence indicates that the accumulation of salt resides in the surface cells whose respiration is markedly augmented. All the experiments indicate that the variables which affect respiration per unit weight also affect in a similar manner the salt absorption per unit volume, as represented by the internal sap concentration. Apparently the principal factor in the salt accumulation of storage tissues is their ability for renewed active metabolism and even growth. Consequently the amount of salt accumulated is closely related to the rate of aerobic respiration, which represents the major source of energy for these tissues.

Hoagland and Broyer (1936) made extensive studies on the accumulation of salts by the excised roots of barley. They obtained absorption ratios of 15 for potassium, 7 for the nitrate ion, and 10 for bromine when the solution in which the roots were placed contained over 7 milliequivalents of the potassium salt per liter. With sufficiently dilute solutions, absorption ratios in excess of 1,000 could be obtained. They found that the accumulation of salts in roots is dependent upon their age and the proportion of active metabolizing cells, a low initial salt content, and available carbohydrate. Apparently the same general factors that influence the accumulation of solutes by the storage tissues of plants also influence accumulation in the roots.

Prevot and Steward (1936) in barley roots found that the potential absorption surface extends back from the apex to the place where secondary roots appear. A pronounced longitudinal gradation in the capacity for salt accumulation was observed, the segments near the apex attaining higher concentrations than those more remote. This gradient is considered to be a consequence of the progressive development of cells from the root apex.

Since the penetration of ions into the cells that have just been mentioned apparently takes place from a solution of low to a solution of high concentration of these ions, the plant cell must do work in absorbing them. The processes involved would seem to be similar to those which occur in the secretion by the glands of animals, by which compounds are taken from a dilute concentration in the blood and excreted to the outside into a much higher concentration. Just how such a process is brought about is not known, so it is only expressing our ignorance when we state that the transference of an ion from a low concentration to a higher one is a process of secretion. The data that have been presented show that little is yet known concerning the means by which absorption of solutes occurs. The ratio of the internal concentration of a given solute in the cell to its external concentration has been termed the "absorption ratio." Some of the factors that influence its value will be discussed in the following topic.

II. FACTORS THAT AFFECT THE ABSORPTION RATIO

A. Light

Since light either directly or indirectly is the source of the energy of the plant cell, Hoagland and Davis (1923, 1929), Hoagland, Hibbard, and Davis (1926), and Hoagland (1930) undertook a study of the influence of light upon the removal of certain ions from the external solution by *Nitella clavata.* The solution and plants were exposed to light for different lengths of time, the amounts of certain elements remaining in the solution were observed, and the absorption ratio was determined. In one experiment the solution employed was 0.01*M* monopotassium phosphate plus potassium chloride in varying amounts. The general plan of the experiment together with certain of the data is shown in the following table:

Experiment number	Length of experiment, days	Concentration of KCl in external solution p.p.m.	Chlorine remaining in solution, p.p.m.					
			Full illumination	Dim light	Light 9 a.m. to 5 p.m.	Light 11 a.m. to 4 p.m	Light 11 a.m. to 2 p.m.	Dark
1	4	17.8	0	1.5	5.0	13.5		
2	4	35.5	0	14	....	27	..	31
3	5	35.5	3	35	16	22	26	
4	4	35.5	2	27	10	15	25	34

In the case of 0.01*M* potassium nitrate, heavy tests for NO_3 were found in the sap for all cells in the light but only slight tests for those in the dark. In every experiment where the absorption of bromine was under consideration, the exposure to light strikingly increased the ability of the cells to accumulate this element. When a concentration of 5 milliequivalents of bromine was used, the concentration of bromine in the cell sap of *Nitella* kept in the dark did not exceed that of the outside solution, while the sap of those exposed to illumination for an equal period of time had a much higher concentration than the surrounding solution. The following figures will serve as an example of the results obtained. These results were noted at a temperature of 20°C.

Illumination	Time, hours	Concentration of bromine in cell sap, milliequivalents
Continuous light	25	6.4
Continuous darkness	25	2.6
Continuous light	50	12.3
Continuous darkness	50	3.3

It was observed by Benedict (1934) that ultraviolet rays between 2900 and 3100 Å caused an increase in the percentage of calcium in all the plants that he examined. He could not, however, formulate any conclusion from his data relative to the effect of these rays on the content of phosphorus in the plants. Ingold (1936) believed that permeability is increased in some cases by illumination, but it was difficult to determine how much absorption was influenced by this factor. Light apparently has a greater effect upon the absorption of cations than of anions.

It is evident, in the case of *Nitella* at least, that light greatly increases the absorption of certain ions. The cells contain only a very limited supply of organic material that might release energy by respiration, so that little kinetic energy is available except that received directly from the sun. It is suggested that if energy is needed for the absorption of solutes by the cells of the roots, it is supplied by the process of respiration. It is worthy of note here that in the absorption of bromine by *Nitella* the temperature coefficient between 10 to 20 and 14 to 24°C. ranged between 2 and 3. This is of such a high value that it would be considered characteristic of a chemical reaction rather than a physical process of diffusion.

B. Concentration of the Exterior Medium

The data on this subject have been obtained, for the most part, by immersing disks of plant tissue of known dimensions and volume in the solution and then either directly or indirectly determining to what degree the external solution had been depleted of the salt or ions in question. The tissues that have been used are carrot, beet, dahlia, and the tubers of *Helianthus* and Irish potato. The solution is either analyzed chemically, or its changes in concentration are determined by those in its conductivity. Since the volume of the tissue is known and since the decrease in the concentration of a given salt or ion in the solution can be determined after any period of absorption, it is possible to obtain values for the amount of salt or ion absorbed per unit of volume of tissue. The values thus obtained give an expression for the internal concentration of the salt. Thus the ratio of the concentration of the solute in the tissue to its concentration in the exterior medium can be obtained.

It has been observed by Nathansohn (1903, 1904) and Stiles and Kidd (1919) that the equilibrium reached in the absorption of a given solute by living tissue depends upon the external concentration of the solute. While the absolute amount of solute absorbed increased with the increasing concentration of the solute in the external solution, the amount absorbed relative to the external concentration nevertheless decreases rapidly from the lowest concentration upward. When the external concentration of the solute is low, the absorption ratio becomes more than unity, while if the external solution is relatively high, the absorption ratio is always less than unity. The following examples will illustrate these statements: Nathansohn (1903) found in the case of the marine alga *Codium* in twentieth-normal to normal solutions of sodium nitrate that the absorption ratio varied from 0.41 to 0.68, but it did not take up the nitrate ion in sufficient amount to produce equal concentrations inside and outside the tissue. He observed, however, that, in sea water where the concentration of the nitrate ion was low, its concentration was greater in the tissue than in the exterior medium. The effect of different concentrations and different salts upon the absorption ratio of carrot tissue as found by Stiles and Kidd (1919) is shown in the following table:

Salt	Initial external concentration	Length of experiment, hours	Absorption ratio
KCl	*N*/5,000	52	25
	N/500	52	17.6
	N/50	52	2.4
	N/10	52	0.8
NaCl	*N*/5,000	48	46.7
	N/500	48	27.0
	N/50	48	3.5
	N/10	48	0.8
$CaCl_2$	*N*/5,000	42.5	15.3
	N/500	42.5	2.8
	N/50	42.5	0.5
	N/10	42.5	0.24

These results show that the degree of absorption depends not only on the concentration of the external solution but also upon the nature of the ions absorbed. Less calcium ions are apparently absorbed than those of potassium and sodium. It is worthy of note that the absorption ratio for dead tissue never becomes greater than unity and that the concentration within the cells is practically identical with that of the external solution. Irwin (1925) noted when living cells of *Nitella* were placed in different concentrations of brilliant cresyl-blue solutions at pH 6.9 that the greater the concentration of the external dye solution the greater was the speed of accumulation of the dye in the cell sap and the higher the concentration of the dye in the sap at equilibrium. This behavior may be explained as a chemical combination of the dye with a constituent of the cell, since this explanation harmonizes with the fact that the temperature coefficient is about 5.0.

C. The Hydrogen-ion Concentration

There are wide differences of opinion in regard to the effect of a variation of the pH value of the cell or its surroundings upon the absorption of solutes. Any change in hydrogen-ion concentration of the cell or its surroundings alters the amphoteric substances in the cell, but it is not known how such a change alters the intake of ions. It was reported by Breazeale and McGeorge (1932) that plants are not able to absorb phosphate or nitrate ions from solutions of greater alkalinity than that represented by a pH of 7.6. Hydroxyl ions depress the absorption of phosphate and nitrate ions and the optimum pH for the absorption of nutrients is at or near neutrality. Osterhout and Kamerling (1934) concluded that any condition that checks the production of acid in the living cell will check the accumulation of nutrients and growth. Jacques and Osterhout (1934) found that increasing the pH value of the sea water, in which plants of *Valonia macrophysa* were growing, would increase the rate of entrance of potassium, and vice versa. It appeared that photosynthesis increases the rate of entrance of potassium by increasing the pH value of the protoplasm. In darkness there is little or no growth or absorption of electrolytes by this plant. It was observed by Jacques (1936) that the lower the pH value the greater is the tendency of potassium to enter the cell sap of *Nitella*. On the other hand, Pirschle and Mengdehl (1931) could find no definite relationship between absorption and the pH value.

III. THE UNEQUAL ABSORPTION OF IONS

The constituent ions of a salt are not necessarily absorbed in equivalent quantities by plant tissue or by the roots, as has been shown by the work of Nathansohn (1903, 1904), Meuer (1909), Ruhland (1909), Pantanelli (1915, 1918), Johnson (1915), Stiles and Kidd (1919), Redfern (1922), Hoagland (1923), Stiles (1924), Rudolfs (1925), Harris, Hoffman, and Lawrence (1925), Pirschle and Mengdehl (1931), and others. This fact has been determined directly by the chemical analysis of the solutions in which the living tissues had been placed or in which the roots of the plants were growing. The three following experiments will serve to illustrate the main facts observed and the chief points of interest in this regard. Redfern (1922) grew plants of corn and peas in a nutrient solution for 2 weeks and then transferred them to various solutions of calcium chloride. After varying periods of time, these solutions were analyzed and the amount of calcium and chloride remaining determined. A sample of the results where the plants were placed in a tenth-normal solution of calcium chloride is shown in the following table:

Corn			Peas		
Duration of experiment, hours	Percentage absorption of calcium	Percentage absorption of chloride	Duration of experiment, hours	Percentage absorption of calcium	Percentage absorption of chloride
23	17.5	6.5	36	17.7	3.6
48	25.3	6.5	60	12.7	4.0
57	27.0	5.9	84	14.6	3.8

The unequal absorption of ions by the barley plant from a variety of different salt solutions was observed by Hoagland (1923). His observations are summarized in the following table:

ABSORPTION OF IONS FROM DIFFERENT SALTS BY THE BARLEY PLANT
Adapted from Hoagland (1923)

Salt used	Initial concentration, milli-equivalent	Ions absorbed, milliequivalents		Ions entering solution, milliequivalents			Length of absorption period, days
		Cation	Anion	Ca	Mg	K	
K_2SO_4	6.32	0.74	0.40	0.32	0.24	....	1⅙
KCl	6.32	1.28	1.30	0.24	0.18	....	
KNO_3	6.45	1.66	3.19	0.19	0.21	....	
KH_2PO_4	6.32	0.90	0.14	0.28	0.21	....	
$KHCO_3$	6.30	1.20	0.84	0.24	0.20	....	
$CaCl_2$	12.50	1.50	2.37	....	0.28	0.02	2
$Ca(NO_3)_2$	12.70	1.66	5.48	....	0.20	0.09	
$Ca(H_2PO_4)_2$	12.30	1.03	0.95	....	0.27	0.20	
$CaSO_4$	12.30	0.88	0.83	....	0.20	0.09	

Stiles (1924) studied the absorption of the constituent ions of single salts by the storage tissues of the carrot, parsnip, beet, and artichoke. Some observations of his work are shown in the following table:

Salt	Concentration, normality	Tissue	Duration of experiments, hours	Absorption, percentage of original amount	
				Cation	Anion
NH_4Cl	1/100	Carrot	48	57.2	32.2
$(NH_4)_3PO_4$	2/100	Carrot	23.5	28.8	6.0
$(NH_4)_3PO_4$	2/100	Parsnip	24.3	34.5	8.2
$(NH_4)_2SO_4$	1/100	Carrot	22.0	3.9	4.2
KCl	1/100	Carrot	22.0	4.6	6.3
NaCl	1/100	Carrot	47.1	18.5	8.5
Na_2SO_4	1/100	Carrot	22.0	15.0	8.0

Jacques (1934) noted that the inorganic constituents of the cell sap, from the cells of rhubarb, field sorrel, and wood sorrel, showed an average of 3.8:1 preponderance of cations over anions.

A. Mechanism

Although it is an established fact that the ions (cations and anions) of a given salt are not absorbed equally by the plant, the mechanism of this unequal absorption is not very clear. According to Stiles (1924), there appear to be two possible consequences. In the first place, the excess of the more absorbed type of ion may be accompanied into the tissue by a quantity of one of the constituent ions of water, either hydrogen or hydroxyl, whichever is of the opposite sign to the more absorbed type of ion. If this theory is granted, then the external medium will be left either alkaline or acid depending on whether the anions or cations are absorbed in the greater amount. Let us consider a solution of sodium chloride as an example in which living tissue is immersed. If the sodium ions are absorbed in excess of the chlorine ions, the former will be accompanied into the tissue by an equivalent quantity of hydroxyl, leaving the same equivalent quantity of hydrogen ions to balance the excess of chlorine ions remaining in the external solution. The external medium would thus become acid. If, however, the chlorine ions are absorbed in excess, hydrogen ions accompany it into the tissue and the external solution becomes alkaline. The acidity and alkalinity of water-cultural solutions have been ascribed in some cases to the result of the unequal absorption of ions.

In the second place, it is possible that the excess of the more absorbed ion is replaced in the external solution by some other ion or ions of the same sign diffusing out from the tissue. Thus in the case of sodium chloride, the excess sodium absorbed might be compensated by the diffusion out from the tissue of magnesium, potassium, calcium, or other cations. If, however, chlorine is the more absorbed ion, its excess absorption could be compensated by the diffusion outward of nitrate, citrate, malate, or other anions.

The action of the protoplasm in the unequal absorption of ions has been considered by various authors. Höber and Höber (1928), in considering the cell sap of *Valonia macrophysa*, concluded that the protoplasmic membrane may consist of separate anion- and cation-permeable areas so that it might be impermeable to a given electrolyte although permeable to both its ions.

Brooks (1929) considered that the protoplasm consists of a mosaic of anion- and cation-permeable areas that are of the nature of charged porous films. These films may exaggerate differences between the diffusion velocities of the ions to which they are permeable. The diameter of these pores may vary according to the environment surrounding the cell. The penetrabilities of different ions are characteristic functions of their own diameters and of the diameter of the pores in the membrane. Briggs (1930) stated that the protoplasmic membrane exhibits alternating phases of ionic permeability. He considered that since the protoplasmic membrane is amphoteric in its nature it can be changed from an anionic to a cationic permeability by changes in the hydrogen-ion concentration. Collander (1930), and Collander and Barlund (1933), however, could not prove different degrees of permeability for different portions of the cell surface.

B. Reaction of the Exterior Solution

The factors that may influence the reaction of the solution in which a plant is growing are numerous and complex. Some of these are the type and concentration of the salts used, the rate of growth, and type of plant, in addition to the unequal absorption of ions. The intention is to state here some of the observations that have been made, in order to show the great variation of results and the complexity of the problem.

One of the earliest observations along this line was made by Skene (1915), who found that the cell walls of sphagnum moss caused the solutions of neutral salts in which they were placed to become acid. The amounts of acid hydrogen thus liberated varied with the salt. In the case of sodium chloride, potassium sulphate, calcium chloride, and sodium nitrate, it amounted to from 0.012 to 0.025 g.; in the case of sodium sulphate, sodium formate, and sodium butyrate, approximately 0.07 g. per 100 g. of dry sphagnum used. Skene found further that this property of sphagnum is exhibited generally by the cell walls of vascular plants but not to such a marked degree, since the number of grams of acid hydrogen liberated ranged only from 0.015 to 0.035 per 100 g. of dry matter. The cell walls of the sphagnums and other plant parts thus have the power to break up salts, removing the base and liberating the acid. It is supposed that this reaction is due to the negatively charged colloidal substances of the cell walls which adsorb the cations and set free the anions, thus causing an acid reaction in the exterior medium. Skene found that the different types of sphagnum with which he worked actually did utilize in growth the cations thus absorbed. Since plant cell walls in general exhibit this same property, although to a slighter degree, he suggested that the absorption of inorganic compounds by the root hairs and the epidermal cells of the root might take place in a similar manner (see also Dufrenoy, 1920).

Denny and Youden (1927) observed that when tissues of potato tubers, carrot roots, apple, whole seeds of corn, rye, and wheat came in contact with a tenth-normal solution of calcium chloride the solution soon became acidified and within an hour might increase as much as twentyfold. They showed that this increase in hydrogen ions was not due to the unequal role of absorption of anions and cations by the tissue, since they found that similar acidifications were produced by adding the salts to the water extract obtained by bringing the tissues in contact with water instead of the salt solution. In these cases, the salt came in contact not with the tissue itself but only with the soluble substances that leached out of the tissue. It was further found that the noncoagulable, noncolloidal, and nonprotein portions of the water extract underwent acidification upon the addition of salts, giving results analogous to those obtained by bringing the salts in contact with the tissues themselves. They further found that increases in hydrogen-ion concentration were obtained when salts were added to various organic acids such as malic, oxalic, succinic, aspartic, and others. The precipitation of insoluble salts of these acids with the resulting increase of the ionization of the remaining acid was shown to account in part for this development of acidity on adding solutions to organic acids, although acidifications were observed in cases where no precipitation occurred. They concluded, therefore, that the observed changes in hydrogen-ion concentration cannot be interpreted as indicating an isoelectric point for the tissue as a whole or furnishing proof that reaction has occurred between the ions of the salt solution and proteins with characteristic isoelectric points. The findings of Denny and Youden are in contrast to those of Rudolfs (1925), who found that when seeds were immersed in representative salt or acid solutions they were able to change the hydrogen-ion concentration of the solutions to definite points, and that a certain equilibrium was reached in all solutions after the seed had been immersed for a sufficient period of time. He considered that the chemical properties of the chief protein constituent of the seeds seemed responsible for the changes in the hydrogen-ion concentrations of the solutions.

The effect of the unequal absorption of ions upon the external medium has been especially studied in the case of cultural solutions. Redfern (1922) noted that the cultural medium in which corn and peas were grown remained neutral, although the cations were absorbed to a much greater degree than the anions. Hoagland (1923) found that single solutions of ammonium salts, potassium sulphate, sodium sulphate, and potassium bisulphate in which barley, peas, and cucumbers were grown generally increased in acidity. When barley, however, was grown in solutions of calcium nitrate, calcium chloride, magnesium nitrate, sodium nitrate, monocalcium phosphate, monomagnesium phosphate, or monosodium phosphate the actual acidity tended to decrease. In the case of any nitrate (except ammonium) the solution generally attained a reaction very close to neutrality. The changes of reaction of certain salt solutions are not always the same in different experiments. Thus, for example, Hoagland (1923) noticed in one experiment that even a slight shading of barley cultures markedly diminished the rate of increase of acidity in a potassium sulphate solution. After a short period of growth, the cultures nearest the window had a pH value of 4.0 and those at the opposite end of the row a pH value of 5.7. Haas and Reed (1926) considered that the changes in reaction of cultural solutions in which citrus seedlings were grown could be attributed directly to unequal absorption of ions together with the excretion of certain ions by the plants. In a complete nutrient solution, citrus seedlings in some cases brought about in a comparatively short period such a high concentration of hydrogen ions as to be injurious to the roots, but acidities greater than pH 3.0 to 3.2 were not observed to develop. In the case of walnut seedlings the acidity of the residual solution was entirely dependent upon the absorption rate of the two ions employed. In solutions of potassium phosphate with initial hydrogen-ion concentrations of pH 5.0 and lower, the preferential absorption of potassium by wheat seedlings resulted in increased acidity. In solutions with initial hydrogen-ion concentrations of 6.0 and 7.0, the increase in acidity was slight, owing to the buffer properties of the solutions (Davidson, 1927).

C. Interchange of Ions from the Plant to the Exterior Solution

When the unequal absorption of ions occurs, it frequently happens that the more absorbed ion is replaced by the diffusion outward from the plant of some other ion or ions of the same sign but of different chemical composition. This interchange of ions thus tends to keep the medium in which the plant is growing from becoming either acid or alkaline in its reaction to any degree. Thus Redfern (1922) found that when corn and peas were grown in a solution of calcium chloride it remained neutral in its reaction, although the calcium ions were absorbed greatly in excess of the chlorine ions. Potassium and magnesium, however, were found in appreciable quantities in the external medium, having diffused out from the roots, and it was considered that this interchange of bases caused the solution to remain neutral in spite of the unequal absorption of the calcium and chlorine. In the case of barley growing in single salt solutions (Hoagland, 1923), calcium and magnesium diffused outward as potassium was removed. When calcium salts were used, potassium and magnesium entered the solution from the plant. Thus in one case where the plant absorbed the potassium ions from a potassium sulphate solution to the extent of 1.44 milliequivalents, the calcium and magnesium ions entering the solution amounted to 0.50 and 0.53 milliequivalents, respectively. In another case where a calcium nitrate solution was used, the absorption of the calcium ions amounted to 1.66 milliequivalents, and the potassium and magnesion ions entering the solution amounted to only 0.09 and 0.20 milliequivalents, respectively. The anion HCO_3, according to Hoagland (1923), is the one which most abundantly replaces the anions that have been more abundantly absorbed. For example, in his experiments with barley growing in the solution of calcium chloride, the absorption of the cation amounted to 0.83 milliequivalent and the anion to 2.06 milliequivalents, while the amount of the HCO_3 ions diffusing into the external solution equaled 1.26 milliequivalents. In the case of a solution of calcium nitrate, the cations were absorbed to the extent of 1.66 milliequivalents and the nitrate ion to the extent of 5.48 milliequivalents. The quantity of HCO_3 ions that entered the exterior solution totaled 4.12 milliequivalents. Apparently, there has been no really satisfactory explanation advanced for the mechanism concerned in the interchange of ions between the plant and the external solution.

D. Factors That Influence the Unequal Absorption of Ions

The absorption and exchange of ions by a plant are influenced by a wide variety of conditions. Some of these are the concentration of the solution, the effect of one ion upon the absorption of another, the nature and concentration of the ions present, temperature, light, and numerous others.

1. Concentration of the External Solution.—The concentration of the solution has a marked influence upon the proportion of ions absorbed. For example, Redfern (1922) noted that peas growing in tenth-normal solution of calcium chloride absorbed 17.7 per cent of the calcium and 3.6 per cent of the chlorine in 36 hr., while in a thousandth-normal solution of the same salt during the same time they absorbed 23.1 per cent of the cation and 15.1 per cent of the anion. The relative absorption of ions by storage tissue is also influenced by the concentration of the solution, as is shown by the following experiments with carrot (Stiles, 1924):

Initial concentration of solution, normality	NH_4Cl, percentage of absorption		NaCl, percentage of absorption	
	48 hr.			
	Cation	Anion	Cation	Anion
1/10	12.9	5.9	25.2	8.3
1/100	57.2	32.2	36.1	15.8
1/1,000	97.1	80.7	38.1	22.8

These figures indicate that the two ions of the salts used tend to be absorbed more nearly equally as the concentration of the solution decreases.

2. The Influence of One Ion upon Another.—The topic here under discussion refers only to ions in dilute solutions that are not toxic to the plant and should not be confused with antagonism, which has to do with the influence of one ion on another in overcoming the toxic effect of solutions on plants.

Ions differ greatly in the rapidity with which they are absorbed. The initial rate of absorption of ions may be in a different order from that which prevails when equilibrium is reached. Thus Stiles and Kidd (1919) found that the cations with which they worked were absorbed initially by storage tissue after the following order: potassium, calcium, sodium, lithium, magnesium, but when equilibrium was approached the order of absorption was potassium, sodium, lithium, calcium, and magnesium. The absorption of anions under the same conditions was initially in the order sulphate, nitrate, chlorine, but this later gave way to the order nitrate, chlorine, sulphate.

Of the cations, potassium seems to be most rapidly absorbed in practically all cases observed (Haas and Reed, 1926), and calcium probably is the least rapidly absorbed. Of the anions, nitrate is the most rapidly and abundantly absorbed, while the sulphate ion has been observed by practically all workers to be the most slowly absorbed. In the majority of cases that have been observed, the cations have been absorbed in greater amount than the anions. The main exception has been the nitrate anion and sometimes the chlorine anion.

Pettinger (1932) found that the application of fertilizers containing chlorine increased the chlorine content of the cell sap of corn plants, the increase being partially proportional to the amount of chlorine supplied by the fertilizer. These plants absorbed the chlorine ion much more readily than the sulphate ion. As the chlorine content of the sap increased, the hydrogen-ion concentration also increased.

The effect of one ion or of several sets of ions upon the absorption of other ions has been noted by Hoagland (1923), Haas and Reed (1926), Lagatu and Maume (1927), Brown (1928), McCool and Weldon (1930), Cooper (1930), Larson (1931), Pettinger (1932), Bartholomew, Watts, and Janssen (1933), Colby (1933), Davidson (1933), and Jacobson and Swanback (1933). Apparently this effect occurs regardless of whether the ions have like or unlike electrical charges, and all three types of relations are involved, *i.e.*, cation to anion, cation to cation, and anion to anion. Haas and Hill (1926) noted that the presence of sodium decreased the rate of absorption of other cations, especially potassium and calcium in barley, citrus, and walnut plants, and potassium ions retarded the absorption of calcium in citrus seedlings. When the potassium in the cultural solution was low in amount, more calcium, magnesium, and phosphorus were absorbed from the solution than when potassium was abundant. Brown (1928) in growing wheat in sand cultures reported that high concentrations of potassium decreased the absorption of magnesium and that a high content of calcium inhibited the absorption of nitrates.

The work of Cooper (1930), on pasture grasses from various associations in New York, showed that there is a close correlation between the standard electrode potentials of elements and the amount of the various minerals in the ash of plants. The productive soils contained relatively large amounts of the elements that form strong ions, while unproductive soils were high in the elements that form weak ions. A negative correlation existed between potassium oxide and calcium oxide, and between calcium oxide and phosphorus pentoxide in the ash of these plants. The calcium seemed to be negatively correlated with the strong anionic materials, which suggested that plants growing on alkaline soils may absorb considerable quantities of the hydroxyl and bicarbonate ions.

It was considered by Bartholomew, Watts, and Janssen (1933) that a deficiency of potassium in a nutrient solution may result in an increased absorption of nitrogen and phosphorus by the leaves of the tomato, while an abundant supply of potassium may increase the absorption of phosphorus by the stems. The absorption of calcium and magnesium did not seem to be significantly influenced by the amount of nitrogen, phosphorus, or potassium absorbed by this plant.

Colby (1933) noted in young, French prune trees that a lack or scarcity of either potassium, magnesium, or phosphorus seriously depressed nitrate absorption and even caused a loss of this ion from the roots late in the season. The lack of calcium prevented root growth, and the trees absorbed very little nitrate. The total phosphorus absorbed during the season was depressed more by the lack of magnesium than by the lack of potassium and sulphur. The lack of calcium apparently prevented the absorption of a considerable quantity of any ion.

The work of Haas and Reed (1926) with citrus and walnut seedlings indicated that the amount of chlorine absorbed stood in the following order with respect to the accompanying cation: calcium, potassium, sodium, and magnesium. With barley plants Hoagland (1923) observed that the order of absorption for chlorine was potassium, sodium, magnesium, and calcium. The presence of chlorine seems to retard the absorption of nitrate, but other ions seem to have little effect upon its absorption. It has been observed that the nitrate ions penetrate more readily from an acid solution than from an alkaline one in experiments with *Nitella* (Hoagland and Davis, 1923). True and Bartlett (1915, 1916) found that the seedlings of *Lupinus albus* absorbed more salts from the mixtures of the nitrates of potassium, calcium, and magnesium than from equally concentrated solutions containing only one or two of these nitrates. The solutions, however, of the three nitrates that were most favorable to absorption were much inferior to corresponding solutions in which the three anions, dihydrogen phosphate, nitrate, and sulphate were present. It was observed by Hoagland, Davis, and Hibbard (1928) that the accumulation of bromium ions in the cell sap of *Nitella* was significantly retarded by chlorine or iodine ions also present in the cultural medium but was not retarded by sulphate, nitrate, or phosphate ions. The accumulation of bromium ions was definitely influenced by the nature of the cation, being most rapid when solutions of potassium bromide or rubidium bromide were used and least rapid with solutions of lithium bromide, calcium bromide, strontium bromide, and magnesium bromide.

Lagatu and Maume (1927) found in the grape that the assimilation of phosphorus checked the utilization of nitrogen, while Larson (1931) noted that in apple trees, with a higher percentage of nitrogen, the percentage of phosphorus was low, but if the percentage of nitrogen was low the percentage of phosphorus was high. McCool and Weldon (1930) grew plants of barley, oats, rye, and wheat in soils to which various amounts of sodium nitrate had been applied. The highest applications of sodium nitrate usually decreased the potassium content of the plants, while the effect of the smaller applications was variable. The calcium content decreased more with the greater application of nitrate, while the magnesium content was not greatly influenced by the amount of salt used. Davidson (1933), however, found that the potassium content of wheat plants grown in soils, to which sodium nitrate had been applied, was consistently higher than in the plants grown in the untreated soil. It was observed by Jacobson and Swanback (1933) in the tobacco plant that an increased proportion of nitrate nitrogen over ammoniacal nitrogen resulted in a large percentage of total calcium in the plant material. Bartholomew, Watts, and Janssen (1933) found that an abundant supply of nitrogen may increase the amount of potassium in the stems of tomatoes.

In the case of barley, Hoagland (1923) found that the slowly-absorbed sulphate ion decreased the rate of absorption of the associated cation. Thus potassium was absorbed much more rapidly from the nitrate and the chloride salt than from the sulphate salt of potassium. The absorption of cations was always greatest when they were accompanied by an anion like NO_3 that is rapidly absorbed (Haas and Reed, 1926). It was reported by Colby (1933), for French prune trees, that the lack of the sulphate ion had a less depressing effect on nitrate absorption than did the lack of any other of the six major elements of a complete nutrient solution.

Thomas (1933) grew Stayman winesap apple trees for six years in soil in large cylinders to which were added different combinations of nitrogen, phosphorus, and potassium nutrients. After this time, the amounts of these three elements that had been assimilated were determined. It was found that the percentage and absolute amount of each of the elements absorbed, and also the ratio in which it was absorbed, varied greatly with the nutrient treatments and to a lesser degree with the cultural system. Each of the respective treatments indicated that the omission of any one of the elements nitrogen, phosphorus, and potassium from the complete fertilizer is followed by a decreased absorption of the remaining elements. This decreased absorption resulted in a nutritional lack of balance as exhibited in the reduction of growth and of flowering. There appeared to exist a specific ratio in which nitrogen, phosphorus, and potassium are absorbed by the trees. The optimum ratio appeared to be 6:1:4, respectively.

The amount of any ion absorbed is not strictly related to the amount present in the plant, since in certain cases the phosphate and nitrate ions are rapidly absorbed, although the plant parts concerned are very rich in these ions. The rate of absorption of ions is not related to their velocities or any other of their physical chemical properties so far as they are known. Since the absorption of ions depends upon light; temperature; humidity; concentration and reaction of the solution; the amphoteric character, amount, and kind of protein present; and the kinds of ions present; it is a complex problem to handle experimentally, and no two experiments will give the same results unless every condition is identical.

IV. TRANSPIRATION AND THE INTAKE OF SALTS

Previous to 1914 the evidence was very confusing in regard to the relationship between the rate of transpiration of a plant and the amount of solutes taken up from the soil. It was considered by Schloesing (1869), Kohl (1886), Sachs (1887), Wollney (1898), Burgerstein (1897), Pfeffer (1900), Jost (1907), and others that the amount of solute taken up by the plant from the soil is proportional to the amount of water transpired. Their conclusions, however, were in most cases based on purely theoretical grounds, although in a few cases they were based on very meager and fragmentary experimental evidence. On the other hand, Ilienkoff (1865), Fittbogen (1873), and Haberlandt (1892) concluded also from a small amount of experimental evidence that there is no relation between transpiration and the quantity of mineral nutrients absorbed from the soil. The example of plants growing in the tropics has been frequently presented to substantiate the view that the transpiration rate under the conditions existing there is very low as compared to that of the temperate regions, yet the intake of salts is apparently abundant. There seems, however, to be insufficient evidence to prove that such a low rate of transpiration exists, as is evidenced by the contentions of Haberlandt (1897), who considered that the rate of transpiration in the tropics is lower than in central Europe, and Stahl (1894), Burgerstein (1897), and Giltay (1897), who claimed that the plants in the tropical rain forests transpire more than those in a temperate region such as Europe.

Beginning with Hasselbring (1914), well-controlled and rather extensive work has been conducted by Kiesselbach (1916), McLean (1919), Muenscher (1922), and Mendiola (1922) to determine the influence of the amount of transpiration upon the quantity of minerals in the plant. The method used, in general, has been to determine the ash content of plants whose rate of transpiration has been made to vary by changing the environmental conditions.

Hasselbring (1914) working in Cuba grew tobacco in large containers both under field conditions and under the shelter of cheesecloth after the method that is used in growing certain qualities of tobacco. Under the cheesecloth shelter the total light was reduced by one-third from that of the open. The temperature of the two stations showed practically no difference, but during the daylight hours the relative humidity under the shade was higher than that in the open. The shading used in this experiment did not diminish the production of total dry matter, but the leaves of the shade-grown plants had a much greater total area than those of the plants grown in the open. The following table gives a summary of the results obtained by the experiment, the figures representing the average for six plants:

Total dry weight, grams	Total water absorbed, cubic centimeters	Total ash in plants, grams	Water absorbed per gram of ash, cubic centimeters
	Plants grown in the open		
188.4	46,344	18.3	2,548
	Plants grown in the shade		
188.1	36,187	21.1	1,718

The plants in the open transpired nearly 30 per cent more water than those growing under the shelter, while the transpiration rate per unit of leaf surface was nearly twice as great for the plants in the sun as those growing in the shade. As indicated by the table, however, the plants that transpired the greater quantity of water contained the smaller amount of ash in percentage of the total dry matter and in absolute amount. Hasselbring, therefore, concluded that the absorption of salts is independent of the absorption of the water and that an accelerated transpiration rate does not exert an accelerating effect on the entrance of salts.

Kiesselbach (1916) studied the relation of transpiration to the ash content of corn grown under different climatic conditions in soils of

Content	Popcorn			Hogue Yellow Dent corn		
	Dry greenhouse	Humid greenhouse	Ratio dry to humid	Dry greenhouse	Humid greenhouse	Ratio dry to humid
Dry weight per plant, grams....	67.9	71.9	......	100.9	90.5	
Ash content per cent, dry basis...	12.28	11.91	100:97	13.6	12.22	100:94
Total weight of ash, grams...........	8.34	8.56	100:102	13.18	11.06	100:84
Water transpired per plant, cc.....	24,279.0	14,304.0	100:59	33,640.0	19,067.0	100:57
Water transpired per gram of ash, cc.	2,910.0	1,670.0	100:57	2,550.0	1,720.0	100:67

different degrees of fertility and moisture content. The data obtained from growing plants in a dry and a humid greenhouse may be taken as an example to represent the effect of climate upon the relation of transpiration to the intake of salts. If the evaporation in the dry greenhouse is taken as 100 per cent, then that in the humid greenhouse would be only 54 per cent. Popcorn and Hogue Yellow Dent were the two varieties used.

These data indicate that there is no relationship between the amount of water transpired and the amount of ash in the plant. Kiesselbach noted that the greater the availability of the soil solutes the greater was the total amount of solute taken into the plant and the greater the amount taken in per unit of water transpired. Less water was transpired per gram of ash content in a relatively low soil saturation than where an abundance of water was present. Kiesselbach observed considerable variation in the ash and transpiration relationships of different corn varieties but considered that there is no absolute correlation among the percentage of ash, the quantity of water transpired per gram of ash, or the transpiration per gram of dry matter, a conclusion also reached by Mendiola (1922) in regard to tobacco plants grown in Pfeffer's nutrient solution under humid and dry conditions. Milo and Black Amber sorghums transpired a slightly smaller quantity of water per gram of ash than did corn, and sunflower plants transpired much more water per gram of ash content. McLean (1919) found that leaves taken from plants growing in the tropical rain forests of Brazil showed a higher ash content than sun plants. He assumed that the transpiration rate is lower for the leaves of the rain forest and considered, therefore, that the absorption of the mineral salts is independent of foliar transpiration.

Muenscher (1922) grew barley plants for 5 weeks in Knop's solution under conditions of high and low transpiration. He reduced the transpiration rate by increasing the humidity of the atmosphere, by reducing the light intensity by shading, and by increasing the concentration of the nutrient solution. It was found that the ash content expressed in percentage of total dry weight of the plants varied but slightly, regardless of whether the plants were grown under conditions of high or low transpiration and irrespective of how transpiration was reduced. The effect of reduced transpiration upon the total ash content, however, expressed in grams, depended upon how transpiration was reduced. With a uniform concentration of nutrient solution the ash content of the barley plants varied but slightly, even though the quantity of water transpired was reduced to less than one-half by increasing the humidity of the atmosphere. In the case of plants, however, in which the transpiration was reduced to less than one-half by shading and the photosynthetic activity was also reduced, the total ash content was correspondingly reduced.

When the total transpiration was reduced by increasing the concentration of the nutrient solution, the total ash content was only slightly reduced. Muenscher considered that his results indicated that there is little or no relation between transpiration and the absorption of salts in barley plants. He considered, however, that the amount of food available for growth, in which salts are used, is a very important factor in determining the quantities in which and the rapidity with which the essential salts enter the plant. It seems to be well established that the solutes enter a plant independently of the entrance of water.

Hoagland grew two sets of barley plants under such conditions that the amount of water absorbed by one set was practically twice that absorbed by the other. The results may be summarized by the following table:

Concentration	Milliequivalents						
	K	Ca	Mg	NO_3	PO_4	SO_4	Water absorbed, cc.
Initial	4.5	8.0	4.4	11.4	1.3	7.2	
Final	1.5	12.5	7.3	9.5	0.4	18.5	505
Final	3.2	9.6	5.2	9.7	1.4	13.1	265

The cultural solution thus became more concentrated with respect to certain ions and more dilute with respect to others as the absorption of water was increased. The ions that regularly increased in concentration were calcium, magnesium, and sulphate, while potassium and the nitrate ions decreased in concentration. The ions thus are not removed from the solution by the plant in the same proportion as water. Under some conditions at least, potassium, nitrate, and phosphate ions may be removed from the solution at a greater rate than water, while at the same time calcium, magnesium, and sulphate ions may be removed less rapidly than water. On this account, therefore, as the plant absorbs water and ions, the solution may become more dilute with respect to certain ions and more concentrated with respect to others.

A simple observation by Scofield (1927) will serve to illustrate this point further. He found that barley and wheat seedlings absorbed during a 24-hr. period a quantity of water equivalent to the original volume of the cultural solution, while at the end of that period the residual solution contained from 78 to 91 per cent of the salts originally present (see also Hoagland, 1923). Since plants do not absorb water and dissolved substances from the soil in the same proportion as these constituents occur in the solution, it is difficult to understand how transpiration may in any way influence the amount of salts in the plant. It is considered by many that the water current set up through the conducting elements in the stem is a factor in carrying the dissolved salts to the leaves. Presumably then, the more rapid the transpiration, the more rapid the current and the more salts thus translocated. However, as Muenscher (1922) stated, it must be borne in mind that although an upward current of water apparently exists, the existence of a transpiration stream in the sense of mass movement of a solution in normally growing plants has not yet been demonstrated.

Gračanin (1932) concluded from studies of the plants of barley, wheat, corn, and peas that no direct relation exists between the absorption of ions and the rate of transpiration.

Freeland (1935, 1936) studied the relationship between the amount of transpiration and the amount of various elements in the tomato, bean, *Coleus*, and sunflower. These plants were grown in sand, watered with Shive's nutrient solution. The rate of transpiration and absorption of water were varied by growing one set of plants under conditions of low relative humidity and the other set in air with high relative humidity. At the conclusion of the experiment, the plants were divided into leaves, stem, and roots. For the tomato the amounts of water absorbed from the two environments were 7,798 cc. in the low humidity and 3,087 cc. in the high humidity. The total quantity of ash, calcium, phosphorus, and potassium was greater in the plants that had the greater amount of transpiration. This increase in mineral content was mainly in the leaves and roots. In the plants of *Coleus*, the amounts of water absorbed were 19,430 cc. in low humidity and 4,679 cc. in high humidity. The ash, calcium, phosphorus, and potassium were greater in all three parts of the plants with higher rates of transpiration. Under the latter conditions, the potassium showed the greatest increase in the leaves, while the increase of calcium was greatest in the roots. Practically the same results were obtained for bean as for *Coleus*. However, the increase in the mineral elements and ash content was seldom directly proportional to the increase in dry weight or to the increase in water absorption. The data show, however, that low humidity, which increased the absorption of water from two to four times, resulted directly or indirectly in an increase in absorption of mineral salts. It is evident that available food and growth cannot account for this increase in mineral content. Freeland believed that the rate of movement of water through the plant may result in differences in the translocation of certain mineral ions. It is not possible, however, to ascribe the translocation results solely to the different amounts of water moving through the plants under high and low humidities, since there were some structural differences between the plants grown in these two environments.

V. NUTRIENT SOLUTIONS

It is impossible to study in any definite manner the nutrient requirements of a plant when it is growing in the soil, since by present known methods the soil solution cannot be controlled, analyzed, or varied with any degree of satisfaction. A water solution offers the best known means to study in a quantitative way the nutritional needs of the plant. Nearly all the cereals and beans, peas, buckwheat, and many other plants lend themselves favorably to growth in such solutions, in spite of the unusual conditions in which their roots are placed. The solutions that are thus used have been termed "nutrient" or "cultural solutions" or "water cultures." When plants are grown in pure sand to which the solution has been added, the medium is called a "sand culture." For quantitative work, however, the water culture has proved the most satisfactory. A nutrient or cultural solution is generally defined as one that contains all the essential elements in the form of salts that a plant absorbs from the soil. A solution that does not contain all the essential elements is sometimes termed a cultural solution, but it is generally inferred when the term is used that the solution contains all the known essential elements that the plant obtains from the soil.

A. Historical

The earliest recorded experiment with water cultures was that of Woodward (1699) who grew plants in spring, river, rain, and distilled water to determine whether it was the water or the solid particles of the soil that nourished the plants. Very little additional study, however, was made in this field until Sachs, Knop, and Nobbe from 1859 to 1865 developed the general procedure in water cultures that is in use at the present time. Sachs (1860) published the first standard formula for a culture solution for plants, and in 1865 Knop proposed a solution that has been one of the most widely used in the study of plant nutrition. Other nutrient solutions have been proposed by Tollens (1882), Schimper (1890), Pfeffer (1900), Crone (1902), Tottingham (1914), Shive (1915), Hoagland (1920), and many others. From 1859 to 1900 the preparation of nutrient solutions and the methods for their use were well established. By their use during that period 10 of the essential elements for plant growth were determined. It was found that the concentration of the standard solutions that were used could vary in concentration of total salts from 0.1 to 0.6 per cent and yet produce an optimum growth of the plants provided that the solutions were frequently renewed. The proportions of the different components of these solutions varied greatly, and it was considered that these had little apparent influence upon the growth and development of the plant, provided that a sufficient quantity of salt was present for the plant's needs.

From 1900 to the present time the investigations of cultural solutions have been characterized by a study of the effects of the physical as well as the chemical properties of the nutrient solutions upon the growth of the plant (Livingston, 1900). The osmotic properties of the solutions have been especially considered. The relationship to growth of the proportion of salts in the solution has been extensively studied, as has also the effect of light and temperature upon the relative value of different solutions for plant development. It has also been found that the unequal absorption of ions of a given salt may have a marked effect upon the reaction of the nutrient solution. Finally, a thorough reinvestigation is in progress to determine the number of elements that are essential to the life of the plant and the function they subserve in its metabolism.

A concise, reliable history of water cultures was published by Johnston (1932), and Sherman (1933) gives a résumé of the theories of plant nutrition from Aristotle to Liebig.

B. Kinds

It is the intention to present here only a few of the many formulas for nutrient solutions. Others may be obtained by consulting Tottingham (1914), Tollens (1882), and Schimper (1890).

Sach's solution (1860), grams		Knop's solution (1865), grams		Pfeffer's solution (1900), grams		Crone's solution (1902), grams	
KNO_3	1.00	$Ca(NO_3)_2$	0.8	$Ca(NO_3)_2$	0.8	KNO_3	1.00
$Ca_3(PO_4)_2$	0.50	KNO_3	0.2	KNO_3	0.2	$Ca_3(PO_4)_2$	0.25
$MgSO_4$	0.50	KH_2PO_4	0.2	$MgSO_4$	0.2	$MgSO_4$	0.25
$CaSO_4$	0.50	$MgSO_4$	0.2	KH_2PO_4	0.2	$CaSO_4$	0.25
$NaCl$	0.25	$FePO_4$	Trace	KCl	0.2	$FePO_4$	0.25
$FeSO_4$	Trace	H_2O	1,000 cc.	$FeCl_3$	Small amount	H_2O	1,000 cc.
H_2O	1,000 cc.			H_2O	1,000 cc.		

The nutrient solutions used prior to 1914 varied greatly in their total concentration and in the proportion of the ions present (Tottingham, 1914). The total concentration of salts ranged from 0.1 to 0.5 per cent, and the ratio of potassium to magnesium varied from 1:1 to 5:1, that of calcium to magnesium 2:1 to 3:1, that of nitrate to phosphate from 3:1 to 10:1, and that of nitrate to sulphate from 1.3:1 to 7:1.

In 1914, Tottingham, using Knop's solution as a basis, began an investigation to determine the influence of a wide range of variations in the proportions of the components of this solution. Three different total concentrations were used, having on a percentage basis a concentration of 0.01, 0.6, and 2 per cent and an osmotic pressure of 0.05, 2.5, and 8.15 atmospheres, respectively. In each of these concentrations the approximate osmotic pressure was divided into 10 equal parts, and the 10 parts were distributed among the four component salts of the solution. This arrangement gives for any given concentration of solution 84 solutions, each having the same osmotic pressure but no two of them containing the same relative proportion of salts. Tottingham found that the solution having the total concentration of 0.6 per cent or 2.5 atmospheres osmotic pressure gave the best growth for wheat in the seedling stage. The best combination of the 84 solutions in this series as judged by the dry weight of tops produced in the experiment was *T3R1S4*, in which the salts were distributed as follows:

Tottingham's best combination of Knop's solution

$$KH_2PO_4 \longrightarrow 0.0130M \qquad Ca(NO_3)_2 \longrightarrow 0.0144M$$
$$MgSO_4 \longrightarrow 0.0145M \qquad KNO_3 \longrightarrow 0.0049M$$

For iron, 2 drops of a suspension containing 0.0024 gr. of ferric phosphate per cubic centimeter was added to each liter of solution at the time of renewal. The nutrient solutions used prior to 1915 contained, excluding the iron, at least four different salts.

When the effect upon growth of the various proportions of these salts in solutions of this type is to be studied, the number of possible combinations becomes large and thus renders the experiment very tedious and cumbersome. Thus, as has been mentioned, if the osmotic pressure of a four-salt solution is distributed in tenths, there are 84 different combinations possible. If the number of salts in a solution could be reduced to three and the osmotic pressure distributed in tenths, the experiment would employ only 36 combinations. If this type of solution could be used, the experimental work would be greatly simplified and more rapid determinations could be made. If a two-salt solution could be used, only nine combinations would be possible on the basis that has just been mentioned. This, however, is impossible, since six elements at least are required in every nutrient solution and only two of these can be carried by any single salt.

In 1915 Shive carried on experiments with a three-salt solution. The potassium nitrate was omitted from the Knop's formula, the solution being composed of the three salts: monopotassium phosphate, calcium nitrate, and magnesium sulphate and a trace of ferric phosphate, thus containing all the elements of the four-salt solution. Shive experimented with three different total concentrations of this three-salt solution, the osmotic pressures of which amounted to 0.1, 1.75, and 4.0 atmospheres, respectively. For each of these three total concentrations, 36 different proportions of the three component salts were tested. It was found that the solution with an osmotic pressure of 1.75 atmospheres was the optimum one of the three for the growth of wheat and buckwheat plants. The production of dry matter by plants growing in this solution was equal to or greater than that produced by plants growing in any of the four-salt solutions. For the growth of young wheat plants, Shive found that the solutions *R5S2* and *R3S3* were the best as determined by the production of dry matter. The following are the formulas for these best three-salt solutions:

Best solution[1]		Next best solution[1]	
R5S2		*R3S3*	
KH_2PO_4	0.0180*M*	KH_2PO_4	0.0108*M*
$Ca(NO_3)_2$	0.0052*M*	$Ca(NO_3)_2$	0.0078*M*
$MgSO_4$	0.0150*M*	$MgSO_4$	0.0020*M*

[1] Ferric phosphate 0.0044 g. per liter of solution.

According to Loomis and Shull (1937), stock solutions for Shive's *R5S2* nutrient solution may be prepared as follows:

Salt	G. in 1 l.
KH_2PO_4	49.01
$Ca(NO_3)_2.4H_2O$	24.56
$MgSO_4.7H_2O$	73.95

Fifty cubic centimeters of each of the three stock solutions, plus 4 or 5 mg. of ferric phosphate, made to 1 l. form a full nutrient solution. When repurified salts are used, it is necessary to add some of the minor elements to obtain the proper development of plants. Ten cubic centimeters of a stock solution consisting of 0.10 g. manganese chloride, 0.05 g. zinc chloride, 0.05 g. boric acid, 0.01 g. copper chloride in 1 l. of water, are added at frequent intervals to the cultural solution.

Because of the simplicity of their preparation and the ease of manipulation in experiments, three-salt solutions of various types have practically superseded the four-salt solutions in the study of plant nutrition. A three-salt solution, however, has the disadvantage that if the proportion of one ion or radical of a salt is altered, the other ion of the salt is changed in a like proportion, so that the effects of an individual ion cannot be so well studied as in a four-salt solution. Six types of three-salt solutions are possible in order to supply potassium, phosphorus, calcium, magnesium, sulphur, and nitrogen to plants. These have been numbered and catalogued by Livingston and Tottingham (1918) and are now commonly called by their serial number when discussed in experimental work in plant nutrition. They are arranged as follows:

I	II	III	IV	V	VI
$Ca(NO_3)_2$ KH_2PO_4 $MgSO_4$	$Ca(NO_3)_2$ K_2SO_4 $Mg(H_2PO_4)_2$	$Ca(H_2PO_4)_2$ KNO_3 $MgSO_4$	$Ca(H_2PO_4)_2$ K_2SO_4 $Mg(NO_3)_2$	$CaSO_4$ KNO_3 $Mg(H_2PO_4)_2$	$CaSO_4$ KH_2PO_4 $Mg(NO_3)_2$

Solution III was tested by Livingston and Tottingham and found to be as good as solution I (Shive's) for the growth of wheat seedlings.

In the soil, many of the plant nutrients are present only in relatively small amounts in the soluble form, although there is generally a considerable reserve in the form of compounds that are only slightly soluble. In order to have a nutrient solution in which the available nutrients would more nearly approximate the concentration in the soil, Duggar (1920) used as nearly as possible salts that are quite insoluble. As a source for calcium he used calcium sulphate, calcium carbonate, tricalcium phosphate, monocalcium phosphate; as a source for magnesium, magnesium carbonate and trimagnesium phosphate were used, and iron was supplied in the form of iron phosphate and ferric citrate. The plants grew in the cultures with these salts just as vigorously as when the more soluble salts were used. Hoagland (1919) used various concentrations of nutrient solutions that were so constituted as to give approximately the same relation between the more important ions as that found in the water extract of soils at the time when the crop was actively absorbing. It was found that solutions with total concentrations ranging from 0.48 atmosphere to 1.45 atmospheres gave very similar results if renewal was frequent. The following solution was found by Hoagland (1920) to be as favorable for the growth of young barley plants as Shive's *R5S2* solution:

Ions	P.p.m.	Ions	P.p.m.
K	190	PO_4	117
Ca	172	NO_3	700
Mg	52	SO_4	202

C. Technique

It is the intention here to discuss only in a broad and general manner the methods of procedure in experimental work with nutrient solutions. The detailed directions can be found by consulting Livingston, Crocker, and Kellerman (1919) and the other references mentioned herein. Various methods can be used for obtaining the seedlings necessary for the experiment, the method of procedure depending to a great extent upon the kind of seeds that are to be germinated. For the cereals and some other seeds the following method has given very satisfactory results: The seeds are soaked in water for 5 or 6 hr. and then placed upon a germination net, which consists of ordinary mosquito netting that has been dipped quickly into melted paraffin of a high melting point so that the threadwork of the network is covered but the meshes remain open. The treated netting is tied as tightly as possible over the top of any wide-mouthed vessel or jar so that when the jar is filled with water or solution the surface practically coincides with the plane of the net. For ordinary demonstration experiments tap water may be used in the germinating vessel but for exact work a cultural solution should be used in order to avoid too abrupt a change when transferred from the germination container to the cultural vessel. The solution in the germination vessel should be changed frequently, preferably daily, by adding fresh solution through a glass tube that extends to the bottom of the container. The control of the temperature during germination is desirable and may be obtained by a water bath or any other means available to the experimenter. The germination jar should be placed in diffuse daylight until the seediings have reached the desired stage of growth (see also Livingston, 1906). After the choice of the seedlings for the cultural experiment has been made, they are lifted from the net, preferably by paraffined forceps, and placed in a glass vessel containing a germination solution 2 cm. in depth so that the roots are in the solutions and the shoots above its surface. The seedlings are then taken from the vessel one by one as needed to be placed in position in the cultural jars, care being taken to avoid contamination.

The best cultural vessels are wide-mouthed glass jars, the ordinary Mason fruit jars of pint, quart, or 2-qt. capacity, being especially adapted for that purpose. Many different arrangements have been used for closing the cultural vessels and for supporting the plants growing therein. One of the most common methods is to use flat corks about ½ in. thick which have been thoroughly impregnated and lightly coated with paraffin. The desired number of holes about 3 cm. in diameter are made through the cork at convenient locations, as described by Tottingham (1914). The young plants are placed in position in these holes and held firmly just above the seed by means of ordinary cotton, which remains dry and becomes compressed as the stem enlarges. The cultural vessels should be filled with nutrient solution so that when the corks are tightly forced into position they will be within 1 to 2 cm. of the surface of the solution (see Livingston, 1906). If the growing plants require support, uprights of light glass or wooden rods anchored to the cork may be used. It is desirable that the uprights be light but substantial and easily attached, and they interfere as little as possible with the handling of the cultural vessels. McHargue (1923) has devised two methods of supports for the larger sand- or water-cultural vessels. In one case, four sockets made from galvanized iron into which wooden uprights may be fitted are placed equidistant around the top of the jar and held tightly in place by means of wire tightly twisted around it. In the other case, the earthenware cover is perforated in the central portion in numerous places in which the plants are held by cotton, and around the edge of the cover equidistant from each other are four holes in which light wooden supports can be tightly fitted. Perforated paraffin disks may also be used, according to Hammett (1928).

In order to make a comparative study of water cultures it is necessary that all plants be exposed to as nearly the same atmospheric conditions as possible. If the cultures are grouped in a given area, the plants shade each other so that they do not all receive the same amount of light. The transpiration loss of some plants is thus reduced and perhaps also their photosynthetic rate. The unequal atmospheric conditions may be overcome in part by changing the position of the plants at numerous intervals. This, however, requires close attention and a great amount of labor and is at the best unsatisfactory. One of the most convenient methods so far devised for placing the cultures under the same conditions is that used by Shive (1915). The plants are placed on the edge of a slowly revolving circular table approximately 4 ft. in diameter which makes a revolution every 4 min. and is continuously rotated during the time of the experiment except when stopped for short intervals for the renewal of solutions. By this means all the plants in the cultural solution are exposed to approximately the same light, heat, and cultural conditions. Trelease (1925) has devised a chamber with thermostatic control and a rotating table for plant cultures.

Gericke and Travernetti (1936) reported an experiment with tomatoes that is of marked importance in the practical application of cultural solutions. They used sheet-metal tanks, each of which was 2.5 ft. wide, 10 ft. long, and 8 in. deep; over them were wooden frames containing a wire screen that was held just above the top of the tanks. Two to three inches of excelsior and sawdust were placed on the screen to provide support for the plants. Of the five tanks, four were heated by means of electrical equipment and one was used as a control. In two of the four tanks the temperature was kept between 75 and 80°F. for the first 6 weeks and then raised to between 80 to 85°F.; the other two tanks were kept at 65 to 70°F., and, after 6 weeks, at 70 to 75°F. The control tank was kept at 55 to 65°F.

The tanks were filled with water to within 1 in. of the top, and each tank received 1 lb. of nutrient salts at the beginning of the experiment. During the 12-month period, six additional applications of salt were made per tank. The plants were started in soil and transplanted as 3-week-old seedlings. Two rows of 10 plants each were placed in each tank, thus giving to each plant an area of about 1.25 sq. ft. The plants in the heated tanks began to ripen fruit about 4 months after transplanting, while those in the unheated tank did not ripen fruit until 6 weeks later. In the heated tanks, the plants grew to a height of 25 ft. and produced fruit for at least 8 months.

The plants in the two tanks that were heated at 70 to 75°F. produced a total of 598.3 lb. of fruit, while those heated at 80 to 85°F. produced 625.6 lb. The control produced only 57.0 lb. of ripe fruit. The authors considered that on 100 sq. ft. of basin area 1 ton of tomatoes could be grown in 12 months, provided that ample room were provided for upward growth.

1. Renewal.—A common practice in water-culture experiments is to renew the solution at intervals of 3 days or, more generally, to remove the plants from the solution and place them in a fresh solution in clean containers. The vessels that have been used are emptied, cleaned, and then refilled at the next change. This method, however, is subject to criticism because it introduces several errors that could be entirely or at least partially overcome if the nutrient solution could be continually renewed. In the first place, if the nutrient solution is relatively dilute, the lack of an optimum amount of a given nutrient might be a limiting factor to growth if the solution were not renewed except at the intervals that have been mentioned. If a study is being made of the effect of a certain strength of solution upon plant growth, it is evident that a certain strength of concentration will not long continue unless a constant renewal is made, since the loss of water by transpiration continually increases the concentration of certain salts, at least in the solution, because they are not absorbed in the same relative amount as the water. In the third place, the influence of the organic compounds that arise from the cells that are lost from the root cap and the epidermis may have a detrimental influence upon the growing plants if allowed to accumulate for any period of time.

The depletion of oxygen and accumulation of carbon dioxide in the solution due to the respiration of the roots are factors to be considered as well as the disturbance of the physiological balance of the solution due to the unequal absorption of salts. Stiles

(1915) observed with rye plants growing in cultural solutions that if the solutions were changed frequently so as to maintain more nearly a constant composition, the solution could vary over a considerable range without producing much effect upon the rate of growth as measured by the amount of dry matter produced within a given time. In the case of barley and wheat seedlings (Brenchley, 1916) growing in nutrient solutions, the improvement in growth was more marked in concentrations up to fifth normal when the solutions were changed frequently. Trelease and Free (1917) investigated the effect of the renewal of solution upon the production of dry weight of the tops of wheat growing in Shive's best solution. The cultures ran for 41 days, and some of the results obtained were as follows:

Change	Dry weight, grams	Change	Dry weight, grams
Daily	1.243	After 2 weeks, then every 3 days	1.131
Every 3 days	1.012	After 1 month	0.654
After 1 week, then every 3 days	0.995	Not at all	0.621
Every 2 weeks	0.780		

In another experiment the culture solution was allowed to flow at the rate of 1 l. per day through the cultural jars, which had a capacity of 250 cc. The results obtained were as follows:

	Dry weight, grams
Continual renewal	1.678
Changed every 3 days	1.222
Not changed at all	0.666

Soybeans, grown in sand or water cultures with a continuous solution renewal of 1 l. of new solution per culture during each 24-hr. interval throughout the growing period, always produced plants that were superior in every respect to those grown in cultures with intermittent solution renewal (Allison and Shive, 1923).

It was shown by Marsh (1935) that plants grew better and more nearly uniform in cultural solutions if the excess solution was drained from the bottom of the container rather than siphoned from near the top of the cultural jar. This is apparently due to the elimination of precipitates and plant wastes that ordinarily collect near the bottom. Methods for supplying a continuous renewal of nutrient solution for water cultures have been described by Clark (1916), Trelease and Livington (1922), Johnston (1927), and Trelease and Thomson (1935), and methods for renewal of plant nutrients in sand cultures have been published by McCall (1916), Zurbicki (1933), and Mehrlich (1935).

2. Aeration.—Various methods have been devised for the aeration of water and sand cultures, some of which are described by Weatherwax (1916). Andrews (1920), and Allison (1921). One method that is frequently used is to pull a stream of air through the medium by use of a filter pump. Another is the drip method in which the cultural solution is replaced drop by drop, the fresh air supply being carried in, with, or by the drops. In the third case, partial aeration, at least, is secured by allowing a slow but steady flow of renewal solution into the cultural vessel. The benefits derived from aeration as measured by the amount of dry matter produced, height of plants or their general vigor have been rather varied, which, however, is to be expected, since plants differ greatly in their oxygen demands when growing under natural conditions. Hall, Brenchley, and Underwood (1914) greatly increased the root and top growth of barley and lupine in water cultures by pulling a stream of air through the medium. Free (1917), however, in experiments with buckwheat growing in Shive's nutrient solution found that the degree of aeration of the cultural solution had no important influence on the growth of these plants. His results were verified by Stiles and Jorgensen (1917), who found that aeration of the nutrient solution increased the rate of growth of barley growing in water cultures but that it had no effect on the growth of buckwheat under the same conditions. Pember (1917) noted that barley did not respond to aeration when grown in nutrient solutions that were renewed every two weeks. In experiments with corn plants, Beal (1918) and Andrews and Beal (1919) obtained much better growth in the aerated sand and water cultures than in the unaerated ones. In one case the plants growing in the aerated cultures were 65 cm. in height as compared to a height of 46 cm. for the plants in the unaerated vessels. Aeration showed its effect upon the anatomy of the plant, since the intercellular spaces were smaller and the tissues more compact in those plants growing in the aerated cultures. Oats, mustard, and peas growing in nutrient solutions respond very favorably to aeration, as is shown by the work of Andrews (1921), who found that the height of plants growing in aerated solutions was twice as much as the dry weight, in some cases five times greater than plants growing in solutions that were not aerated. In the case of soybeans, Allison (1921) and Allison and Shive (1923) stated that the best plants with respect to health, vigor, and yield were always secured in the cultures with continuous solution renewal and with constant and thorough aeration of the medium. They also found that the continuous renewal of the nutrient solution did not alone maintain the supply of dissolved oxygen necessary for the maximum development of the plants and that the roots showed a much greater response to aeration than the aerial portions of the plants.

Loehwing (1932) found that the ratio of the weight of tops to the weight of roots of sunflower plants, grown in aerated sand cultures, was much lower than that of the controls. The aerated plants not only showed accelerated growth, greater height, and larger tops and roots than did the controls, but they contained more calcium, phosphorus, sugars, hydrolyzable polysaccharides, and total nitrogen. The expressed sap of the aerated plants contained more total solids and sugars but had less ash and total nonnitrate nitrogen than the nonaerated ones.

Clark and Shive (1932) noted that the roots of the tomato plant grew throughout the container in the aerated nutrient solutions, while those of the nonaerated cultures grew only near the surface. They found that aeration had a more pronounced influence upon the growth of tops of this plant than upon the growth of the roots. The plants in the nonaerated cultures bloomed and fruited earlier than did those in the aerated cultures.

Bryant (1934), in studies of the roots of barley plants grown in aerated and nonaerated water cultures, found that after 60 days the nonaerated plants had 225 roots, while those in the aerated solution had 75 roots. The average length of the roots in the aerated cultures was 37.4 cm., as compared to 10.9 cm. for the nonaerated and the roots of the former were about 15 per cent greater in diameter than those of the latter. The cortex of the roots grown in the aerated cultures consisted of compact parenchyma with few and small intercellular spaces, while that of the roots grown in the nonaerated solution showed large air passages separated by narrow strands of parenchyma. The tissues of the nonaerated roots began to differentiate nearer the root tips than did those in the aerated cultures. In the mature regions of the aerated roots the cell walls were approximately twice as thick as the corresponding cell walls of the roots grown in the nonaerated solutions.

3. Preparation.—Since the method to be followed in the preparation of nutrient solutions will depend upon the type of solution to be used and the nature of the experiment, it will be impossible to give here anything but general methods of procedure. This perhaps can be best accomplished by giving in detail the methods to be followed in one or two specific cases. The first step in the preparation of nutrient solutions, especially if quantitative work is to be done, is to obtain salts of known purity either dry or containing a known amount of water. The next step is the preparation of stock solutions from which solutions of the desired composition and concentration may be made when desired. One of the greatest difficulties that has been experienced in the preparation of four-salt solutions and to some extent in three-salt solutions is to obtain a solution of the desired concentration and proportions and avoid any precipitate. In ordinary demonstration experiments with water or sand cultures, the formation of some precipitate is of small matter, but in a quantitative study it becomes an important factor and must be avoided. Tottingham (1914) in his work with Knop's solution found it best to employ the component salts in stock solutions, one part, *A*, containing only calcium nitrate and the other, *B*, containing the other three salts. The maximum possible concentration without precipitation of part *B* for a period of 14 days at temperatures of 20 to 23°C. was found to be 10 per cent when monopotassium phosphate was used and 9 per cent when dipotassium phosphate was used. This limits the concentration of solution *A* to 12.33 and 12 per cent, respectively. Tottingham found the maximum possible concentration of the complete Knop's solution without precipitation for a period of from 3 to 4 days at a temperature of 20 to 26°C. to be 2.8 and 2.2 per cent when monopotassium phosphate and dipotassium phosphate were used, respectively. It was found that the precipitate formed in unstable Knop's solution consisted almost entirely of calcium sulphate.

For a supply of iron a suspension of ferric phosphate of approximately 0.02 g. molecular concentration is generally recommended. A suspension of this type contains about 3 g. of ferric phosphate per liter or 3 mg. per cubic centimeter. Enough of this preparation may be made to last for several years. Enough of this suspension should be added to each liter jar at each filling to give 3 mg. of ferric phosphate.

Of interest in this regard is the work of Horner, Burk, and Hoover (1934) in devising methods for the preparation of synthetic and natural humates containing iron, aluminum, manganese, zinc, nickel, and copper. One or two applications of those humates will ordinarily supply sufficient iron, or similar metal, to water or sand sultures to last throughout the entire period of growth. The metals in the form of humates are stable in alkaline, neutral, and moderately acid media, and are not precipitated by phosphates. The general consideration of the iron supply in nutrient solutions is discussed under the heading of Iron in Chap. VI.

Five characteristics of a nutrient solution may be distinguished when its osmotic, ionic, and molecular relationships to plant response are considered: (1) the solution type or kinds of salt entering into the solution; (2) osmotic salt proportions, the division of the osmotic pressure of the solution into a definite number of parts, and their distribution among the salts in all the various proportions; (3) volume molecular salt proportions; when considered in its relationship to osmotic salt proportions it would mean that part of a volume molecular concentration of each salt which must be added to each solution to produce the osmotic proportion that is desired for that particular salt; (4) total volume molecular concentration or the total number of molecules of all kinds per liter; and (5) the total osmotic concentration, which is the resultant of the number of molecules and ions in the solution per liter. This calculated osmotic value or potential-osmotic pressure is commonly expressed in terms of atmospheres. Owing to ionization and in some cases also to hydration, the osmotic salt proportions and the volume molecular salt proportions of a solution will not vary proportionally from solution to solution. If a series of solutions are planned, therefore, so as to vary by definite increments of osmotic salt proportions, the volume molecular concentration will vary irregularly. The preparation of nutrient solutions for the study of their osmotic and salt relationships to plant growth, therefore, involves a rather tedious procedure and entails much care and labor if accurate and satisfactory results are to be obtained. In the case of a four-salt solution, if the osmotic pressure is divided into ten equal parts and distributed among the four salts in all the various proportions, 84 different solutions must be prepared. With a three-salt solution the same procedure requires the preparation of 36 different solutions. Let us here consider a yet more simple case in which the total osmotic pressure of 1 atmosphere of a three-salt solution of type I is distributed in eighths among the three component salts monopotassium phosphate, calcium nitrate, and magnesium sulphate in all the possible proportions. The molecular proportions of each of the salts and the partial molecular concentrations necessary to produce an osmotic pressure of 1.0 atmosphere for the 21 different solutions that are possible under this combination are given in the following table. This is taken from the *Report of the Committee of Biology and Agriculture of the National Research Council* (1919).

The method by which the figures presented in this table were obtained for each solution was somewhat as follows: For example, in the case of solution *R*1*S*1, a solution was prepared containing ⅛ g. molecule of monopotassium phosphate, ⅛ g. molecule of calcium nitrate, and

⅝ g. molecule of magnesium sulphate. This solution was then diluted until its freezing point indicated a total osmotic pressure of 1.0 atmosphere. In a solution that has thus been diluted, the molecular proportion of each salt can easily be calculated from the original volume, and since

Partial Molecular Concentrations and Molecular Proportions (Osmotic Proportions) of KH_2PO_4, $Ca(NO_3)_2$, and $MgSO_4$ (Type I) in 21 Solutions All Having a Calculated Osmotic Value of Approximately 1.00 Atmosphere at 25°C. but Differing by Increments of One-eighth in Salt Proportions

Solution	Molecular proportions			Partial molecular concentrations necessary to produce the fractional pressures desired		
	KH_2PO_4	$Ca(NO_3)_2$	$MgSO_4$	KH_2PO_4	$Ca(NO_3)_2$	$MgSO_4$
*R*1*S*1	1	1	6	0.0027	0.0027	0.0161
*S*2	1	2	5	0.0025	0.0049	0.0123
*S*3	1	3	4	0.0024	0.0071	0.0094
*S*4	1	4	3	0.0022	0.0089	0.0067
*S*5	1	5	2	0.0022	0.0108	0.0043
*S*6	1	6	1	0.0020	0.0122	0.0120
*R*2*S*1	2	1	5	0.0053	0.0027	0.0132
*S*2	2	2	4	0.0049	0.0049	0.0099
*S*3	2	3	3	0.0047	0.0071	0.0071
*S*4	2	4	2	0.0045	0.0090	0.0045
*S*5	2	5	1	0.0041	0.0104	0.0021
*R*3*S*1	3	1	4	0.0076	0.0025	0.0101
*S*2	3	2	3	0.0072	0.0048	0.0072
*S*3	3	3	2	0.0068	0.0068	0.0045
*S*4	3	4	1	0.0065	0.0086	0.0021
*R*4*S*1	4	1	3	0.0099	0.0025	0.0074
*S*2	4	2	2	0.0094	0.0047	0.0047
*S*3	4	3	1	0.0090	0.0068	0.0022
*R*5*S*1	5	1	2	0.0123	0.0024	0.0049
*S*2	5	2	1	0.0118	0.0047	0.0023
*R*6*S*1	6	1	1	0.0145	0.0024	0.0024

the degree of ionization of each of the salts at that dilution can be obtained from prepared tables, the fractional molecular concentration of each salt necessary to produce the fractional osmotic pressure desired can thus be calculated. If, for example, at a given concentration the dissociation value is 50 per cent for a given salt, then the osmotic value of that salt is 1.5 times its molecular concentration, so that the molecular concentration divided by 1.5 would give the molecular concentration necessary to produce the fractional osmotic pressure that is desired (see table above).

In the next table are given the strength of stock solutions and the proportion of distilled water to make 1,000 cc. of each of the 21 nutrient solutions that have just been described.

NUMBER OF CUBIC CENTIMETERS OF DISTILLED WATER, AND OF THE SINGLE SALT-STOCK SOLUTIONS REQUIRED TO MAKE 1 L. OF EACH OF THE 21 NUTRIENT SOLUTIONS OF TYPE I WITH A TOTAL CONCENTRATION OF 1.00 ATMOSPHERE

Adapted from recommendations of Research Council, 1919

Solution	H_2O	KH_2PO_4, $\frac{M}{3.5}$	$Ca(NO_3)_2.4H_2O$, $\frac{M}{3.5}$	$MgSO_4.7H_2O$, $\frac{M}{3.5}$
R1S1	924.7	9.45	9.45	56.4
S2	931.1	8.75	17.15	43.0
S3	935.85	8.4	24.85	32.9
S4	937.7	7.7	31.15	23.45
S5	939.45	7.7	37.8	15.05
S6	943.3	7.0	42.7	7.0
R2S1	925.8	18.55	9.45	46.2
S2	931.05	17.15	17.15	34.65
S3	933.85	16.45	24.85	24.85
S4	937.0	15.75	31.50	15.75
S5	941.9	14.35	36.4	7.35
R3S1	929.3	26.6	8.75	35.35
S2	932.8	25.2	16.80	25.20
S3	936.55	23.8	23.80	15.75
S4	939.80	22.75	30.10	7.35
R4S1	930.7	34.65	8.75	25.9
S2	934.2	32.9	16.45	16.45
S3	937.0	31.50	23.80	7.7
R5S1	931.4	43.05	8.4	17.15
S2	934.2	41.3	16.45	8.05
R6S1	932.45	50.75	8.4	8.4

4. The Triangle System in the Studies of Nutrient Solutions, Sand Cultures, and Fertilizer Experiments.—In the study of nutrition experiments dealing with a large number of cultures of different salt combinations it is desirable to employ some diagrammatic scheme or arrangement to represent concretely the various salt proportions and to interpret more easily the results obtained. The triangle method, which was suggested by Schreinemocher (1893) and Bancroft (1902) for certain percentage experiments in physical chemistry, was first adapted by Schreiner and Skinner (1910) to nutrition experiments. Later the method was used by Tottingham (1914), Shive (1915), and McCall (1916) in both water- and sand-culture experiments and has now become the general method for interpreting the data obtained in that type of work. A summary of the method and its various applications has been given by

Schreiner and Skinner (1918), and a critical review of it has been made by Espino (1919).

As an example of the application of the triangle system let us here consider first the distribution of the 36 different combinations that are possible when the osmotic pressure of a three-salt solution is divided into tenths and proportioned in all the possible ways among the component salts. In Fig. 11, the distribution of the various proportions of type-I solution consisting of monopotassium phosphate, magnesium sulphate, and calcium nitrate is represented on the plant of an equilateral triangle according to the arrangement now in use for all three-salt solutions. The base of the triangle is considered as the potassium side, the left side as the calcium side, and the right as the magnesium side. Each

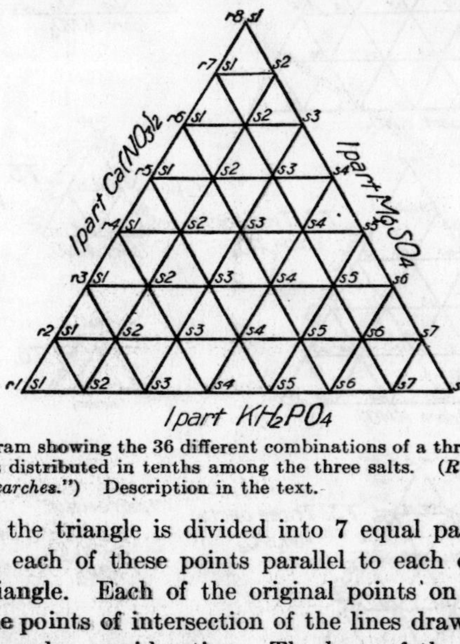

FIG. 11.—Diagram showing the 36 different combinations of a three-salt solution if the osmotic pressure is distributed in tenths among the three salts. (*Redrawn from Shive in "Physiological Researches."*) Description in the text.

of the sides of the triangle is divided into 7 equal parts and lines are drawn through each of these points parallel to each of the other two sides of the triangle. Each of the original points on the sides of the triangle and the points of intersection of the lines drawn represents one of the cultures under consideration. The base of the triangle and the lines parallel with it are numbered from the base, which is row 1 (*R*1), to the apex, which is row 8 (*R*8). The points of intersection are numbered as series (*S*) from the line on the left *S*1 to the extreme point on the right *S*8. All the cultures on the base line in the case of type-I solution have one-tenth of their osmotic pressure due to monopotassium phosphate, all the cultures on *R*2 have two-tenths of their osmotic pressure due to monopotassium phosphate, and so on until *R*8, which has eight-tenths of its osmotic pressure due to that salt. In the first line on the left of the triangle *S*1, all the cultures have one-tenth of their osmotic

pressure due to calcium nitrate, while in *S*2, two-tenths of the osmotic pressure is due to that salt, and so on until *S*8, where eight-tenths of the pressure is due to the calcium nitrate. The proportion of the osmotic pressure due to magnesium sulphate at any points can be obtained by subtracting from ten the sum of the osmotic proportions of monopotas-

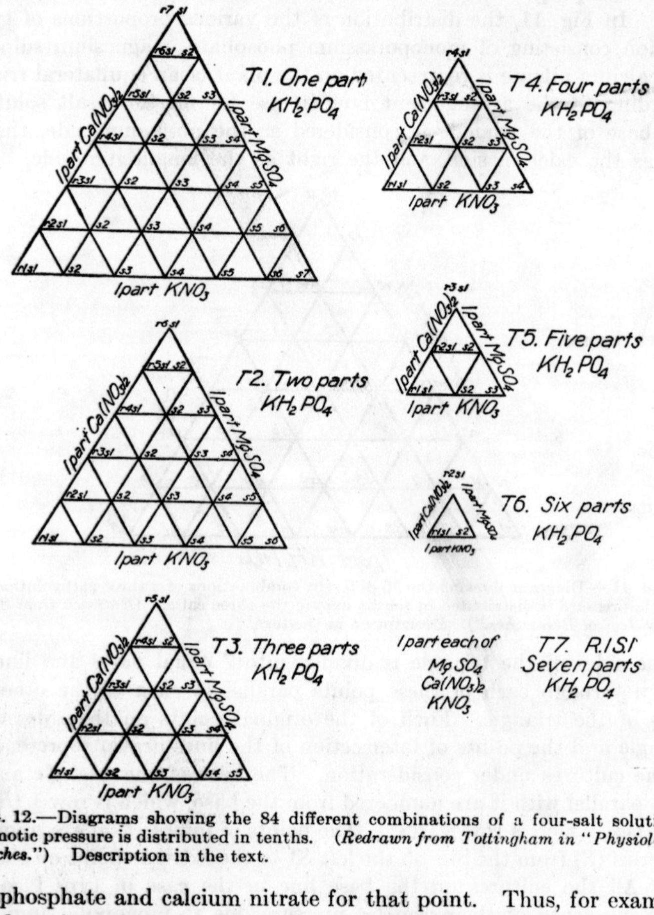

Fig. 12.—Diagrams showing the 84 different combinations of a four-salt solution if the osmotic pressure is distributed in tenths. (*Redrawn from Tottingham in "Physiological Researches."*) Description in the text.

sium phosphate and calcium nitrate for that point. Thus, for example, if a culture is at the point *R*3*S*4 it has three-tenths of its osmotic pressure due to monopotassium phosphate four-tenths to calcium nitrate, and three-tenths to magnesium sulphate. It is also customary to designate the type of solution being used by placing the type number before the symbol of concentration. Thus II*R*4*S*3 signifies that the solution in

question is type II and that four-tenths of the osmotic pressure is due to potassium sulphate, three-tenths to calcium nitrate, and three-tenths to monomagnesium phosphate.

When a four-salt solution is under consideration and the osmotic pressure is distributed in tenths among the various salts, the representation by the triangle method becomes much more complicated, since 84 different combinations are possible. Figure 12 represents the schematic arrangement of the 84 combinations of Knop's solution as used by Tottingham (1914). For a procedure of this type seven triangles are necessary, one, however, forming only a single point. The triangles are designated as *T*1 to *T*7 corresponding to the tenths of the total osmotic pressure due to monopotassium phosphate. The base of each of these triangles is the potassium nitrate side, while the left and right sides are the calcium nitrate and magnesium sulphate sides, respectively. Taking any point as *T*2*R*3*S*4, for an example, we know that two-tenths of the total osmotic pressure of the nutrient solution at that location is due to monopotassium phosphate, three-tenths to potassium nitrate, four-tenths to calcium nitrate, and one-tenth to magnesium sulphate.

D. Factors That May Determine the Relative Value of Nutrient Solutions

One of the main aims in studying nutrient solutions has been to determine the best proportions, concentrations, and means of presenting the nutrient salts in order to secure the maximum response in the growth of the plants. It has frequently been stated that the aim is to determine or find the "best solution" with the idea of applying the proportions and concentrations of solutes thus found for a maximum response in agricultural practice. It is important, then, to consider some of the factors that may determine whether or not a nutrient solution will be the "best solution" for plant growth.

1. Temperature and Light.—It was reported by Gericke (1921) that the nutrient solutions which produced the best growth of Marquis wheat in the early growth phases varied with the temperatures under which the experiment was conducted. The plants were grown in the six types of three-salt nutrient solutions in which the total molecular concentration was distributed in eights among the component salts in all possible proportions. The temperatures used were 28 and 17°C., the former being considered about the optimum and the latter distinctly below the optimum for the early growth of wheat. The "good or best solutions" were considered to be those in which the growth values obtained lay within the upper one-fourth of the total range of values for the same temperature and the same solution type. The following table shows some of the best solutions for the two temperatures observed:

Temperature, degrees Centigrade	Type I KH_2PO_4 $Ca(NO_3)_2$ $MgSO_4$	Type II K_2SO_4 $Ca(NO_3)_2$ $Mg(H_2PO_4)_2$	Type III KNO_3 $Ca(H_2PO_4)_2$ $MgSO_4$	Type IV K_2SO_4 $Ca(H_2PO_4)_2$ $Mg(NO_3)_2$
28............	*R1S3* *R1S4* *R3S4* *R4S1*	*R1S3* *R1S4* *R2S1*	*R1S3* *R1S4* *R1S6* *R2S1*	*R1S4* *R1S5*
17............	*R3S2* *R3S3* *R4S1* *R5S1*	*R3S3* *R5S1* *R5S2*	*R2S5* *R3S4* *R4S1* *R5S1*	*R4S1* *R5S1* *R5S2*

These figures indicate that there is generally a marked difference between the proportions that give the best growth at higher temperatures and those that give the best growth at the lower temperatures. It appears, for example, that the proportion of potassium may be less for high than for low temperatures.

Trelease and Trelease (1926) grew wheat plants in nutrient solutions at temperatures of 14, 19, and 30°C. and studied the relation of temperature to the physiological values of these solutions as indicated by the growth of the roots. There was little or no difference at the first two temperatures, but at a temperature of 30°C. there were marked differences in the physiological values of the different solutions.

From a comparison of the relative value of a series of solutions for the growth of wheat plants under different degrees of illumination, Trelease and Livingston (1924) concluded that if a number of markedly different nutrient solutions are repeatedly compared by tests made under different climatic complexes, the relative physiological values assigned to the several solutions may be either the same or more or less different for the several tests according to the nature and magnitude of the climatic differences dealt with. They considered that the comparative physiological values of different salt solutions cannot be stated with any degree of precision without having described in an adequate manner the climatic condition under which the tests were made.

2. Concentration.—The experimental evidence in regard to the effect of the concentration of the nutrient solution on the growth of plants is confusing. Since the information is not very definite, it is the intention to present here only the results that have been reported and the criticisms that have been offered upon the question. Breazeale (1905) grew wheat seedlings in nutrient solutions with concentrations, respectively, of 15, 75, 155, 750, and 1,550 p.p.m. The seedlings were grown for 28 days and the solution changed daily. The plants increased in growth and vigor with the increasing concentration of the solutions up to 155 p.p.m., but above that point they were small and resembled those grown in the weaker solutions. Breazeale concluded that there is an optimum physical concentration of the nutrient solution in which plants thrive best aside from variations in the amounts present of the different nutrient materials. The work of Breazeale was substantiated by that of Hall, Brenchley, and Underwood (1914) who concluded from their experiments with barley, which was grown in both water and sand cultures with a standard nutrient solution of four concentrations, that the growth made by plants in the soil solution is, in the main, determined by the amount of nutrients that are available. They also considered that the concentration of the solution within certain wide limits, irrespective of the total amount of plant food available, is a factor in the rate of plant growth. Ayres (1917), Tottingham (1914), and Shive (1915) furnished evidence which indicates that the effect upon growth of any given set of salt proportions varies with the total concentration of the solution. Shive found that, in experiments with solutions of 0.01, 1.75, and 4.00 atmospheres osmotic pressure, the best growth of tops of young wheat was obtained in the solution with an osmotic pressure of 1.75 atmospheres, the poorest growth in the weakest solution, and the medium growth in the most concentrated solution. Trelease (1917) observed young wheat plants growing in a solution consisting of monopotassium phosphate, magnesium sulphate, potassium chloride, and calcium nitrate in the molecular proportions of 1.000, 0.587, 0.485 and 0.341, respectively, for these salts in which the concentration of the solution varied by six increments from 0.5 atmosphere to 7.0 atmospheres. The maximum yield of tops was obtained when the nutrient solution had a total osmotic concentration of 1.6 atmospheres. Between the concentrations of 1.6 and 7.0 atmospheres the dry weight of the tops was approximately a linear function of the concentration, decreasing as the concentration increased. When, however, the molecular proportion of the salts in the solution was changed to 1.000, 1.155, 7.282, and 0.699, the optimum concentration for growth was between 4.5 and 5.5 instead of at 1.6 atmospheres (see also Hoagland, 1919). Cameron (1911) and Stiles (1915) working with solutions similar to some of those used by the authors just mentioned were unable to obtain any marked variations in the amount of growth over a fairly wide range of concentrations. They considered that the slight differences observed might have been due to the depletion of some one element and that this, perhaps, could be overcome by the more frequent renewal of the solution.

Hoagland and Martin (1927) considered that all essential ions, except perhaps nitrate, are always present in the soil solution, and that no matter how rapidly the ion is absorbed, its concentration is not reduced to zero.

They believed that no fixed critical concentration exists for any single element. Thus they found that 1 p.p.m. of phosphate in a cultural solution was necessary to maintain a good growth of barley, while with other plants a much lower concentration would produce an adequate growth. They believed that the absorbing-power characteristic of different plants must be ascribed to their different abilities to draw indirectly on mineral supplies not initially present in the soil solution. Thus with buckwheat and barley, no difference between the two plants was manifested in their reduction of the amount of nitrate, calcium, magnesium and potassium in the soil solution. The content of those constituents, however, varied greatly, the buckwheat withdrawing from an equal volume of soil much more calcium, magnesium, phosphorus, potassium, and nitrogen than did the barley.

Nightingale and Farnham (1936) grew plants of sweet peas in two nutrient solutions with a concentration of 0.5 to 1.0 atmosphere and of 2 to 3 atmospheres. As compared with the more concentrated solution, the plants in the dilute solution grew more vigorously, were more succulent, and abscissed more buds. The leaf blades of the plants in the weaker solution were thin, somewhat circular in outline, and of a dark green color, while those of the plants in the concentrated solution were thick, oval or elliptical, and light green. The young, active cells were more abundant in the roots and tops of the plants grown in the more dilute solutions than they were in the plants of the more concentrated solution. In the former, the tissues differentiated and matured more slowly than in the latter. Gračanin (1935), in experiments with corn and wheat, found that the amount and rate of growth did not depend upon the amount of nutrient solution but upon its concentration.

The conflicting results obtained in regard to the effect of the concentration of the solution on growth may, in part at least, be due to factors other than the ones under consideration, which have been overlooked by those conducting the experiments.

3. Water vs. Sand Cultures.—A comparative study of sand and water cultures as media for plant growth has been made by Stolhane (1914), Hall, Brenchley, and Underwood (1914), McCall (1915), Shive and Martin (1918), Shive (1920), Bakke and Erdman (1923), Hoagland and Martin (1923), and others. A sand culture furnishes a medium that is apparently more nearly comparable to soil than is a water culture. It has been found by practically all investigators that the growth of roots as judged by the production of dry matter as well as by the profuse branching and general extent is much higher in sand cultures than in water cultures. This has generally been interpreted as due to better aeration of the sand cultures. In many cases, however, the best growth of tops as measured by both green and dry weight occurs in the water cultures. One of the first questions that arises is what effect the sand has upon the nutrient solution that is added to it. Shive (1920) with three-salt nutrient solutions showed that different grades of washed sand gave no evidence of the adsorption of salt or ions in sufficient amount to alter materially either the reaction or the total salt concentration of the solution in contact with the solid particles. The adsorptive capacity, however, of the unwashed sand was sufficient to reduce the total concentration of the solution from 1.76 to 1.61 atmospheres, an average reduction of 8.5 per cent, during the first 24-hr. period. The reaction of the nutrient solution was not markedly altered by its contact with the unwashed sand. By the renewal of the nutrient solution the initial adsorptive effect of the unwashed sand was soon eliminated, apparently by saturating its adsorptive capacity. As an example of the contrast between water and sand cultures the work of Bakke and Erdman (1923) may be mentioned. In the case of Marquis wheat grown to the heading stage in type III three-salt solution with a concentration of 1.00 atmosphere, they found, in general, a greater dry weight of tops produced in the water cultures and a greater dry weight of roots in the sand cultures. The solutions producing the maximum yield of tops for the water and sand cultures showed marked variations in salt proportions. The best water-culture solution was *R1S1*, while the best sand-culture solution was *R3S3*. The amount of water transpired was with but few exceptions greater from the water cultures than from the sand cultures. The sand cultures that transpired more water than the water cultures were characterized by a high proportion of monocalcium phosphate with only a small proportion of potassium nitrate and magnesium sulphate.

It was reported by Dunlap (1936) that, for the purpose of transplanting, seedlings can be grown to better advantage in sand cultures than in soil. The seedlings so grown withstand removal better than those grown in the soil, do not damp off, and can be kept for long periods of time without becoming too large for successful transplanting.

4. Physiological Balance.—It has already been noted that one of the outstanding lines of recent investigations in plant nutrition has been the attempt to determine whether certain combinations and ratios of salts or ions in nutrient solutions are more beneficial to plant growth than others.

The experimental data obtained have indicated that the question of physiological balance is an intricate and complex one, since it varies with the type and concentration of the solution, the type of plant under consideration, and the general conditions of the environment. Certain phases of this question are treated under other topics and it is the intention here to consider only a few of the experiments that have been conducted along this line.

The results obtained by Tottingham (1914) with wheat seedlings are very striking. He grew these plants in ordinary Knop's solution and in 84 different combinations of this solution in which the osmotic pressure was distributed in tenths among the component salts. He found, for example, that with a concentration of 0.01 per cent and an osmotic pressure of 0.05 atmosphere the solution producing the greatest dry weight of tops was 39 per cent superior to the ordinary Knop's solution of the same concentration. This solution contained 0.6, 3.1, 0.5, and 1.4 times as much of monopotassium phosphate, magnesium sulphate, calcium nitrate, and potassium nitrate, respectively, as occurs in the corresponding Knop's solution. In the case of roots, the best solution was 15.4 per cent superior to Knop's solution of the same total concentration. This solution contained 1.8, 0.8, 0.2, and 3.4 times as much of the respective salts as occurs in the corresponding Knop's solution. In the solutions in which the concentration was 0.6 per cent and the osmotic pressure approximately 2.5 atmospheres, the one that produced the greatest dry weight of tops was 11 per cent better than Knop's solution of the same concentration. The greatest dry weight of roots obtained in this series was 15.4 per cent better than that obtained in the same concentration of the ordinary Knop's solution.

Evidence that different plants require different physiological balances for their best development was furnished by Shive (1915, 1917). In a series of 36 combinations in a three-salt solution with an osmotic pressure of 1.75 atmospheres, it was found that the solution which gave the best growth for buckwheat was 1*R*4*S*2, while the best solution for the growth of young plants of Marquis wheat was 1*R*5*S*2. Johnston (1925) in a study of the relative value of the six types of three-salt solutions for the growth of Irish Cobbler potatoes found that types I and II were the best suited if the dry weight of the plants is used as a criterion.

Experimental evidence secured by Gericke (1922) showed that the pairing of ions or the type of salt in which they are made available to the plant exerts a marked influence upon growth. He grew wheat seedlings for 1 day in a solution of potassium nitrate with a concentration of 1.0 atmosphere and then removed the plants, rinsed the roots in distilled water, and placed them in a solution of monomagnesium phosphate of the same concentration. The next day the plants were transferred to a like solution of calcium sulphate and after 24 hr. again transferred to the potassium nitrate solution. It was found that the plants which were thus changed by this continued rotation at 24-hr. periods grew as large and produced approximately the same dry weight as those grown in the complete nutrient solutions of the same salts or in other three-salt solutions which supplied the same ions. It was further found that plants grown in these three single-salt solutions above mentioned were much superior to those grown in other types of single-salt solutions which were likewise changed every 24 hr.

In an experiment in which 10 different salts were selected to give six different types of nutrient solutions, it was found that the order of change from one salt to another was of only minor importance. It was also found that the differences in the pH value of the single-salt solutions were within the range of good plant growth and thus exerted little or no influence but that the kind of salt used in making the change was of great importance. Thus when potassium nitrate, monomagnesium phosphate, and calcium sulphate were used singly as above mentioned they gave much better yields than did the salts of any other type solution. Gericke considered that the cause for this might be the way in which the two essential ions are paired. Thus the best results were obtained when nitrate and potassium were present together, since the absence of the other essential elements for a period of 48 hr. was apparently of little consequence. Likewise, monomagnesium phosphate seemed to be the best salt with which to supply the magnesium and phosphorus to the plant, since if the plants, for example, were placed in magnesium sulphate or magnesium nitrate alone for 24 hr. they were decidedly injured. It seems apparent, then, that the pairing of nutrient elements or ions of opposite charge is an important factor in securing the maximum growth of plants in water cultures. This does not necessarily mean that a definite cation-anion ratio must exist, since ionic ratios can cover a considerable range of values without any apparent physiological effect. Results of a somewhat similar nature were obtained by Sewell (1924), who found that when wheat is growing in a soil with a sufficient supply of nitrogen, the plant will produce a maximum yield with less of the salts containing potassium and phosphorus than if the nitrogen supply is limited. On that account, if these elements are deficient in the soil a smaller application is required if there is a liberal supply of nitrogen present.

Trelease and Trelease (1925), from a study of the growth of wheat roots, concluded that the roots are slightly sensitive to small differences in salt proportions except when the partial concentration of calcium nitrate was below 5 per cent of the total concentration. In 1925, they grew wheat in 29 different cultural solutions with a wide range of proportions of the constituent salts. The plants grew vigorously in all these combinations, but, when inoculated with the spores of mildew (*Erysiphe graminis*), pronounced differences in susceptibility were found to exist. They considered from this that susceptibility can be influenced to a high degree by the composition of the solution supplied to the plants and that the tests gave no support to the view that low vigor of the host is an index to its susceptibility to invasion by a parasite.

Gericke (1929) found that a wide range of salt proportions was equally satisfactory for the early growth of Marquis wheat. Duncan (1931) found in growing corn in sand cultures that high nitrogen produced stocky, sturdy plants giving good yields, while low nitrogen gave tall, slender, late-maturing plants. High potassium produced large plants with a relatively extensive root growth, while low potassium produced a small top growth and a very limited formation of roots. It has been reported by Bartholomew, Watts, and Janssen (1933) for tomato, by Spencer and Shive (1933) for *Rhododendron ponticum*, and by Wynd (1933) for orchid seedlings that a wide range of salt proportions can be used equally well in the growth of those plants. Beckenbach, Wadleigh, and Shive (1936) studied statistically the effects on the growth of corn of varying the proportion of the different ions in the nutrient solutions. Within the limits of the concentrations employed, no correlation was found between the amounts of sulphate or phosphate in the solution and the growth of the plants. Increasingly high concentrations of nitrate, potassium and calcium gave the most marked correlation with increased growth, while increasingly high concentrations of magnesium within the same limits resulted in decreased growth. None of the three cation ratios—calcium/magnesium, potassium/magnesium, or potassium/calcium—showed any significance in the experiment.

The problem of physiological balance in the soil under agricultural practice is of vast importance if the application of fertilizers is to be placed upon a scientific basis. The most thorough work along this line has been done by Hibbard (1927). The plants that he especially considered were Marquis wheat and Worthy oats, and the salts used as fertilizers were monocalcium phosphate, sodium nitrate, and potassium sulphate. The wheat plants were grown in soil in the greenhouse, while the oat plants were grown in the field, but in each case the nitrogen, phosphorus pentoxide, and potassium oxide were varied by increments of 10 per cent so that there were 36 combinations or ratios. The plants were grown to maturity and the yields of grain taken as a measure of the results. Hibbard concluded that the data obtained indicate that the yields of cereals such as wheat, oats, and corn may be identical under somewhat wide variations of nutrient ratios. He considered that his work suggests that nothing can be gained by attempting to find a definite ratio or physiological balance in soil cultures through the triangular system of fertilizer tests. He considered further that many of the crop plants have become accustomed to wide variations in environmental conditions and for that reason are not suitable indicators for studies of this kind. Under field conditions the proportions of various salts available to the plant must vary from time to time because of the fluctuations of the salt content of the ordinary soils. Similar conclusions were drawn by Eisenmenger (1928) in studying the rate of elongation of wheat roots supplied with simple solutions of monopotassium phosphate, calcium nitrate, and magnesium sulphate and with solutions containing pairs of these salts. He considered that the wide range of salt proportions that allow nearly optimum growth is one of the conspicuous features of his results. Optimum salt proportions probably could be defined precisely if a sufficiently detailed study were made, but the growth rate in the optimum solution would be only slightly higher than in many solutions differing considerably with respect to the relative proportion of their constituent salts. Hibbard found, however, under the conditions of his experiment that the use of a single fertilizer ingredient did not give so good results in soil cultures as a combination of the three and suggested that only under conditions where the price for a mixture of the three fertilizer ingredients is prohibitive would the use of single fertilizers be advisable.

The relationship of the moisture content of the soil to the physiological balance has been studied by Shive (1919 to 1922). He found that good physiological balance and optimum total concentration of a nutrient solution are not alone sufficient to produce the best growth of which the solution is capable unless it is diffused as a film on the particles of a solid substratum. An optimum supply of moisture is thus necessary to impart to the soil its maximum physiological value so that the actual plants producing value of any fertilizer treatment is largely determined by the moisture conditions of the soil.

E. The Hydrogen-ion Relations in Nutrient Solutions

In discussing this subject a few facts should be noted concerning the effects of soil acidity on plant growth. The acidity of the soil has many general influences on fertility due to its effects on the physical, chemical, and biological conditions of this medium. According to Truog (1918) an acid condition of the soil usually lowers the availability of nearly all the essential elements, and favors the accumulation of toxic substances. Loehwing (1930) found, however, that the application of lime and potassium to sterile muck soils often proved harmful due largely to the low hydrogen-ion concentration of the cell sap which caused the precipitation of iron in the lower parts of the plant.

The data relating to the effect of the plant on the hydrogen-ion concentration of the culture solution and the effect of this reaction upon the plant are very confusing. This is to be expected, since the reactions under consideration are influenced by so many different factors. Among these are the type of plant, the rate of ion absorption, exosmosis from the roots, temperature, and the nature of the nutrient solution. It is the intention here to state only a few experimental results to show the general complexity of the question. Tarr and Noble (1922) grew seedlings of wheat, soybeans, and corn in nutrient solutions which were maintained at constant hydrogen-ion concentrations extending over a range from pH 3 to 8 at definite gradations of 1 pH They found that a reaction of pH 3 was prohibitive to growth in all cases. The maximum growth of wheat seedlings occurred at a reaction of approximately pH 4, while a lesser concentration exerted no harmful effect until a pH of 6 was reached, when chlorosis began to appear. The maximum growth of soybeans and corn was obtained at a reaction of pH 5. In both these cases chlorosis occurred at a reaction of pH 6 due to the insolubility of iron at all hydrogen-ion concentrations less (more neutral) than pH 6. According to Hoagland (1926), slight acidity pH 5 to 7 is not itself injurious to various types of plants, but alkaline solutions above pH 8 in general prove to be less favorable, and pH values above 9 are injurious to most agricultural plants. Bermuda grass, however, can grow vigorously when the medium in which it is growing has a pH value of 9. The effects of alkalinity on the plant are both direct and indirect, so that they are difficult to interpret, chiefly because it is difficult or impossible to maintain the desired concentration of the essential salts, since some of these precipitate out under alkaline conditions. Hoagland (1923) obtained some indications that barley, cucumbers, and peas grow more rapidly when a certain degree of acidity is maintained than when the reaction of the solution is allowed to change under the influence of the plant. In no case did he observe as good a growth at pH 7 as when the nutrient solution was slightly acid. It seems therefore to be uncertain whether the reaction to which a solution is brought by a plant growing in it is the most favorable for the growth of that plant. Duggar (1920) found in the case of wheat, corn, and Canada field peas growing in modified Shive and Livingston-Tottingham solutions among others that these plants under favorable conditions gave excellent growth even though the pH value of the solutions varied from 4.5 to 7.1. He concluded from these observations that there is a considerable range of pH values within which plants will grow well.

It was observed by Kachioni-Val'ter (1927) that the optimum growth and development of buckwheat in the presence of chlorine were at pH 5.0 to 6.5; that, at a pH of 7.0 to 9.0, growth and development were hindered; and that pH 4.5 had an intermediate effect. In the absence of chlorine, the optimum region of the pH value became narrower and approached a pH of 6.0.

It has been observed by numerous investigators, in the study of the relationship of the hydrogen-ion concentration to growth of plants, that there is frequently, between the extreme limits of acidity or alkalinity that are injurious and interfere with growth, a third region where the reaction affects the plant injuriously. Thus, if the growth of the plant or plant part is plotted against the hydrogen-ion concentration of the solution in which the plant is grown, the curve has a double maximum with a minimum located between the two maxima. That this phenomenon has not been observed in all cases should not be considered (Robbins, 1923) as necessarily indicating the nonexistence of such a curve, since there are many conditions that might conceal or obliterate the minimum located between the two maxima. The pH values of the solutions in which the plants were grown may have been so separated that the minimum point for growth could not have been observed. Then such factors as temperature, water supply, and salt content may have limited growth so that the maximum and minimum points could not be distinguished. It may be well to give a few examples of this. Salter and McIlvane (1917) noted a double maximum curve for the growth of wheat seedlings grown in nutrient solution with different pH values and found the minimum between these two maxima to lie at a pH value of 6, the same minimum point as that observed by Cole (1922) for corn seedlings. Hixon (1920) found a double maximum for growth in the case of peas, corn, wheat, and oats, a pH value of 5.0 representing the minimum point for peas, and pH 6.0 the minimum point for the other three plants. Arrhenius (1922) grew numerous crop plants in soils in which the hydrogen-ion concentration ranged from pH 3.0 to 10.0. The two hydrogen-ion concentrations giving the highest yield were as follows:

Dale oats	pH 7 and 5	Alfalfa	pH 8 and 7
Rubin wheat	pH 9 and 4	Flax	pH 6 and 4
Red clover	pH 6 and 5	Sugar beets	pH 8 and 6

The data indicate clearly the presence of a double maximum and also show that the same hydrogen-ion concentration may show wide variations in plant production.

A double-maximum curve also apparently results when the swelling, osmotic pressure, viscosity, and electrical conductivity of certain colloids are plotted against the pH value of the solution, and it has been found that the minimum between the two maxima is located at the isoelectric point of these substances. These facts led Robbins (1923) to consider that there may be an isoelectric point for plant tissue that may correspond to the pH value of the minimum point of growth. By means of experiments with the absorption of water and dyes by potato tuber tissue in contact with solutions of different hydrogen-ion concentrations, he obtained a double-maximum curve with a minimum between for water absorption. His results indicated that an ampholyte, possibly a protein, plays a chief role in this process, and its isoelectric point is approximately pH 6.0. He considered that this assumption of an isoelectric point for plant tissue and its influence on water absorption offers a satisfactory explanation for the double-maximum curves in the growth of plants and other physiological functions when the plants are grown in media of different hydrogen-ion concentrations.

The effect of the nature of the nutrient solution upon the rate of change of the pH value when plants are growing therein has been studied by Jones (1921), Jones and Shive (1922), McCall and Haag (1920), Loo (1927), Zinzadzé (1932, 1933), and Trelease and Trelease (1933, 1935). In the case of wheat growing in three-salt nutrient solutions, the rate of change of the hydrogen-ion concentration decreased with an increase in the total osmotic value of the solution and with its volume. The following table shows the change of the pH value of different types of nutrient solutions in which wheat had been grown for 52 hr. (Jones and Shive, 1922).

Kind of solution	pH value	
	Original	After 52 hr.
Crone's	6.6	6.6
Sach's	6.7	6.6
Knop's	4.6	5.7
Shive's *R5S2*	4.5	5.3
Tottingham's *T3R1S4*	4.6	5.5
Detmer's	4.7	5.9

They considered that the resistance offered to reaction change resulting from contact with the roots of growing plants was largely dependent upon the volume molecular proportions of the soluble phosphate salts contained in the solution. McCall and Haag (1920) stated that the hydrogen-ion concentration of various types of three-salt solutions is a function of the volume molecular proportion of the dihydrogen phosphate salt.

Loo (1927), and Trelease and Trelease (1933, 1935) considered that the type or proportion of the nitrogen salts used determined the pH value of the cultural solution. Wheat plants were grown in culture solutions with initial pH values adjusted to 4.3, 5.1, and 6.0 by suitable proportions of phosphoric acid, monopotassium phosphate, and dipotassium phosphate, respectively. A wide range of nitrate/ammonium ionic ratios was obtained by varying the proportions of potassium nitrate and ammonium sulphate, potassium nitrate and ammonium nitrate and calcium nitrate and ammonium nitrate. With low nitrate/ammonium ratios, the pH value of the solution decreased rapidly under the influence of the plants and approached in extreme cases a pH of 3.0. With high ratios, on the other hand, the pH values increased rapidly, tending to reach a limiting value of pH 6.5. When a nitrate/ammonium ratio of suitable value was used, a physiologically balanced solution was obtained in which the hydrogen-ion concentration tended to remain approximately constant during the 8-day period between the renewal of the solution.

Zinzadzé (1934) found that slightly soluble phosphate plus ammonium nitrate is efficient in stabilizing the reaction in nutrient solutions.

The majority of observations indicate that plants growing in nutrient solutions whose reactions are either above or below the neutral point tend to change the hydrogen-ion concentration toward neutrality, although this will depend in part upon the nature of the solution (Toole and Tottingham, 1918; Hoagland, 1917; and Meier and Halstead, 1921). Thus Bakke and Erdman (1923) noted that wheat plants at the heading stage changed the reaction of type-III salt solution in which they grew for 3½ days from pH 3.75 to 5.94, when water cultures were used, and to 6.66 when sand cultures were used. An example of the different reactions brought about by different plants is afforded by the work of Jacobson (1925) with wheat and rice plants. With Marquis wheat plants 97 days old placed in quart jars that contained solution type III *R2S1* with the molecular proportions distributed in eighths, the pH value increased from 3.9 to 6.3 during the first 12 hr., after which the reaction of the solution remained fairly constant for the 72-hr. period of observation. The nitrogen concentration of the solution decreased from 85 to 24 p.p.m., a loss of practically 72 per cent during the first 12 hr., and by the end of the 36-hr. period only a trace of nitrogen remained. The results indicated that the potassium of the potassium nitrate was not taken up nearly so rapidly as the nitrate, and it combined with other elements after being liberated from its anion so that the combination thus formed was capable of partly neutralizing the acidity of the nutrient solution. In the case of rice plants, however, growing in four different types of nutrient solutions, the pH value decreased from 5.0 to about 3.5 during the first 24 hr., after which it remained practically constant for the 3 days that the experiment was continued. It was considered that the cations of the salts containing sulphate might have been removed from the solution more rapidly than the anion so that the acidity was increased. This has also been suggested by Hoagland (1918). It was observed in both the rice and wheat cultures that the excretion of carbon dioxide by the roots tended to decrease the pH value of the solution.

F. Criticism of the Data of Plant Nutrition

The basic purpose of the study of plant response in water and sand cultures is to determine the nature of the nutrients, their concentration, their proportions, and the method of their application that will give the best response when applied in agricultural practice. The question then arises as to how far the observations that have been made with sand and water cultures are comparable and applicable to plants growing under conditions in the field.

One of the chief objections to the use of water and sand cultures from the practical standpoint is that the plant is growing in a very unnatural medium. The differences between the plants grown in sand and in water cultures illustrate the fact that the medium in which the roots grow may affect the plant response.

In cultural solutions the nutrient ions are free to move, while in soils they are adsorbed on colloidal particles or fixed in crystals and thus exist under conditions of restraint (Jenny and Cowan, 1933). While many plants grow to maturity and produce seed in water and sand cultures, the production is, with but few exceptions, less than that of plants growing in the soil. The seed produced, however, has in the cases which have been tested shown little difference in its germinating power from that produced by plants growing in the soil.

Dunn (1921) found that the cortex of roots, grown in nutrient solutions, showed intercellular spaces, the time and place of their appearance varying with the kind of plant and solution used. These spaces were formed apparently by the disintegration of cells. The concentration of the solutions used in most cultural work is much higher than the soil solution upon which a plant normally has to depend (Crocker, 1919). Apparently, too, the solutes in cultural solutions are more readily available than they are in the soil, from both the standpoint of solubility and the rate of diffusion. Neither the frequent renewals nor continuous renewals of the solution in cultural work furnish a medium like that of the soil, although the exact status of the soil solution in regard to its variability of concentration, solubility, and composition is as yet quite uncertain. On that account, therefore, in the words of Stiles (1915), it is a very uncertain procedure to argue from a comparatively simple medium like a nutrient solution of mineral salts to a complex substance like the soil.

Jenny and Cowan (1933) grew plants of soybeans in different systems containing calcium as the only nutrient ion. In one system it was free in solution, in another it was adsorbed on colloids, while in a third it was fixed in the interior of crystals. They found that with a low content of calcium the plant yields were much higher for the free-calcium system than for the adsorbed one, indicating that the plant encounters considerable difficulty in obtaining the adsorbed ions. With higher concentrations of calcium, however, the plants grew best in the adsorbed system. They believed their work indicated that excreted hydrogen ions replaced the calcium ions on the clay, thus enabling calcium to enter the plant. They considered it probable that, inasmuch as the plant can feed successfully on adsorbed ions, the significance of the "soil solution" has been overestimated. According to them the solubility concept is not wholly adequate to account for plant growth in soils of humid regions, and they considered that ionic exchange must be given consideration in a discussion of the uptake of ions by plants. As pointed out by Davis (1921), a very great defect in most plant-nutrition work has been the comparatively small number of plants and cultures used in each experiment. The variation of plants growing even under field conditions is very great, so that unless a large number of individuals is considered in cultural work the results may be of no significance. Davis selected three combinations of Shive's three-salt solution (1.75 atmospheres) which had been reported, respectively, as the best, medium, and poor solution for the growth of wheat and grew in them Sonora wheat in 33 to 50 triplicate cultures for each of the three selected solutions. He found that when the probable error was obtained for the number of cultures used, there was no significant difference between the cultural solutions which had been reported as the best and medium, respectively, but that the solution reported as poor showed a decided inferiority.

The wide variation in cultures was illustrated by the chance selection of duplicate cultures. Thus where the mean of the dry weights of the series was 2.15 g., the mean of 16 pairs selected at random ranged from 1.82 to 2.55 g. Since in most of the experimental work that has been published on plant nutrition the cultures were grown only in duplicate or triplicate, considerable doubt is thrown upon the value of the data that have been reported.

One of the chief aims of nutrition work with plants has been to find the "best solution" for plant growth. The criterion for judging this has been the best response as measured by growth obtained from a series of solutions under conditions that too frequently have not been well defined. The standards by which the best response has been measured are the length and weight of tops, the amount of transpiration, the total amount of dry matter produced, and others. In outlining the standards by which the so-called "best growth" is to be judged, the purpose for which the plant is to be grown must obviously be taken into consideration. If a plant is to be grown for forage, a different standard must be considered from that which would have to be considered if only a grain crop were desired. Clements (1928) made chemical analyses of plants grown in the various positions in the triangle for starch, total sugars, hemicellulose, nitrate nitrogen, total organic nitrogen, and ash in the various parts, with the idea of determining the internal relations of the plant to the various salt combinations. It was found, however, that these criteria, for the most part, are unreliable for determining the best response of plants to nutrient solutions. Hoagland and Martin (1923) have pointed out that we are not justified in stating that a normal plant is the one making the best growth under a given set of environmental conditions. No solution has a fixed value so far as its ionic, atomic, or molecular properties are concerned for growth, since the relation of a plant to a cultural medium is dependent upon the environmental conditions. On that account, therefore, no statement of any value can be made in regard to the "best solution" unless a strict account of the various environmental conditions is presented—a requirement that is practically impossible. The stage of growth of the plant must also be taken into consideration when considering the value of a given nutrient solution, since the nutrient requirements of a plant vary with its stage of development, as will be discussed in the following chapter.

The consensus of opinion among workers in plant nutrition at the present time is that there is a wide range of solutions in regard to concentration, composition, and proportion of components in which plants will thrive equally well and that the response of plants to any solution will depend to a great extent upon the environmental conditions under which they are growing. In the words of Stiles (1916), we may state that the value of water- and sand-culture work is limited because of the difficulty of analyzing results due to complex factors that are not under control and because of the further difficulty of controlling in some cases even the factor whose action is being investigated.

Gericke (1930) considered that the actual salt requirement cannot be determined by growing plants to maturity in any so-called complete nutrient solution. Apparently the conditions that determine the relation between the magnitude of yield and the absorption of any given elements from a cultural solution involve not only the amounts of the elements absorbed but also the time required for absorption, as well as the time required for the various absorbed elements to function in growth.

On the other hand, it must be remembered that nutrient solutions furnish the best means that are available at present to study the nutrition of plants in a quantitative way, and much of our knowledge of plant nutrition has been gained by their use. Care should nevertheless be exercised in interpreting the results obtained with nutrient solutions in terms of the behavior of plants under natural conditions.

VI. ANTAGONISM AND BALANCED SOLUTIONS

It has long been known that solutions of certain single salts are toxic to living organisms placed therein but that in a mixed solution of these salts the organisms function normally and are not injured (True, 1915). The best example is the organisms living in sea water. They grow and develop in this mixture of salt solutions, but if placed in a solution of any one of the single salts that compose it they perish, in many cases in a relatively short time. The term "antagonism" is used to designate the hindrance that a given salt has upon the toxic action of another salt. Thus antagonism means that one toxic substance acts as an antidote to another.

As has been mentioned above, it is possible to mix two or more toxic substances in solution in such proportions that they lose their individual toxic properties. Such a solution has been termed a "physiologically balanced solution" (Loeb, 1906) and may be defined as a solution in which the toxic effects that each salt would have were it alone present in the solution are hindered by one or more antagonistic salts in the solution. It has also been defined as a solution containing salts in such proportions that they have none of the toxic properties of the individual salts. The term, however, has been used in a very broad sense and has referred in many cases to the proportional arrangement of the component parts of nutrient solutions regardless of their toxic properties. A physiologically balanced solution may contain only two salts, so that it is not necessarily a nutrient solution, but on the other hand a nutrient solution to be of the best advantage must be a balanced one.

The question of antagonism and balanced solutions in regard to plants has been extensively investigated by Osterhout and his pupils

(1906 to 1922). The intention here is not to discuss this problem in any detail but only to mention a few examples to illustrate the importance of this subject in relation to plant life. Osterhout (1914) has pointed out that a mixture of two equally toxic solutions must have precisely the same effect on growth as the pure solutions themselves, provided the effects of the salts are additive. By the additive effect is meant the effect produced when each salt acts independently of every other so that its toxicity is neither augmented nor diminished by the presence of other salts. If antagonism between the salts exists, there will be better growth in the mixed solutions than in the single ones. The amount of this increased growth expressed as percentage of the growth obtained in the pure solution is the most satisfactory measure of antagonism. Precautions must be taken in the determination of antagonism that only those parts of plants that come in contact with the solution should be used for measurement. Trelease and Trelease (1926) and Eisenmenger (1928) have demonstrated that the roots of young wheat seedlings can be used satisfactorily for demonstrating the growth-retarding effects of single-salt solutions and the antagonistic action of mixed solutions. The following table shows the relative effect of mixed- and single-salt solutions on the growth of wheat roots therein (Osterhout, 1907):

Medium	Growth of roots in 40 days, millimeters	Medium	Growth of roots in 40 days, millimeters
NaCl 0.12M	60	1,000 cc. NaCl 0.12M 78 cc. $MgCl_2$ 0.12M 10 cc. $CaCl_2$ 0.12M	327
$MgCl_2$ 0.12M	7		
KCl 0.12M	68		
$CaCl_2$ 0.12M	70	Dilute artificial sea water 0.12M NaCl 1,000 cc. 0.12M $MgCl_2$ 78 cc. 0.12M $MgSO_4$ 38 cc. 0.12M KCl 22 cc. 0.12M $CaCl_2$ 10 cc.	360
1,000 cc. NaCl 0.12M 10 cc. $CaCl_2$ 0.12M	254		
1,000 cc. NaCl 0.12M 22 cc. KCl 0.12M 10 cc. $CaCl_2$ 0.12M	324		
		Distilled water	740

Osterhout and others have found antagonism between the following cations:

NH_4 *vs.* Ca	NH_4 *vs.* Na	Na *vs.* Ba
K *vs.* Ca	NH_4 *vs.* K	K *vs.* Ba
Na *vs.* Ca	Na *vs.* K	Na *vs.* Sr
Mg *vs.* Ca	Mg *vs.* K	K *vs.* Sr
Mg *vs.* Sr	Fe *vs.* Mn	

Raber (1920) from experimental work with sodium acetate and sodium sulphate concluded that antagonism existed between anions, so any theory of the antagonistic action of salts must take into account the action of the anions as well as the cations.

Kearney and Harter (1907) noted the effect of calcium sulphate in overcoming the toxic effect of the magnesium and sodium salts that are found in alkali soils. The following table shows the critical concentrations for plants of lupine and corn grown in pure solutions of the various salts and in the same solutions after an excess of calcium sulphate had been added. The term "critical concentration" as here used was the concentration of solution that allowed the root tips of any one-half of the total number of seedlings tested to retain their capability of further elongation after being in the solution for 24 hr.

Salt	Critical concentrations			
	In pure solutions		After adding $CaSO_4$ in excess to solutions	
	Lupine	Corn	Lupine	Corn
$MgSO_4$	0.007*M*	0.25*M*	0.40*M*	0.60*M*
$MgCl_2$	0.007*M*	0.08*M*	0.20*M*	0.30*M*
Na_2CO_3	0.010*M*	0.02*M*	0.03*M*	0.05*M*
Na_2SO_4	0.040*M*	0.05*M*	0.30*M*	0.40*M*
NaCl	0.040*M*	0.04*M*	0.20*M*	0.25*M*
$HNaCO_3$	0.030*M*	0.05*M*	0.05*M*	0.10*M*

The neutralizing effect of calcium sulphate was much greater toward the magnesium salts than toward the sodium salts, while the order of toxicity of the different salts was also changed by its addition.

The causes of antagonism are little understood, and it is not the intention here to enter into any discussion of the various theories that have been proposed. The relationship of the changes in permeability produced by salts and their antagonistic relations as suggested by Osterhout (1915) may give some insight into the causes of antagonism. As has been mentioned in Chap. II, the substances that alter the permeability of the protoplasm may be divided into those which cause an increase and those which cause a decrease in its permeability. Osterhout considered that substances belonging to one class will antagonize those belonging to the other class, so that in order to predict which substances will antagonize each other it is necessary only to determine to which of these classes the substances belong. The amount of antagonism between salts may also be predicted, at least to a certain extent, since the greater their effect on the permeability the greater will be their antagonistic action. This theory of antagonism assumes that the protoplasmic membrane possesses a certain degree of permeability at which it functions normally, and when this permeability is increased or decreased beyond a certain limit the activities of the protoplasm are impaired. If an organism is placed, for example, in a solution of sodium chloride, the permeability of the protoplasm is increased to a point that is injurious to it; and if it is placed in a single solution of calcium chloride, the permeability will be decreased to the point of injury. If, however, these two salts are mixed in the proper proportion, their effects on the protoplasm will counteract each other, and the permeability of the protoplasm will be unaltered so that the cell or organism will function normally.

Bibliography

ALLISON, R. V., 1921: The relation of aeration to the development of the soybean plant in artificial culture, *N. J. Sta. Ann. Rept.*, **1921**: 338–345.

——— and J. W. SHIVE, 1923: Studies on the relation of aeration and continuous renewal of nutrient solution to the growth of soybeans in artificial culture, *Am. J. Bot.*, **10**: 554–567.

ANDREWS, F. M., and C. C. BEAL, 1919: The effect of soaking in water and of aeration on the growth of *Zea mays*, *Bull. Torrey Bot. Club*, **46**: 91–100.

———, 1920: Apparatus for aerating plants, *Proc. Ind. Acad. Sci.*, **1919**: 165–167.

———, 1921: The effect of aeration on plants, *Proc. Ind. Acad. Sci.*, **1920**: 147–148.

ARRHENIUS, O., 1922: Hydrogen ion concentration, soil properties, and growth of higher plants, *Arch. Bot.*, **18**: 1–54.

———, 1923: Soil reaction and plant development, *Medd. Centralanst. Jordbruksomradet.* 245, 14 pp., *Expt. Sta. Rec.*, **50**: 532.

ASPREY, G. F., 1933: Studies on antagonism. I, The effect of the presence of salts of monovalent, divalent, and trivalent kations on the intake of calcium and ammonium ions by potato tuber tissue, *Proc. Roy. Soc.*, B, **112**: 451–472.

———, 1933: Studies on antagonism. II, The effect of previous immersion of potato tuber tissue in solutions of monovalent, divalent, and trivalent cations on the subsequent absorption of the ammonium ion, *Proc. Roy. Soc.*, B, **113**: 71–82.

AYRES, A. H., 1917: Influence of the composition and concentration of the nutrient solution on plants grown in sand cultures, *Univ. Calif. Pub.*, **1**: 341–394.

BAKKE, A. L. and L. W. ERDMAN, 1923: A comparative study of sand and solution cultures of Marquis wheat, *Am. J. Bot.*, **10**: 18–32.

BANCROFT, W. D., 1902: Synthetic analysis of solid phase, *J. Phys. Chem.*, **6**: 178.

BARTHOLOMEW, R. P., V. M. WATTS, and G. JANSSEN, 1933: The effect of variations in the nutrient media upon the nitrogen, phosphorus, and potassium content on plants, with special reference to the tomato, *Ark. Sta. Bull.* 288.

BEAL, C. C., 1918: The effect of aeration on the roots of *Zea mays*, *Proc. Ind. Acad. Sci.*, **1917**: 177–180.

BENEDICT, H. M., 1934: Effect of ultraviolet radiation on growth and on the calcium and phosphorus contents of plants, *Bot. Gaz.*, **96**: 330–341.

BERRY, W. E., and F. C. STEWARD, 1934: The absorption and accumulation of solutes by living plant cells. VI, The absorption of potassium bromide from dilute solution by tissue from various plant storage organs, *Ann. Bot.*, **48**: 395–410.

BLINKS, L. R., and A. G. JACQUES, 1930: The cell sap of *Halicystis*, *J. Gen. Physiol.*, **13**: 733–738.

BREAZEALE, J. F., 1905: Effect of the concentration of the nutrient solution upon wheat cultures, *Sci.*, **22**: 146–149.

———, 1923: Nutrition of plants considered as an electrical phenomenon, *J. Agr. Res.*, **24**: 41–55.

——— and W. T. MCGEORGE, 1932: Nutritional disorders in alkaline soils as caused by deficiency of carbon dioxide, *Ariz. Sta. Tech. Bull.* 41.

BRENCHLEY, W. E., 1916: The effect of the concentration of the nutrient solution on the growth of barley and wheat in water cultures, *Ann. Bot.*, **30**: 77–91.

BRIGGS, G. E., and A. H. K. PETRIE, 1928: On the application of the Donnan's equilibrium to the ionic relations of plant tissues, *Biochem. J.*, **22**: 1071–1083.

———, 1930: The accumulation of electrolytes in plant cells. A suggested mechanism, *Proc. Roy. Soc.*, B, **107**: 248–269.

———, 1932: The absorption of salts by plant tissues, considered as ionic exchange, *Ann. Bot.*, **46**: 301–322.

BROOKS, S. C., 1929: The accumulation of ions in living cells. A non-equilibrium condition, *Protoplasma*, **8**: 389–412.

———, 1931: The accumulation of strong electrolytes in living cells, *Sci.*, **73**: 589–590.

BROWN, D. D., 1928: The effect of different proportions of calcium nitrate, and potassium dihydrogen phosphate on the growth of wheat in sand cultures, *Soil Sci.*, **26**: 441–446.

BRYANT, A. E., 1934: Comparison of anatomical and histological differences between roots of barley grown in aerated and in non-aerated culture solutions, *Plant Physiol.*, **9**: 389–391.

BURGERSTEIN, A., 1897: Ueber die Transpirationgrösse von Pflanzen feuchter Tropengebiete, *Ber. bot. Ges.*, **15**: 154–165.

CAMERON, F. K., 1911: "The Soil Solution," The Chemical Publishing Co., Easton, Pa.

CASALE, L., 1921: A physical-chemical study of the absorptive power of soils and of the method whereby plants absorb nutritive materials from the soil, *Staz. Sper. Agrarie Ital.*, **54**: 65–113; *Bot. Abst.*, **10**: 115.

CLARK, H. E., and J. W. SHIVE, 1932: Influence of continuous aeration upon the growth of tomato plants in solution cultures, *Soil Sci.*, **34**: 37–42.

CLARK, O. L., 1916: A method for maintaining a constant volume of nutrient solutions, *Sci.*, **44**: 868–869.

CLEMENTS, H. F., 1928: Plant nutrition studies in relation to the triangular system of water cultures, *Plant Physiol.*, **3**: 441–459.

COLBY, H. L., 1933: Seasonal absorption of nutrient salts by the French prune grown in solution culture, *Plant Physiol.*, **8**: 1–34.

COLE, A. M., 1922: Effect of variations in H-ion concentration on the growth of seedlings, *thesis*, *Univ. Mo.*

COLLANDER, R., 1930: Permeabilitätsstudien an *Chara ceratophylla*. I, Die normale Zusammensetzung des Zellsaftes, *Acta Bot. Fennica*, No. 6, 1–20. Helsingforsiae.

——— and H. BARLUND, 1933: Permeabilitätsstudien an *Chara ceratophylla*. II, Die Permeabilität für Nichtelektrolyte, *Acta Bot. Fennica*, No. 11, 1–114. Helsingforsiae.

CONRAD, J. P., 1934: Physiological acidity and alkalinity of inorganic nitrogenous compounds in solution cultures, *J. Am. Soc. Agron.*, **26**: 364–372.

COOK, R. L., 1930: Effect of soil type and fertilizer on the nitrate content of the expressed sap and the total nitrogen content of the tissue of the small grains, *J. Am. Soc. Agron.*, **22**: 393–409.

Cooper, H. P., 1930: Ash constituents of pasture grasses, their standard electrode potential and ecological significance, *Plant Physiol.*, **5**: 193–215.

Cooper, W. C., M. J. Dorcas, and W. J. V. Osterhout, 1929: The penetration of strong electrolytes, *J. Gen. Physiol.*, **12**: 427–433.

——— and W. J. V. Osterhout, 1930: The accumulation of electrolytes. I, The entrance of ammonia into *Valonia macrophysa*, *J. Gen. Physiol.*, **14**: 117–127.

Crocker, W., 1919: Review of plant nutrition work, *Bot. Gaz.*, **67**: 174–178.

Crone, G., 1902: Ergebnisse von Untersuchungen über die Wirkung der Phosphorsäure auf die höhre Pflanzen und eine neue Nahrlösung, *Sitzungber. Niederrhein Ges. Nat. Heilkundee Bonn*, **1902**: 167–173.

Davidson, J., 1927: Effect of hydrogen-ion concentration on the absorption of phosphorus and potassium by wheat seedlings, *J. Agr. Res.*, **35**: 335–346.

———, 1933: The possible effect of hydrogen-ion concentration on the absorption of potassium and phosphorus by wheat plants under field conditions, *J. Agr. Res.*, **46**: 449–453.

Davis, A. R., 1921: The variability of plants grown in water cultures, *Soil Sci.*, **11**: 1–33.

Denny, F. E., and W. J. Youden, 1927: Acidification of unbuffered salt solutions by plant tissue in relation to the question of tissue isoelectric points, *Am. J. Bot.*, **14**: 395–414.

Dufrénoy, J., 1920: The biological significance of selective adsorption, *Am. Midland Nat.*, **6**: 159–165.

Duggar, B. M., 1920: The effect of conditions on the relation of seed plants to H-ion concentration of nutrient solutions, *Sci.*, **52**: 416.

———, 1920: Hydrogen ion concentration and the composition of nutrient solutions in relation to the growth of seed plants, *Ann. Mo. Bot. Gard.*, **7**: 1–50.

———, 1920: The use of insoluble salts in balanced solutions for seed plants, *Ann. Mo. Bot. Gard.*, **7**: 307–327.

Duncan, G. H., 1931: Influence of nutritional balance upon the development of corn plants, *Trans. Ill. State Acad. Sci.*, **23**: 143–148; *Exp. Sta. Rec.*, **74**: 32.

Dunlap, A. A., 1936: Sand culture of seedlings, *Conn. Sta. Bull.* 380.

Dunn, G. A., 1921: Note on the histology of grain roots, *Am. J. Bot.*, **8**: 207–211.

Eisenmenger, W. S., 1928: Toxicity, additive effects and antagonism of salt solutions as indicated by growth of wheat roots, *Bull. Torrey Bot. Club*, **55**: 261–305.

———, 1931: Factors modifying the toxicity of phenol, *Plant Physiol.*, **6**: 325–331.

Espino, R. B., 1919: Critical review of article by Schreiner and Skinner on the triangle system for fertilizer experiments, *Plant World*, **22**: 53–54.

Fittbogen, J., 1873: Untersuchungen über das für eine normale Production der Haferpflanze notwendige Minimum von Bodenfeuchtigkeit sowie über die Aufnahme von Bestandteilen des Bodens bei verschiedenem Wassergehalt desselben, *Landw. Jahrb.*, **2**: 252–371.

Free, E. E., 1917: The effect of aeration on the growth of buckwheat in water cultures, *Johns Hopkins Univ. Circ.*, **3**: 198–199.

Freeland, R. O., 1935: Effect of transpiration upon the absorption and distribution of mineral salts in plants, (Reprint) *Abstract, Doct. Diss. No. 17, Ohio State Univ. Press.*

———, 1936: Effect of transpiration upon the absorption and distribution of mineral salts in plants, *Am. J. Bot.*, **23**: 355–362.

Friedemann, W. G., 1929: A method for comparing the value of ammonia nitrogen and nitrate nitrogen, *Sci.*, **70**: 43–44.

Furnstal, A. H., and B. Johnson, 1936: Preparation of sintered pyrex glass aerators for use in water culture experiments with plants, *Plant Physiol.*, **11**: 189–194.

Gericke, W. F., 1921: Influence of temperature on the relations between nutrient salt proportions and the early growth of wheat, *Am. J. Bot.*, **8**: 59–63.

———, 1922: Water culture experimentation, *Sci.*, **56**: 421–422.

———, 1922: On the physiological balance in nutrient solutions for plant cultures, *Am. J. Bot.*, **9**: 180–182.

———, 1929: Some relations of maintained temperatures to germination and the early growth of wheat in nutrient solutions, *Philippine J. Sci.*, **38**: 215–240.

———, 1930: Plant food requirement of rice, *Soil Sci.*, **29**: 207–226.

——— and J. R. Travernetti, 1936: Heating of liquid culture media for tomato production, *Agr. Eng.*, **17**: 141–142.

Giltay, E., 1897: Vergleichende Studien über die Stärke der Transpiration in den Tropen und in mitteleuropäischem Klima, *Jahrb. wiss. Bot.*, **30**: 615–644.

Gračanin, M., 1932: La concentration des ions comme facteur de la résorption, *C. R. Acad. Sci. Paris*, **195**: 1311–1313.

———, 1932: The effect of light on the resorption of salts by plants, *Preslia Bull. Soc. Bot. Tchecoslovaque à Prague*, **11**: 1–5.

———, 1935: Die Konzentration der Nährlösung als Faktor des Wurzelwachstums, *Lab. Plant Nutrition Univ. Zagrebensis.*

Gregory, F. G., 1928: The differential effect of the ions of three-salt solutions on the growth of potato plants in sand culture, *Proc. Roy. Soc.*, B, **102**: 311–327.

Grunder, M. S., 1933: Yield and chemical composition of certain pasture crops, fertilized and unfertilized, *J. Agr. Res.*, **46**: 375–386.

Haas, A. R. C., and H. S. Reed, 1926: The absorption of ions by citrus and walnut seedlings, *Hilgardia*, **2**: 67–106.

Haberlandt, G., 1892: Anatomisch-physiologische Untersuchungen über das tropische Laubblatt, *Sitzungsber. Akad. Wiss. Wien.*, **101**: 785–816.

———, 1897: Ueber die Grösse der Transpiration im feuchten Tropenklima, *Jahrb. wiss. Bot.*, **31**: 273–288.

Hall, A. D., W. E. Brenchley, and L. E. Underwood, 1914: Soil solution and mineral constituents of the soil, *Phil. Trans. Roy. Soc.*, **204**: 179–200.

———, ———, and ———, 1914: The soil solution and the mineral constituents of the soil, *J. Agr. Sci.*, **6**: 278–301.

Hammett, F. S., 1928: A convenient device for plant solution culture work, *Sci.*, **67**: 106–107.

Harris, J. A., W. F. Hoffman, and J. V. Lawrence, 1925: Differential absorption of anions by varieties of cotton, *Proc. Soc. Exp. Biol. Med.*, **22**: 350–352.

Harvey, R. B., 1927: Ionization and adsorption isoelectric points, *Plant Physiol.*, **2**: 220.

Hasselbring, H., 1914: The relation between the transpiration stream and the absorption of salts, *Bot. Gaz.*, **57**: 72–73.

———, 1914: The effect of shading on the transpiration and assimilation of the tobacco plant in Cuba, *Bot. Gaz.*, **57**: 257–286.

Henricksen, H. C., 1933: Absorption of ash constituents by citrus trees, *Ann. Rept. Puerto Rico Sta.*, **1932**: 18–19.

Hibbard, R. P., 1917: Physiological balance in the soil solution, *Mich. Sta. Tech. Bull.* 40.

———, 1927: Negative results on physiological balance in soil cultures, *Plant Physiol.*, **2**: 1–38.

Hixon, R. M., 1920: The effect of the reaction of a nutrient solution on germination and the first stages of plant growth, *Medd. Vetenskapsak. Nobelinstitut*, **4**: 1–28.

Hoagland, D. R., 1917: The effect of hydrogen and hydroxyl ion concentration on the growth of barley seedlings, *Soil Sci.*, **3**: 547–560.

———, 1918: The relation of the plant to the reaction of the nutrient solution, *Sci.*, **48**: 422–425.

———, 1919: Relation of the concentration and reaction of the nutrient medium on the growth and absorption of the plant, *J. Agr. Res.*, **18**: 73–117.

———, 1919: Relation of nutrient solution to composition and reaction of cell sap to barley, *Bot. Gaz.*, **68**: 297–304.

———, 1920: Optimum nutrient solutions for plants, *Sci.*, **52**: 562–564.

———, 1923: The effect of the plant on the reaction of the culture solution, *Cal. Sta. Tech. Paper* 12.

———, 1923: The absorption of ions by plants, *Soil Sci.*, **16**: 225–246.

——— and A. R. Davis, 1923: The composition of the cell sap of the plant in relation to the absorption of ions, *J. Gen. Physiol.*, **5**: 629–647.

——— and ———, 1923: Further experiments on the absorption of ions by plants, including observations on the effect of light, *J. Gen. Physiol.*, **6**: 47–62.

——— and J. C. Martin, 1923: A comparison of sand and solution cultures with soils as media for plant growth, *Soil Sci.*, **16**: 367–388.

———, 1925: Physiological aspects of soil-solution investigations, *Hilgardia*, **1**: 227–257.

——— and A. R. Davis, 1925: Suggestions concerning the absorption of ions by plants, *New Phytol.*, **24**: 99–111.

———, 1926: Some phases of the inorganic nutrition of plants in relation to the soil solution, *Sci. Agr.*, **6**: 141–151, 177–189.

———, P. L. Hibbard, and A. R. Davis, 1926: The influence of light, temperature, and other conditions on the ability of *Nitella* cells to concentrate halogens in the cell sap, *J. Gen. Physiol.*, **10**: 121–146.

——— and J. C. Martin, 1927: Recent experiments concerning the adequacy of artificial culture solutions and of soil solutions for the growth of different types of plants, *Proc. First. Int. Cong. Soil Sci.*, Ithaca, 1926, **3**: 1–12.

———, A. R. Davis, and P. L. Hibbard, 1928: The influence of one ion on the accumulation of another by plant cells with special reference to experiments with *Nitella*, *Plant Physiol.*, **3**: 473–486.

——— and A. R. Davis, 1929: The intake and accumulation of electrolytes by plant cells, *Protoplasma*, **6**: 610–626; *Biol. Abst.*, **5**: 2655.

———, 1930: Observations on the nature of the process by which mineral elements are accumulated by plant cells, *Sci.*, **72**: 403–404.

———, 1931: Absorption of mineral elements by plants in relation to soil problems, *Plant Physiol.*, **6**: 373–388.

——— and T. C. Broyer, 1936: General nature of the process of salt accumulation by roots with description of experimental methods, *Plant Physiol.*, **11**: 471–508.

Höber, R., and J. Höber, 1928: Beobachtungen über die Zusammensetzung des Zellsaftes von *Valonia macrophysa*, *Pflügers Arch. Ges. Physiol.*, **219**: 260–272; *Biol. Abst.*, **5**: 2655.

Hollenberg, G. J., 1932: Some physical and chemical properties of the cell sap of *Halicystis ovalis*, *J. Gen. Physiol.*, **15**: 651–652.

Horner, C. K., D. Burk, and S. R. Hoover, 1934: Preparation of humate iron and other humate metals, *Plant Physiol.*, **9**: 663–670.

Ilienkoff, P. A., 1865: Einige Versuche zur Bestimmung des Einflusses, welchen die Bodenfeuchtigkeit auf die Vegetation ausübt, *Ann. Chem. Pharm.*, **136**: 160–165.

Iljin, W. S., 1924: The influence of salts on the alternation of concentration of cell sap in plants, *Studies in Plant Physiol. Lab. Charles Univ., Prague*, **2**: 5–25; *Bot. Abst.*, **15**: 799.

INGOLD, C. T., 1936: The effect of light on the absorption of salts by *Elodea canadensis*, *New Phytol.*, **35**: 132–141.

IRWIN, M., 1923: The behavior of chlorides in the cell sap of *Nitella*, *J. Gen. Physiol.*, **5**: 427–428.

———, 1925: On the accumulation of dye in *Nitella*, *J. Gen. Physiol.*, **8**: 147–182.

———, 1926: Mechanism of the accumulation of dye in *Nitella* on the basis of the entrance of the dye as undissociated molecules, *J. Gen. Physiol.*, **9**: 561–575.

JACOBSON, H. G. M., 1925: Changes in hydrogen-ion concentration in nutrient solutions. I, In a culture with wheat. II, In cultures with rice, *J. Am. Soc. Agron.*, **17**: 577–583; 583–586.

——— and T. R. SWANBACK, 1933: Relative influence of nitrate and ammoniacal nitrogen upon intake of calcium by tobacco plants, *Plant Physiol.*, **8**: 340–342.

JACQUES, A. G., and W. J. V. OSTERHOUT, 1930: The kinetics of penetration. II, The penetration of CO_2 into *Valonia*, *J. Gen. Physiol.*, **13**: 695–714.

——— and ———, 1932: The accumulation of electrolytes. IV, Internal versus external concentrations of potassium, *J. Gen. Physiol.*, **15**: 537–550.

——— and ———, 1934: The accumulation of electrolytes. VI, The effect of external pH, *J. Gen. Physiol.*, **17**: 727–750.

———, 1934, 1935: The accumulation of electrolytes. VII, Organic electrolytes, *J. Gen. Physiol.*, **18**: 235–242; 283–300.

———, 1936: The kinetics of penetration. XIII, Effect of pH on the entrance of potassium into *Nitella* at low concentrations, *Proc. Nat. Acad. Sci.*, **22**: 191–193.

JENNY, H., and E. W. COWAN, 1933: The utilization of adsorbed ions by plants, *Sci.*, **77**: 394–396.

JOHNSON, H. V., 1915: The absorption of ions by living and dead roots, *Am. J. Bot.*, **2**: 250–254.

JOHNSTON, E. S., 1925: Comparative study of the "six types" of nutrient solutions in relation to the growth of potato plants in sand cultures, *Soil Sci.*, **20**: 397–403.

———, 1927: An apparatus for controlling the flow of nutrient solutions in plant cultures, *Plant Physiol.*, **2**: 213–215.

———, 1932: Growing plants without soil, *Smithsonian Rept.*, **1931**: 381–387.

JONES, L. H., 1921: Some factors affecting the rate of change of hydrogen-ion concentration in nutrient solutions, *N. J. Sta. Rept.*, **1921**: 330–333.

——— and J. W. SHIVE, 1922: Influence of wheat seedlings upon the hydrogen ion concentration of nutrient solutions, *Bot. Gaz.*, **73**: 391–400.

JOST, L., 1907: "Lectures on Plant Physiology," Eng. ed., pp. 43, 79, Clarendon Press.

KACHIONI-VAL'TER, L. S., 1927: Influence of pH on development and growth of buckwheat, *Mém. Inst. Agron. à Lenigrade*, **4**: 151–174; *Biol. Abst.*, **4**: 351.

KEARNEY, T. H., and L. L. HARTER, 1907: The comparative tolerance of various plants for the salts common in alkali soils, *U. S. Bur. Plant Ind. Bull.* 113.

KIESSELBACH, T. A., 1916: Transpiration as a factor in crop production, *Nebr. Sta. Res. Bull.* **6**: 170–184.

KNOP, W., 1865: Quantitative Untersuchungen über den Ernährungsprocess der Pflanze, *Landw. Versuchs-Stat.*, **7**: 93–107.

KOHL, F. G., 1886: "Die Transpiration der Pflanzen," 124 pp., H. Bruhn.

LAGATU, H., and M. L. MAUME, 1927: Méthode de contrôle chimique du mode d'alimentation de la vigne, *C. R. Acad. Agr. France*, **13**: 439.

LAPIQUE, L., 1925: Sur l'absorption des sels par les cellules végétales épictèse et sélection, *Bull. Soc. Chim. Biol.*, **7**: 621–637; *Bot. Abst.*, **15**: 1161.

LARSON, C. A., 1931: The effect of continued application of irrigation water and commercial fertilizers on Wenatchee orchard soils, *unpub. thesis, Wash. State College.*

LIVINGSTON, B. E., 1900: On the nature of the stimulus which causes the change of form in polymorphic green algae, *Bot. Gaz.*, **30**: 289–317.

———, 1906: A simple method for experiments with water cultures, *Plant World*, **9**: 13–16.

——— and W. E. TOTTINGHAM, 1918: A new three-salt nutrient solution for plant cultures, *Am. J. Bot.*, **5**: 337–347.

———, 1919: Review of Gericke's work on the influence of temperature on the physiological balance of nutrient solutions, *Plant World*, **22**: 26–27.

———, K. F. KELLERMAN, and W. CROCKER, 1919: A plan for cooperative research on the salt requirements of representative agricultural plants, *Rept. Com. Div. Biol. Agr. Nat. Res. Council.*

LOEB, J., 1906: "Dynamics of Living Matter," Leincke & Buechner, Agents Columbia Press, New York.

LOEHWING, W. F., 1930: Sap hydrion concentration as a factor in plant metabolism, *Proc. Iowa Acad. Sci.*, **35**: 135–139.

———, 1932: Soil aeration in relation to plant metabolism, *Abst. Am. J. Bot.*, **19**: 844.

LOO, T. L., 1927: The influence of hydrogen ion concentration on the growth of the seedlings of some cultivated plants, *Bot. Mag. (Tokyo)*, **41**: 33–41; *Exp. Sta. Rec.*, **61**: 627.

———, 1928: The effect of renewal of nutrient solutions upon the growth of culture plants and its relation to aeration, *Jap. J. Bot.*, **4**: 71–98.

LOOMIS, W. E., and C. A. SHULL, 1937: "Methods in Plant Physiology," xviii + 472 pp., McGraw-Hill Book Company, Inc., New York.

LUNDERÅRDH, H., and H. BURSTRÖM, 1933: Atmung und Ionenaufnahme, *Planta*, **18**: 683–699.

MACHT, D. I., and M. E. DAVIS, 1935: Toxicity of alpha- and beta-nicotines and nornicotines for *Lupinus albus*, *Am. J. Bot.*, **22**: 329–332.

MARSH, R. P., 1935: A modified cultural jar, *Sci.*, **82**: 256.

MCCALL, A. G., 1915: A new method for the study of plant nutrients in sand cultures, *J. Am. Soc. Agron.*, **7**: 249–252.

———, 1916: A method for the renewal of plant nutrients in sand cultures, *Ohio J Sci.*, **16**: 101–103.

———, 1916: Physiological balance of nutrient solutions for plants in sand cultures, *Soil Sci.*, **2**: 207–255.

——— and J. R. HAAG, 1920: The hydrogen-ion concentration of certain three-salt nutrient solutions for plants, *Soil Sci.*, **10**: 481–485.

MCCOOL, M. M., and M. D. WELDON, 1930: The effect of sodium nitrate on the composition of the expressed sap of the small grains, *J. Am. Soc. Agron.*, **22**: 434–437.

MCGEORGE, W. T., 1932: Chemical equivalent base exchange reactions in plants, *Plant Physiol.*, **7**: 119–124.

MCHARGUE, J. S., 1923: Convenient supports for plants in pot or water culture experiments, *Soil Sci.*, **16**: 359–363.

MCLEAN, R. C., 1919: Studies in the ecology of tropical rain forests, with special reference to the forests of Brazil, *J. Ecol.*, **7**: 5–54; 121–172.

MEHRLICH, F. P., 1935: A mechanism for controlled continuous flow of nutrient solutions, *Plant Physiol.*, **10**: 169–177.

MEIER, H. F. A., and C. E. HALSTEAD, 1921: Hydrogen-ion concentration relations in a three-salt solution, *Soil Sci.*, **11**: 325–353.

MENDIOLA, N. B., 1922: Effect of different rates of transpiration on the dry weight and ash content of the tobacco plant, *Philippine J. Sci.*, **20**: 639–655.

MEUER, R., 1909: Über die regulatorische Aufnahme anorganischen Stoffe durch die Wurzeln von *Beta vulgaris* und *Daucus Carota*, *Jahrb. wiss Bot.*, **46**: 503–567.

Muenscher, W. C., 1922: The effect of transpiration on the absorption of salts by plants, *Am. J. Bot.*, 9: 311–330.

Nathansohn, A., 1903: Ueber Regulationserscheinungen im Stoffaustausch, *Jahrb. wiss. Bot.*, 38: 241–290.

———, 1904: Über die Regulation der Aufnahme anorganischen Salze durch die Knollen von *Dahlia*, *Jahrb. wiss. Bot.*, 39: 607–644.

———, 1904: Weitere Mitteilungen über die Regulation der Stoffaufnahme, *Jahrb. wiss. Bot.*, 40: 403–442.

Nightingale, G. T., and R. B. Farnham, 1936: Effects of nutrient concentration on anatomy, metabolism and bud abscission of sweet pea, *Bot. Gaz.*, 97: 477–517.

Osterhout, W. J. V., 1906: On the importance of physiologically balanced solutions for plants, *Bot. Gaz.*, 42: 127–134.

———, 1907: On the importance of physiologically balanced solutions for plants, *Bot. Gaz.*, 44: 259–272.

———, 1909: The nature of balanced solutions, *Bot. Gaz.*, 47: 148–149.

———, 1909: The penetration of inorganic salts in living protoplasm, *Z. physikal. Chem.*, 70: 408–413.

———, 1912: The permeability of protoplasm to ions and the theory of antagonism, *Sci.*, 35: 112–115.

———, 1914: Quantitative criteria of antagonism, *Bot. Gaz.*, 58: 178–186.

———, 1914: The measurement of antagonism, *Bot. Gaz.*, 58: 272–276.

———, 1915: On the nature of antagonism, *Sci.*, 41: 255–256.

———, 1915: The determination of additive effects, *Bot. Gaz.*, 60: 228–234.

———, 1916: Antagonism and Weber's law, *Sci.*, 44: 318–320.

———, 1916: The penetration of balanced solutions and the theory of antagonism, *Sci.*, 44: 395–396.

———, 1917: Antagonism and permeability, *Sci.*, 45: 97–103.

———, 1918: The basis of measurement of antagonism, *J. Biol. Chem.*, 34: 363–368.

———, 1919: Decrease of permeability and antagonistic effect caused by bile salts, *J. Gen. Physiol.*, 1: 405–409.

———, 1922: "Injury, Recovery and Death in Relation to Conductivity and Permeability," pp. 124–184, J. B. Lippincott Company, Philadelphia.

———, 1922: Some aspects of selective absorption, *J. Gen. Physiol.*, 5: 225–231.

———, 1925: Is living protoplasm permeable to ions? *J. Gen. Physiol.*, 8: 131–146.

——— and W. M. Stanley, 1932: The accumulation of electrolytes. V, Models showing accumulation and a steady state, *J. Gen. Physiol.*, 15: 667–689.

———, S. E. Kamerling, and W. M. Stanley, 1934: The kinetics of penetration. VI, Some factors affecting penetration, *J. Gen. Physiol.*, 17: 445–467.

———, ———, and ———, 1934: Kinetics of penetration. VII, Molecular *versus* ionic transport, *J. Gen. Physiol.*, 17: 469–480.

——— and S. E. Kamerling, 1934: The kinetics of penetration. VIII, Temporary accumulation, *J. Gen. Physiol.*, 17: 507–516.

Pantanelli, E., 1915: Assorbimento elettivo di ion nelle piante, *Bull. Orto Bot. R. Univ. Napoli*, 5: 1–54.

———, 1915: Über Ionenaufnahme, *Jahrb. wiss. Bot.*, 56: 689–733.

———, 1918: Decorso dell' assorbimento di ioni nelle piante, *Bull. Orto Bot. R. Univ. Napoli*, 6: 1–37.

———, 1929: Über Ionenaufnahme, *Protoplasma*, 7: 129–137.

Pearsall, W. H., and J. Ewing, 1924: The isoelectric point of some plant proteins, *Biochem. J.*, 18: 329–339.

——— and ———, 1924: The diffusion of ions from living plant tissues in relation to protein isoelectric points, *New Phytol.*, 23: 193–207.

PEMBER, F. R., 1917: Studies by means of both pot and solution culture of the phosphorus and potassium requirements of the barley plant during its different periods of growths, *R. I. Sta. Bull.* 169.

PETRIE, A. H. K., 1933: The intake of ions by the plant and its relation to the respiration of the root, *Australian J. Exp. Biol. and Med. Sci.*, **11**: 25–34; *Biol. Abst.*, **8**: 45.

PETTINGER, N. A., 1932: Effect of fertilizers on the chlorine content of the sap of corn plants, *J. Agr. Res.*, **44**: 919–931.

PFEFFER, W., 1900: "Physiology of Plants," Vol. I, 420 pp., Eng. ed., Ewart, Clarendon Press.

PIERRE, W. H., and G. G. POHLMAN, 1933: Preliminary studies on the exuded plant sap and the relation between the composition of the sap and the soil solution, *J. Am. Soc. Agron.*, **25**: 144–159.

PIRSCHLE, K., and H. MENGDEHL, 1931: Ionenaufnahme aus Salzlösung durch die höhere Pflanze, *Jahrb. wiss. Bot.*, **74**: 297–363.

POTAPOV, N. G., and N. T. STANKOV, 1934: Periodicity of mineral nutrition within the twenty-four hours, *C. R. Acad. Sci. U. S. S. R.*, **2**: 43–45; *Chem. Abst.*, **28**: 5099.

PREVOT, P., and F. C. STEWARD, 1936: Salient features of the root system relative to the problem of salt absorption, *Plant Physiol.*, **11**: 509–534.

RABER, O. L., 1920: The antagonistic action of anions, *J. Gen. Physiol.*, **2**: 541–544.

REDFERN, G. M., 1922: On the absorption of ions by the roots of living plants, *Ann. Bot.*, **36**: 167–174.

REED, H. S., and A. R. C. HAAS, 1924: Some effects of certain calcium salts upon the growth and absorption of citrus seedlings, *Am. J. Bot.*, **11**: 15–18.

——— and ———, 1924: Some relations between the growth and composition of young orange trees and the concentration of the nutrient solution employed, *J. Agr. Res.*, **28**: 277–285.

RIPPEL, A., 1926: Quantitative Untersuchungen über Kationenaustausch in der Pflanze, *Jahrb. wiss. Bot.*, **65**: 819–885.

ROBBINS, W. J., 1923: An isoelectric point for plant tissue and its significance, *Am. J. Bot.*, **10**: 412–440.

ROSENFELS, R. S., 1935: The absorption and accumulation of potassium bromide by *Elodea* as related to respiration, *Protoplasma*, **23**: 503–519.

RUDOLFS, W., 1925: Effect of seeds upon hydrogen ion concentration equilibrium in solution, *J. Agr. Res.*, **30**: 1021–1026.

———, 1925: Selective absorption of ions by seeds, *Soil Sci.*, **20**: 249–252.

RUHLAND, W., 1909: Zur Frage der Ionenpermeabilität, *Z. wiss. Bot.*, **1**: 747–763.

SACHS, J., 1860: Vegetationsversuche mit Ausschluss des Bodens über die Nährstoffe und sonstigen Ernährungsbedingungen von Mais, Bohnen, und anderen Pflanzen, *Landw. Versuchs-Stat.*, **2**: 219–268.

———, 1887: "Lectures on the Physiology of Plants," Eng. ed., Clarendon Press.

SALTER, R. M., and T. C. MCILVAINE, 1917: The effect of reaction of solution on germination of seeds and on growth of seedlings, *J. Agr. Res.*, **19**: 73–97.

SCHIMPER, A. F. W., 1890: Zur Frage der Assimilation der Mineralsalze durch die grüne Pflanze, *Flora*, **73**: 207–261.

SCHLOESING, T., 1869: Tabac sous cloche et à l'air libre, *Ann. Sci. Nat. Bot.*, V, **10**: 366–369.

SCHREINEMOCHER, F. A. H., 1893: Konzentrierung oder Verdünnung einer Lösung bei konstanter Temperatur, *Z. physikal. Chem.*, **11**: 81–85.

SCHREINER, O., and J. J. SKINNER, 1910: Ratio of phosphate, nitrate and potassium on absorption and growth, *Bot. Gaz.*, **50**: 1–30.

——— and ———, 1910: Some effects of a harmful organic soil constituent, *U. S. Bur. Soils Bull.* 70.

——— and ———, 1918: The triangle system for fertilizer experiments, *J. Am. Soc. Agron.*, **10**: 225–246.

Scofield, C. S., 1927: The effect of absorption by plants on the concentration of the soil solution, *J. Agr. Res.*, **35**: 745–757.

Sewell, M. C., 1924: Relation of the molecular proportions in the nutrient solution to the growth of wheat, *J. Agr. Res.*, **28**: 387–393.

Sherman, C. C., 1933: Theories on the nutrition of plants, from Aristotle to Liebig, *Yale J. Biol. and Med.*, **6**: 43–60.

Shive, J. W., 1915: A study of physiological balance in nutrient media, *Physiol. Res.*, **1**: 327–399.

———, 1915: A three-salt nutrient solution for plants, *Am. J. Bot.*, **2**: 157–160.

———, 1917: A study of physiological balance for buckwheat grown in 3-salt solutions, *N. J. Sta. Bull.* 319.

——— and W. H. Martin, 1918: A comparison of salt requirements for young and for mature buckwheat plants in water cultures and sand cultures, *Am. J. Bot.*, **5**: 186–191.

———, 1919: The influence of the moisture content of sand cultures upon the physiological salt balance for plants, *N. J. Sta. Rept.*, **1919**: 358–363.

———, 1919: The influence of sand of different degrees of fineness upon the concentration and reaction of a nutrient solution, *N. J. Sta. Rept.*, **1919**: 363–366.

———, 1920: Relation of moisture in solid substrata to physiological salt balance for plants and to the relative plant-producing value of various salt proportions, *J. Agr. Res.*, **18**: 357–378.

———, 1920: The influence of sand upon the concentration and reaction of a nutrient solution for plants, *Soil Sci.*, **9**: 169–179.

———, 1920: The influence of varying degrees of soil moisture upon the physiological salt balance for plants, *N. J. Sta. Rept.*, **1920**: 387–395.

———, 1922: The relation of soil moisture to physiological salt balance for plants, *Soil Sci.*, **14**: 391–411.

——— and A. L. Stahl, 1927: Constant rates of continuous solution renewal for plants in water cultures, *Bot. Gaz.*, **84**: 317–323.

Sideris, C. P., B. H. Krauss, and E. Masunaga, 1928: The effect of salt concentration of the culture solution on the growth and composition of pineapple plants, *Am. J. Bot.*, **15**: 353–372.

Skene, M., 1915: The acidity of sphagnum and its relation to chalk and mineral salts, *Ann. Bot.*, **29**: 65–87.

Sorauer, P., 1880: Studien über Verdünstung, *Forschungen auf dem Gebiete der Agr. Physik.*, **3**: 351–490.

Spencer, E. L., and J. W. Shive, 1933: The growth of *Rhododendron ponticum* in sand cultures, *Bull. Torrey Bot. Club*, **60**: 423–440.

Stahl, E., 1894: Einige Versuche über Transpiration und Assimilation, *Bot. Ztg.*, **52**: 117–146.

Steward, F. C., 1932: The absorption and accumulation of solutes by living plant cells. I, Experimental conditions which determine salt absorption by storage tissue, *Protoplasma*, **15**: 29–58.

———, 1932: The absorption and accumulation of solutes by living plant cells. II, A technique for the study of respiration and salt absorption in storage tissue under controlled environmental conditions, *Protoplasma*, **15**: 497–516.

———, R. Wright, and W. E. Berry, 1932: The absorption and accumulation of solutes by living plant cells. III, The respiration of cut discs of potato tuber in air and immersed in water, with observations upon surface: volume effects and salt accumulation, *Protoplasma*, **16**: 576–611.

———, 1932: The absorption and accumulation of solutes by living cells. IV, Surface effects with storage tissue. A quantitative interpretation with respect to respiration and salt absorption, *Protoplasma*, **17**: 436–453.

———, 1933: Salt absorption and respiration of *Valonia*, *Carnegie Inst. Yearbook*, **32**: 281–283.

———, 1933: The absorption and accumulation of solutes by living plant cells. V, Observations upon the effects of time, oxygen and salt concentration upon absorption and respiration by storage tissue, *Protoplasma*, **18**: 208–242.

——— and W. E. Berry, 1934: The absorption and accumulation of solutes by living plant cells. VII, The time factor in the respiration and salt absorption of Jerusalem artichoke tissue (*Helianthus tuberosus*), with observations on ionic interchange, *J. Exp. Biol.*, **11**: 103–119.

——— and ———, 1936: The absorption and accumulation of solutes by living plant cells. VIII, The effect of oxygen upon respiration and salt accumulation, *Ann. Bot.*, **50**: 345–366.

Stiles, W., 1915: On the relation between the concentration of the nutrient solution and the rate of growth of plants in water culture, *Ann. Bot.*, **29**: 89–97.

———, 1916: On the interpretation of the results of water culture experiments, *Ann. Bot.*, **30**: 427–436.

——— and I. Jorgensen, 1917: Observations on the influence of aeration of the nutrient solution in water culture experiments with some remarks on the water culture methods, *New Phytol.*, **16**: 182–197.

——— and F. Kidd, 1919: The influence of external concentration on the position of the equilibrium attained in the intake of salts by plant cells, *Proc. Roy. Soc.*, B, **90**: 448–470.

——— and ———, 1919: The comparative rate of absorption of various salts by plant tissue, *Proc. Roy. Soc.*, B, **90**: 487–504.

———, 1924: The absorption of salts by storage tissue, *Ann. Bot.*, **38**: 617–635.

Stolhane, A. A., 1914: A comparative study of nutritive solutions in sand and water cultures, *Rec. Trav. Lab. Agron. Moskov. Selsk. Khoz. Inst.*, **10**: 289–336; *Exp. Sta. Rec.* **36**: 630.

Tarr, L. W., and S. C. Noble, 1922: The effect of hydrogen ion concentration upon the growth of seedlings, *Del. Sta. Bull.* 131.

Thomas, W., 1930: The effect of colloidal silica on the absorption of phosphoric acid by plants, *Sci.*, **71**: 422–423.

———, 1933: Absorption, utilization, and recovery of nitrogen, phosphorus and potassium by apple trees grown in cylinders and subjected to differential treatment with nutrient salts, *J. Agr. Res.*, **47**: 565–582.

Tollens, B., 1882: Ueber einige Erleichterungen bei der Cultur von Pflanzen in wasserigen Lösungen, *J. Landw.*, **30**: 537–540.

Toole, E. H., and W. E. Tottingham, 1918: The influence of certain added solids upon the composition and efficiency of Knop's nutrient solution, *Am. J. Bot.*, **5**: 452–461.

Tottingham, W. E., 1914: A quantitative chemical and physiological study of nutrient solutions for plant cultures, *Physiol. Res.*, **1**: 133–245.

Trelease, S. F., 1917: The relation of the concentration of the nutrient solution to the growth of young wheat plants in water cultures, *Johns Hopkins Univ. Circ.*, **3**: 225–227.

——— and E. E. Free, 1917: The effect of renewal of culture solutions on the growth of young wheat plants in water culture, *Johns Hopkins Univ. Circ.*, **3**: 227–228.

——— and B. E. Livingston, 1922: Continuous renewal of nutrient solution for plants in water culture, *Sci.*, **55**: 483–486.

——— and ———, 1924: The relation of climatic conditions to the salt-proportion requirements of plants in solution cultures, *Sci.*, **59**: 168–172.

———, 1925: Chamber with thermostatic control and rotating table for plant cultures, *Bull. Torrey Bot. Club*, **52**: 389–395.

——— and H. M. Trelease, 1925: Growth of wheat roots in salt solutions containing essential ions, *Bot. Gaz.*, **80**: 74–83.

——— and ———, 1926: Relation of temperature to the physiological values of salt solutions as indicated by growth of wheat roots, *Bull. Torrey Bot. Club*, **53**: 605–609.

——— and ———, 1926: Toxicity and antagonism in salt solutions as indicated by growth of wheat roots, *Bull. Torrey Bot. Club*, **53**: 137–156.

——— and ———, 1928: Susceptibility of wheat to mildew as influenced by salt nutrition, *Bull. Torrey Bot. Club*, **55**: 41–68.

——— and ———, 1933: Physiologically balanced culture solutions with stable hydrogen ion concentration, *Sci.*, **78**: 438–439.

——— and J. R. Thomson, 1935: Regulating the flow of solution for plant cultures, *Sci.*, **81**: 204.

——— and H. M. Trelease, 1935: Changes in hydrogen-ion concentration of culture solutions containing nitrate and ammonium nitrogen, *Am. J. Bot.*, **22**: 520–542.

True, R. H., and W. J. Gies, 1903: On the physiological action of some of the heavy metals in mixed solutions, *Bull. Torrey Bot. Club*, **30**: 390–402.

———, 1915: Antagonism and balanced solutions, *Sci.*, **41**: 653–656.

——— and H. H. Bartlett, 1915: The exchange of ions between the roots of *Lupinus albus* and cultural solutions containing one nutrient salt, *Am. J. Bot.*, **2**: 255–278.

——— and ———, 1915: The exchange of ions between the roots of *Lupinus albus* and cultural solutions containing two nutrient salts, *Am. J. Bot.*, **2**: 311–323.

——— and ———, 1916: The exchange of ions between the roots of *Lupinus albus* and culture solutions containing three nutrient salts, *Am. J. Bot.*, **3**: 47–57.

Truog, E., 1918: Soil acidity. I, Its relation to growth of plants, *Soil Sci.*, **5**: 169–195.

Wean, R. E., 1935: Automatic flowmeter for drip solutions in plant nutritional studies, *Sci.*, **82**: 336.

Weatherwax, P., 1916: An apparatus for aerating culture solutions, *Proc. Ind. Acad. Sci.*, **1914**: 157–160.

Wendt, G. V., 1930: Stimulationsversuche mit Jod, *Gartenflora*, **79**: 213–218; *Biol. Abst.*, **7**: 791.

Wodehouse, R. P., 1917: Direct determinations of permeability, *J. Biol. Chem.*, **29**: 453–458.

Wollney, W., 1898: Untersuchungen über den Einfluss der Luftfeuchtigkeit auf das Wachstum, *Bot. Cent.*, **76**: 249.

Woodward, J., 1699: Thoughts and experiments on vegetation, *Phil. Trans. Roy. Soc.*, **21**: 382–398.

Wynd, F. L., 1933: The sensitivity of orchid seedlings to nutritional ions, *Ann. Mo. Bot. Gard.*, **20**: 223–237.

Zinzadzé, C., 1932, 1933: Nutrition artificielle des plantes cultivées. I, Mélanges nutritifs à pH stable, *Ann. Agron.*, Nov.-Dec. **1932**: 809–853; Jan.-Feb. **1933**: 53–72.

———, 1934: Nutrient media with stable hydrogen ion concentration, *Sci.*, **79**: 480–481.

Zurbicki, Z., 1933: New method of arranging experiments with vegetables in sand cultures with flowing solution, *Plant Physiol.*, **8**: 553–558.

CHAPTER VI

THE ELEMENTS ABSORBED BY THE PLANT

I. NUMBER AND KIND

According to Palladin (1927), 35 elements have been detected in the analysis of plants. These are carbon, hydrogen, oxygen, nitrogen, sulphur, phosphorus, potassium, magnesium, calcium, iron, chlorine, bromine, iodine, fluorine, boron, silicon, sodium, lithium, rubidium, strontium, barium, zinc, mercury, aluminum, thallium, titanium, tin, lead, arsenic, selenium, manganese, cobalt, nickel, copper, and silver. Robinson, Steinkoenig, and Miller (1917) analyzed a large number of plants from different regions for the rarer elements. In addition to those just mentioned, they detected chromium, vanadium, and cesium. Vanadium was found, however, in only six cases and then only in traces. Cesium was detected in only three cases and in two of them cesium beryls were known to be present in that region. A general idea of the number and relative amounts of the elements that compose plants may be had by considering a given plant, although, as will be mentioned later, the elemental composition of a plant will depend upon the species and upon the conditions under which it is grown.

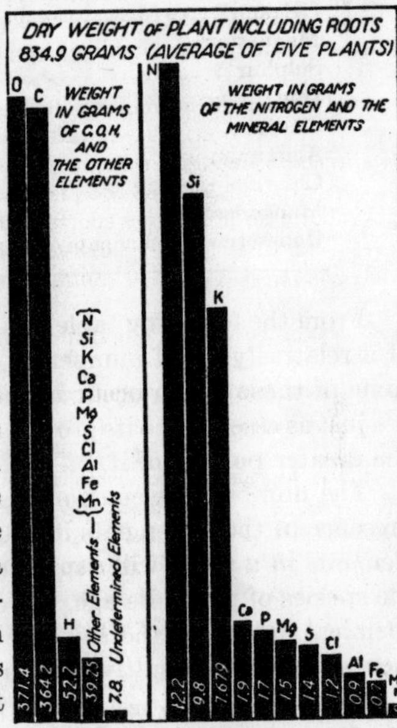

FIG. 13.—Diagram showing the elemental composition of a corn plant at the late dough stage.

An elemental analysis of the corn plant (Latshaw and Miller, 1924) may be taken as this example. These plants were analyzed at a stage when the grain was fully dented and in the late dough stage. The leaves were all intact, and the plants were evidently at a stage where absorption of solutes from the soil had practically ceased. The proportion of these different elements is shown in Fig. 13 and in the table on page 284.

Weight of the Elements That Composed the Stem, Leaves, Cob, Grain, and Roots of Pride of Saline Corn Plants Grown at Manhattan, Kans., in 1920
(Average of Five Plants)

Average dry weight of stems, leaves, cob, grain, and roots, 835.9 g.	Weight, grams	Percentage of the total dry weight
Oxygen	371.4	44.431
Carbon	364.2	43.569
Hydrogen	52.2	6.244
Nitrogen	12.2	1.459
Phosphorus	1.7	0.203
Potassium	7.7	0.921
Calcium	1.9	0.227
Magnesium	1.5	0.179
Sulphur	1.4	0.167
Iron	0.7	0.083
Silicon	9.8	1.172
Aluminum	0.9	0.107
Chlorine	1.2	0.143
Manganese	0.3	0.035
Undetermined elements	7.8	0.933

From the foregoing table it is seen that the bulk of a plant is composed of a relatively small number of elements, but as will be mentioned later some of those which occur in the plant in only relatively small amounts are just as essential to its growth and development as those which compose the greater portion of it.

The number of elements in a plant will depend principally upon the presence of these elements in the soil. The amount or proportion of the elements in a plant will depend upon many factors. Some of these are: the species of plant, its age, the distribution of its roots, the physical and chemical nature of the soil, the proportion and distribution of the elements, methods of cultivation, and the general climatic conditions.

A. Influence of the Plant

It is common-knowledge that plants of different species grown under identical conditions differ in their elemental composition, especially in regard to the proportionate amount of the various elements that compose them.

Observations have been made by Parker and Truog (1920), Richardson (1920), Headden (1921), Skinner and Reid (1921), Greaves and Nelson (1925), Wherry (1926), von Wrangell (1927), Newton (1928), Gregory and Crowther (1928, 1931), Haley, Longenecker, and Olson (1931), Daniel

(1934), Harper, Daniel, and Murphy (1934), Davidson and Chambliss (1932), Maume and Dulac (1934), Archibald and Bennett (1935), Dadswell (1935), Hurd-Karrer (1935), Trelease and Martin (1936), and others in regard to the different proportions of the elements contained in plants of different species grown under the same conditions. Richardson (1920) determined the composition of the ash of certain dune plants and the sand in which they grew. Approximately 92 per cent of this sand was silica, while the remainder consisted of iron and aluminum oxides 4.3 per cent, calcium oxide 1.4 per cent, magnesium oxide 0.7 per cent, potassium oxide 1.0 per cent, and less than 1 per cent of organic matter. The composition of the ash of the sand cherry (*Prunus pumila*), artemisia (*Artemisia caudata*), *Andropogon* (*sp.*), and the scouring rush (*Equisetum hyemale*) growing in this sand will suffice to show the specific differences in the amount of certain elements absorbed by these plants.

ANALYSIS OF THE ASH OF PLANTS GROWING IN SAND DUNES

Constituent	Percentage of the ash			
	Prunus pumila	*Artemisia caudata*	*Andropogon*	*Equisetum hyemale*
Silica	1.5	12.1	65.4	49.4
Iron oxide	0.7	1.7	2.5	1.0
Aluminum oxide	0.02	0.4	2.6	2.3
Calcium oxide	44.1	35.5	10.2	13.4
Potassium oxide	10.9	11.6	6.7	6.0
Magnesium oxide	4.2	6.4	3.2	3.7

The most striking differences to be observed in the proportion of elements absorbed by these plants is in the amounts of silicon, calcium, and aluminum. *Andropogon* and *Equisetum* absorbed and accumulated large amounts of silicon, while *Prunus* and *Artemisia* accumulated only a relatively small amount. *Prunus* and *Artemisia*, on the other hand, accumulated from three to four times as much calcium as did *Andropogon* and *Equisetum*, while the amount of aluminum is from six to ten times greater in some plants than in others.

Headden (1921) noted that alfalfa growing in soils which contained both lithium and strontium had little lithium in the ash but strontium was present in considerable quantity. In tobacco, on the other hand, growing in the same soil, lithium was present in considerable quantity, but only a trace of strontium could be detected. Newton (1928) grew sunflowers, beans, wheat, barley, peas, and corn in the same nutrient solution in the same container. He harvested the wheat and barley at the heading stage, the corn at the time of tasseling, and the peas and beans at the blooming stage and determined the percentage of the various elements present in the plant. The similarities and differences in absorption which are characteristic of these plants are shown in the following data:

Plant	Percentage of the dry weight of				
	Ca	K	Mg	N	P
Sunflower	2.2	5.0	0.64	3.6	0.56
Beans	2.1	4.0	0.59	3.6	0.55
Wheat	0.8	6.7	0.41	4.5	0.49
Barley	1.9	6.9	0.54	4.7	0.52
Peas	1.6	5.3	0.50	4.5	0.19
Corn	0.5	3.9	0.40	2.9	0.39

The sunflower plant contained 2.7 times as much calcium as the wheat plant and 4.4 times as much as the corn. The sunflower also contained approximately 1.5 times as much magnesium as wheat and corn. Wheat and barley contained a percentage of potassium considerably higher than that for any of the other plants. Newton found that absorption from the cultural solution of the various elements was higher than by plants grown in a loam soil and that there was much less variability in absorption from the solution than from the soil. An observation of von Wrangell (1927) is worthy of note in considering the different rates of absorption by plants. He noted the differences in the amount of iodine contained in different types of plants growing in the same soil and found that lettuce and spinach contained 100 times more iodine than other agricultural plants under the same conditions.

Hurd-Karrer (1935) grew different species of plants in greenhouse plots in which the soil had been treated with 5 p.p.m. of selenium as sodium selenate. The amounts of selenium absorbed by the various plants are shown in the following table:

Plant	P.p.m. of dry weight	Plant	P.p.m. of dry weight	Plant	P.p.m. of dry weight
Mustard	1,240	Alfalfa	560	Proso millet	285
Broccoli	1,180	Peas	560	Corn	275
Sunflower	790	Oats	535	Crested wheat grass	255
Flax	685	Wheat	470		
Sweet clover	645	Barley	450	Brome grass	200
German millet	590	Spinach	315	Sorgo	130

Trelease and Martin (1936) reported that western wheat grass growing on various soils accumulated 1 to 60 p.p.m. of selenium, while *Astragalus bisulcatus*, growing on the same soil, accumulated 200 to 4,300 p.p.m. On soil containing 2.1 p.p.m. of selenium, *A. bisulcatus* accumulated 1,250 p.p.m., while *A. missouriensis* stored only 3.1 p.p.m. Daniel (1934) found that ordinarily legumes contained 3.9 times as much calcium, 1.7 times as much phosphorus, and 2.6 times as much nitrogen as the grasses. Plants that are low in calcium and phosphorus remained low in these elements when grown on a fertile soil. Those which are relatively high in these elements always contain relatively large amounts even though growing on a poor soil. The legumes have low nitrogen/phosphorus and high calcium/phosphorus and nitrogen/calcium, while the grasses have high nitrogen/phosphorus and low calcium/phosphorus and nitrogen/calcium.

Maume and Dulac (1934) showed that the percentages of nitrogen, phosphorus, and potassium, varied among the varieties of wheat at a like stage of development, when grown under the same conditions.

Since different plants are composed of different proportions of the various elements, the statement is frequently made that plants exercise "selective absorption" or "feeding power." Since these terms have been so frequently used and since they have been so differently interpreted, they should be mentioned here. It has been assumed by some that the differences in the composition of plants are due primarily to differences in the specific absorbing powers of the roots. They consider thus that the absorbing cells of the roots differ in their ability to exclude certain elements more than others. On the other hand, some consider that the difference in the amount of certain elements in plants is due to the specificity of the protoplasm, which requires different proportions of materials for its development in different plants. The protoplasm of the developing cells is thus considered to regulate the amount of any given element in the plant rather than the regulatory action of the absorbing cells of the root. The speed of absorption of a given compound in this case would thus be determined by the rate at which it was utilized by the cells in the interior of the plant. As has been mentioned in Chap. V, the absorption of solutes is apparently not so simple a process and the phenomenon of the heaping up of salts in the ionic form on the interior of the absorbing cells cannot be explained by either of the above-mentioned theories. Davis, Hoagland, and Lipman (1923) also point out that the differences in the extent of the root systems and the differences in the rate of production of carbon dioxide by these roots are factors that may determine to a considerable degree the differences in the amount of minerals absorbed by plants.

The relationship between the amount of nutrients absorbed by a plant and its development is a complex one and is very difficult of interpretation. In this regard the statement of Thomas (1934) is appropriate—"A comparison of the quantitative relations between the elements during the growth of a particular species, subjected to different growth conditions, cannot be made without a key of interpretation. Consequently comparisons between the composition of different species with respect to any of the dominant or accessory mineral elements cannot be made except by such a key."

B. Influence of Environment

The amount of any element that is found in the plant is influenced by the abundance and location of that element in the soil. Thus Scharrer and Strobel (1927) and Orr, Kelly, and Stuart (1928) found that they could greatly increase the iodine content of sugar beets, peas, oats, and alfalfa by the addition of this element to the soil in which these plants were growing. Parker and Pierre (1928) grew corn in various concentrations of phosphate (PO_4) from 0.05 to 50 p.p.m. During the younger stages of growth, increased concentration increased absorption; but in later periods, the absorption was not proportional to the concentration of the external solution in many cases. In case of potassium, the maximum growth of corn was secured with 2 p.p.m. The concentration necessary for growth, however, varies with the species. Thus soybeans and tomatoes give maximum growth at 3 to 5 p.p.m. of potassium, while maximum growth is secured for oats at 2 p.p.m. and alfalfa at only 1 p.p.m.

Godden (1926) reported that pastures in England responded to the application of artificial fertilizers by a considerable modification of the mineral contents of the plant. The constituents that showed the greatest variations were calcium and potassium, while phosphorus varied to a slightly lesser extent. Brown (1929) studied the mineral composition of apples that had been grown on 10 different soils. The amount of potassium in the apples showed a definite relationship with the amount in the soil, and there were indications of a similar relationship between the percentage of phosphorus, magnesium, and calcium. It was found by Vandecaveye and Baker (1932) in Washington that the application of lime, phosphate, or potassium, either alone or in combination, resulted in a higher content of calcium, phosphorus, and potassium in pasture grasses and hay. Thomas (1932) considered that the plant from the early stages of growth to blooming absorbs nutrients in approximately the same proportion as they are present in the medium in which it grows. Under some conditions, however, the amount of an element in a plant is not greatly influenced by the relative amount in the soil. Thus Brown (1932) found that the influence of fertilizers on the total ash content was not very evident in white clover, bluegrass, wild oats, and other plants growing in permanent pastures of Connecticut. Daniel and Harper (1934) studied the total calcium and phosphorus in the plant and soil and concluded that the amount of any single element in the soil will not give an accurate indication of the amount of that element which will be found in the plant.

The location of the various nutrients with respect to the roots may have a marked effect upon the amount of elements contained in the plant, as indicated by the work of Gile and Carrero (1917, 1921). They grew corn plants in such a way that the roots were divided among three solutions each of which lacked two elements so that the nitrogen was in one flask, the phosphorus in another, and the potassium in a third. It was found that the amount of these three elements in the plants thus grown was lower than under normal conditions, although the rate of absorption of nitrogen, for example, was greatly increased by the roots in that solution. The cause of the smaller amount is seemingly due to the slowness of distribution of these elements within the plant, since they have to be transferred over the plant by and from a comparatively small number of fibrovascular bundles.

The amount of a given element in a plant may be influenced by the proportion and kind of other elements contained in the soil or culture solution. Thus Ginsburg (1925) found that soybeans growing in a complete nutrient solution had an ash content of 13.3 per cent. When phosphorus was withheld from the solution, the ash content was 18.6 per cent and amounted to 18.4 per cent when magnesium was lacking. The plants growing in the incomplete cultural solutions, with the exception of those grown in solutions lacking sulphur, always absorbed higher percentages of calcium and lower percentages of both nitrogen and magnesium than did the checks. In the case of spinach, which was grown on soil that was heavily fertilized with different fertilizers or fertilizer mixtures, it was observed that the calcium, magnesium, and sulphur content of the plant soon reached an equilibrium regardless of the quantity or kind of nutrients applied but that the silicon, potassium, and sodium content showed wide fluctuation with the various applications of nutrients (True, 1919). Leaves of orange trees growing in cultures to which no calcium salts were added were found to be rich in potassium, while those of plants receiving no potassium salts in the nutrients were high in calcium (Reed and Haas, 1923).

Carolus (1933), in experiments with tomatoes, found that the use of 60 lb. per acre of ammonia produced a marked absorption of nitrogen, stimulated the absorption of magnesium, and slightly depressed the intake of potassium. The application of 120 lb. per acre of phosphorus pentoxide resulted in a large increase in the absorption of phosphorus and calcium, and in an increased intake of magnesium. A slight deficiency or excess of potassium in the plant caused a condition of physiological unbalance that resulted in both decreased growth and reduced yields, and impaired the quality of the fruit.

The amount of the various elements absorbed by plants is evidently influenced to a marked degree by the condition of the soil. Thus Brown (1929) considered in the case of the apple that the mechanical structure of the soil had a marked influence on the composition of the fruit. Greaves and Hirst (1929) reported that the ash content of growing wheat, oats, barley, and corn in Utah was increased 46, 31, 36, and 8 per cent, respectively, by the application of optimum quantities of irrigation water. Daniel and Harper (1934) also considered that the amount of moisture in the soil is an important factor in determining the amount of the absorption of various elements.

It was observed by Parker, Hester, and Carolus (1933) that the maximum amount and percentage of calcium in cabbage, peas, and beans occurred when the pH reaction of the soil was between pH 5.5 and 6.2. Magnesium was absorbed to a greater extent at the lower pH and was replaced by calcium as the neutral point was approached. Hester (1934) grew lima beans in soils with different pH values and in different calcium-ion concentrations, and found that at the pH values favorable to maximum crop production the soil carried a desirable nutrient balance for this crop.

Wynd and Fuller (1931) observed that tomato and cucumber plants which had been treated with ultraviolet radiation showed a decided increase in calcium and a decrease in phosphorus as compared to the controls. Steward and Arthur (1933, 1934) found that various species of plants, when grown under conditions of low light intensity, exhibited an increase in ash content upon irradiation. This increase in ash appeared 42 to 72 hr. after exposure to the rays. The effective wave lengths were within the range of 2900 to 3130 Å. The application of irradiated ergosterol in olive oil to the basal leaves brought about an increase in the ash of the plants.

Street (1934) found in the field pea that the crude ash was higher in plants grown in a 10-hr. than in a 13-hr. light period. Under the former condition, the plants were high in potassium and low in calcium and magnesium, while under the latter condition they were lower in potassium and higher in calcium and magnesium. Plants exposed to light for 17 hr. daily were markedly low in potassium and usually highest in calcium and magnesium.

If there is an abundance of the soluble inorganic salts present in the soil, most plants will absorb a greater quantity of them than is necessary for their maximum growth. This has been especially noted for the salts of phosphorus, potassium, and sulphur. Jordan (1913) in nutrition experiments with barley, peas, tomatoes, tobacco, buckwheat, and rape noted that both phosphorus and potassium compounds increased in amount in these plants after the maximum increase in growth had been reached. Similar observations have been made by Hoagland (1919 and 1926), who found that the concentration of the cell sap in regard to electrolytes is much higher than its concentration in the soil. A further indication that many of the elements are present in the plant cells in excess of their needs is the fact that a large proportion of them in many cases can readily be extracted with water. Thus Ames and Boltz (1912) found in the alfalfa plant that 90 per cent of the potassium, 85 per cent of the magnesium, 75 per cent of the nitrogen, and 50 per cent of the sulphur were soluble in water. A considerable portion of the nitrogen is known to be in simple soluble protein compounds, but apparently the other elements mentioned are only loosely combined and probably not at all.

Canals, Canaye, and Cabanes (1930) showed in the dialysis of the expressed saps of radish, carrot, turnip, beet, and other plants that most of the minerals occur in the dialyzable form. The loss of sodium and potassium was complete, while the loss of phosphorus and magnesium ranged from 95 to 100 per cent during dialysis. The loss of calcium was highly variable ranging from 30 to 94 per cent. The calcium of the leaves was predominantly in the dialyzable form, while in the roots a considerable portion was not in that condition. Lausberg (1935) found that leaves of *Ricinus* and *Vicia* lost large amounts of ash constituents when sprinkled or when placed in water for varying periods of time. Sprinkled leaves lost 9 to 32 per cent of their ash constituents, while the loss of potassium was 23 to 60 per cent, and that of calcium 6 to 10 per cent. When both sides of the leaves were sprinkled equally, twice the amount of minerals was lost from the upper as from the lower side. Halma and Haas (1931) noted that some of the inorganic constituents of citrus cuttings underwent marked changes in solubility during the rooting process.

It is a general observation that a plant will respond favorably to an increasing amount of a nutrient in the soil provided this nutrient was not originally present in an amount sufficient for maximum growth. Above a certain limit, however, the plant will cease to respond to increased application of the nutrient. It will generally continue, however, in such cases to absorb increasing amounts from the soil as indicated by an analysis of the tissues. The occurrence in a plant of a greater quantity of an element than is necessary for its needs is termed "luxury absorption" or "luxury consumption."

According to Thomas (1932) the optimum ratio of the constituents of a fertilizer may be defined as that ratio which on the addition of any one of its constituents—nitrogen, phosphorus, or potassium—results in no increased utilization by the plant of any of the other elements as determined by the time-absorption method.

The amount of a given element in a plant may be sufficient for its growth and development, but it may not be sufficient for the needs of the animal that consumes the plant. This condition prevails in regions where there is a scarcity of certain elements in the soil, or because they are in a form that is unavailable to the plant. The elements iodine, phosphorus, and copper are the ones that are the most generally lacking in the proper amount in the soil and consequently in the plant. The lack of iodine in the drink or diet of animals produces a type of goiter. The lack of sufficient phosphorus or copper in forage and grain causes the so-called "salt sickness" of animals. Certain areas of Florida, Minnesota, and Montana are especially lacking in available copper or phosphorus, and the herbivorous animals in those regions are not sufficiently supplied with these elements (Scott, 1929; Bryan and Becker, 1935; and Dove, 1934).

C. Indicators of the Need

Even after analyzing the soil for various constituents, it is practically impossible to interpret these data in relation to the influence of the constituents on the growth of the plant. A chemical analysis of the soil indicates more of its potential fertility than it does of the available nutrients that are actually concerned in crop production (Thornton, 1933). Many investigations have been made in recent years to diagnose the needs of plants by noting their general appearance, the characteristics of the cell sap, and by applying certain microchemical tests to the various tissues. A large number of workers, including Dastur and Cooper (1932); Emmert (1930–1935); Emmert and Ball (1933); Fraps (1931); Gilbert (1926); Gilbert and Hardin (1927); Gilbert, McLean, and Adams (1927); Gilbert and Smith (1929); Gilbert and Pember (1935); Hoffer (1926); Jones (1931); Kraybill (1930); Laurie and McElwee (1934); Lauritzen (1934); McMurtrey (1933); Mitchell (1934, 1935); Murneek and Gildehaus (1931); Pettinger (1931); Poehlman (1935); Pohlman and Pierre (1933); Smith, Crandall, and Frear (1932); Shaw (1934); Thornton (1932, 1933); Thornton, Conner, and Frasure (1934); Van Haarlem (1935); and others; have made the investigations. The work that has been done shows that with some plants, and under certain conditions, the three types of observations may be indicators of the fertilizer needs of the plant. Much more information, however, must yet be acquired before such observations can be considered as reliable and of practical application.

1. General Appearances.—In corn plants growing in the field, a scarcity or lack of potassium is evidenced by the premature death of the plants, chaffy ears borne on premature broken shanks, and badly decayed roots (Hoffer, 1926). In some plants, dead tissue around the margins and below the veins of the leaves is an indication of potassium deficiency. In tobacco, McMurtrey (1933) mentioned that typical potassium deficiency is distinguished by small, necrotic spots or specks at the tips or margins of the chlorotic leaves. The leaves curl upward at the tips and margins. In the tomato when the supply of phosphorus is restricted, growth is retarded, and the leaves acquire a typical brown or purplish tinge. In tobacco a shortage of phosphorus is indicated by yellowing or drying of the lower leaves; the roots are relatively long, little-branched, and reddish.

According to Hoffer (1926), the common symptoms of nitrogen starvation in corn are light yellow-green leaves and stalks. McMurtrey (1933) stated that in the absence of nitrogen, the plant became light-green with more or less yellowing, drying, or firing of the lower leaves. The roots were long, little-branched, and white.

Under this topic the so-called "deficiency diseases" of plants should be mentioned. These are characteristic abnormal conditions that result from the unavailability of certain elements to a plant. A comprehensive review of these disorders is given by Humphrey (1934). In this text these "diseases" are mentioned for the most part under the discussions of the various elements. Some of these diseases are: chlorosis, sand drown of tobacco, gray speck of oats, hollow heart of turnip, heart rot of beet, little leaf or rosette of fruit trees, white bud of corn, tea yellows, leaf scorch of apples, and frenching of the leaves of various plants. In some cases the lack of an element may not cause a definite disease, but its scarcity may make the plant more susceptible to the invasion of various fungi (Heck, 1934).

2. Characteristics of the Sap.—According to Pettinger (1931), the quantity of sap that can be extracted from the cornstalk tissues is closely related to the diameter of the stalk and to soil productivity. There is a general relationship between the color of the expressed sap and the productivity of the soil on which the plants are grown. Those saps that are colorless or only light brown after clarification indicate a fertile soil and a sufficiency of potassium. A clarified sap with a dark-brown color is associated with an infertile soil and a scarcity of potassium.

If the expressed sap contains less than 100 p.p.m. of nitrates, the soil is deficient in available nitrogen, but if it contains above 300 p.p.m. of nitrates, there is an ample supply in the soil for all the needs of the plant. If the sap contains less than 0.20 mg. of phosphorus pentoxide per cubic centimeter, the soil is deficient in available phosphorus, but if there is present more than 0.20 mg. of phosphorus pentoxide per cubic centimeter, the soil supply is abundant. When the concentration of potassium oxide in the sap is lower than 1.0 mg. per cubic centimeter, potassium is deficient in the soil. If potassium oxide is present in excess of 2.0 mg. per cubic centimeter of sap, the soil contains an abundance of potassium for the growth and maturation of the plant.

3. Microchemical Tests.—In corn the presence of iron in the tissues of the joints indicates a lack of potassium (Hoffer, 1926). The presence of iron is indicated by a red color when a few drops of a solution of potassium thiocyanate is applied to the fresh tissue and followed with a drop or two of dilute hydrochloric acid. The relative amount of potassium can be estimated by treating sections with a 10 per cent solution of platinic chloride and examining the crystals that are formed. The relative amount of potassium in the plant may be determined by treating a portion of macerated tissue with a special solution of cobalt nitrite. A heavy precipitate indicates abundant potassium, a medium precipitate indicates a doubtful supply, and no precipitate, or only the trace of one, indicates that the soil is deficient in that element (Thornton, 1933). The application of 2 or 3 drops of a special solution of diphenylamine to freshly cut tissue gives a characteristic blue color if nitrates are present. A dark-blue color indicates that there is available to the plant an abundant supply of nitrates, a pale-blue color indicates an adequate amount, and no color suggests that the supply in the soil is inadequate. If a portion of crushed plant tissue is treated with ammonium molybdate solution and a small portion of stannous chloride is added and the whole shaken, the inorganic phosphorus concentration is indicated by a series of color intensities and shades ranging from colorless or yellowish to a dark blue. A medium- to dark-blue color indicates a sufficiency to an abundance of this element in the soil, while it is doubtful if a sufficiency is present when the color is green to light blue.

D. Distribution of the Elements in the Plant

The distribution of the various elements in the different parts of the plant is shown in the following table. A knowledge of their distribution is of importance in interpreting the role that they serve in the functions of the plant.

PERCENTAGE DISTRIBUTION OF THE DIFFERENT ELEMENTS IN VARIOUS PARTS OF CORN PLANTS GROWN AT MANHATTAN, KANS., IN 1920

Organ	C	O	H	N	P	K	Ca	Mg	S	Fe	Si	Al	Cl	Mn
Leaves	26.6	27.7	26.3	25.0	28.6	45.2	58.2	32.3	39.8	23.0	62.3	19.5	42.8	27.9
Stems	24.5	23.7	22.7	13.8	10.5	32.2	18.0	21.0	22.7	14.6	8.6	2.9	36.9	12.6
Grain	32.0	31.8	34.8	46.0	52.3	14.2	3.4	34.2	25.8	15.7	0.4	6.7	7.1	35.7
Roots	7.0	7.1	9.5	6.3	4.2	3.6	19.5	6.8	10.7	44.1	27.6	66.3	5.4	14.9
Cobs	9.8	9.7	6.6	8.9	4.4	4.7	0.9	5.6	1.1	2.7	1.1	4.6	7.8	8.0

The distribution of the various elements in the plant parts varies to some extent with the stage of growth. The distribution of the elements in barley will suffice as an example (Pfeiffer, Rippel, and Pfotenhauer, 1919). The total ash in this plant increased up to the time of the milk stage. The total phosphorus increased up to harvest time, being highest in the leaves before heading and in the stems at heading time. The total calcium increased up to the milk stage and was highest in the leaves at heading time and in the stems during the milk stage. The total potassium increased up to the milk stage and was highest in the leaves before heading. The total nitrogen increased up to harvest time, being highest in the leaves before heading and in the stem at heading time.

II. FORM

A green plant obtains practically all its carbon and a part of its oxygen from the air and all the other elements from the soil. With the exception of some of the oxygen, all the elements that enter a plant must be presented to it in the form of soluble compounds. Oxygen is the only element that the plant can utilize when presented in elemental form, but it, too, enters the plant dissolved in water.

A. INORGANIC COMPOUNDS

It is a definitely established fact that the green plant is able to grow and develop to maturity if it is supplied with the necessary inorganic compounds in sufficient quantity for its needs. Under the proper conditions of light, climate, and soil fertility, the green plant is capable of manufacturing from the simple inorganic compounds that it obtains from the soil and the air all the numerous and varied organic substances that enter into its composition. The kind of inorganic compounds that best serve the needs of the plant will be discussed later in this chapter.

B. ORGANIC COMPOUNDS

The concentration of the soluble organic substances in the soil may be as high as that of the inorganic compounds, so it is of interest to know to what extent the green plant can absorb and utilize organic material.

Funchess (1916) found that large amounts of coumarin and vanillin applied to the soil depressed the yield of oats when they were planted immediately after the application of these compounds. Pyridine and quinoline were very beneficial to both oats and corn in all the soils tested. The addition of potassium and phosphorus greatly increased the beneficial effects of these two compounds. Asparagine and nucleic acid were beneficial to plants, and dihydroxystearic acid when applied alone had no effect on either oats or corn but, when added in combination with fertilizer, it produced a larger yield than when fertilizer alone was applied. Stokes, Leukel, and Barnette (1930) utilized the sewage from septic tanks for the irrigation of Napier grass and Japanese cane. The yield of green material from the irrigated plots was as much as 3.3 times that from the nonirrigated ones. Tanaka (1931) found that urea proved a good source of nitrogen for some plants. The sugars he used were mostly favorable to plant growth, while organic acids except in low concentrations gave mostly negative results. As a source of phosphorus, lecithin was more readily assimilated than phytin. Of the organic sulphur compounds, only cystine was assimilable to any extent. Malyschev (1932) found that excised roots of various plants could be grown in solutions of various salts to which sugar had been added. The most extensive experimental work has been done with the carbohydrates, organic acids, and nitrogenous compounds, although numerous other substances have been used.

1. Carbohydrates.—Acton (1889) supplied various carbohydrates and glycerin to *Acer pseudoplatanus*, *Phaseolus vulgaris*, *Quercus robur*, and *Euphorbia hirsuta* among other plants growing in water cultures. He found that glucose, saccharose, and glycerin were absorbed and utilized in the dark by these plants in the formation of starch but that glycogen, dextrin, and soluble starch were not so utilized. Laurent (1904) showed that corn, beans, and other plants could absorb glucose through their roots and utilize it, as indicated by the formation of starch in their leaves. The same results were also noted by Mazé and Perrier (1904) for corn grown in water cultures.

Knudson (1916) studied the influence of carbohydrates on plant growth by growing them under sterile conditions in an agar medium to which Pfeffer's nutrient solution and a sugar solution varying in concentration from 1 to 2 per cent had been added. The assimilability of the sugar was judged by noting the influence on the dry weight of the plants. Knudson found that corn thus grown was able to absorb by means of its roots and utilize in its metabolism glucose, fructose, saccharose, and maltose, the beneficial effects being in the order that the sugars are here named. Canada field peas responded in growth to the presence of saccharose, glucose, maltose, and lactose in the order stated. Timothy, vetch, and radish were also benefited by the presence of sugar. In the case of radish, the average dry weight of 71-day-old plants was as follows:

WEIGHT IN GRAMS

Case	Plants grown in Pfeffer's solution plus				
	Check	Sucrose 2 per cent	Maltose 2 per cent	Glucose 2 per cent	Lactose 2 per cent
I..........	0.791	1.237	1.572	1.713	1.128
II..........	1.208	.....	1.950	2.461	1.900

The beneficial effects of sugar when supplied to the roots of the radish were also observed by Molliard (1907). Robbins (1918) grew the moss plant (*Ceratodon purpureus*) on sterile culture media and found that organic carbon in the form of levulose, glucose, galactose, lactose, sucrose, and maltose was absorbed and utilized, as was evidenced by the formation of starch in the plant in the dark. Mannite, glycerin, and starch, however, could not be utilized. He observed that the amount of growth with levulose as a source of carbon was from two to seven times greater than that with glucose as a source of carbon, and that with the presence of levulose the greater amount of growth occurred in the dark, while with glucose the greater amount occurred in the light. Light, however, seems to be necessary for the continued growth of this moss plant, even though a supply of available carbohydrates is furnished.

Brannon (1923), using the methods of Knudson, obtained data that indicate that different kinds of plants differ as to the sugar they are able to use most efficiently. Alfalfa grew best for him in solutions containing glucose, while timothy did better in solutions containing fructose. The radish used both sugars equally well. It seems that all the plants tested can use glucose and sucrose to some degree, while galactose, mannose, and fructose have been shown to be toxic to some plants under certain conditions. Roach (1926) studied the relation of soluble carbon compounds to the growth of certain soil algae. He found the relative average rates of growth in the different media to be as follows: glucose in the light, 100 per cent; maltose, 100 per cent; galactose, 94 per cent; sucrose, 84 per cent; fructose, 73 per cent; glycerin, 43 per cent; and mineral salts alone, 60 per cent. He also observed (1927) that the five species examined were capable of growing in complete darkness provided that a suitable organic compound was present in the medium.

Mayer (1929) obtained optimum root growth in a cane-sugar solution, while Wynd (1933) studied the growth of orchid seedlings on a series of different sugars over a period of 8 months. The order of excellence of growth on the different sugars was *d*-mannose > *d*-glucose > maltose >

d-fructose > sucrose > raffinose. No growth was obtained on *d*-galactose or the pentoses: 1-arabinose, 1-rhamnose, and 1-xylose.

The evidence that green plants can absorb and utilize carbohydrates is based in many cases upon the behavior of plants grown in the dark. The abnormal development and behavior of plants when grown in darkness, however, prevent one from drawing a general conclusion as to the response of green plants to organic nutrients under normal conditions of plant growth. The process of photosynthesis is a factor, too, that prevents a clear conception of the response of plants to organic nutrients, and this process cannot be eliminated in green plants when grown in the open in daylight even if the atmosphere is depleted of carbon dioxide. The process of respiration releases within the cells a considerable quantity of this gas, which is utilized at once by the chloroplasts in the process of photosynthesis. Excellent material for a study of organic nutrition would seem to be the albino plants, which frequently appear as sports in numerous species. These plants are apparently comparable in every regard to the green plants with the exception that they lack chlorophyll and functional plastids so that they cannot carry on photosynthesis.

Knudson and Lindstrom (1919) studied the effect of sugars upon the duration and growth of albino corn plants. Plants that were grown in Pfeffer's solution to which was added a two-tenth molecular solution of sucrose showed a very appreciable gain in weight during a 55-day experiment, while those grown in Pfeffer's solution alone showed a loss and perished 25 days sooner than those supplied with sugar. These authors attribute the failure of the albino plants to make a sustained growth to the slow diffusion of the sugar so that the plants were unable to absorb a sufficient amount of organic material to supply the needs of the entire plant. Their hypothesis is strengthened by the fact that after the leaves are dead, the roots continue to remain alive in the sugar solution several months after the check plants are dead. Robbins (1924) observed that excised root tips of peas, corn, and cotton made considerable growth in the dark in solution cultures containing mineral salts and glucose and fructose but that they made little growth under the same conditions when carbohydrates were lacking. The growth of these roots was markedly greater in the cultures containing glucose than in those containing fructose.

The roots under these conditions, however, do not grow to the extent that they do in the soil. Robbins (1929) considered that this lack of growth might be due to the toxicity of the nutrient solution, the failure to absorb sufficient glucose, lack of some essential element, lack of growth substances, or to a disturbance in the respiratory process.

2. Organic Acids.—Ravin (1913) investigated the influence of certain organic acids and their neutral and acid salts on the growth of the radish.

His results as summarized in the following table by Knudson (1916) indicate that the radish can absorb and make use of in growth such acids as malic, tartaric, succinic, and citric and their salts:

Culture medium	Dry weight grown in free atmosphere, milligrams	Dry weight grown in confined atmosphere, milligrams
Knop's solution	36.3	11.7
+ $KHSO_4$	41.0	16.9
+ K_2SO_4	42.0	17.0
+ Glucose	54.9	30.5
+ free organic acid	52.5	30.5
+ acid potassium salts of organic acids	53.9	33.0
+ neutral potassium, salts of organic acids	53.6	28.7

3. Organic Nitrogenous Compounds.—The utilization of organic nitrogenous compounds is discussed in detail under The Nitrogen Metabolism of the Green Plant (Chap. IX).

The absorbing system of the green plant is apparently adjusted for the absorption of the rapidly diffusing inorganic compounds and cannot readjust itself for the intake of sufficient quantities of the more complex organic compounds to supply all its needs for growth and repair. The experimental evidence indicates that green plants are capable of absorbing by their roots and utilizing in their metabolism a sufficient amount of organic matter to have a beneficial influence upon their growth. It seems evident, therefore, that the soluble organic matter of the soil may play a direct role in the nutrition of green plants.

III. ESSENTIAL VS. UNESSENTIAL ELEMENTS

From 1860 to 1900 much experimental work was done by Sachs, Knop, Nobbe, Pfeffer, and others to determine the elements that are essential to plant life. Their work indicated that 10 elements are absolutely necessary for all green plants. These are carbon, oxygen, hydrogen, nitrogen, phosphorus, potassium, calcium, magnesium, sulphur, and iron and have been considered until quite recently as the only ones necessary for the normal growth and development of plants. The other elements found in plants were considered of little or no physiological importance, and their presence was attributed to the fact that they entered the plant simply because of their occurrence in the soil in the soluble form; if comparatively large amounts were present, it was attributed to a specific power of the plant to accumulate them.

The experimental work by which the so-called "essential" elements were determined was considered to be above criticism until quite recently, when it was discovered that there are three important sources of error that had been overlooked: (1) many at least of the so-called "chemically pure" chemicals then used contained impurities that might supply elements other than the ten in sufficient amount to influence plant growth; (2) the various elements in the seed were not taken into consideration, and it is now known that the seed contains a sufficient amount of certain elements to supply the needs of the plant even up until the time of fruiting; (3) no precautions were taken to avoid the contamination of the cultures by certain elements that may dissolve in small amounts from the walls of the containers.

Haas (1930) noted that citrus plants grew normally in earthenware containers or in Swedish enamel mixing bowls without the addition of boron to the cultural solution. In sand cultures in galvanized iron cans, such solutions, however, gave poor results. Boron is present in the walls of many of the containers used in nutrition experiments, because borax is widely used in the preparation of pottery, enamel, agate ware, and certain kinds of glass. Silicon also is readily obtained by plants from the walls of the vessels commonly used in cultural studies. Consequently, Furnstal and Johnson (1936) devised sintered, pyrex glass aerators for water cultures to prevent any contamination by some of the rarer elements, which might invalidate results.

The elements needed by plants subserve two main functions: (*a*) Some of them are component parts of the cell structure. The amount of the different elements used for these purposes is relatively large, and any deficiency of them is soon noticed in the general growth of the plant. (*b*) Some of the elements apparently act as carriers of other ions, catalyzers of reactions, or antidoting agents. For these purposes the amount of some of the elements needed is very small and could easily be supplied in sufficient amount by impurities from the three sources that have just been mentioned. The observation that small amounts of certain of the so-called "unessential" elements improve the growth of certain plants and prevent certain deficiency diseases, and the discovery of the relation of vitamins in animal nutrition have perhaps given the impetus to renew the investigations concerning the role of many of the elements that have been considered unessential.

Mazé (1914–1919) has led the field in this work. He considered that zinc, silicon, chlorine, aluminum, boron, and perhaps other elements are essential to the growth and development of green plants. McHargue (1919–1935) in extended experiments, which will be discussed later in the chapter, has furnished very convincing data to prove that manganese is an essential element for plant growth. He considered that plants obtain such factors as are necessary for their growth from the mineral elements contained in a fertile soil, taking up small amounts of iron, manganese, copper, zinc, boron, nickel, and others with the aid of which are synthesized complex organic compounds in the plant. He considered that these organic compounds function as enzymes, catalysts, and vitamins and that when they are consumed by animals they are resynthesized into catalases, oxidases, hormones, and animal vitamins. Warington (1923, 1926), Brenchley (1914), Brenchley and Thornton (1925), Sommer and Lipman (1926), and many later investigators have shown that *Vicia faba* (Windsor bean) will not grow normally in water cultures unless small quantities of boron are added. A histological study showed that there is also a marked change in the anatomy of the root and nodules in the absence of this element (see Boron).

The increased study of the effects of the so-called unessential elements has given rise to the terms "rarer elements," "minor elements," and "less common elements." This does not mean that they are of infrequent occurrence in the soil, but that they are so named because apparently they are needed in only small amounts by the plants. These rarer elements, according to Thatcher (1932), and Young (1935), may affect the growth of plants in a number of ways. They may be necessary in very small amounts in certain physiological functions. Some may act as positive catalysts and others as negative catalysts. The latter type may depress certain reactions that are injurious to the plant. Some of these elements may inactivate toxins in the soil or may aid in oxidizing or reducing certain compounds in the soil to a form more readily available to the plant. The effects of manganese in maintaining the correct ratio of ferrous iron to ferric iron is given as an example of this. The rarer elements may have an effect upon colloids in overcoming the detrimental effects of certain ions. Thus the potassium ions in a pure aqueous solution are so strongly adsorbed that they injure the colloidal condition of the cell. When aluminum is added, neither the aluminum nor the potassium ion can monopolize the cell surface, and the effective action of both ions is decreased. These minor elements may have an action similar to vitamins in that the plant may grow to maturity and produce seed in their absence, but its general health will be greatly improved and its resistance to deficiency diseases greatly increased by their supplementary effects. Some of these elements may play an indirect role in the nutrition of the plant by balancing the nutrient solution in which it is growing. The concentration of these elements needed to produce the various effects mentioned above is low and variable, rarely below 1/500 p.p.m. and never greater than 400 p.p.m.

Thatcher (1934) proposed a classification of the chemical elements with respect to their functions in the metabolism of the green plant as follows: group I—hydrogen and oxygen—energy-exchange elements; group II—carbon, nitrogen, sulphur, and phosphorus—energy storers; group III—sodium, potassium, calcium, and magnesium—translocation regulators; group IV—manganese, iron (cobalt and nickel), copper and zinc—oxidation-reduction regulators. The other elements were placed into four other groups, but their function in the plant was not suggested. Haas and Reed (1927) found that young orange trees grown in sand cultures with the ordinary nutrient solutions showed marks of decline, while those receiving cultural solutions made with tap water did not show these symptoms. The leaves curled, became spotted and corky, and split along the veins. In other cases, the leaves were shed prematurely, and multiple buds were formed in the leaf axils or above the former leaf scars, but none of them was able to develop. They discovered that these plants could be rejuvenated by the addition to the sand culture solution of a suspension so as to give a concentration of 0.2 p.p.m. of aluminum, iodine, titanium, bromine, strontium, lithium, manganese, boron, and ammonia, respectively. The salts used were aluminum sulphate, potassium iodide, titanium sulphate, potassium bromide, strontium nitrate, lithium nitrate, manganese sulphate, ammonium nitrate, and boric acid. Within a week after this application, signs of rejuvenation were observed, new growth was in evidence, and none of the symptoms previously mentioned remained.

Cotton (1930) found that plants of buckwheat growing in a full nutrient solution containing minute quantities of manganese sulphate, zinc sulphate, boric acid, aluminum sulphate, sodium chloride, cobalt nitrate, copper sulphate, and lithium nitrate made decidedly better growth than those grown in a solution containing only the three main salts and iron. In growing oats in sand cultures, McHargue and Shedd (1930) noted that the yield of straw was increased 47 per cent by the addition of manganese to the culture, 53 per cent by the addition of manganese and copper, 65 per cent by the addition of manganese, copper, and zinc, 73 per cent by adding manganese, copper, zinc, and boron, and 63 per cent by the addition of manganese, copper, zinc, boron, and arsenic. The yields of grain increased in the same order 1,659, 2,134, 1,825, 1,637, and 1,157 per cent. Schropp and Scharrer (1933) showed that wheat, rye, maize, oats, peas, beans, clover, alfalfa, and tobacco growing in culture solutions were favorably influenced in their growth by the addition to the regular solutions of small amounts of a solution containing boron, manganese, copper, zinc, aluminum, nickel, cobalt, titanium, tin, iodine, and bromine. Meyer (1931) considered that the beneficial effects of the addition of kaolin to sand cultures was due to some of the rarer elements that it contained.

McHargue, Roy, and Pelphrey (1932) determined the amount of iron, copper, manganese, and zinc in various crop plants as shown below:

Plant	Ounces per ton of dry crop			
	Fe	Cu	Mn	Zn
Soybean	8.0	0.26	3.9	1.6
Sweet clover	4.2	0.29	1.6	1.6
Alfalfa	4.8	0.32	2.5	1.6
Bluegrass	10.2	0.45	2.6	0.5
Corn stover	6.4	0.22	5.1	1.9
Wheat straw	3.5	0.10	2.2	0.5

These data show that copper, manganese, and zinc are not absorbed in nearly so great an amount as the iron but yet in amounts great enough to be significant in the metabolism of the plant.

It was observed by Hoagland and Snyder (1933) that a supplementary solution of aluminum, iodine, bromine, titanium, tin, lithium, manganese, boron, zinc, copper, nickel, and cobalt caused strawberry plants to grow much better than in its absence. Pettinger, Henderson, and Wingard (1932) noted that the addition of magnesium, zinc, copper, boron, and arsenic to a basal nutrient solution increased the frost resistance of young corn plants.

Brenchley (1932), however, obtained only neutral or negative results by adding copper, lithium, titanium, and aluminum to cultures of barley, mustard, peas, or maize.

These results indicate that some elements other than those in the regular nutrient solutions exerted a beneficial influence upon the growth of these plants. As our knowledge now stands, we are not justified in saying that only 10 elements are necessary for plant life but must acknowledge that apparently others are essential. At least we are safe in saying that many of the so-called "nonessential" elements have a marked influence on the growth and development of plants.

IV. OCCURRENCE AND FUNCTION

A. Carbon, Hydrogen, Oxygen, and Nitrogen

Carbon, hydrogen, and oxygen enter into the composition of practically all organic compounds of the plant and make up approximately 95 per cent of its dry weight. The carbon is practically all obtained from the air in the form of carbon dioxide. The hydrogen and a considerable portion of the oxygen are obtained from the water absorbed by the plant. The remainder of the oxygen is obtained in the elemental state from the atmosphere. Nitrogen enters into the composition of the proteins, which in turn are apparently the active components of the protoplasm. The subject of nitrogen nutrition is discussed in detail in Chap. IX and will not be further mentioned here.

B. Phosphorus

1. Occurrence.—The amount of phosphorus in plants ranges from 0.2 to 0.8 per cent of the total dry substance. It is generally present in somewhat greater quantities than sulphur and magnesium and in smaller quantities than calcium. A considerable amount of the phosphorus in plants is soluble in water. In alfalfa (Ames and Boltz, 1912), 75 per cent; in immature cabbage (Peterson and Peterson, 1926), 61 per cent; and in the turnip, 80 per cent (Hartwell, 1917) of the total phosphorus present is soluble in water. The phosphorus in plants occurs in the greater amounts in the seeds and fruits. In the mature corn plant, 29 per cent of the total phosphorus is in the leaves, 11 per cent in the stem, 56 per cent in the grain and cobs, and 4 per cent in the roots (Latshaw and Miller, 1924). This is in accordance with the observation of MacGillivray (1926) that one-half the total phosphorus in the tomato plant is located in the fruit. The amount of phosphorus found in plant tissue will depend upon the amount present in the soil, and Jordan (1913) has shown that up to a certain point the production of plant substance increased with the increased phosphorus supply but that beyond this the phosphorus was not utilized in an increased plant growth. Hartwell (1917) found that young turnips which had been deprived of phosphorus and later placed in a nutrient solution well supplied with it showed crystals of ammonium-magnesium-phosphate within 24 hr. after the addition of the magnesium mixture. An examination for the relative abundance of the phosphorus showed it to be most abundant in the bundles and surrounding tissues near where the taproot merges into the thickened root. Next in amount is inside the cambium in the more active xylem elements and next outside the cambium in the cortex, and only traces are in the pith and leaf petioles. MacGillivray (1927) found that the percentage of phosphorus in tomato plants receiving an ample supply of phosphorus showed a steady increase in amount from the bottom to the top regions of the stem. Plants, however, grown in nutrient solutions lacking phosphorus did not show such an increase in phosphorus content but had approximately the same percentage phosphorus content in all regions of the stem from the bottom to the regions near the tip where it was higher. This indicates that high percentages of phosphorus are associated with regions of rapid growth. Phosphorus is apparently redistributed in plants to a greater extent than most of the other elements, and this property of redistribution may be one reason for plants having a smaller total amount of phosphorus than nitrogen, potassium, or other elements.

Webster (1928, 1929) found that the total phosphorus, in percentage of dry weight, was 0.54, 0.47, 0.42, 0.39, 0.37, and 0.36, for the seeds of soybean, cowpea, wheat, barley, kafir, and corn, respectively. The inorganic phosphorus represented only 2 to 5 per cent of the total phosphorus. The inorganic phosphorus in the roots of yellow corn, mung beans, alfalfa, and green onions only amounted to 0.012, 0.024, 0.136, and 0.189 per cent, respectively. According to Thoday and Evans (1932), the inorganic phosphorus in *Kleinia articulata* is localized mainly in the bundle zone. It was reported by Cockefair (1931) that phosphorus is distributed in the general meristematic tissues, in the synthetically active tissues, and in the fruiting organs. Storage organs are rich in phosphorus only in those parts where organic phosphorus has accumulated. The characteristics of plants suffering from a deficiency of phosphorus vary strikingly for different plants. Karraker and Bortner (1933) noticed in the plants of tobacco deprived of this element that the leaves had a dusky tinge, and the lower ones developed characteristic spots. The plants had a spindling appearance due to small petioles and stems in proportion to the size of leaves and the height of the plant. Fisher (1935) found that tomato plants lacking phosphorus appeared purplish in color, and the stems were covered with small deep-blue spots. The roots were long and brown with but few lateral branches.

2. Form.—Plants absorb and utilize phosphorus for the most part in the form of phosphates, although they can to a limited extent utilize some organic phosphorus, especially nucleic acid (Schreiner, 1923). Certain species of plants possess marked differences in their ability to acquire phosphorus from different sources (Baguley, 1912; Jordan, 1913; Truog, 1916; and Bauer, 1923). Cabbage, rape, sweet clover, alfalfa, and buckwheat were found to utilize to good advantage the phosphorus of phosphate rock, while barley, millet, corn, and oats were relatively less efficient in this regard.

Truog (1916) has proposed that according to the laws of mass action and chemical equilibrium, the reaction making the phosphorus of phosphate rock available to the plant may be represented as follows

$$Ca_3(PO_4)_2 + 2H_2CO_3 \rightleftharpoons Ca_2H_2(PO_4)_2 + CaH_2(CO_3)_2$$

According to this theory, if the soluble phosphate is removed, but only a portion of the soluble calcium bicarbonate, an equilibrium will finally be established in which little or no phosphorus will be brought in solution. Truog thus considered that plants containing a relatively high amount of calcium oxide have a relatively high feeding power and plants of a relatively low calcium oxide content have a relatively low feeding power for the phosphorus of phosphate rock (see also von Wrangell, 1922).

Hartwell and Damon (1927) found under conditions of an insufficient amount of phosphorus accompanied by liberal amounts of nitrogen and potassium that over long periods of time beets, cabbage, rape, and rutabagas are high phosphorus-response crops. In contrast to this, carrots, millet, tomatoes, and spring wheat are low phosphorus-response crops, and sorghums, soybeans, and oats occupy an intermediate position in this regard.

3. Amount.—It was noted by Tidmore (1930) that the quantity of phosphorus necessary for the maximum growth of corn, sorghum, and tomato depended upon the nature of the medium in which they grow. Thus, while these plants would not make a satisfactory growth in culture solutions with a phosphate concentration less than 0.1 p.p.m., they grew satisfactorily in a soil where the displaced solution contained less than 0.05 p.p.m. of phosphate. Teakle (1929) found that wheat plants showed vigorous growth and tillering when grown in nutrient solutions containing 1.0 p.p.m. to 50 p.p.m. of phosphate, and only small growth and greatly-reduced tillering when the phosphate content was between 0.05 and 0.1 p.p.m.

The amount of phosphorus in the sap of the plant generally varies directly with the amount in the soil. Thus by heavy application of phosphorus to the soil, Mather (1929) obtained an increase of this element in the plant sap of 34, 89, 100, 130, 135, 150, and 175 per cent for alfalfa, timothy, barley, rape, sweet clover, wheat, and alsike clover, respectively. In one instance, rape grown to maturity showed a 300 per cent increase. This effect has also been observed by Teakle (1929) for wheat; by Blair and Prince (1932) for corn, oats, rye, soybeans, hay, and potatoes; by Ames and Kitsuta (1932), Murphy (1933), and Neller (1935) for various crops; by Hartwell and Hammett (1911) for turnips; and by Jacob and others (1935) for millet, cabbage, and Sudan grass.

4. Physiological Role.—Phosphorus plays both a direct and indirect role in the metabolism of the plant, as indicated by the following topics.

a. Component of Compounds.—There are various organic compounds in plants in which phosphorus is one of the component elements.

1. *Phytin.*—This is an acid calcium-magnesium salt of inosite phosphoric acid. According to Posternak (1903) from 80 to 90 per cent of the phosphorus of seeds exists in this form. Averill and King (1926) reported that phytin makes up from 1 to 3.3 per cent of the air-dry material of seeds. Andrews and Bailey (1932) found that the amounts of phytin phosphorus in wheat bran, embryo, and endosperm were 1.4, 0.59, and 0.00 per cent, respectively. It was observed by McCance and Widdowson (1935) that the phytin phosphorus of corn, peanuts, peas, potatoes, and carrots amounted to 58.0, 57.5, 10.8, 23.0, and 15.8 per cent, respectively, of the total phosphorus of these parts, while apples, celery, and turnips were entirely lacking in this compound.

Knowles and Watkin (1932) found that three-fourths of the phytin in the mature wheat plant was in the spike. At maturity approximately

50 per cent of the total phosphorus of the grain was in the form of phytin, while in the straw 95 per cent of the total phosphorus was in this compound. The phytin phosphorus in the entire plant increased from 25 per cent of the total phosphorus in early spring to 60 per cent at harvest time. They observed in all parts of the wheat plant that the phytin phosphorus and the protein phosphorus rose and fell concurrently, which indicates a relationship between phytin and protein.

DeTurk, Holbert, and Howk (1933) noted that phytin is absent from the corn plant in all parts prior to pollination. Within two weeks after pollination, however, phytin appears in the developing seeds but is not present in any other part of the plant. They suggested that the activation of the zymogen of the enzyme responsible for the synthesis of phytin may be furnished by the pollen. Phytin disappears from corn seeds at a fairly rapid rate during germination.

2. *Phospholipides.*—These compounds are substituted fats containing phosphoric acid and nitrogen. Lecithin is the most abundant of the phospholipides in plants and is found in practically every living cell. Reed (1907) found that when phosphorus is withheld from plants the formation of lecithin does not occur, and that as a consequence there is an accumulation of fat in the cells. Webster and Dalblom (1930) stated that the lipoid phosphorus is only a very small fraction of the total phosphorus in the mung bean. Knowles and Watkin (1932) in wheat observed that the lipoid phosphorus makes up only 2.5 per cent of the total phosphorus of the spike, and from 6.5 to 4.0 per cent of the total phosphorus of the straw, decreasing in amount as maturity approaches.

3. *Hexose Phosphates.*—These are compounds in which the hexose molecule is combined with phosphoric acid. These compounds are apparently associated with respiration.

4. *Nucleic Acid.*—This compound is apparently associated with the nucleus. When phosphorus is withheld, no nuclear division occurs and meristematic activity is greatly diminished.

5. *Phosphoproteins.*—The evidence that such proteins are formed rests wholly on theory. The phosphorus evidently plays a role in protein formation since in its absence sugars increase in amount and coagulable proteins decrease. Under such conditions, plants have the general appearance of those that have been deprived of nitrogen.

b. Reduction.—It was found by Kraybill (1930) that nitrates are not reduced by tomato plants in the absence of phosphates. Eckerson (1929, 1931) noted in the depletion of phosphates by the tomato plants that the nitrate-reducing substances decrease rapidly in succession in the leaves, stem, and roots, and the nitrates accumulated in the same order. Jashnova (1930), as quoted by Nicol (1934), found that the nitrification of large amounts of ammonia was slowed up, if phosphates were present in insufficient quantity.

c. Respiration.—Lyon (1923, 1927) reported that phosphates exert an accelerating effect upon potato oxidase, so that carbon dioxide is produced by an oxidation of some component in a solution of glucose. Phosphate also catalyzes the slow oxidation of pyrogallol and of tannic acid by atmospheric oxygen. It also increases the rate of production of carbon dioxide by the anaerobic phase of respiration through its action as a catalyst toward oxidases. These accelerated actions are equally pronounced when enzymes are contained within the cells of *Elodea* or wheat seedlings.

According to Eckerson (1929), phosphorus, in all forms of life thus far examined, seems to stand on the threshold between potential and kinetic energy relations. If hexose phosphates are formed in plants it is supposed that they hold the same relation to respiration as in animals. It is considered that in the fermentation of sugars intermediary esters in a nascent state are formed. Sugars may be phosphorylated by the aid of coenzymes, through which action a readily decomposable form of sugar is formed. This type of sugar is susceptible to the action of destructive enzymes which transform its energy from the potential to the kinetic form.

Chatterjee (1933) noted that the acceleration of the outgo of carbon dioxide from the leaves of *Eugenia jambolana* and *Allium tuberosum* was greatest when the leaves were injected with a mixture of phosphate and glucose. The injection of phosphate alone, however, had little effect if the amount of glucose was small.

d. Root Growth.—In certain regions the application of phosphates is especially beneficial in promoting the maximum production of root crops. In the absence of phosphorus, root crops like turnips, mangolds, and rutabagas do not enlarge but remain permanently dwarfed. The general impression that the development of the roots of all plants is augmented by the application of phosphorus is probably due to the marked returns obtained from its use as a fertilizer (Russell, 1915; and Noll, 1923).

It was observed by Turner (1929) with barley, wheat, and cotton growing in water cultures, that the ratio of tops to roots decreased as the concentration of phosphate increased. This increase in phosphorus, however, retarded the development and growth in length of the lateral roots. He considered that the decreasing ratio of tops to roots under these conditions may be explained by assuming that the increase in volume and the storage of materials in the roots are stimulated by the presence of phosphorus. Sommer (1936) found in cultural solutions of tomato, wheat, cotton, and corn that high concentrations of phosphorus did not stimulate the development of roots. Apparently there is little or no experimental evidence to show that the proportion of roots to tops of ordinary crops is increased by the application of phosphorus.

e. Nitrogen Relations.—It has previously been mentioned that plants growing in a medium lacking sufficient phosphorus resemble nitrogen-starved plants. Such plants, however, contain more total nitrogen than those grown in complete nutrient solutions (MacGillivray, 1926). According to Kraybill (1930), the stems of phosphorus-deficient tomato plants contained three times as much water-soluble nitrogen and four times as much ammoniacal nitrogen, nitrate nitrogen, amide nitrogen, and α amino acid nitrogen as the stems from plants growing in complete nutrient solutions. McCool and Weldon (1930) noted that the phosphorus content of the plant sap was decreased by the application of sodium nitrate. In prune trees, Colby (1933) found in the leaf tissues that phosphorus starvation resulted in a relatively high content of nitrogen. Carolus (1935) found with collards, cowpeas, and tomatoes, growing under greenhouse conditions, that in general low availability of phosphorus was associated with a very high content of nitrogen in the plant, and a high content of phosphorus with a low content of nitrate nitrogen. Thus, in black cowpeas, the absence of adequate nitrogen increased the phosphate in the cell sap from 44 to 1,040 p.p.m. In collards a low content of phosphorus increased their nitrate nitrogen from 148 to 970 p.p.m. The same general behavior in the phosphorus and nitrogen relationship was also observed by Haas (1936) in experiments with citrus plants.

Emmert (1934) found that relatively large amounts of nitrogen decreased and small amounts of this element increased the phosphate content of the vascular bundles. He considered that this interrelation between nitrogen and phosphorus is probably caused by their utilization by the plant. The presence of an abundant supply of nitrogen as compared with phosphorus causes a utilization of this element and, consequently, a decrease in its amount in the conducting tissues. If nitrogen is deficient, then the phosphorus may accumulate because it is not used in the synthesis of nitrogenous, organic compounds.

f. Carbohydrate Relations.—It was observed by Reed (1907) that the transformation of starch into water-soluble carbohydrates was seriously impaired in the absence of phosphorus. Hartwell (1917) noted that, simultaneously with the increase in the absorption of phosphorus by the turnip root, which previously had been deprived of it, the leucoplasts containing the starch grains shrank in size as the grains were corroded and dissolved until finally all the starch had disappeared from the root tissue only to reappear when phosphorus was again withdrawn from the nutrient solution. MacGillivray (1926) noted in the tomato that the absence of phosphorus greatly increased both reducing and nonreducing sugars in the plant. This increase of sugars in the absence of phosphorus has also been noticed by Eckerson (1929) and Kraybill (1930). Eidelman (1930) stated that for average lengths of day and with relatively high temperatures photosynthetic activity tends to show a positive correlation with the content of phosphorus. The maxima, however, do not coincide under the conditions of a shortened day and lower temperatures.

g. Maturity.—It has long been observed that phosphate fertilizers induce earlier ripening of the grain crops growing on soil low in phosphorus. The same effect has also been noticed for cabbage and cotton. The effects seem to vary with the different types of phosphorus, the more soluble ones hastening maturity to a greater degree. Increasing the rates of application, however, above the needs of the crop as indicated by yields has not been accompanied by further increases in earliness (Noll, 1923). Phosphorus, however, is not the only element that promotes earliness, since moderate applications of nitrogen and potassium and sometimes calcium when they are present in insufficient amounts have similar effects on maturity to those of phosphorus.

h. General Effects.—In oats, McClelland (1931) found that the application of phosphorus increased the number of tillers, panicles, number of spikelets per panicle, yield, and size of seed. Eckerson (1929) noted that in the absence of phosphorus the protoplasm of young cells became less refractive, and that the chloroplasts disintegrated first in the stem and lower leaves and finally in the upper ones. Owen (1931) noted in tomatoes that if phosphates were omitted the potassium content of the foliage was depressed, and that if they were added to a medium deficient in this radical the content of potassium in the leaves was increased. Hurd-Karrer (1935) found that the presence of phosphorus greatly inhibited arsenic injury of plants.

C. Calcium

Calcium is essential to the growth and development of all green plants with the exception of some of the lower algae. Nonchlorophyllous flowering plants must also be supplied with this element, although such plants as a whole contain less calcium than green plants. Calcium, however, is not necessary for the development of the fungi.

1. Occurrence.—A considerable portion of the calcium found in plants is in forms that are soluble in water. In cabbage that is ready for storage, 60 per cent of the calcium present is soluble in cold water (Peterson and Peterson, 1926), but after the plants have been in storage for six weeks, the soluble calcium amounts to only 42 per cent of the total amount present. Ames and Boltz found that 40 per cent of the calcium in alfalfa can be extracted with water. In the tomato plant, Nightingale and others (1931) found that all the calcium of the fresh tissues of calcium-deficient plants was insoluble. The leaves contain the greatest amount of calcium, and the seed or grain contains the lowest amount. In the case of the corn plant (Latshaw and Miller, 1924), 58 per cent of the calcium was found in the leaves, 18 per cent in the stems, 20 per cent in the roots, and only 4 per cent of it in the grain and cob.

The amount of calcium in the plant varies greatly with the species and with environmental conditions. Thus, for example, the percentage of calcium based on dry weight has been found to vary from 3 to 8 per cent in the vines of the cucurbits, to be 2.7 per cent in mature tobacco, and to amount to only 0.4 per cent in timothy hay and wheat straw. As a rule, the legumes, lettuce, cabbage, and tobacco have a high percentage of calcium on a dry basis, while the members of the grass family have a relatively small amount. In the case of cucurbit vines, the amount of calcium increases from the early to the late stage of development, while the potassium, phosphorus, and nitrogen percentage decreases (Wilkins, 1917).

Cowell (1932) found the calcium content of the outermost leaves of cabbage may be 20 to 30 times as great as that of the inner leaves. In *Kleinia articulata*, Thoday and Evans (1932, 1933) observed crystals of calcium oxalate in the collenchymatous, hypodermal layers. This element was present in abundance in solution in the sap of cells of the pith and in patches of cells in the inner cortex, two-thirds of the total calcium being in the pith cells. Yanovsky, Nelson, and Kingsbury (1932) found that practically one-half of the ash of the hackberry was calcium oxide. It was found by Harper and Daniel (1934) that the calcium content of *Elodea* was as much as 8 to 10 per cent, while in *Typha latifolia* it occurred only in relatively small amounts. They considered that the absorption of calcium by aquatic plants may be an important factor in the clarification of the water of ponds and lakes. Larson (1928) found that the concentration of soluble calcium favorable for normal plant growth is apparently about 32 to 64 p.p.m., while 16 to 32 p.p.m. seems to be the minimum concentration. He considered that 32 p.p.m. is the most economical concentration, since above that point, although the yield is increased, the increases become successively smaller.

2. Physiological Role.—According to Bamford (1931), the necessity of calcium for plant growth was first demonstrated by Salm-Horstmar (1856). Numerous functions have been ascribed to calcium, although some of these are little understood. The roles may be classified under three main headings: (*a*) antidoting agent, (*b*) component part of the plant, and (*c*) general effects.

a. Antidoting Agent.—Calcium may function as an antidoting element in two ways: (1) it may play a part through the calcium/magnesium ratio, and (2) it may function in the neutralization of organic acids in the plant.

1. *The Calcium/Magnesium Ratio.*—Calcium reduces the toxic effects of single-salt solutions of sodium, potassium and magnesium, as was discussed under Antagonism and Balanced Solutions. Since calcium or magnesium salts in single solutions are toxic or injurious to plants and since either will neutralize, in part or in whole, the deleterious effects of the excess of the other, Loew (1903) proposed the hypothesis that one of the principal functions of calcium in plant metabolism is to neutralize the toxic action of magnesium and that a certain calcium-magnesium ratio varying with the type of plant is necessary for the proper growth and development of the plant. An enormous amount of experimental work was carried on by Loew and his pupils and by many others in an attempt to prove this hypothesis and to find the proper proportion of calcium and magnesium for various crop plants. Lipman (1916), however, after a thorough résumé of the literature on the subject, concluded that the experimental work presents no evidence in support of the hypothesis that a specific lime-magnesium ratio exists for any plant or group of plants. He considered that there is no more reason to assume that there should be a proper ratio in the soil between calcium and magnesium than that there should be a proper ratio between calcium and potassium or between calcium and any other essential element.

The conclusions of Lipman are further upheld by the work of Wyatt (1916) with wheat, soybeans, cowpeas, and alfalfa. The crop yields of these plants and the ratio of calcium to magnesium in them bear no direct relations to the ratio of the carbonates applied. Different ratios of calcium to magnesium within rather wide limits produced no marked differences in yield, although increasing the size of the applications increased the calcium and magnesium content of the plants. Prianischnikow (1923) found that weak solutions of the salts of magnesium, calcium, barium, and strontium in concentrations of $0.0005N$ to $0.005N$ increase the resistance of cells to free acid, and, of the cations used, calcium showed the strongest antagonism.

Moser (1933) could find no significant correlation between the calcium/magnesium ratio and crop yields. He believed that the significant factor in determining yields is the amount of active calcium in the soil. The addition of lime to the soil is beneficial not because it changes the calcium/magnesium ratio, but because it increased the replaceable calcium therein.

2. *Neutralization of Organic Acids.*—It has long been known that oxalic acid is one of the most abundant and the strongest of the acids in the plant and is supposed to be a by-product of carbohydrate and protein metabolism. Schimper (1888, 1890) suggested that one of the principal functions of calcium is to precipitate oxalic acid and soluble oxalates in the form of calcium oxalate, thus preventing the injurious effects that they might exert upon the plant. Calcium, for this purpose, however, is apparently not necessary in certain cases, for neither oxalic acid nor oxalates have been observed in some plants. The theory of Schimper was further weakened by the work of Bruch (1902), who showed that the grain plants tolerate relatively large amounts of oxalates, since wheat plants growing in nutrient solutions containing 18 to 278 p.p.m. reached the blooming stage. He further found that calcium oxalate could replace other calcium salts as a source of calcium in nutrient solutions, and Ames and Boltz (1912) found in the alfalfa plant that one-third of the calcium present was in an insoluble form other than oxalates, which indicates that calcium is essential for the formation of some of the organic compounds. Parker and Truog (1920) have more recently revived the theory that calcium plays an important role in the neutralization of acids within the plant. They observed that there is a close relationship between the calcium and nitrogen content in plants. In regard to the ratio of calcium to nitrogen, plants can be divided into two groups. The first group has a calcium nitrogen ratio of about 0.3 and contains almost entirely members of the grass family that have low calcium requirement and are more or less tolerant to soil acidity. The second group has an average calcium nitrogen ratio of about 0.55 and includes the legumes and those plants which require a large amount of calcium and are sensitive to soil acidity. These authors considered then that the larger the amount of nitrogen the more protein is formed and the more oxalic, acetic, succinic, and formic acids are produced in protein metabolism. Larger amounts of calcium are thus necessary to neutralize these acids than if a smaller amount of protein were manufactured. They assumed that, if calcium is lacking, the above acids accumulate in such quantity that they become injurious to the cells and inhibit further

protein production. Hobart (1924), however, considered that the neutralization of these acids can be performed by other bases and that the necessity of calcium can be explained by assuming that it acts as a stimulant to the protoplasm.

b. Component of Plant.—It is definitely known that calcium is one of the components of the middle lamella of the cell wall. It is considered by some that it may enter into the composition of the protoplasm and of certain types of proteins in the cell.

1. *Cell Wall.*—It was stated in Chap. I that calcium pectate is one of the compounds of the middle lamella. It was noted by Reed (1907) that when calcium was withheld from the plant, new cell walls were imperfectly formed if at all, although the nuclei of the cell divided mitotically. The most important observations on the function of calcium in the cell wall have been made by True and his workers (1918, 1921, 1922). They found that plants growing in various concentrations of a single salt of potassium or sodium were not able to carry on a sustained absorption but in the end gave up more ions to the medium than they were at any time able to take up. In solutions of calcium and magnesium salts, there was a concentration below which the roots were not able to absorb, and ions leached out into the medium, although, in solutions above this concentration, absorption took place in a greater or lesser degree. In mixtures containing calcium and other nutrient salts, the total quantity of ions absorbed by the plants exceeded by far the total number of calcium ions present. This indicates that in a solution of mixed salts, calcium in some way produces conditions that bring about the absorption of ions which without the presence of calcium would not be absorbed or would cause the leaching of ions from the plant. This behavior of the diffusion relations can, in part at least, be explained by the following microchemical observations:

It was found that when seedlings were grown in potassium solutions, ions readily entered the cells of the root, but that within 24 hr. calcium ions began to diffuse out of the calcium pectate middle lamella of the cells. Potassium pectate was formed in the cell walls instead of calcium pectate and soon dissolved out, since it is readily soluble in water. At this stage organic matter and salts began to leach rapidly from the cell vacuole. Approximately two-thirds of this material was organic and nonelectrolytes and consisted principally of sugars and amino acids. This action of the potassium salt upon the cell wall not only changed its character but also greatly changed the permeability of the protoplasmic membrane next to it. These changes are seen also by the inward passage of materials. Corn seedlings after 5 days in a potassium nitrate solution were placed in a one per cent solution of copper sulphate and within 1 hr. the copper ions had penetrated all the root tissue. Similar seedlings after 5 days in a calcium nitrate solution showed the penetration of copper ions only after 24 hr. in a similar copper solution. In solutions of magnesium salts, the calcium pectate in the middle lamella is replaced by magnesium, which is also an insoluble compound but more permeable than the calcium pectate. The fatal results from growing plants in magnesium solutions do not occur until the calcium of the middle lamellae has been wholly replaced by magnesium, after which sufficient magnesium ions are apparently free to penetrate the protoplasm and injure it. True considered that the integrity of the calcium pectate forming the middle lamella is maintained when a sufficient quantity of calcium ions are present in the external medium and with it the normal retention of the cell contents. When according to the laws of mass action this quantity of calcium ion falls below the equilibrium concentration, other cations replace the calcium in the middle lamella. No cation apparently can replace calcium in this regard, although magnesium comes more nearly to doing it than any other.

2. *Proteins.*—It is considered by some that calcium enters into the composition of proteins, and Loew (1903) believed that calcium protein compounds enter into the organized substance of the nucleus and plastids. He considered that these calcium compounds have a well-defined power of imbibition, which if disturbed by replacement substances like potassium-protein compounds leads to fatal results.

3. *Protoplasm.*—Sorokin and Sommer (1929) found that plants of *Pisum sativum* grown in culture solutions without calcium usually die within 2 or 3 weeks. The death of the plants is associated with the degeneration that takes place in the protoplasts of the meristematic cells of the root tips. The first visible effect of the absence of calcium is a decrease in the amount of cytoplasm and the appearance of large vacuoles in the meristematic cells, which thus resemble old cells. With the gradual depletion of the amount of calcium, the cells of the meristematic regions of the roots show transitional stages from interrupted and irregular mitoses to almost typical amitoses. Sorokin and Sommer considered that the absence of calcium does not have a direct effect on the disintegrating tissues due to the separation of the cells because of the absence of calcium pectate. According to these investigators, the effect of the absence of calcium is on the meristematic cells, either because it is a necessary chemical constituent of the protoplast or because its absence so affects the physical condition of the colloidal system that normal action of the protoplast is not possible.

Bamford (1931) studied the influence of a deficiency of calcium on the growth and cytological structure of the root tips of young wheat and corn seedlings growing in water cultures. Under such conditions, the cells of the root apex gradually lost all their stainable contents, but the cell walls remained intact. The protoplasm appeared to undergo a gradual disintegration, until it completely disappeared. At first the nucleus remained normal, but eventually it degenerated into a heterogeneous mass and disappeared. This process of disintegration at first affected only the epidermis and the peripheral layers of cells in the cortical region, but later progressed to the interior and finally upward in the roots. However, Lutman (1934) found that the lack of various essential elements, other than calcium, was reflected in the cytology of the cells by a premature vacuolization and by indications of maturity and senescence.

c. General Effects.—Under this heading are considered (1) the effects of calcium on the general appearance of the plant, (2) translocation of carbohydrates, and (3) the response of roots.

1. *General Appearance.*—In a deficiency of calcium in the soil or water culture, the lower leaves in the case of the cereals show an inrolling of the margins, which Loew (1903) and Tottingham (1914) ascribe to a high magnesium calcium ratio. Dickson (1918) observed this behavior and noticed that later brown spots appeared on these inrolled leaves, which finally became dry and twisted around the stem. Day (1929, 1935) working with *Pisum sativum* observed that, with plants deprived of calcium, the lower leaves were chlorotic, and the younger leaves were curled and tough; the plants under these conditions were shorter and less succulent. McMurtrey (1932, 1933) noted that the first indication of a deficiency of calcium in the tobacco plant was a peculiar hooking downward of the tip of the young leaves of the bud followed by the death of the tissues at their tips and margins. If later growth appears, the tips and margins show a cut-out appearance. At later stages of growth, calcium deficiency is manifested by the cessation of terminal growth and the development of chlorosis and necrotic spots in the older leaves (Davis, 1930). Mann (1924) found that young apple trees in sand cultures deficient in calcium produced leaves that were larger than those borne by trees supplied with a complete cultural solution and that in the case of the gooseberry a deficiency of calcium resulted in leaf water content higher than that shown in the complete nutrient. Resistance to evaporation in bright light was greatest in the case of the plants grown in a deficiency of calcium.

2. *Translocation of Carbohydrate.*—It was observed by Boehm (1875), that an abnormal accumulation of starch occurred in *Phaseolus vulgaris* grown in water cultures lacking that element—an observation that has since been verified for numerous plants by many investigators. The movement of starch from any tissue in a plant will depend primarily upon the diastase activity or the power of digestion and the depletion of the sugar in the surrounding tissues. It has been found (Groom, 1896) that potassium oxalate retards the action of diastase, and it has been considered that, with a sufficient supply of calcium, calcium oxalate is formed instead of the potassium compound, and this salt is not detrimental to diastatic action. Hartwell (1916) suggested that an accumulation of starch may accompany any general disturbance of the metabolism of the plant, so in the absence of calcium the accumulation of starch may be only an indirect result. It has also been observed (Grafe and Portheim, 1906) that the addition of levulose, dextrose, and saccharose to calcium-free solutions enables bean plants to live in them a longer period than where no sugar is added. The cause for the growth of plants not being inhibited at once by the absence of calcium is attributed to the fact that considerable carbohydrate is present in the form of sugar, and that this is sufficient for its immediate needs. In some cases, a deficiency of either magnesium or calcium causes an increased growth, but under such conditions the weight of the grain is less than where a sufficient supply of calcium is available, so that the total dry weight of the plant is less. This observation is held to indicate that a deficiency in calcium causes a disturbance in the translocation of carbohydrates and proteins and in their storage during seed formation rather than a hindrance to their synthesis. It was observed by Nightingale and others (1931) in the tomato that the calcium-deficient plants accumulated carbohydrates in large quantities apparently because absorption and accumulation of nitrates did not take place. When calcium-deficient plants were placed in the light and given a supply of calcium, absorption and assimilation of nitrates occurred within a few hours.

3. *Root Response.*—When calcium is withheld from water cultures, the size of the roots of the plants growing therein is greatly reduced. Reed and Haas (1923) showed

that the roots of citrus seedlings grown in the absence of calcium were injured, as indicated by the gelatinization of the superficial layers and the ultimate death of the roots. If injury has not progressed too far, the addition of calcium induces the production of lateral rootlets, the lowest of which mark the boundary between the dead and living portions of the root.

It was noted by Bamford (1931) that no lateral roots were developed by wheat and corn when calcium was absent from the culture solution. Albrecht and Davis (1929) considered that calcium plays some physiological role favoring nodule formation of soybeans.

D. Magnesium

Magnesium is usually present in the plant in somewhat smaller amounts than calcium. It is relatively more abundant in seeds and leaves than in the other parts of the plant. In the case of the corn plant, 34 per cent of the magnesium is in the grain, 32 per cent in the leaves, 21 per cent in the stems, 7 per cent in the roots, and 6 per cent in the cobs.

Daniel (1935) found that the average content of magnesium in 19 species of mature grasses was 0.156 per cent, and in 45 types of mature legumes it was 0.379 per cent. The magnesium content of the grasses ranged from 0.059 to 0.316 per cent, while the legumes varied from 0.329 to 1.024 per cent. Most soils are not deficient in magnesium. Wheeler and Hartwell (1904), however, showed that the application of magnesium salts to certain Rhode Island soils was beneficial to various plants and that a single application of Mg lime would cover the deficiency of Mg for many years to come.

Magnesium, as first shown by Willstätter (1906), is one of the constituents of chlorophyll, and when this element is lacking a disturbance in the formation of chlorophyll is soon noticed.

According to Beaumont and Snell (1935), plants differ markedly in their response to a deficiency of magnesium in the soil. Thus buckwheat and spinach are very sensitive, and turnips, mangels, corn, and tobacco considerably so. The small grains, grasses, clover, and potatoes, however, are only slightly affected. Dickson (1918) found that oat plants grown in a nutrient solution which had only approximately one-tenth of the requisite magnesium supply developed numerous broad leaves which at first were very bright green but which when older showed marked stripings between the veins because of the disappearance of the chlorophyll. Sand drown, a chlorosis of tobacco involving both the green and yellow pigments of the chloroplasts, has been shown (Garner and McMurtrey, 1923) to be due to a deficiency of magnesium in the soil.

According to Garner, McMurtrey, Bowling, and Moss (1930), the outstanding features of sand drown in tobacco in the field are:

1. A disintegration of both the green and yellow pigments of the chlorophyll, beginning at the tips of the lower leaves. The symptoms

of magnesium deficiency differ from those of potassium deficiency in that with magnesium deficiency the leaf surface ordinarily remains smooth, with no downward curvature of the tips and margins, and spotting, resulting from localized disintegration of the tissues, only rarely occurs.

2. An increase in the water content of the leaf, and a decided decrease in the amount of dry matter per unit of area.

According to Beaumont and Snell (1935), the symptoms of magnesium deficiency for plants in general are: (1) the development of chlorosis in the intervascular tissue of the older leaves, which produces a striped appearance in leaves with parallel veins and a mottled appearance in those with netted veins; and (2) (in severe cases) the browning of the margin of the leaves and the dropping of some of the leaves from the plant.

Reed and Haas (1924) observed that chlorosis of young citrus and walnut trees resulted from an insufficient application of magnesium. Brown, Houghland, Smith, and Carolus (1933) found in a number of potato-producing regions that a characteristic, chlorotic condition of the lower leaves developed, which apparently was due to a magnesium deficiency in the soil. Mameli (1918) found that within certain limits the development of chlorophyll is proportional to the magnesium supplied in the solution. Jones (1929) concluded that chlorosis of corn in certain regions of Massachusetts was due to the lack of magnesium, since the normally green plants contained a larger amount of this element than did the chlorotic ones.

Parbery (1935) noted that chlorotic orange leaves in Australia were characterized by marked deficiencies in magnesium. Lutman and Walbridge (1929) believed that lack of magnesium may lead to early chlorosis of the older leaves of the potato plant.

It was found by Garner, McMurtrey, Bowling, and Moss (1930) that a minimum content of 0.25 per cent of magnesium, or 0.4 per cent of magnesium oxide, in the leaf of the tobacco is required to prevent magnesium deficiency symptoms. In practically all cases the troubles arising from magnesium deficiency have been remedied by the application to the soil of various magnesium salts as fertilizers.

Magnesium appears to function in many cases as a carrier of phosphorus. The salts of magnesium undergo dissociation very easily and thus readily give up the anions that they carry. Phosphorus enters into the formation of nucleoproteins and lecithins, two compounds that enter into the composition of protoplasm. If magnesium functions as a carrier of phosphorus, then the content of magnesium should be relatively high in those regions where these compounds are being formed. This apparently is the case, since magnesium is more abundant and frequently increases in amount in the tissues—as the tips of stem and roots—where rapid development is taking place.

Magnesium is more abundant in oily seeds like those of flax and cotton than it is in starch-containing seeds like the cereals. The average proportion of magnesium in starchy seeds to that in oily seeds is about 2:5 (Loew, 1903). Since the formation of oil is preceded or accompanied by the formation of lecithin, it would appear that magnesium in the case of oil formation serves as a carrier of phosphorus, and this supposition is further supported by the fact that, when both lecithins and nucleoproteins are present in the cell, magnesium occurs in greater amounts than if only one of these compounds is present. In regard to oil formation, Reed (1907) observed that the filaments of *Vaucheria* growing in magnesium-free solutions were devoid of oil globules, although globules were present in those plants growing in the control solutions. He considered that there is an intimate relationship between the presence of magnesium and the formation of vegetable oils. Oat plants growing in a nutrient solution deficient in magnesium produced inferior grain, and the ratio of grain to straw was decreased (Dickson, 1918). Apparently only a small amount of magnesium is needed to satisfy the needs of the plant, and seemingly it can be used over and over again as a carrier. It has been observed that beans can reach their full vegetative growth with no supply of magnesium other than that stored in the seed.

When magnesium in the form of magnesium sulphate is present in relatively large amounts in nutrient solutions, the leaves of cereals growing therein show characteristic magnesium injury. The tips first wilt and then dry up and drop off. This injury occurs in a much greater degree under conditions conducive to high transpiration than to low. The addition of calcium salts is beneficial in overcoming this injury, and the results have been considered as proof that one of the functions of calcium is to overcome the toxic action of magnesium, but Gericke (1922) has shown that this injury can also be prevented by having a proper proportion of phosphorus present in the solution.

Trelease and Trelease (1931) reported that magnesium injury of seedling wheat is marked by a rolling and curling of the young leaf during its emergence from the sheath. This effect is first visible after the young wheat plants have grown in the cultural solution from 9 to 12 days and the third leaf is just appearing. Later, numerous tillers develop prematurely, growth of the seminal roots is inhibited, and a number of adventitious roots are produced. The plants of oats, rye, and barley exhibit the same general characteristics. Carmin (1931) noted that magnesium sulphate was more toxic to the roots than to the tops, and more toxic to the main roots than to the lateral ones.

E. Potassium

Potassium is absolutely essential for plants and cannot be entirely replaced by any other element. Sodium may be substituted for potassium to a certain degree, but by no means entirely (Colla, 1929; Tincker, 1932; Alten and Gottwick, 1933; McCalla, 1934; and Hartt, 1934). The plant can apparently absorb from the soil and utilize in its metabolism any soluble inorganic compound of potassium. The form in which it is presented to the plant apparently makes little difference in some cases, and in other cases the differences observed may be attributed to the nature of the soil, environmental conditions, or the nature of the plant.

1. Occurrence.—The range of distribution of potassium in the plant has been studied by McCallum (1905), Weevers (1911), Dowding (1925), Lukaszewicz (1926), Penston (1931), and Thoday and Evans (1933).

All meristematic tissues are rich in this element, as indicated by strong positive microchemical tests obtained in the growing tips and cambium. Potassium appears especially abundant in the growing parts in those cells in the region of secondary root outgrowths and the formation of new leaves. McCallum (1905) noticed that the cell in the *Equisetum* spore which will give rise to the primary root hair is rich in potassium, and as the tip elongates the potassium keeps in contact with it. The same thing holds true in the development of the pollen tube. The potassium in the pollen grain migrates into the tube and keeps in contact with the growing tip. Potassium is in the mesophyll cells of the leaf, the cells of the medullary rays, the phloem, and to some extent in other parenchymatous tissue, but little or none is found in the older xylem portions of the stem. In the spruce when the bud elongates, there is a translocation of the potassium within the embryonic cone to the next year's meristem and to the embryonic leaves (Dowding, 1925). This author noted also that in the wheat grain the cells of the aleurone layer contain a large amount of potassium but that the endosperm has only a trace. After 15 days of germination the potassium disappears from the aleurone layer, having apparently leached out into the surrounding medium, since there is no evidence that it is transferred into the endosperm. All parts of the embryo contain potassium, but the greatest amounts are contained in the root and stem apexes. Potassium is thus present in regions of meristematic division, photosynthesis, translocation, storage, and other special, physiological activities.

Chemical analysis of the corn plant at early maturity (Latshaw and Miller, 1924) shows that 45 per cent of the potassium in that plant is located in the leaves, 32 per cent in the stems, 14 per cent in the grain, approximately 5 per cent in the cob, and 4 per cent in the roots. Dickson (1918) found by analysis that potassium is localized first in the growing region of the young wheat seedling but that later it appears in large quantities in the region of grain formation. Potassium is well distributed over the plant, and André and Demoussy (1927) considered that it is the predominant soluble salt in plants because of its mobility.

The potassium present in potassium-deficient plants is apparently withdrawn from the older regions and transferred to the meristematic tissues (Janssen and Bartholomew, 1929; Nightingale, Schermerhorn, and Robbins, 1930; and James and Penston, 1933). It was observed by Hartt (1934) in sugar cane that potassium migrated from the dying leaves to the living ones and that under such conditions it was the only element that did so. Penston (1935) noted a pronounced variation in the potassium content of potato leaves throughout the day. It tended to reach a maximum about 3 to 4 P.M. and to fall at night. The greatest variation between the maximal and minimal content during the day amounted to 73.2 per cent. He concluded that potassium is being continually brought into the leaf, presumably in the transpiration stream, and removed through the phloem, since a loss of potassium becomes apparent in the evening when the rate of transpiration is decreased. In the expressed sap of perennial rye grass, taken from July to December, the absolute amount of potassium rose to a maximum at maturity or during late senescence, and then declined (Petrie, 1934).

In the individual cells, potassium seems to be absent from the nucleus. Stocklasa (1911) could find none in the chloroplasts, and Dowding (1925) stated that, in the spruce leaves during the summer, potassium occurs as a network of granules between the chloroplasts. He observed that in the epidermal cells of the hyacinth and onion potassium occurs in the corners of mature guard cells where oil and starch can be found, but that in the initial cells that have not yet divided there is no localization of this element. McCallum (1905) observed potassium in the cytoplasm, but Stocklasa could not positively recognize it there in many cases, but considered that the greater portion of it is in the vacuole of the cell.

James and Penston (1933) found potassium in the cytoplasm and vacuole but only on the surface of nuclei and plastids. In perennial rye grass, Petrie (1934) considered that potassium occurs entirely in the ionic form in the cell sap and cytoplasm. The amount of cell sap and cytoplasm probably rises to a maximum at maturity and then declines. The absolute amount of potassium is probably correlated with this drift.

The potassium present in the plant is for the most part in forms readily soluble in water (Janssen and Bartholomew, 1929). Kostytschew and Eliasberg (1920) reported that when plant tissue is extracted with cold water, the tissue yields potassium-free ash and lead acetate, and tannin precipitates were also found to contain no potassium. Weevers (1911) by water extraction removed 99 per cent of the total potassium from the pollen tubes of *Pinus sylvestris* and leaves of the horse-chestnut.

Mann and Wallace (1925) found that potassium leached readily from apple leaves when they were immersed in water. It was reported by Anderson, Swanback, and Street (1932) for corn and tobacco; Morris (1933) and Morris and Sayre (1935) for corn; and Tyner (1935) for rape, sweet clover, buckwheat, alfalfa, corn, and the sorghums that the potassium in plant tissue is in a form that is readily soluble in water. It would appear that if it is combined at all with the protoplasm, it is easily dissociated from it. Then apparently it is present in the cells chiefly in the form of inorganic compounds and as salts of the organic acids.

Practically all investigators including Bartholomew and Janssen (1929), Johnston and Hoagland (1929), Wallace (1931), Janssen and Bartholomew (1932), Wallace and Proebsting (1933), and Hartt (1934) have found that the amount of potassium in the plant is increased by its application to the soil. In many of these observations the luxury absorption of this element was also apparent.

It was observed by Hoagland and Martin (1933) that in many cases large amounts of potassium are fixed in a nonreplaceable form almost immediately after placing in the soil. They found in experiments with barley on widely different soils that if the amount of replaceable potassium is high, all or the major portion of the potassium absorbed by the plant may be balanced by the loss of replaceable potassium from the soil. High-replaceable potassium tends to induce luxury absorption of this element by the crop. As cropping continues, the proportion of potassium derived from nonreplaceable forms grows larger until a point is reached at which no further loss of replaceable potassium can occur. At that point the solubility of the nonreplaceable forms determines the supplying power of the soil. Proebsting (1933), however, with apple trees, buckwheat, and tomatoes concluded that neither the exchangeable nor the water-soluble potassium is an accurate means of measuring the ability of the soil to supply the plant with this element.

Plants vary widely with respect to the amount of potassium needed for their maximum growth. Janssen and Bartholomew (1930) noted with soybeans, cowpeas, oats, sweet clover, and other plants, that the greatest growth in water cultures was obtained when there were 2 to 3 p.p.m. of potassium in the solution. This was below the concentration at which the maximum absorption of potassium occurred, indicating that more potassium is absorbed by plants than is needed for their use. It was found by Bartholomew and Janssen (1929) that alfalfa and Hubam clover made optimum growth in a solution containing 0.5 p.p.m. of potassium, while oats, cowpeas, soybeans, and cotton made the best growth at a concentration of 2 p.p.m. Johnston and Hoagland (1929) found that the optimal growth of the tomato plant was obtained in a flowing solution with a concentration of approximately 5 p.p.m. of potassium.

Tyner (1935) observed that corn, rape, buckwheat, peas, and sorghums were poor feeders on feldspathic potassium, while alfalfa, alsike clover, red and sweet clovers could appropriate potassium to good advantage from that source. For buckwheat growing in flowing cultures, the minimum concentration of potassium necessary for good growth was approximately 3 p.p.m., while 1 p.p.m. was sufficient for good growth of red clover. Proebsting (1933) also observed that plants vary in their ability to withdraw potassium from the soil, and therefore he considered that the only reliable method for determining the ability of a plant to obtain adequate potassium from the soil was to grow the plant upon it.

Apparently plants can use to advantage the potassium that has been placed on the leaves. Domontovitsch (1930) found that plants which had been grown in a potassium-free medium responded favorably if the leaves were painted with a 2 to 3 per cent solution of some potassium salt.

Loring and Druce (1930) found that the potassium content of the ash of the potato plant had a greater proportion of the isotope with the higher mass than did the potassium in the soil. They considered that this indicated a highly selective ability of the plant. Heller and Wagner (1931), however, could find no accumulation of the isotopes of potassium in beet tops.

2. Physiological Role.—The exact role of potassium in the plant is not known. Its function is only surmised by noting the effects on the plant in its absence.

a. Formation of Carbohydrates.—It was first pointed out by Nobbe and his workers (1871) that carbohydrates are formed normally only when the potassium supply is sufficient. The production of sugar by sugar beets, mangolds, potatoes, and other crops that form a large amount of carbohydrate is reduced by a deficiency in potassium before any diminution in the vegetable growth of leaves and stems can be observed. In an experiment with mangolds, Russell (1915) reported that when potassium was deficient in the soil 7,255 lb. of leaves produced 14,684 lb. of root, while plants growing on the same soil with a sufficient supply of potassium and with very little more leaf weight (8,508 lb.) produced 40,128 lb. of root, or nearly three times as much as the plants growing under a deficiency of potassium. In all the green plants studied by Reed (1907), potassium was apparently essential for starch formation. The lack of this element gives rise so far as can be observed to no pathological condition of the chloroplasts, and under conditions of potassium deficiency they remained normal for 2 months and even increased in number. The loss of the power of synthesis by green plants in the absence of potassium has also been observed by Stocklasa (1908, 1912) and by Weiszmann (1923), but Smith and Butler (1921) could not sustain these general observations in their work with bluestem wheat growing in a solution deficient in potassium.

As has previously been mentioned, potassium is present in comparatively large amounts in the growing tips. It was first shown by Schimper (1890) that a potassium supply is necessary for the normal development of the growing apexes. Dickson (1918) grew oat plants in a nutrient solution containing only one-tenth of the optimum potassium supply. Plants grown after this manner produced only about one-half the amount of total dry matter of stem and leaves as did those growing in a solution with the proper amount of potassium. The total dry weight of the plants including grain, however, was two-thirds that of the normal plants. The decrease in grain was thus not proportional to the decrease in the dry matter of the vegetative parts of the plant. This result is evidently due to a lack of potassium for meristematic development, and as a consequence the vegetative growth of the plant was checked.

The data obtained from later experiments on the influence of potassium on the metabolism of carbohydrates are contradictory and confusing. Nightingale, Schermerhorn, and Robbins (1930) considered that potassium either directly or indirectly is essential for carbon dioxide assimilation. Carbohydrates, however, frequently accumulate in potassium-deficient plants because nitrogen assimilation is retarded. They considered that the translocation of sugars and digestion of starch can occur freely in plants that are extremely low in potassium. Janssen and Bartholomew (1930) found a high correlation between the percentage of potassium and the total weight of the starch and sugar in the plant. Hartt (1929) found in sugar cane that potassium-deficient plants had a greater percentage of total sugars, reducing sugars, and sucrose. Greater lignification occurred in the potassium-deficient plants, while greater cutinization was found in plants supplied with potassium. Tobler (1929) found that flax fertilized with potassium sulphate had more fiber than the control plants. The fibers were larger, more angular, had more wall laminations, and had intervening, intercellular spaces. Ramie, oats, and willow showed similar modifications, but hemp was only slightly affected. Tincker (1932) and Tincker and Darbishire (1933) found in dahlias, artichoke, and *Phaseolus multiflorus* that potassium accumulated in the roots and tubers as they increased in weight. In plants the evidence indicates that in general a potassium deficiency may diminish carbohydrate and protein synthesis. Richards (1932) found a positive correlation between the amount of potassium supplied and the rate of respiration in potassium-deficient plants. This is apparently due to the increase in the amount of carbohydrate resulting from an increased potassium supply. Hibbard and Grigsby (1934) found in plants of the pea, which had been deprived of potassium and calcium, that simple sugars, sucrose, starch, hemicellulose, and protein were produced but in less quantity than by plants growing in complete nutrient solutions. They believed that the deficiency of the two elements disturbed and retarded protoplasmic activity.

On the other hand, Lilleland (1930) could find no indication that the content of potassium in the soil had any influence on the sugar content of prunes. Janssen and Bartholomew (1932) found that cowpeas grown in a low-potassium nutrient solution generally were higher in reducing sugars, and total sugars, but lower in starch than plants grown in high-potassium nutrient solutions. Thick cell walls were associated with low potassium in the plant. The percentage of protein was usually greater for the potassium-deficient plants. Houghland and Schricker (1933) concluded that there is only a slight possibility of altering the starch content of the potato by varying the potassium supply. Phillips, Smith, and Dearborn (1934) noted that the potassium-deficient plants of tomato were high in solids and reducing sugars.

b. Relation to Nitrogen.—Potassium is considered to play a rôle in the formation of proteins in the plant. Stocklasa (1916) considered that in the presence of sunlight and with a sufficient supply of carbohydrate and other material, protein can be formed in either the presence or the absence of potassium. In the dark, however, the formation of protein occurs only in the presence of potassium, provided the necessary materials are at hand.

Loew (1934) considered that potassium is essential to the manufacture of the protein in the plant cell. The increase of protein reserves in seeds, in conjunction with an increase in potassium content, supports this view and explains the higher potassium content of leguminous seeds as compared with that in the seeds of grains and grasses. It has been observed by Gildehaus (1931) for apples, Janssen and Bartholomew (1932) for cowpeas and sugar beets, Colby (1933) for French prune trees, and Rippel, Behr, and Meyer (1933) for potatoes that the amount of nitrogen is markedly higher in those plants that have been grown in a potassium-deficient medium than in those that have grown in a sufficient supply of this element. Nightingale, Schermerhorn, and Robbins (1930) considered that potassium is apparently essential either directly or indirectly for the initial stages in nitrate reduction in the plant and probably for the synthesis of proteins of meristematic tissue. Janssen, Bartholomew, and Watts (1934) believed that they found evidence in tomato plants that a deficiency of potassium causes an accumulation of amino forms of nitrogen. The plants that were deficient in potassium had a lower percentage of nitrogen in a water-soluble form than those with a high potassium content.

c. Formation of Oils.—Potassium apparently plays a role in the formation of oils in plants. Soybean plants grown with a deficient potassium supply produce small and immature seeds which have an oil content lower than that of the normal seeds (Schuster, 1927). It is assumed that the carbohydrates have not under these conditions become fully transformed into oil.

d. Catalyst.—Most authors are agreed that potassium plays the role of a catalyst or condensing agent. Loew (1903) believed that the superiority of potassium over sodium in plant nutrition is due to its superiority as a condensing agent. He cites numerous cases in organic chemistry where condensation processes take place only in the presence of potassium or much more rapidly than with sodium or related elements. Stocklasa (1908, 1920) considered that potassium acts as a catalyst in the formation of organic compounds. He is of the opinion that potassium is concerned through its radioactivity in photosynthesis and that there is a relationship between this process and the beta and gamma rays that potassium emanates.

Bruno (1928) considered that potassium may influence plant growth and carbohydrate formation as a result of a photoelectric effect originating in the radioactivity of potassium. Loew (1934) believed that the specific action of potassium is due to the emission of beta rays by a potassium compound in the protoplasm. These rays promote the physiological oxidation processes in the protoplasm and transform the energy in the living cell. Further evidence that potassium acts as a condensing agent is furnished by the work of Reed (1907). In *Spirogyra*, which ordinarily can be stimulated to divide, mitotic nuclear division did not proceed in the absence of a potassium supply. The cells elongated but the nucleus did not divide. The assumption for this behavior is that nucleoprotein cannot be synthesized in the absence of potassium even if the necessary constituents are present. When such elongated cells are placed in a solution of potassium, normal mitotic division proceeded in some cases, but in a majority of them it did not. This behavior may be due to an insufficient supply of nucleoprotein, but potassium may also play some important function in the initial steps of mitotic division and, being absent at that time, could not function at a later stage.

Loew (1934) considered that, in the cell nuclei, potassium promotes synthetic chemical processes.

e. Action on Enzymes.—The effect of potassium on diastatic activity has been studied intensively. In most cases, potassium appears to accelerate the action of diastase. James (1930) believed that potassium aids in increasing the effectiveness of the catalytic surface, but the manner by which it does this is not known. It may increase the surface of the enzyme by induction of a greater dispersion, by increasing its absorbing and combining power, by a change in the degree of dissociation, or by the formation of new enzyme (Briggs, 1922). Cattle (1933) found that by whatever method the plants were compared, the lack of potassium depressed the diastatic activity of the leaves. The amount of this depression varied with the age of the leaf. James and Cattle (1935) considered that the effect of potassium is to increase the amount of diastase formed. It apparently does not enter into combination with the enzyme, since colloidal precipitates from extracts of plant materials usually contain diastase, but they may be entirely free of potassium. White (1936), studying the growth of colonies of *Lemna major* in solutions deficient in potassium, found that they were characterized by a high content of starch but by a low assimilation rate. The enzyme-containing extract of colonies grown in potassium-deficient mediums showed a diminished capacity for hydrolyzing starch.

Hartt (1929, 1934) found, however, in sugar cane that the potassium-deficient plants had the greatest diastatic activity in the tops, while in the roots it was the same for all conditions of potassium nutrition. She found further that potassium stimulated the action of invertase and catalase. Cattle (1933) noted that the catalase activity is decreased in potassium-starved plants.

3. General Response of Plants to Potassium.—Plants differ to a considerable degree in their response to potassium. Hartwell (1927) found that the yield on an area with insufficient potassium in comparison with the yield on an area with optimum fertilizer ranged from 16 per cent for onions to 88 per cent for oats. He grouped the plants in reference to their response to the applications of potassium as follows: low potassium-response crops—oats, rye, wheat, millet, and carrots; medium potassium-response crops—barley, rutabagas, parsnips, potatoes, and cabbage; high potassium-response crops—tomatoes, mangels, buckwheat, corn, and onions.

Potassium has been reported to modify the general appearance of the plant; to affect the quality of fruits, grains, and vegetables; to prevent certain diseases; and to influence the intake of other elements.

a. General Appearance.—It was reported by Hartt (1929, 1934) for the cuttings of sugar cane that the symptoms of potassium starvation are decreased growth, "dieback," and a deficiency of chlorophyll. Nightingale, Schermerhorn, and Robbins (1930) noted that potassium-deficient plants die prematurely if fruit is present, because most of the potassium is withdrawn from the vegetative parts and translocated to the fruit. Kraybill (1930) considered that dead tissue around the margins and between the veins of the leaves is an accurate indicator of potassium deficiency in the tomato plant. Johnston and Hoagland (1929) in the same plant noticed a characteristic spotting of the leaves that indicated a lack of potassium, while Fisher (1935) stated that symptoms of potassium deficiency appeared after 7 to 10 days as yellowish-brown spots on the leaves near the margin, and that the spots gradually spread until they coalesced and the entire leaf became yellow. The margins of the leaves turned under, and the lower leaves were pimpled between the veins. Wallace (1930) and Davis (1930) reported that the lack of potassium increased shoot growth, leaf scorch, and the defoliation of shoots.

b. Quality.—The quality of fruits and grains has been attributed to an optimum potassium supply. Schuster (1927) considered that the effect of the application of potassium on the quality of the crop will depend upon the amount of potassium in the soil and the property of the soil to correct the effect of the acid derived by the liberation of the applied potassium. He considered that the literature on potassium nutrition produces evidence which indicates that potassium aids in producing quality in potatoes, sugar crops, tobacco, soybeans, and wheat.

Lanham (1927) found that the application of potassium had no influence on the carrying quality of tomatoes grown in Texas. He considered that the carrying quality is mainly varietal and is influenced to some extent by the time of harvesting. Both the fruit and the plants from the fertilized plots contained more potassium than the checks, but the tomatoes containing the highest amount of potassium were not always those from the plots having the highest application of potassium. Robbins, Nightingale, Schermerhorn, and Blake (1929) stated that an adequate supply of potassium appears essential for the production of "chunky" sweet potatoes. It apparently functions in the formation of a type of protein necessary for the rapid development of the cambium, which is chiefly responsible for the thickness and chunkiness of the potato.

c. Relation to Other Elements.—Brown (1927) found that in the apple low values for the hydrogen-ion concentration are associated with total ash and potassium content so that potassium and acidity are inversely correlated. A correlation seems also to exist in the apple between potassium and phosphorus. Johnston and Hoagland (1929) found that a low supply of potassium in the soil cause an increased absorption of calcium, magnesium, and phosphorus. Fonder (1929) noted in alfalfa that there seemed to be an inverse correlation between the content of potassium and calcium. Colby (1933), with French prune trees, noted under conditions of potassium starvation that the leaf tissues were higher in phosphorus, calcium, and magnesium but lower in ash than those trees that were supplied with sufficient potassium. Phillips Smith and Dearborn (1934) found that the effects of an early deficiency of potassium upon the tomato plant are a lower ash content and a higher calcium, magnesium and phosphorus content than are found in plants growing in a complete nutrient solution.

McCalla and Woodford (1935) found that the limiting of potassium to wheat plants resulted in a marked increase of calcium and magnesium in the dry matter.

d. Other Effects.—According to Brown (1929) good keeping quality of apples is associated with a high percentage of available potassium and phosphorus in the soil and with the amount of each in the fruit. Beaumont and Chandler (1933) found indications that a lack of potassium tended to make apples and peaches firmer at harvest but to make them softer during storage. Young, Janssen, and Ware (1932) and Ware and Young (1934) found that a plentiful application of potassium, alone or in combination with salts containing nitrogen and phosphorus, gave definite control of cotton wilt. Warne (1934) found that apples, gooseberries, and currants are especially susceptible to "leaf scorch" under conditions of potassium deficiency. The percentage of dry matter is higher in the scorch leaves, but the potassium content is almost always lower in the scorched than in the unscorched leaves.

Davis, Hill, and Johnson (1934) noted in strawberries that lack of hardiness was associated more markedly with deficient potassium than with any other factor. Wohack (1930) claimed that the frost-protecting effect of potassium is very evident even when it is applied late in the season. There seems to be some evidence that potassium tends to prevent lodging by stiffening the straw. Mann (1924) noted that young apple trees in sand cultures deficient in potassium had small leaves subject to leaf scorch, and that the gooseberry under similar conditions had leaves with lower water content which were less resistant to water loss than those growing in the controls. James (1931) found a significant, positive correlation between the amount of potassium and the water content of the plant throughout the growing season.

F. Sulphur

It has long been recognized that sulphur is necessary for plant development, but the amount needed by the plant was until recently underestimated. Much sulphur was lost by the ash-analysis method, which until comparatively recently was used in its determination. By more accurate methods of analysis it has been found that plants need from two to one hundred times as much sulphur as had formerly been supposed. Hart and Peterson (1911) stated that cereal crops remove from the soil about two-thirds as much sulphur as phosphorus, the legumes as much sulphur as phosphorus, and in the case of alfalfa a greater amount, while such crops as turnips and cabbage remove two to three times as much sulphur as phosphorus. In mustard, potato, onions, and chives, also, the sulphur content of the plant greatly exceeds that of the phosphorus (Crocker, 1923).

1. Occurrence.—Sulphur occurs in plant tissues chiefly in the form of proteins, volatile compounds, and sulphates. In the proteins it exists in the form of cystine. Some of the various volatile forms of sulphur are mustard oil, allyl sulphide, vinyl sulphide, and mercaptans. Those volatile compounds are most abundant in the mustard family, but some of them are to be found in the *Liliaceae* (onion and garlic), *Tropaeolaceae*, and other families. The mustard oils do not occur as such in the plant but are found in the form of glucosides, the best known of which are sinigrin found in black mustard and sinalbin found in white mustard. Sulphur occurs in the inorganic form in plants principally as sulphates, and as much as 65 per cent of the sulphur found in certain plants is in the form of soluble sulphates. Plants cannot utilize elemental sulphur; apparently it must be oxidized to the form of sulphates before it is of any benefit to them. According to Reimer and Tartar (1919), all the sulphates produce similar beneficial results when applied as fertilizers.

It was noted by Frazer (1935) that certain coniferous plants obtained several times as much elemental sulphur from localities where there was distinct atmospheric contamination from factories as when they grew in regions not subjected to this contamination.

It is also of interest to note that, when sulphur is applied as a fertilizer in the soil, a marked increase in the amount of sulphates in the plant is soon noted. In the case of rape, radishes, and clover that had been grown on a soil with only the normal amount of sulphur present, the plants grew vigorously but showed little or no sulphates in their tissues. This indicates that a considerable portion at least of the sulphur that is found in the plant in the form of sulphates is simply an excess supply of sulphur.

The following brief table adopted from Peterson (1914) shows the relative amount of the various types of sulphur compounds in certain plants:

DISTRIBUTION OF THE SEVERAL FORMS OF SULPHUR IN PERCENTAGE OF THE TOTAL SULPHUR

Material	Soluble sulphate *A*	Volatile sulphur *B*	Soluble sulphur *C*	Insoluble sulphur *D*	Sum of fractions *B, C, D*	Loss on drying	Total unoxidized sulphur $(C - A) + D$
Rutabagas, average 4 samples	37	12	73	11	96	19	47
Cabbage	24	9	74	15	98		65
Sugar beets	34	..	62	38	100	11	66
Alfalfa hay	50	..	70	30	100	4	50
Rape, dried[1]	64	..	85	8	93	..	29

[1] Potassium, nitrogen and phosphorus, and sodium sulphate added to soil.

Sulphur is fairly evenly distributed throughout the plant. In a mature corn plant, 40 per cent of the total amount was in the leaves, 23 per cent in the stem, 26 per cent in the grain, and 11 per cent in the roots. Marsh (1922) observed that the amount of sulphur in the tomato plant increased from the tip toward the base, the oldest tissues having the highest sulphur content. In the case of the apple, the highest percentages of sulphur are found in the leaves and then in descending order in spurs, young bark, and old bark. The percentage of sulphur is at a minimum in the apple spur at the time of fruit-bud differentiation.

2. Physiological Role.—Sulphur functions as building material for the formation of protein and other constituents of the plant. The proteins of plants contain from 0.003 to 7.2 per cent, and it has been noted that plants with a high protein content or especially rich in organic sulphur compounds respond most markedly to the application of sulphur as a fertilizer. In Washington (Olson and St. John, 1921; Peterson, 1914) alfalfa grown on soils treated with sulphur was found to contain a higher percentage of protein than alfalfa on soils not so treated. This indicates that the sulphur under these conditions was used by the plant in the production of protein. This supposition is strengthened by the work of Tottingham (1918) on red clover, which indicated that a deficiency of sulphur restricted growth in that plant by limiting the synthesis of protein.

3. Formative Effects.—Aside from functioning as building material, sulphur produces certain formative effects in plants. These are manifested especially in three ways.

(*a*) An increased root or absorbing system. The effect of sulphur fertilizer on the root system of plants has been noted by Hart and Tottingham (1915), Pitz (1916), Miller (1919), Reimer and Tartar (1919), and others. It was found in the case of alfalfa that the root systems of plants which have been fertilized with various sulphur fertilizers were from two to three times as large as those of the unfertilized plants. This holds true also for clover and rape. It was suggested by Crocker (1923) that the increased root development which has been so frequently observed from the application of superphosphates and attributed to phosphorus might be due, in part at least, to the sulphates therein rather than entirely to the phosphates.

(*b*) The appearance of chlorophyll. It has been observed that plants grown on soils deficient in sulphur have a pale-green color, and this can be remedied by the application of sulphur as fertilizer. The plants thus treated show a dark, vigorous, green color (Eaton, 1922; Demolon, 1912). Powers (1930), in working with potatoes, corn, and tomatoes that had been dusted with land plaster or grown in soils that had been treated with sulphur, found that they developed a darker green color than the controls. He found that the amount of chlorophyll in alfalfa plants was increased as much as 18 per cent by the application of sulphur. The application of this element tended to increase the length of stems of this plant, to thicken the leaves, and to prevent their shattering in haying. Storey and Leach (1933) reported that the "tea-yellows" disease is produced by a deficiency of sulphur in relation to other elements. This disease may be prevented in the field by an application of either sulphates or elemental sulphur.

Just what role sulphur plays in this regard, however, is not known. This element is apparently necessary for the formation of chlorophyll, but it must act in an indirect manner, since it does not enter into the composition of the pigment.

(*c*) Effect on the nodules of the legumes. Duley (1916) was the first to note an increase in the number of nodules on the roots of sweet clover by the application of flowers of sulphur as a fertilizer. Pitz (1916) found that calcium sulphate when added to soil apparently had no marked effect on the total number of bacteria that grow on agar plates and did not increase to any extent ammonification or nitrification. The application of sulphur did, however, increase the growth of the legume bacteria and the number of nodules on red clover. Powers (1923) noticed in clover that the plants fertilized with sulphates had a higher nitrogen content than the controls. Since the number of nodules on the roots was increased, and since rape, which assimilated a large amount of sulphur, did not show any relationship between the content of nitrogen and sulphur, he attributed this increased nitrogen content of clover to the stimulating effect of the sulphates on the action of the legume bacteria.

4. General Effects.—It was noted by Nightingale, Schermerhorn, and Robbins (1932) for tomato, apple, narcissus, and asparagus, and by Eaton (1935) for soybeans, sunflower, rape, kale, and mustard that plants deficient in sulphur have the general appearance of those which have been completely deprived of nitrogen. These plants are extremely high in carbohydrates and contain much more nitrate than the plants receiving a sufficient sulphur supply. The rate of nitrate reduction and the oxidation of sugars is reduced but not totally inhibited. The digestion of starch and the translocation of sugars take place freely. Such plants have thick cell walls and a relatively high proportion of fibers and lignified tissue. The protoplasm appears much like that of plants lacking nitrogen in that it is limited in amount but not noticeably injured as it is when phosphorus, calcium, or potassium is deficient.

5. As a Fertilizer.—Sulphur is present in the soil in the form of either sulphates or organic matter. Most of the sulphates are quite soluble and are not readily adsorbed, so they are leached out rapidly, and only small amounts are found in the soil. In a large number of soils analyzed by Woodard (1922), the sulphur content was less than the phosphorus content. He estimated that the highest sulphur content found in these soils was sufficient for 39 years of alfalfa, 355 years of wheat, and 232 years of corn, while the lowest sulphur was sufficient for only 5 years of alfalfa, 46 years of wheat, and 30 years of corn. Good to excellent crops of alfalfa remove annually from 37 to 90 lb. of sulphur per acre from the soil, and, in general, hay from fields with the heaviest crops have the highest percentage of sulphur (Hall, 1922). The response of plants to the application of sulphur as a fertilizer will depend upon the type of plant and upon the soil. In many regions the sulphur supply is not a limiting factor in production, as shown by Bruce (1925) for Kansas soils and by Stewart (1920) for Illinois soils.

Hart and Tottingham (1915) working in Wisconsin found that plants most affected by the application of sulphates to the soil were members of the legume and mustard families. They observed, however, a noticeable stimulation to seed production in barley and oats but no increase in the quantity of straw. In the case of clover, they obtained an increase in dry matter due to calcium sulphate of about 23 per cent. With rape, the greatest increase occurred where the calcium sulphate was applied with a complete fertilizer, giving an increase of 17 per cent over the complete fertilizer. In general, the calcium sulphate was more effective than the more soluble sodium sulphate.

Shedd (1914, 1917) reported that wheat, corn, oats, timothy, alfalfa, rape, beets, potatoes, turnips, radishes, beans, tobacco, and grapes were benefited by the application of sulphur to certain Kentucky soils. He found that the sulphur content of 32 varieties of tobacco averaged 0.458 per cent of the total dry matter as compared to a phosphorus content of 0.302 per cent. Olson and St. John (1921) in Washington obtained increases of twofold or more in alfalfa yields by treating the soil with gypsum, acid phosphate, and sulphuric acid, while Powers (1923) from Oregon showed that marked results could be obtained by the application of sulphur for alfalfa, red clover, and alsike, but only a moderate response in yield was had with wheat and potatoes and little or no increase for corn, kale, or sunflowers.

G. Iron

1. Occurrence.—Iron is necessary for the growth and development of green plants and cannot be replaced by any other element. Although iron is universally distributed in plants, its detection microchemically is difficult, since it occurs in very small quantities. Molisch (1892) found that combined iron occurs in plants in the greatest amount and that very little of the iron is in a form that is readily soluble in water. He observed combined iron in the protoplasm, the vacuole, and the globoids of the aleurone bodies. Jones (1920) stated that, as a whole, plant tissues give the iron reaction more rapidly and intensely than animal tissues. The iron stain occurred in the chloroplasts, nuclei, and in large masses scattered throughout the cytoplasm. The leaves of some plants did not show the iron stain, while others did with a very definite reaction.

Plants vary widely in the amount of iron that they contain. Thus Stiebeling (1932) found that apples, cucumber, and watermelons had the least amount, containing less than 0.0004 per cent of this element. Bananas, celery, carrots, onions, and tomatoes had from 0.0004 to 0.00079 per cent, asparagus, beets, dock, and sorrel had from 0.0008 to 0.00159 per cent, while cowpeas, dandelion leaves, parsley, spinach and water cress contained over 0.0016 per cent. The iron in plants apparently occurs in two forms designated variously as "active and inactive," "free and combined," "soluble and insoluble," and "available and nonavailable." Sayre (1930) believed that the iron in corn stems is in the form of an oxide or hydroxide in a solid solution in some host substance.

2. Physiological Role.—If plants are deprived of iron, they do not normally produce chlorophyll, a fact that was first noted by Gris (1844). The new leaves and stems formed by plants deprived of iron are pale or yellow, a condition that has been termed "chlorosis." It has been definitely proved, however, that iron does not enter into the composition of chlorophyll (Willstätter and Stoll, 1913). The quantity needed by the plant at any time is very small. Gile (1916) estimated the available iron in the nutrient solution in which he grew normal healthy rice plants to be always less than 1 part per 10,000,000 parts of solution. The use to which iron is put in the plant is not known. Wolff (1913) and others considered that it acts in plants as a catalyzer, since such marked results are obtained with such small amounts, while Warburg (1925) has proposed the theory that iron is the oxygen-carrying component of the respiration ferment.

Hopkins (1930) believed that the concentration of the iron ions plays an important role in the cellular processes involving biological oxidation.

Oddo and Pollacci (1920) grew plants of *Zea mays*, *Solanum nigrum*, *Datura stramonium*, *Euphorbia sp.*, and *Aster sinensis* in a standard nutrient solution lacking iron but to which was added the magnesium salt of pyrrolecarbonic acid that has the formula

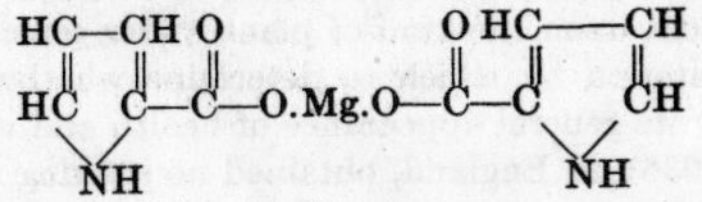

in a concentration equivalent to 0.0232 g. magnesium in 1 l. of water. These plants, although deprived of iron, when supplied with this organic compound formed chlorophyll. Since iron under ordinary conditions is indispensable for green plants, these authors suggested that the formation of chlorophyll in the presence of iron is due to the catalytic action of this element in the formation of the pyrrole nucleus, which in itself is the center of the chlorophyll complex. They believed that if this nucleus is already formed, the presence of iron is not necessary for the greening of the plant. This observation is in line with the statements of Willstätter (1906) and Mameli (1918) that the function of magnesium in the greening of the plastids is directly proportional to the presence of pyrrole. Deuber (1926), however, was unable to confirm the work of Pollacci and Oddo in this regard. He grew plants of corn, cowpea, soybean, and *Spirodela* in nutrient solutions containing a magnesium salt of pyrrolecarbonic acid substituted for iron. In no case did this compound prevent chlorosis of the leaves of the plant. He found further that the pyrrole salt in concentrations of 0.001 to 0.250 g. per liter was toxic to the plants used and that when the salt was applied as a paint to the leaves of cowpeas it had the same injurious effects as in solution.

Pollacci (1935), however, reported that *Chlorella vulgaris*, and other algae, thrived and remained green in cultural solutions devoid of iron but containing magnesium pyrrolecarboxylate instead. He considered that his results confirmed the previous conclusions of Oddo and Pollacci that the magnesium compound can replace iron in the nutrition of green plants. Oserkowsky (1933) noted that iron may be present in larger amounts in chlorotic leaves than in green ones. He believed this indicated that a specific kind of iron, which he designated as "active iron," is concerned in chlorophyll formation. He found that the chlorophyll content of leaves from chlorotic plants is proportional to the amount of active iron in the leaves. He considered that this active iron is in the ionic form or is in a compound that readily yields ionic iron.

3. Form and Amount Necessary for Plant Growth.—This is a subject that formerly was considered of little import in the nutrition of plants. All the directions for nutrient solutions stated that only a trace of iron was necessary and that it mattered little in what form it was presented to the plant, provided it was soluble in water. Quite recently, however, much work has been done on the question of iron nutrition, which shows that not only the quantity of iron available but the form in which it is presented to the plant can exert a marked influence upon plant growth and development. These two factors in turn are dependent upon the composition of the medium or nutrient solution—its hydrogen-ion concentration, the nature of the general environment—and to some extent upon the kind of plant under consideration. There is at the present time no criterion by which to determine whether a plant is receiving sufficient iron except by its general appearance of health and vigor.

Moore and Jones (1936), in England, obtained no striking differences in yield by the application of iron to potatoes and beets, but the plants on the treated plots had darker green leaves and a healthier appearance than those of the control plots. Gile (1916), however, remarks in this regard that although the color of the leaves will indicate a marked deficiency of iron, a slight deficiency may materially diminish the yield without appreciably affecting the appearance of the plant.

Jones and Shive (1921) used ferric phosphate and ferrous sulphate as a source of iron in Shive's nutrient solution (*R5C2*, 1.75 atmospheres) in which Marquis wheat was grown. The solution was renewed at intervals of about 4 days for a period of 90 days. The results obtained are shown in the following table:

Iron per liter of solution, milligrams	Series I $FePO_4$, dry weight, grams	Series II $FeSO_4$, dry weight, grams
0.01	4.55	10.95
0.10	5.75	14.65
0.25	8.20	15.95
0.75	12.05	27.40
1.00	14.90	18.75
1.50	16.25	19.15
2.00	20.50	23.50
3.00	18.00	20.90
5.00	16.80	18.45

Control no Fe..........8.40

In series I the plants showed more or less chlorosis even when large amounts of iron were supplied, a condition which could easily be remedied in a day or two by the introduction of ferrous sulphate. There was no chlorosis in series II except when small amounts of iron were used. Ferrous sulphate, however, was apparently toxic in higher concentrations.

That plants may differ in regard to their utilization of iron was shown by Corson and Bakke (1917). They found a marked difference in the growth of wheat plants when the source of iron was ferric phosphate and ferrous sulphate, the former being the more efficient. Little or no difference, however, was observed in the behavior of Canada field peas to these two sources of iron. They considered that the iron stored in the seed might be a determining factor, a fact which was also suggested by Tottingham and Beck (1916) in their observations on the response of wheat plants to the salts of iron and manganese. Hartwell and Pember (1908) noted that ferrous sulphate was equally effective as a source of iron for barley and rye seedlings growing under the same conditions.

Marsh and Shive (1925) studied the relation of the growth of soybean to the amount and form of iron in several types of cultural solutions. The iron compounds used were ferric glycerophosphate, soluble ferric phosphate, ferric tartrate, and ferrous tartrate. Considering that the availability to the plant of iron in cultural solutions is closely correlated with the solubility of the iron compound in the solution, their data suggest that in a general way ferric glycerophosphate could be used to advantage in solutions in which the hydrogen-ion concentration is kept at a high level. Soluble ferric phosphate and ferrous sulphate are not very usable when the pH values approach the neutral point. In solutions of this type they suggest that ferric tartrate should prove the most effective source of iron for the plant. Marsh and Shive further found that if the iron supply is adjusted from day to day, large healthy plants are produced regardless of the type of cultural solution or the iron compound used. The iron in healthy plants was present in a smaller percentage, based on dry weight, than in the chlorotic plants or those suffering from iron toxicity. The iron in healthy plants, although low in concentration, seems to be uniformly distributed throughout the stem and leaves, while in the case of those suffering from toxicity of iron it is high in concentration. The iron content of chlorotic plants is low in the leaves but high in the stems. These authors suggest that in order to maintain the plant in healthy green condition, the supply of soluble iron in the cultural solution must be kept at as low a concentration as possible without inducing chlorosis from lack of available iron. A concentration of soluble iron slightly above this optimum may result in toxicity due to a too high content throughout the plant. Chlorosis, however, may occur due to the lodgment of iron in the roots and stems, thus preventing the translocation of this element to the leaves in sufficient amount (Hoffer, 1920). Marsh and Shive considered that the general appearance of the plant must serve as an index of the iron supply in each case, since the changing plant environment has pronounced influence upon the iron requirement.

Gines (1930) used six different types of iron salts in growing rice in nutrient solutions and found that each of these was capable of furnishing sufficient amounts of iron for the needs of the plant. A striking observation in this regard was noted by Gericke (1925) in the effects of two different light intensities on the growth of wheat in nutrient solutions devoid of iron. It appeared from the general behavior and appearance of the plants that the need for iron is much greater per unit of growth of plants exposed to high light intensities than to low ones. According to Ingalls and Shive (1931), the soluble-iron content of plants varies directly with the variation in the hydrogen-ion concentration brought about by changes in light intensity from day to night. Low hydrogen-ion concentration corresponds to high light intensity and high hydrogen-ion concentration to low light intensity. Plants in which the tissue fluids have low hydrogen-ion concentration values show high total and relatively

Plant	Leaf		
		Iron per gram of dry matter	
	Acidity, pH	Total	Soluble
Buckwheat	5.10	0.278	0.0477
Tobacco	5.74	0.325	0.0390
Tomato	5.87	0.297	0.0277
Asparagus	5.87	0.373	0.0256
Soybean	6.01	0.469	0.0246
Clover	6.10	0.571	0.0281

low soluble-iron content. Those plants in which the tissue fluids have high hydrogen-ion concentration show low total and relatively high soluble-iron content, as shown by the table on page 331.

The same conditions that in the presence of iron in the cultural medium are decidedly beneficial to growth become relatively harmful in the absence of iron. This observation is an important one, for it not only indicates that there can be no fixed minimum amount of iron for a unit yield of product, but it suggests that this may apply to other elements as well.

4. Relation to Hydrogen-ion Concentration.—Jones (1920) and Jones and Shive (1924 to 1923) replaced potassium nitrate with ammonium sulphate in Tottingham's solution and noted that the solution became more acid when plants were growing therein, changing from pH 4.8 to 4.2 during a 3½-day period, while the regular Tottingham solution became less acid during that time. One milligram of iron per liter in the form of ferric phosphate was not sufficient to supply the needs of plants in Tottingham's solution, but one-half milligram was ample when ammonium sulphate replaced the potassium nitrate. In Tottingham's solution 0.25 to 0.50 mg. of iron in the form of ferrous sulphate was sufficient, but in the modified solution this amount was toxic to the plants. McCall and Haag (1921) observed that wheat plants growing in solutions having pH values ranging from 4.02 to 7.0 suffered from a lack of available iron or from a change in metabolism that resulted in the immobility of iron within the plant. This observation is in line with that of Hoffer (1920), who considered that the rate of growth of young plants of corn and the degree of chlorosis may be determined by the relative ability of the plant to develop and maintain a sufficiently high hydrogen-ion concentration to keep the iron compounds mobile as occurs in normal green plants.

The solubility of ferric citrate, ferric phosphate, ferric sulphate, and ferrous sulphate in Tottingham's solution within a pH concentration of 4.2 and 6.0 was investigated by Tottingham and Rankin (1923). Using the growth measurements of young wheat plants as an indicator, the results showed that ferric citrate was the most favorable form of iron used. It remains soluble over a considerable range of pH values of the nutrient solution, and, when supplied at the rate of 10 mg. of iron per liter of nutrient solution, although it is not completely dissolved, it provides abundant iron for plant growth when the solution is renewed every 4 days. Reed and Haas (1924) found that the addition of certain organic compounds to an alkaline nutrient solution, in which ferric tartrate was the source of iron, increased the amount of soluble iron in the solution. The sodium and potassium salts of organic acids, *e.g.*, citrates and tartrates, effect marked solubility of iron without appreciably altering the pH of the solution. Starch and sugars added to the solution also dissolved appreciable amounts of iron. Deuber (1926) in using potassium ferrocyanide in concentrations equal to 0.033 and 0.066 p.p.m. of iron noted that the higher concentration produced a stoppage of growth in soybeans. Ferric ferrocyanide was a satisfactory source of iron when the solution had a reaction of pH 5.0, but, at a less acid reaction, growth of these plants and chlorophyll development were restricted.

A direct correlation was found by Barnette (1921) and Barnette and Shive (1923) between the decrease in hydrogen-ion concentration of the cultural solution and the appearance of chlorosis in plants when ferric phosphate and ferric sulphate were used as sources of iron. They could partially control the hydrogen-ion concentration of the solution in contact with the roots of the growing plants by the use of a large volume of solution renewed frequently. Hopkins and Wann (1926, 1927) also observed that iron was unavailable to *Chlorella* in nutrient solutions at the higher pH values. The addition of sufficient sodium citrate, however, kept the iron in solution even at alkaline reactions for an indefinite period. If too high an amount of sodium citrate is present, the ionization of the ferric citrate is decreased, so that although there may be a large total amount of iron present little or no growth will occur because of the low concentration of the ferric ion. The availability of iron is determined apparently in as large a measure by the composition, concentration, and general reactions of the cultural solutions, as it is influenced by the action of the plants growing therein.

It was noted by Rogers and Shive (1932) that the xylem always showed the lowest and the phloem the highest pH values of the plant tissues. The sclerenchymatous tissue had pH values that were the same as or slightly higher than the xylem. Small amounts of iron were always present in the xylem parenchyma, but in no instance did pronounced accumulations of iron occur there. Whenever marked accumulations of iron were present, they usually occurred in tissues with a high pH value, such as the phloem where it bordered on the xylem or other tissues with a low pH value. The iron in these accumulations can be regarded for the most part to be in the precipitated form and immobile, so that it does not function in the metabolism of the plant. It seems logical, therefore, to conclude that the xylem is the tissue in which the upward translocation of iron occurs. No accumulation of iron was found in plants with low pH tissues throughout. In such plants, the content of iron is low and uniformly distributed. The roots seemed more acid than the stem. They should thus readily absorb iron if it is available in the soil since most of them showed pH values at or above pH 6.0, the critical value for the precipitation of iron. The iron may be precipitated before it reaches the chlorophyll-bearing parts. Certain plants with a high pH of tissues may show little tendency toward chlorosis because apparently they contain certain organic solvents in the presence of which iron in the mobile form resists the precipitating action of high pH reactions. Oxalic, tartaric, and citric acids are some of the compounds that will prevent the precipitation of iron.

Oserkowsky (1932) in pear trees found no significant difference between the hydrogen-ion concentration and the iron concentration in the extracted tracheal sap of green and chlorotic branches from the same environment. Bennett and Oserkowsky (1933) found that the tracheal sap in several instances contained more copper than iron but no special relation of copper to the functioning of iron was indicated.

5. Absorption from the Soil.—Certain plants, especially rice and pineapple, on certain calcareous and manganese soils are affected with chlorosis and do not make a normal growth. This phenomenon has been encountered and studied most extensively on the calcareous soils of Porto Rico and the manganese soils of Hawaii. Gile (1911) applied a solution of ferrous sulphate to calcareous soils, but this showed no effect in reducing the chlorosis of the leaves of the pineapple. Crystals of the salt, however, applied in the soil in close proximity to the roots gave very evident results in reducing chlorosis. In another experiment (1916), rice plants were grown in acid, neutral, and alkaline solutions with 0.008 g. of iron per liter in various forms. The plants grown in the acid solution contained the highest percentages of iron. Plants in the neutral solution contained more iron than the alkaline solution when some forms of iron were used and an equal amount when others were used. The percentage of nitrogen-, phosphorus-, calcium-, magnesium-, and carbon-free ash remains the same in all plants regardless of the reaction of the solution. From these and other similar experiments, Gile (1911, 1916), Gile and Carrero (1920), Mazé (1912), Sidorin (1914), and Willis and Carrero (1921) agreed in considering that lime-induced chlorosis is due to a depression in the available iron. They are inclined to believe that the carbonate of lime in the soil reacts with the iron forming ferric carbonate and thus depressing the availability of the iron for the plants. In the manganese soils of Hawaii, Johnson (1924) has shown that it is present mainly in the dioxide form. He considered that the manganese dioxide either present as such or formed from manganese salts keeps the iron that is present in the soil oxidized to the ferric form, which is not readily available to the plants. Allyn (1927) believed that lime-induced chlorosis is not a result of iron becoming locked up in the soil, but rather that it is the result of a disturbance in the metabolism of iron after it has entered the plant.

The lack of iron available in the soil for pineapples has been overcome by spraying the plants with iron solutions. The most effective and economical spray that has been used in Hawaii is an 8 per cent solution of ferrous sulphate, although other forms of iron may be used. Plants that are already yellow are sprayed three or four times at intervals of a week and then as often thereafter as is necessary to keep them green and healthy. The cost of the iron spray above mentioned amounts to about 60 cts. per acre per spraying (Johnson, 1916). It is worthy to note that leaves nicked or injured recover from chlorosis more quickly than do the intact ones after being sprayed.

Burke (1932) reported that spraying trees with ferrous sulphate or applying it to the soil did not improve their chlorotic condition, while treatment internally with iron salts in the solid state or in solution gave beneficial results. Wallace (1935) obtained good results in overcoming chlorosis by using dry, powdered citrate according to the method of Burke. Finch, Albert, and Kinnison (1933, 1934) remedied chlorosis in certain plants in Arizona by the use of iron in various ways.

6. Immobility within the Plant.—It has been observed that when plants which have their needed iron supply are deprived of it, the new stem and leaves that develop are chlorotic or almost pure white so that there is a marked and abrupt contrast between the green leaves and the colorless ones. This phenomenon was observed by Molisch (1902) in squash, sunflower, corn, pineapple, and rice. The leaves of plants growing in an iron-free solution begin to die at the top in marked contrast to the behavior of plants that are deprived of or have a limited supply of nitrogen, potassium, or phosphorus, since in that case the lower leaves dry up while the uppermost ones are the last to perish. These observations point to the fact that iron is immobile in plants and is not readily withdrawn from where it occurs to newer organs. The immobility of iron was further demonstrated by Gile and Carrero (1916). The young leaves of rice plants growing in a solution lacking iron were brushed at the tips with a 0.2 to 0.4 per cent solution of ferrous sulphate. The tips of the leaves just emerging from the stalk were used, and, after they grew out, the part that had been brushed was a normal green, while the lower unbrushed part of the leaf was strongly chlorotic and remained so until treated with iron. Gile and Carrero found that spraying with iron compounds completely restored the green color of the pineapple leaves but that the new leaves formed after the treatment were chlorotic.

H. Silicon

1. Occurrence.—Silicon is an element that has an almost universal occurrence in plants, but the amount present varies greatly for different plants and for different parts of the same plant. Richardson (1920) considered that, for plants of all types and for all parts of plants, silicon is the greatest variable of all the elements that compose them. It is especially abundant in the aerial parts of *Equisetum*, the *Gramineae*, and in the bark of trees. Silicon has been found to compose 71 per cent of the ash of *Equisetum telemantia* and 51, 67, and 46 per cent, respectively, of the straw of barley, wheat, and oats. On the other hand, the silicon content of the stem and leaves of red clover and alfalfa equals only 2.7 and 9.5 per cent, respectively, of their ash content. The distribution of silicon in the different parts of the plant is shown in the analysis of the corn plant at the dent stage (Latshaw and Miller, 1924). The percentage of silicon present in the different parts was as follows:

Plant part	Per cent of silicon in dry matter	Plant part	Per cent of silicon in dry matter
Leaves	2.50	Grain	0.02
Stem	0.45	Cob	0.14

The deposits of silica are more frequent in tropical plants than in those of the temperate zone. As a result of the higher soil temperatures, silicic acid is dissolved and taken into the plant with the water of transpiration, but it is not used in plant metabolism (Frey-Wyssling, 1930). He considered that the deposits of silica and calcium in plant tissues are not excretions but are accumulations of materials that are of no immediate use to the plant.

2. Physiological Role.—The large amounts of silicon present in many plants led early to the belief that it might be an essential element in the nutrition of these plants. It has been shown, however, that those plants which normally contain such large amounts of silicon can be grown to maturity in nutrient solutions that have only a trace present. Sachs (1887) and Knop (1861 to 1862) grew corn and Kreuzhage and Wolff (1884) grew oats in nutrient solutions lacking silicon and the resulting plants contained only a fraction of the normal content of this element. Jodin (1883) grew corn through four generations without the application of silicon. Gile and his workers (1915 to 1916) found that rice that normally contains about 10 per cent of silicon in the dry matter contained only 0.2 per cent when grown in a silica-free nutrient solution. It is now known that these plants were not grown in the absence of silicon for the amount of silicon in the seed, accruing from the glass container, or occurring as impurities in the water or chemicals used, may be sufficient to supply the physiological demands for this element.

Although plants can be grown to maturity in the absence of silicon or at least when only a trace is present, certain benefits that we shall now discuss accrue to the plant by its presence. Palladin (1927) considered that silicon plays an important role in protecting plants from the attacks of various parasites. He stated that wheat and rye grown in nutrient solutions deficient in silicic acid often suffer very severely from rusts. He also considered that silicated cell walls are a protection against the attacks of plant lice and mentioned an experiment with *Lithospermum arvense* grown in a nutrient solution lacking silicic acid. Plants in this solution suffered much more than control plants grown in solutions in which an ample amount of silicon was present. Sommer (1926) reported that seeds of millet grown without silicon were badly infected by fungi and germinated very poorly with and without disinfection, while the seeds of plants grown with silicon were not infected by fungi and germinated well.

King and Davidson (1933) obtained an increased growth of diatoms by the addition of silicate to the cultural solution, but a high concentration had an unfavorable effect upon their growth. It was found by Germar (1934) that silica is deposited for the most part in the epidermis of the leaves of rye, wheat, and barley. If the transpiration is low, however, the silica is deposited for the most part in the stems. Strong illumination, a deficiency of nitrogen, and an excess of potassium favor the accumulation of silicon but the supply of phosphorus does not influence its intake. The leaves of cereals well supplied with silicon are more resistant to infection by mildew, *Erysiphe graminis*. This resistance is correlated with silicon deposition in the epidermis, which makes the layer more resistant to the enzymic penetration of fungi. Resistance to fungi that enter through the stomata is not increased by the infiltration of silicates.

The effect of silica upon the phosphorus nutrition of plants has been extensively studied. Kreuzhage and Wolff (1884) in cultural-solution experiments with oats observed that the total growth of these plants was not much increased by the presence of soluble silicates but that the proportion of grain formed was constantly raised. This effect is very similar to that brought about by the addition of phosphoric acid to cultural solutions which are deficient in that element, so they concluded that the action of silicon and phosphoric acid is in some way related.

Hall and Morison (1906) reported the results obtained from plots at the Rothamsted station which had been treated with soluble silicates during a period of more than 40 years. One series of plots received sodium nitrate with various combinations of fertilizers so as to provide plots receiving (1) nitrogen alone, (2) nitrogen and phosphoric acid, (3) nitrogen and potassium, and (4) a complete fertilizer. From 1864 to 1904, one-half of each of these plots was cross-dressed with sodium silicate. The effect of sodium silicate on the yield of barley for a period of 41 years is shown in the following table:

Plot	Fertilizer per acre					Yield per acre			
	$NaNO_3$, pounds	Acid phosphate, pounds	K_2SO_4, pounds	Na_2SO_4, pounds	$MgSO_4$, pounds	Grain, bushel		Straw, hundredweight	
						Without silicate	With silicate	Without silicate	With silicate
1	275	...	...	...	...	27.3	33.8	16.2	19.8
2	275	350	...	...	...	42.2	43.5	24.6	25.8
3	275	...	200	100	100	28.6	36.4	17.9	21.7
4	275	350	200	100	100	41.2	44.5	25.3	27.6

The table shows that the effect of the sodium silicate is chiefly on those plots which are low in phosphoric acid, and it appears to ripen the grain in a similar but not so striking manner as the application of phosphorus.

Hall and Morison concluded that the increased and earlier grain formation observed in the presence of silicates is due to an increased assimilation of phosphoric acid by the plant, but that there is no evidence that the function of phosphorus can be replaced by silicon. Jennings (1919) increased the dry weight of wheat seedlings from 18 to 29 per cent by the addition of 1 per cent silica to the culture solution in which they were growing. He also showed that the silicon content of these plants was increased as much as 10 to 15 per cent over those grown in the solutions containing no silica. Lemmerman and his workers (1922 to 1925) found that yields were increased by the application of silicates particularly where there was an absence of phosphorus but where potassium and nitrogen were applied in the requisite amounts. They found that colloidal silicic acid was the most favorable form in which to apply the silicon. These workers considered that silicon does not replace phosphoric acid in the nutrition of the plant but increases the amount of phosphoric acid that can be taken up from the soil and the efficiency with which it is utilized. Gile and Smith (1925) observed in experiments with millet grown in sand cultures that the addition of silica gels greatly increased the growth of plants supplied with acid phosphate. The growth of the plants was approximately proportional to the quantity of phosphoric acid in the plants but bore no relation apparently to the quantity of silicon taken up. These authors ascribe the beneficial action of the silica gels to those plants supplied with rock phosphate to the increased quantity of phosphoric acid. Schollenberger (1922), however, found that plots treated with various silicon compounds and soluble phosphates furnished no data on which to base conclusions as to the action of silica in phosphorus assimilation. Brenchley, Maskell, and Warington (1927) found that, under control conditions in water cultures in the absence of phosphorus, the addition of soluble silicates brought about a significant increase in the dry weight of barley. This addition of silicates caused an increase in the height of the main stem, hastened the development of leaves, and was the most marked in phosphate-free solutions. The soluble silicates were more active in causing the increase in dry weight than the glass silicates. In the heaviest application of silica the shoots contained ten times as much silica as they did in solutions that did not contain it. These authors considered that there is a variation in the response of mustard and barley to silicates on different types of soil. They suggested that silica might act within the plant by unlocking phosphate from relatively quiescent parts of the plant and enabling it to be transferred to regions where assimilation and growth are active.

Weihe (1931), although admitting that cultivated plants required silicic acid in small amounts, considered that the increases in yield obtained by its application are not to be explained entirely from the favorable effects of silicon inside the plant. He considered it probable that the finely divided silicic acid, as a result of the large surface exposed, causes adsorption phenomena that increase the water capacity of certain soils and thus improve the nutrition of the plant. Sreenivasan (1934) believed that the peptizing action of colloidal silica may facilitate the uptake of phosphorus from the more insoluble, mineral phosphates.

3. Lodging.—The term "lodging" refers especially to the plants of the small grains that have completely or partially fallen over. This condition is generally caused by a bend in the stem which usually occurs near one of the lower nodes. This bending occurs below the node if lodging takes place early in the season while the leaf sheaths are yet closely enveloping the stem; it occurs above the node if lodging develops after the sheaths have withered and become loose around the stems. The part immediately above the node is the smallest in diameter of any portion of the stem and during the period of development it is usually the highest in moisture content. The plant thus lodges because under certain environmental conditions some portion of the lower stem does not possess sufficient mechanical strength to support the parts above it. In the last analysis, lodging results from too low an amount of dry matter per unit length of culm.

Liebig considered that lodging is caused by the lack of sufficient silicon to properly strengthen the straw. Sach observed that lodging usually developed when the stand of plants was too thick, so that the culms partly shaded one another and took on some of the characteristics of etiolation. He believed that one of the causes of lodging is the lack of sufficient lignification of the stems. Pfeffer (1900) considered that the stiffness of straw is not due to silicon but that it depends upon the development of the internodes under the influence of illumination and exposure.

The causes of lodging are now known to be much more complex than formerly supposed, because lodging is the resultant of numerous and varied factors. Some of these are: fertility of the soil, rate of seeding, climate, heredity, and fertilization.

Lodging is frequently associated with hypernutrition, especially of a nitrogenous nature, and is very prevalent on fertile soils. Under the conditions of hypernutrition there is more tillering, a high proportion of straw growth, and thus more shading. This results in a relatively small amount of dry matter per unit of culm length and a low carbohydrate/nitrogen relation. Under such conditions there is a relatively low amount of polysaccharides per unit of stem. The amount of lignin present seems to vary with different plants or under different conditions. Welton (1928) considered that the stems which lodged were relatively low in lignin, while Davidson and Phillips (1930), and Phillips, Davidson, and Weihe (1931) found that the plants which lodged contained more lignin than those which did not lodge. Prutzkova (1932) and his workers, however, reported that the varieties of wheat which were subject to lodging contain a lower amount of lignin than those which did not lodge. There is much disagreement, however, as to the significance of lignin in contributing to the strength of tissues, so that its importance in relation to lodging may have been overestimated.

A high rate of seeding or the use of varieties that tiller prolifically results in a relatively thick stand that induces shading. Shading has to some extent the same general effects as hypernutrition. A season that is warm, cloudy, and rainy tends to induce a low carbohydrate/nitrogen relation in the plants and thus causes lodging. A relatively infertile soil, a low rate of seeding, and a cool, clear season are factors that tend to reduce the vegetative growth of plants and thus produce a relatively large amount of dry matter per unit length of the culms. Under such conditions a relatively high carbohydrate/nitrogen relation occurs, which results in thick cell walls and a relatively wide band of hypodermal tissue that stiffens the straw and makes the plants less susceptible to lodging. Gassner (1910) reported that if certain strains of oats were germinated at 25°C. they lodged very markedly in about 5 weeks after they had been transplanted in the field. Prutzkova (1932) found that varieties of wheat not subject to lodging contained higher amounts of ether-soluble substances, hemicelluloses and celluloses, and smaller amounts of sugar, than the lodging varieties. There was no difference, however, in the quantity of ash and nitrogen contained in the two types of plants.

Heredity plays some part in the lodging of grains. It is well known that the "hard wheats" have a soft straw, relatively easy to break, that they tiller profusely, and are subject to lodging. The "soft wheats," on the other hand, have hard, relatively stiff straw and only under exceptional conditions do they lodge. Salmon (1931) observed in general that the straw of soft wheats which showed resistance to lodging was more difficult to break than that of hard wheats which lodge easily. The correlations between lodging in various seasons were not statistically significant but those between the breaking strength of straw in different seasons were both high and highly significant. This is explained by the fact that the percentage of lodging is governed by weather conditions and the stage of growth to a much greater extent than is the breaking strength of the stem. Clark and Wilson (1933) could find no correlation between the tillering rates of wheat varieties and their lodging behavior. Brady (1934) observed that the number of vascular bundles, and width of lignified tissue and sclerenchyma were all closely associated with varietal differences in resisting lodging. All these characters, however, were so sensitive to soil variation that their use for the isolation of strains resistant to lodging could be employed only on a relative basis.

Hall (1934) with corn could find little or no relationship between the amount of lodging and the height of ear, cross section of stalk, size of brace roots, number of tillers, or weight of ears. However, in the selfed lines certain phases of the root development were to some extent associated with lodging.

Fertilization may have a marked effect upon lodging. Thus an excess of nitrogen produces an abundant vegetative growth with relatively weak culms. Tubbs (1930), however, noted that a large amount of nitrogen can be safely applied, provided the proper potassium nitrogen ratio is maintained. The plants that had a deficient supply of nitrogen and phosphorus showed lignification equal to that of fully manured plants. Anatomical sections from potassium-deficient plants, however, showed considerably less depth of staining, indicating less lignification. The thickness of the lignified parenchyma was quite narrow in these plants as compared to the fully fertilized ones. The role of potassium as a condensing agent in the formation of carbohydrates and possibly in lignification offers a possible explanation for the preceding observations. A deficiency of potassium had no influence upon the number of tillers produced, while a deficiency of phosphorus and nitrogen caused a marked reduction in that regard.

It was observed by Davidson and LeClerc (1923) and Phillips, Davidson, and Weihe (1931) that the application of sodium nitrate caused a distinct decrease in the silica content of the straw of wheat. Since this application was followed by lodging, they considered that the original view of Liebig for the necessity of silicon for strength of straw received new support. These observations also show clearly the superiority of potassium over sodium as a condensing agent in plant nutrition.

I. Boron

1. Occurrence.—Boron is very widespread in the plant kingdom, and Agulhon (1910) has determined its presence in the fungi, algae, ferns, and various members of the flowering plants. In the plants examined by him the amount of boron ranged from 1.2 per cent of the ash in the case of *Betula alba* to as low as 0.12 per cent of the ash in *Cannabis sativa.* Brenchley (1914) stated that, in general, annual plants seem to have the least boron in the composition of their ash and that the bark and wood are richer in this element than the leaves. The work of Cook (1916) showed that the absorption and distribution of boron vary greatly with different plants. In growing plants in soil to which manure treated with borax and calcined colemanite had been added, wheat and oats absorbed very little boron, while leguminous and succulent plants absorbed comparatively large amounts. The amount of boron in the entire plants of oats and peas grown in the same kind of soil varied in the case of oats from no detectable amount to 0.025 per cent of the total dry matter and in the case of the pea from 0.0005 to 0.04 per cent of the dry matter, the amounts depending on the amount and form of the boron applied. The fruit of the tomato plant contained only traces of boron, while the seed of the cowpea contained large amounts, from 0.0094 to 0.0222 per cent of boric acid on an ash-free basis. Wheat, beets, cowpeas, and potatoes contained boron principally in the tops and, with the exception of beets, comparatively little or none in the roots. String beans, soybeans, and cowpeas showed a more equal distribution of boron among the roots, tops, and fruits than any of the other plants tested.

McLean and Hughes (1936) determined the content of boron in *Vicia faba* and *Gossypium herbaceum* grown in solutions containing known quantities of boron. The highest percentage per unit of dry matter was in the leaves, and it increased with the age of the leaf. The roots had the smallest amount while, in the seeds, boron was confined to the cotyledons. In *Vicia faba* the percentage content in the cotyledons was 50 per cent above that of the stems.

2. Need.—The influence of boron on the growth and development of plants has been more extensively studied in recent years than that of any other element. The plants used have included the cereals, legumes, algae, cotton, flax, mustard, citrus, tomato, walnut, tobacco, grape, prune, peach, apricot, and others. The evidence indicates that boron is essential to the growth and development of the plant. The amount of this element that is necessary for the optimum functioning of plants is relatively low, while a very small excess becomes extremely toxic. Tolerance of this element varies widely with the kind of plant and the conditions under which it is grown. Of all the plants that have been investigated, cotton has the highest boron requirement and the cereals the lowest. It was reported by Haselhoff (1913) that beans were benefited by the application of $\frac{1}{10}$ p.p.m. of boron but corn was injured by that concentration. If boric acid was used as a source of boron, a concentration of 1 p.p.m. was toxic, but if borax was used the plants were not visibly affected. Warington (1923) found that in water cultures a continuous supply of boron was necessary for the healthy growth of *Vicia faba,* and the benefits of this element were evident in concentrations ranging from 1/12.5 p.p.m. to 40 p.p.m. Barley plants, however, were injured by concentrations as low as 1/2.5 p.p.m.

Geigel (1935) found that a concentration of boron above 1 p.p.m. was toxic to *Spirodela polyrhyza,* while with *Chlorella* a concentration of 30 p.p.m. was required to decrease growth, and it could survive a concentration of 140 p.p.m. The addition of dextrose or sucrose made *Chlorella* more tolerant to boron.

According to Haas (1930) and Haas and Klotz (1931), the addition of boric acid to a cultural solution in amounts sufficient to make a concentration of boron of $\frac{1}{10}$ to $\frac{1}{5}$ p.p.m. was sufficient to cause a rapid recovery of citrus plants suffering from a boron deficiency. Johnston and Fisher (1930) found that $\frac{1}{2}$ p.p.m. of boron was sufficient to cause a fourfold increase in fruit production in the tomato. Loomis and Wilson (1933) reported that $\frac{1}{10}$ and 5 p.p.m. of boron were inferior to $\frac{5}{10}$ p.p.m. in the response of this plant, while Fisher (1935) found 1.1 p.p.m. to be the optimum concentration. The maximum growth and fruitfulness of cotton occur only when boron is supplied in a concentration of 10 p.p.m. The boron requirement of cotton is the highest among some 60 plants that were compared. When the content of boron is ideal for cotton, the growth is depressed 25 per cent in beans, corn, and the smaller cereals (Eaton, 1932).

Johnston and Dore (1929) found 5.5 p.p.m. of boron in a nutrient solution to be toxic to tomato plants, while Newell (1930) under the conditions of his experiment noted that 2.75 p.p.m. injured these plants. Morris (1931) in working with wheat reported that a concentration of 50 p.p.m. of boron caused a retardation in growth of 40 per cent, while a concentration of 100 p.p.m. completely inhibited growth.

It was considered by Brenchley and Warington (1927) that the chemical combination in which boron is presented to the plant is immaterial, even the relatively insoluble borates being effective.

The accumulation of boron in the soil from the water used in irrigation has been especially detrimental to citrus plants in certain regions. In the irrigation waters of California the content of boron ranges from 0.15 to 0.30 p.p.m. in most cases, while in waters that have caused trouble, boron ranges from 0.5 to 2.5 p.p.m. In the mature leaves of orange trees that have been irrigated with water low in boron, there are present from 50 to 150 p.p.m. of boron, based on dry weight, while in those on plants irrigated with water high in boron, from 600 to 1,000 p.p.m. are present. There is a distinct relationship between the boron content of the leaves and that of the water supplied to the plant (Scofield and Wilcox, 1930, 1931).

Plants differ greatly in their rate of absorption of boron when grown under the same conditions. Thus Eaton and Blair (1935) observed that, on a dry basis, with plants grown in sand to which was added a solution containing 5 p.p.m. of boron, the amount of boron in p.p.m. absorbed was navy beans, 648; lima beans, 515; wheat, 453; pumpkin, 291; lettuce, 261; turnip, 245; Canadian field peas, 207; tomato, 159; alfalfa, 139; cotton, 123; asparagus, 120; and carrot, 69. This element tends to accumulate in some plant parts more than in others. Under the conditions mentioned above, the leaves of lemons contained 1,232 p.p.m., while the stem and roots contained only 54 p.p.m. The leaves of elm had 943 p.p.m., while there were only 22 p.p.m. in stems. The combined stems and roots of corn contained 60 p.p.m. but there were only 16 p.p.m. in the grain. The boron absorbed by plants is apparently mobile but upon reaching the leaves is changed to the immobile form.

Since plants differ in their rate of absorption of boron under identical conditions, Eaton and Blair (1935) considered that a specific rootstock might influence the amount of boron in the scion. They found that the concentration of boron accumulating in the leaves of the scion was directly influenced by the rootstock upon which it was grown. The concentration in the scion was reduced if grafted to varieties normally accumulating lesser concentrations and increased if grafted to varieties normally accumulating higher concentrations. The amount of boron absorbed, however, is not wholly dependent on the rootstock. Thus Eureka walnuts accumulated more boron in their leaves than the purple variety when both were grown on the same rootstock in the same nutrient solution. With other conditions equal, the rate of absorption of boron is apparently determined by the characteristic of the absorbing root cells and the equilibriums between the mobile and nonmobile forms of boron in the plant. McLean and Hughes (1936) considered that the role of boron in the plant is probably that of an activator or regulator of metabolic processes.

3. Deficiency Effects.—The effects of a deficiency of boron on the growth, development, and general appearance of plants vary with the type of plant, its age, and the general conditions under which it is grown. Some of the more general symptoms shown by various plants as a result of lack of a sufficiency of boron are given below.

Brenchley and Thornton (1925) observed that the absence of boron had a marked effect upon the development of nodules in the roots of *Vicia faba*. The number of nodules of macroscopic size was smaller in those plants that were not supplied with this element. In one case the number of nodules per plant was 353 and 199, respectively, for the plants receiving boron and those deprived of it. In the absence of boron the vascular supply of the nodule was defective, being either entirely absent or when present very weakly developed and running only a short distance into the nodule. The nodules that had no vascular strands remained minute and were usually buried in the cortical tissue. Very little nitrogen was fixed in these abnormal nodules, being in one case only one-tenth that fixed in plants where the nodules were normally developed.

It was found by Warington (1926) that the anatomical structure of *Vicia faba*, growing in a nutrient solution containing boric acid in the proportion of 1:2,500,000, is exactly similar to that of plants growing in the soil. If, however, boron is omitted, frequent disintegration of phloem and ground parenchyma, poor development of the xylem, and hypertrophy, discoloration, and disintegration of the cambial cells occur.

It was noted by Sommer and Sorokin (1928) that the meristematic tissues in the root tip of *Pisum sativum* are the first to show the effects of the absence of boron. In a relatively short time after the plants are transferred to boron-free solutions, the cells cease dividing and a general arrest of growth follows. The existing cells undergo premature development or pathological changes. Isolated xylem elements appear in regions occupied by the meristem or in the region of elongation in normal roots. The root apexes are enlarged owing to hyperplasia of the plerome and hypertrophy of the periblem regions. Fagundes (1934) obtained a marked increase in the growth of lima and broadbeans, and soybeans when boron was added to the cultural solution. More iron is necessary when the proper amount of boron is present, apparently because of the greater amount of material being synthesized in these plants.

McMurtrey (1929, 1935) found with tobacco plants grown in nutrient solutions deficient in boron that growth was greatly reduced, the most characteristic effect being the injury produced on the growing point. The first indication of this deficiency is the light-green color of the bud leaves, which is followed rapidly by a breakdown of the tissues. Frequently the leaves are distorted by growth around the injured areas. The stalk toward the top of the plant may manifest a one-sided or twisted growth. The final manifestation is a "dieback" of the terminal bud.

The effects of the absence of boron on the tomato have been studied by many workers. Johnston and Dore (1928, 1929) found that plants grown in a boron-deficient nutrient solution showed four distinct types of injury: (*a*) death of the terminal growing point of the stem, (*b*) breaking down of the conducting tissues of the stem, (*c*) a characteristic brittleness of the stem and petiole, and (*d*) roots of an extremely poor growth and of a brownish, unhealthy color. The total sugars and starches were more abundant in the leaves and stems of the boron-deficient plants, while a greater amount of benzene-soluble matter was found in the leaves of normal plants and in the stems of the boron-deficient plants. Johnston and Fisher (1930) reported that dark, dead areas appear on the fruit of the tomato plants grown in a solution deficient in boron during the fruiting phase. According to them, boron is fixed in the tissues and cannot be used repeatedly by the plant. It apparently functions as a simple nutrient element, which is in accord with the views of Warington (1923). She believed that it acts in a nutritive way rather than in a catalytic manner, since a continuous supply of it appears to be required throughout the life of the plant. According to Loomis and Wilson (1933), tomato plants did not develop properly when grown in a boron-deficient nutrient solution contained in soft-glass vessels.

It was found by Van Schreven (1935) that tomato plants suffering from a lack of boron were cured by the application of boric acid. The plants showing boron deficiency had up-curled leaf margins in sand cultures and down-curled margins in water cultures. This leaf curling was attributed to the thickening of the interveinal tissues due to the enlargement of individual cells, while the veins remained unchanged. The plastids did not decrease in number but the chlorotic symptom was due to their decrease in size and to the increase in the volume of the cell contents. Fisher (1935) stated that boron deficiency was apparent about 9 days after the plants were transferred to a solution lacking this element. The cotyledons and leaves turned a distinct purple color and within a few days the terminal shoot curled inward and died. The conductive tissue disintegrated and the petioles became extremely brittle. The fruits were covered with darkened or dried areas, which were apparently due to a breaking down of the cells making up this tissue.

In barley and wheat grown outdoors in large sand beds flooded daily with nutrient solution, Eaton (1930) found that shot blotch of barley, caused by *Helminthosporium sativum*, did not develop when boron was omitted from the cultural solution. In the summer plantings of barley, powdery mildew (*Erysiphe graminis*) was abundant on plants without boron and absent when it was applied. In the winter plantings it was abundant on all plants. There was no mildew on summer-grown wheat, but it was equally abundant on all winter wheats studied, regardless of the amount of boron applied.

Haas and Klotz (1931) found that a deficiency of boron in grapefruit and Valencia orange was marked by a gradual reduction in the size of the shoot produced, which in extreme cases resulted in the formation of "multiple buds." When boron was deficient, the cambium and parts of the phloem disintegrated, but the xylem tissues were affected to a lesser extent if at all. A copious amount of gum was formed and appeared in breaks in the cortex. The total amount of sugar and starch was doubled in the leaves of the plants deficient in boron. The addition of this element to the cultural medium caused a reduction in the sugar content of the leaves. The plants grown in the boron-deficient cultures showed a slightly greater diastatic activity than those receiving the boron.

It was reported by Gilbert and Pember (1931) that red clover, when grown in water cultures, is extremely sensitive to a deficiency of boron and manganese, and that the addition of minute amounts of these elements produced a marked increase in the growth of both roots and tops.

McHargue and Calfee (1932) found that the varieties of leaf and head lettuce which they observed required boron for normal growth. A deficiency of boron caused a burning of the leaves and an early death of the growing parts. The functions of boron could not be performed by magnesium, copper, zinc, nickel, cobalt, barium, strontium, iodine, or arsenic, which is in accordance with the report of Brenchley and Warington (1927), who tried to substitute 52 elements but found none that could replace boron.

Bobko and Belvoussov (1933) noted that the absence of boron caused a characteristic disease of the sugar beet known as "heart rot," which could be corrected by the application of small quantities of boron. Hill and Grant (1935) in the growth of turnips in pot sand cultures deficient in boron noted that there was a marginal yellowing of the foliage, followed by a purpling and scorching. The roots were small and shriveled or rotted at their juncture with the top. These symptoms could be remedied by the application of boron. An inverse relationship was found between the amount of boron supplied and the occurrence of hollow heart of the roots. There was also an inverse relationship between the amount of boron and the percentage of ash in the dry matter of the roots. It was found by Hoagland and Snyder (1933) that in the absence of boron the leaves of strawberry plants were deformed, dwarfed, cupped upward, puckered, and generally brown at the tips.

Schmucker (1934) in considering the effect of boron on pollen growth decided that boric acid can be designated as inorganic growth material because of its ability to form strongly acid compounds with certain organic substances. Warington (1933) found that reduced length of day rather than the lower temperature is the factor controlling the delay in appearance of the symptoms of a deficiency of boron during the spring and autumn as compared with the summer months. This behavior is noticeable in *Vicia faba*, *Phaseolus multiflorus*, and *Glycine hispida*. In the absence of boron, the influence of the length of day was less striking than when boron was present. The symptoms of boron deficiency were less pronounced under short-day than under full-day conditions. According to Warington (1934) the absorption of calcium is affected by the presence and quantity of boron. A relationship of some sort apparently exists between boron and calcium, but its exact nature is yet undetermined.

4. Excess Effects.—Boron is toxic to fruit trees when present in relatively small concentrations (Haas, 1929). Citrus and walnut leaves may become thin, mottled, chlorotic, and crinkled when boron is in excess. The addition of various amounts of ferrous sulphate to cultures of lemon seedlings tends to overcome the toxicity of boron. Unless badly defoliated, trees injured by boron may recover if the toxic agent is leached out with water. The leaves with an excess of boron are mottled and burned along the margins. Such leaves of citrus and walnut contain reduced amounts of calcium but increased amounts of potassium and do not mature. When tomato plants are supplied with an excess of boron, Newell (1930) reported a yellowing of the whole leaf, large brown spots developing thereon, and the occurrence of a mottled appearance. Brandenburg (1931) considered the dry rot of beet to be a boron-injury disease. With citrus plants, Webber (1935) found that leaf injury is generally the characteristic manifestation of boron toxicity, the older leaves showing the most pronounced symptoms. These are the yellowing of tips and margins, often followed by marginal or spotted browning. This condition may be followed by premature defoliation and the subsequent production of malformed leaves. Under conditions of excess boron, the trees of stone fruits seldom exhibit such symptoms of the leaf. The petioles and larger veins of the leaves of prunes and apricots may become brown and rough, and occasionally exude gum. Rodriquez (1933) considered that boric acid inhibits the solubility of iron. This suggests that chlorosis, which sometimes appears when boron is present in excess, may be due to a lack of availability of iron and not to the boron itself.

Histological studies of prune, peach, apricot, and grape showed that a toxic concentration of boron in the nutrient solution is not a stimulus to a specific reaction resulting in a specific histological change. The histological evidence of injury by boron is in general similar to that attributable to other causes. It seems probable that an excessive concentration of boron in the cell often injures the protoplasm to such an extent that it undergoes a progressive degeneration. If chloroplasts are present they are the first protoplasmic structures to be affected. In other cases, excess of boron may stimulate the cell to abnormal growth and division.

5. Fertilizer.—The effect of boron compounds in fertilizers and in manures treated with them has received considerable attention in the United States. Cook and Wilson (1917) applied manure, which had been treated with borax and calcined colemanite, to soil at Arlington, Va., and observed the effect of the boron compounds thus applied upon the growth and yield of wheat. The presence of 0.0022 per cent of boric acid added as borax to the first 6 in. of soil or 0.0029 per cent as colemanite was apparently nontoxic to this crop, although, in 2 years out of 3, borax applied at the above rate reduced the yield of grain 10 per cent, while colemanite had little if any effect. Only minute quantities of boron were absorbed by these plants, but more was absorbed from the borax than from the colemanite.

These workers (1918) observed that there is a decided difference in soils in rendering added boron nontoxic to plants. In some soils there is a tendency for the plant to absorb boron in proportion to the quantities added, while in others the same amount of boron is absorbed irrespective of the quantities added. The behavior of plants grown at Arlington, Va., and Bethesda, Md., will serve as an example in this regard. At the latter station, 0.0044 per cent of boric acid as borax and 0.0058 per cent as colemanite in the upper 6 in. of soil caused no injury to lettuce, spinach, or kale; but at the former station these concentrations caused a reduction in the crop of these plants.

Skinner, Brown, and Reid (1923), Neller and Morse (1921), and Blair and Brown (1921) found that the potato can tolerate a greater quantity of borax than corn and beans. The effect of borax on potatoes at Arlington, Va., when used in quantities of less than 5 lb. per acre applied in the furrows, was one of stimulation, but 20 lb. per acre produced injury and depressed the yield. In Maine, however, 5 lb. of borax applied in the drill at the time of planting produced definite injury. In the case of corn, 4 lb. of borax per acre in the drill caused a decrease in yield of both grain and stover. When sown broadcast, 20 lb. per acre was required to produce any decrease in yield, while practically no plant growth took place where the application exceeded 50 lb. of borax per acre. Skinner and Allison (1923) observed on the silt loam at Arlington, Va., that cotton was slightly injured by the broadcasting of 10 lb. of borax per acre and decidedly so by the application of 20 lb. Boric acid and to a lesser degree borax exerted a stimulating effect on plants in nutrient solution (Collins, 1927); the greatest stimulation appearing when 2.5 mg. of boron per liter of cultural solution was used. He, however, does not consider that boron is necessary to the production of mature plants in nutrient solutions. In field experiments, all compounds of borax tested reduced the dry weight of plants when applied in quantities equivalent to 1 to 7.5 lb. of boron acre, depending upon the type of soil.

The evidence indicates that boron is an essential element for plant growth but that only extremely small amounts are necessary to supply the needs of the plant. The application of the compounds of boron, however, as fertilizers must be made in relatively small amounts or they will have a toxic effect upon the plants. In most cases, applications as low as 5 lb. per acre cause detrimental effects on crop growth and production.

J. Manganese

1. Occurrence.—Manganese is universally distributed in plants (Bertrand and Rosenblatt, 1921 to 1922), and, according to Robinson (1917), it fluctuates more in amount than most of the other elements. Most investigators concur that it tends to accumulate in the leaves and that this element reaches its highest concentration in that part of the plant. In corn grown in Kansas, the manganese content was 0.017, 0.037, and 0.043 per cent of the dry matter of the stem, grain, and leaves, respectively (Latshaw and Miller, 1924). According to Kelley (1914), the leaves of the pineapple contain ten times as much manganese as the stalk. In the various plants grown in Kentucky, McHargue (1925) found manganese present in parts per million of air-dry material as follows: soybean leaves, 160; soybean seeds, 32; wheat bran, 125; wheat embryos, 150; corn embryos, 40; corn endosperm, 16; rice polishings, 100; and polished rice, 10. In Oregon, alsike clover utilized manganese in larger amounts than any other legume there grown (Jones and Bullis, 1921). The average manganese content for legumes considered in terms of milligrams per kilogram of air-dry material was alsike clover, 68; vetch, 42; white clover, 34; field peas and red clover, 33; sweet clover, 27; and alfalfa, 23. Gössl (1905) found that swamp plants and water plants, in general, contain more manganese than dry-land plants and that the evergreens have a greater quantity than the deciduous trees. It has been observed that the total amount of manganese in the plant will vary according to the nature of the soil, and McDonnell and Roark (1917) stated that the stems of *Chrysanthemum cinerariaefolium* if grown in Japan contain more manganese than those from other countries.

The amount of manganese in the seeds is an important point to consider, since it is assumed, by those who consider manganese an element essential to the normal growth of plants, that the seed, in some cases at least, contains a sufficient quantity to supply the plant during its growing period. Wester (1921) found that seeds from 13 plant families contained in every case manganese ranging in amount from 2 to 6 mg. for every 100 g. of dry material. McHargue (1923) investigated the iron and manganese content of a wide list of seeds, and the observations may be summed up in the table on page 347.

The average manganese content of the seeds of wheat, oats, and hemp, produced under natural conditions, is equal to or greater in amount than the average amount of

Plant	Per cent iron content on dry basis	Per cent manganese content on dry basis	Plant	Per cent iron content on dry basis	Per cent manganese content on dry basis
Wheat, 10 varieties....	0.0039	0.0047	Alsike clover ...	0.0517	0.0028
Oats, 6 varieties.......	0.005	0.005	Crimson clover.	0.0016	0.0029
Garden peas, 3 varieties	0.0096	0.0012	Tobacco........	0.024	0.007
Beans, 5 varieties	0.0103	0.0018	Hemp..........	0.0085	0.0165
Soybeans, 9 varieties ..	0.0074	0.0028	Sunflower......	0.0034	0.0023
Alfalfa..............	0.010	0.0012			

iron found therein. This observation agrees with that of Headden (1915), who found that manganese is present in the grain of wheat in approximately the same proportion as iron, although the latter greatly predominates in the soil. Although the iron and manganese were present in approximately the same amounts in the grain, the former was present in the straw in amounts from two to six times greater than the latter. Headden examined samples of wheat seed that had been grown over wide areas and under a wide range of cultural conditions, climate, and soils and found that the manganese content ranged from 0.003 to 0.006 per cent of the dry seeds regardless of the conditions under which the crop was grown.

It was observed by Jacobson and Swanback (1932) in the tobacco plant that the highest percentage of manganese was found in the middle leaves, and the lowest in the upper ones. Bolin (1934) determined the amount of manganese in eight species of grass and found it ranged from 78.1 mg. per kilogram for Kentucky bluegrass to 207.5 mg. per kilogram for orchard grass. Alfalfa showed an average of only 46.6 mg. per kilogram.

2. Amounts Needed.—The amount of manganese needed by plants is small and apparently occurs in sufficient quantities as impurities in the various salts used in nutrient solutions. Thus McHargue (1922) found that the salts of calcium, magnesium, and iron generally used for plant nutrients may contain as impurities a supply of manganese sufficient for the plant's needs. This fact has apparently been overlooked by most investigators in plant nutrition. McHargue observed that crocks which were supposed to be perfectly glazed absorbed during the growing season from the soil with which they were filled a sufficient amount of manganese to supply the needs of the succeeding crop grown in them in a manganese-free medium. This experiment emphasizes the need for acid-proof glazed jars in experimental nutrition work. McHargue (1922) observed that when wheat seedlings were grown in Pfeffer's nutrient salt solution from which manganese was excluded, they showed no observable differences from those supplied with manganese during the first 6 or 8 weeks. Shortly after this period, the plants from which manganese was withheld began to give evidence of unequal growth. The first observable difference was the lack in the development of chlorophyll, the leaves becoming a yellowish green. The plants deprived of manganese made a stunted growth and produced no seed, whereas those to which manganese was available grew normally and produced seed. There was an increase of 135 per cent in the dry matter produced in the plants receiving manganese when compared with the dry weight of an equal number of other plants from which it was withheld. It was found that the seeds of radish, soybean, cowpea, field bean, and corn do not contain a sufficient amount of manganese for the growth of these plants to maturity. Samuel and Piper (1929) studied the relation of manganese to the growth of rye and oats. They found that these two plants differ in regard to their requirements for manganese. Thus rye developed to maturity in water cultures on an amount of manganese that did not allow the complete development of oats. A concentration of 1 part in 50,000,000 parts of nutrient solution allowed complete development of rye, while the minimum quantity of manganese that allowed healthy growth of Algerian oats was about 14 parts per 1,000,000 of dry matter at the flowering stage. Hopkins (1930) increased the growth of *Chlorella* 600-fold by the addition of manganese, in a concentration of ⅕ p.p.m., to nutrient solutions that lacked this element. At a pH of 7.0 the increase in growth was 17-fold, while at a pH of 8.0 it was 170-fold. In 1934 he reported that certain unicellular algae and *Lemna minor* showed no growth in the absence of manganese from the nutrient solution. When, however, amounts of manganese ranging from 1/100 to ⅕ p.p.m. were added to this cultural solution, healthy growth occurred. If manganese was added to the solution before the process of disintegration of the plants had progressed too far, recovery occurred. Clark and Fly (1930) found that concentrations of manganese exceeding 1 p.p.m. became toxic for *Lemna minor*, and that the size, weight, reproduction, and ash content of this plant became very irregular. Clark (1933) obtained a fine growth of *Lemna major* in a nutrient solution containing manganese in a concentration as low as 1/3000 p.p.m. Saeger (1933) showed that a manganese content of 1/1000 p.p.m. was sufficient to maintain the vigorous growth of five species of *Lemnaceae*, while a concentration of 1 p.p.m. was excessive for optimum growth.

The reaction of the plant to manganese apparently depends upon the nature of the medium in which it is growing. Thus Jacobson and Swanback (1932) found that the tobacco plant showed toxic symptoms when growing in a nutrient solution containing as low a concentration of manganese as 1 p.p.m. In sand cultures, however, the application of 80 p.p.m. was necessary to produce toxicity, while in the field 486 lb. per acre did not cause injury. An increased acidity of the soil resulted in the presence of large amounts of active manganese and a greater intake by the plant. Leeper (1934) observed that the "manganese deficiency disease" is confined to soils of pH 6.7, or higher, and that it occurs especially on heavily limed, sandy soils. In soils with a sufficient supply of manganese for plant needs, the active manganese exceeded 100 p.p.m., while in the soils associated with manganese deficiency it was lower than 15 p.p.m. Gilbert (1934) considered that a manganese deficiency can be prevented in many soils by maintaining a suitable acidity that allows the manganese to remain available to the plant. Under the same soil conditions, different crops will respond differently to a deficiency of manganese (Odland and Crandall, 1932).

3. As a Fertilizer.—The influence of manganese as a fertilizer in soil and in water cultures has been studied by numerous investigators. The results indicate that the effect of the application of this element will depend upon the concentration of the various salts used and upon the type of plant under observation. Nagaoka (1903, 1904, 1906) showed that the application of 50 kilograms of manganese sulphate per hectare increased the rice harvest over one-third the first year, 16 per cent the second, only slightly the third year, and caused a depression in yield the fourth year, due probably to the increased acidity of the soil. Small quantities of manganese sulphate in soil cultures stimulated the growth of rice, peas, and cabbage according to the observations of Loew and Sawa (1902), while Loew and Honda (1904) report that *Cryptomeria japonica* growing in soil treated with manganese sulphate doubled in 1½ years its cubic contents over plants that had grown in soil not treated with this compound. Katayma (1906) applied both iron and manganese sulphate to soil at the rate of 0.01 per cent and observed an increase of 6.2 per cent straw and 7.2 per cent increase in grain yield of barley but a depression in the yield if larger amounts were applied. Brenchley (1910) stated that at the Rothamsted station concentrations of less than 1:100,000 of manganese sulphate stimulated the vegetative growth of barley but retarded the ripening of the grain. The application of 100 lb. of manganese sulphate per acre increased appreciably both ammonification and nitrification (Brown and Minges, 1916). The increase in activity was lower in applications above this amount, but a decrease occurred if amounts greater than 2,000 lb. per acre were applied.

Schreiner and Dawson (1927) observed that a highly calcareous soil occurring in certain regions of Florida failed to produce a crop even with a heavy application of inorganic fertilizer unless stable manure was applied. Striking results were obtained by the application of 25 to 50 p.p.m. of manganese sulphate in addition to the balanced inorganic fertilizer. In control greenhouse experiments with this soil, plants did not flower, developed chlorosis, and were stunted in growth. These effects rapidly disappeared on the application of manganese. Samuel and Piper (1929) found that on certain soils oats are unable to absorb a minimum quantity of manganese and that they suffer from the gray-speck disease, a manganese-deficiency disease that can be cured by the application of soluble manganese salt to the soil. On soils with abundant available manganese, considerably more manganese is absorbed by oats than is required for normal development.

Conner (1932) reported that applications of manganese to oats growing under neutral and alkaline conditions increased yields, and Skinner and Ruprecht (1930) found manganese sulphate to be beneficial to tomatoes growing in the calcareous soils of Florida which are deficient in manganese. Saeger (1933) statéd that extracts of autolyzed yeast, raw carrots, and spinach will stimulate the growth of plants when added to a nutrient solution deficient in manganese.

McCool (1935) studied the effects of different light intensities on the injurious action of manganese on soybeans, buckwheat, snapdragons, and tobacco. The visible injury to the plants grown in the manganese-treated soil decreased as the light intensity became less. The percentage of manganese in the leaves of the soybean and buckwheat growing in the open decreased consistently as the light intensity increased.

4. Physiological Role.—It has been considered by Bertrand (1897, 1905, 1912), McHargue (1919–1932), Bishop (1928), Bryan (1929), Hopkins (1931), Conner (1932), Clark (1933), and others, that manganese is an essential element for the growth of plants. It has been observed that the absence or scarcity of manganese results in disturbed carbohydrate metabolism, chlorosis, retarded growth, a decrease in the content of ash, and the failure to reproduce.

Our understanding of the function performed by manganese in the metabolism of the plant rests almost entirely on theoretical grounds. Bertrand (1897, 1905, and 1912) considered that the addition of small amounts of soluble manganese increases the oxygen-carrying power of the oxidases. This enzyme is universally present in plants and fulfills a definite function in their metabolism. From their observations on the effect of manganese on the germination of seeds, Michiels and DeHeen (1906) considered that this element has a role in enzymic action. Kelley (1914) from his observations on the effect of manganese on the plants of pineapple believed that manganese may act in two general ways: (1) It may have an effect upon the inert bases of the soil by bringing about a greater degree of mobilization of calcium and magnesium therein, and (2) it may cause a stimulation of the necessary oxidation going on in the plant and in the soil. The first effect is substantiated by the work of Aso (1907), who found that a 41.8 per cent increase in the yield of rice was obtained from the first application of manganese sulphate, while similar applications to the same soil the following year produced only a little better than a 2 per cent increase. McHargue (1922) thinks that manganese is concerned in nitrogen assimilation and the synthesis of proteins and that it plays the role of a necessary catalyst in plant metabolism and also functions with iron in the synthesis of chlorophyll. He also considered that, since the lack of manganese affects the production of dry matter, the indications are that it has some very important function in carbon assimilation.

Bishop (1928) noted that manganese occurred in plants in the regions of most active chemical change, and he considered that it is concerned in chlorophyll formation and hence in carbon assimilation. Hopkins (1930) suggested that manganese functions in an indirect manner in plants by its action upon the oxidation of iron, because in vitro the reduction of iron from the ferric to the ferrous form by sodium acetate is hindered by the presence of manganese. Bertrand and Rosenblatt (1932) found in the leaves of dandelion, lettuce, cabbage, and celery, with different degrees of etiolation, that for a given plant the content of manganese varied directly with the amount of chlorophyll. McHargue and Calfee (1932) reported that chlorotic plants of *Lemna minor* could be restored to their original dark-green color by the addition to the nutrient solution of traces of manganese sulphate, but not by the addition of iron salts.

Miller (1933) found that wheat and tomato plants which were deficient in manganese were much lower in sugar than corresponding plants that had been supplied with a small amount of this element. Thus wheat tops with sufficient manganese contained on a dry basis 8.3 per cent of sucrose and only 1.6 per cent when manganese was lacking. The roots contained 3.3 per cent of sugar when manganese was present, but only a trace of this carbohydrate when it was absent. Pettinger, Henderson, and Wingard (1932) noted that a deficiency of manganese caused a peculiar type of chlorosis in corn growing in sand cultures. The fact that chlorosis is so often one of the criteria of a deficiency of manganese indicates that it is concerned in some manner in the formation of chlorophyll.

Chlorosis of plants seems to occur when they are deprived entirely of a manganese supply, and this may prevail also even when an abundance of this element is in the soil or the cultural solution (McCool, 1913). McGeorge (1923) considered that the chlorosis of pineapple leaves on plants grown in soils rich in manganese is due to a greater assimilation of lime indirectly caused by the presence of manganese in the soil. He believed that the principal physiological disturbance is the greater immobility of the iron in the plant, resulting from the excessive lime content of the leaves and stalks, since iron is present in an equally available form in both manganiferous and nonmanganiferous soil types of the same hydrogen-ion concentration. Rippel (1923) observed that manganese in the form of soluble salts produced chlorosis in barley plants grown in water cultures. This was overcome or cured by increasing the iron supply. Since the iron content of green and chlorotic plants was the same, it was concluded that manganese did not interfere with iron absorption but with the activity of iron within the plant.

Haas (1933) noted that walnut leaves affected with "yellows" contained a higher amount of manganese and iron than the healthy ones. He considered that the availability of these elements and not their total amount is the determining factor in their utilization by plants.

According to Gilbert (1934) the causes of manganese chlorosis have been definitely linked with soil alkalinity. The pH range within which chlorosis may occur appears to vary with the crop and climate and not with the soil type. Crops vary widely in their susceptibility to manganese chlorosis, spinach being very sensitive, while tomatoes are relatively immune. A relatively low temperature, by causing a slower growth rate, may enable the plant to avoid the chlorotic effect. Chapman (1931) noted that chlorotic plants always recovered their normal color uniformly over the whole of the treated leaves when sprayed with a 2 per cent solution of ferrous sulphate. When manganese was added with the iron or injected into the trees before spraying with iron, the leaves greened only in spots—suggesting a reduction in the solubility of iron by manganese in the leaf. Similar observations were made by Haas (1932), who found that excessive concentrations of manganese may produce chlorosis, although iron is present in the cultural solution. When manganese is deficient, apparently less iron accumulates in the leaves.

McLean and Gilbert (1925) were able to effect a complete cure of chlorosis of spinach, growing on heavily limed soils, by spraying the plants with a solution of 8 parts per 1,000,000 of manganese sulphate. When harvested, a 40 per cent increase in yield was obtained from the treated plants as compared with the untreated ones. The injection of solutions containing 5 to 50 p.p.m. of manganese into the leaves through the stomata brought about the recovery of the chlorotic spinach plants.

Bryan (1929), in experiments having to do with the effects of copper and manganese on cowpeas, beans, and sorghums in Florida, found that the responses were as striking when the chemicals were applied to the leaves as when they were applied to the soil. Gilbert (1934) also mentioned that chlorosis could readily be corrected by the application of a solution of manganese sulphate as a spray. Miller (1933) found with plants of wheat and tomato that the manganese necessary for the plant could be injected as a solution into the stems.

K. Chlorine: Its Occurrence, Function, and Value as a Fertilizer to the Plant

With the exception of the conifers, chlorine is present in practically all plants, according to Jung (1920), who studied microchemically the occurrence and distribution of this element in plant tissue. This author considered that chlorine occurs in the plant only in the form of chlorides which are generally dissolved in the cell sap. It usually increases in amount from the roots to the leaves, being most abundant in succulent parenchymatous tissue. It is especially abundant in fleshy roots and rhizomes and is very scant in the epidermis, fibrovascular bundles, pollen, and woody tissue. The mosses, ferns, and epiphytes contain little or no chlorine. All seeds that have been examined contain more or less chlorine. Most soils contain considerable amounts of it in the form of chlorides. Schmidt (1915) found that the entire chlorine content of the sugar-beet seed is soluble in water and can be precipitated directly by silver nitrate. This strongly indicates that the chlorine in this seed exists largely in the form of inorganic chlorides and that if it is involved in any organic compounds they must be very unstable.

In consequence of the results obtained by the earlier experimenters in plant nutrition, chlorine has generally been considered as one of the unessential elements for the growth of plants. Nobbe and Siegert (1862), and Leydhecker (1866), however, observed certain peculiarities in the development of buckwheat in a chlorine-free nutrient solution which led them to conclude that chlorine was necessary for the proper growth and development of this plant. As a result of this work, the buckwheat plant has been repeatedly mentioned in the literature as an example of a plant that requires chlorine for its development. Many other investigators, however, since that time, among whom are Shulov (1909), Tottingham (1919), Johnston (1917), and Trelease (1917, 1920), have concluded that, aside from the small amount of chlorine contained in the seed, the buckwheat plant does not need chlorine for its normal growth and development.

De Peralta (1927 found, however, that the addition of chlorine in the form of sodium chloride or potassium chloride to nutrient solutions was beneficial to the growth of young plants of rice.

A great amount of work has been done to determine the value of the chlorides as fertilizers. The results that have been published are very confusing. In many cases, chlorine is apparently beneficial to plant development, while in others it is deleterious in its influence on growth. The environmental conditions under which the experiments have been conducted have been very poorly defined, and this no doubt has been one of the main reasons for the contradictory data that have been reported. The student is referred to Tottingham (1919) for a thorough review of the literature on this subject. Tottingham (1915, 1917, 1919) carried on extensive and well-controlled experiments in chlorine nutrition. Wheat, buckwheat, radish, carrot, parsnip, sugar beet, and potato were grown in nutrient cultures, in soil cultures in the green-house, and in field plots. The chlorides of potassium and sodium were used chiefly as sources of chlorine, the sulphate or nitrate of potassium being replaced partly or wholly by the potassium chloride, while sodium chloride was applied either alone or supplementary to a complete nutrient mixture.

Skinner (1931) reported that potassium chloride could be used as a source of potassium for the fertilizing of cotton in the southeastern United States without apparent injury from the chlorine. Apparently cotton is more tolerant of this element than are many other plants.

In the case of wheat grown in water cultures, increasing amounts of potassium chloride depressed slightly the yield of dry matter in the tops and roots as well as the length of roots. The addition of sodium chloride, while it depressed the length of roots, led to a slightly increased yield of dry matter as compared with other solutions. Buckwheat grown in similar cultural solutions showed that the chlorine ion in all cases produced disturbances which were shown in a decreased root length, decreased production of dry matter of leaves and roots, and decreased water absorption per culture.

The application of chlorides to radishes growing in soil cultures in the greenhouse increased the dry weight of the roots. The production of dry matter in the roots was decreased in proportion to tops by the application of chlorine in the carrot but increased in the case of the sugar beet. The percentage of glucose in the radish and parsnip was depressed by both potassium chloride and sodium chloride, while in the case of the carrot and sugar beet they induced an increase of this sugar. Sodium chloride increased the amount of sucrose in the radish and carrot and depressed it in the case of the parsnip and sugar beet. The application of the chlorides led to the production of a more watery root of the radish, carrot, parsnip, and sugar beet, a result that has been observed by several other investigators (Bolin, 1916).

In greenhouse experiments with potatoes the percentage of starch in the Triumph variety was not altered by the application of potassium chloride, while the Rural New Yorker showed a decrease in starch under these conditions. Sodium chloride produced more watery tubers than the unfertilized controls. In field experiments no significant variations in the percentage of starch resulted from the varied treatments with the chlorides. Tottingham considered that the depressing effect of chlorine on the starch content of potatoes as reported by Pfeiffer (1898) and his workers does not occur under all conditions of culture. The application of sodium chloride alone in the field, however, produced tubers of an inferior quality that were possessed of a bitter alkaline flavor. Rudolfs (1919), from data secured in determining the effects of sodium chloride on oak, birch, and maple trees, considered that the application of small amounts of the salt exerted a stimulating effect, as shown by the luxuriant growth and the extremely dark color of the leaves. Lomanitz (1923) in growing alfalfa in cultural solutions found that the solutions containing sodium chloride gave higher yields than did the checks. The greatest increase was obtained with the solution in which one-tenth atmospheric osmotic concentration of sodium chloride was substituted for the one-tenth atmosphere of potassium nitrate of Tottingham's solution *T1R1S5*, the roots showing a greater relative increase than did the tops. In another experiment (1924) alfalfa was grown in nutrient solutions with varying amounts of sodium chloride. An increase in yield over the checks was observed in five out of six cultures. He considered that sodium chloride might function as a mineral nutrient.

It has been noted by later investigators that chlorine markedly increases the water content of tissue, affects the metabolism of carbohydrates, and the production of chlorophyll. Thus Garner, McMurtrey, Bowling, and Moss (1930) showed that the application of 20 to 30 lb. of chlorine per acre to certain types of soil stimulated the growth of tobacco plants so that the average increase in production amounted to 10 per cent. The immediate effect of muriate of potash is to reduce decidedly the symptoms of magnesia deficiency in the plants. Chlorine has a marked effect in increasing the water content of the tobacco plant during the growing season, and thus enables it to resist injurious desiccation. This behavior protects the leaves against the type of injury known as "drought spot." An excess of this element, however, may seriously interfere with the metabolism of carbohydrates, and this adverse nutritional effect may result from an application of chlorine in amounts as low as 40 lb. per acre. Wilson (1933) noted that an excess of chlorine interferes with normal carbohydrate metabolism in the tobacco plant so that the leaves have a high content of starch. The leaves become thick and brittle, curl upward at the margins, and have a high water content and a low amount of organic acids. The cell sap of plants grown in the field in the presence of excessive amounts of chlorides had a lower osmotic concentration and less bound water than the sap of plants to which a liberal application of complete fertilizer had been given. Watson (1936) found that the only significant result of the application of potassium chloride, on the carbohydrate metabolism of the potato was a marked reduction in the amount of sucrose in the leaves during midday. The amount of starch and reducing sugars was not noticeably affected. The rate of translocation of carbohydrates during periods of darkness was unaffected, but the rate of the removal of dry matter was increased. By augmenting the application of potassium chloride, the rate of photosynthesis was apparently increased in the afternoon and decreased in the morning. The percentage of water in the plant was also increased by the application of this fertilizer. These variations were evident when the application amounted to 25 to 100 lb. of potassium oxide per acre. Maiwald (1923) noted that the chlorophyll content of potato leaves was sharply reduced by the accumulation of the ions of potassium and chlorine, which apparently inactivated the iron in the plants. Dhein (1929) noticed that large applications of potassium chloride caused a reduction of the chlorophyll in the plants studied. Basslavskaya and Syroeshkina (1936) working with potatoes found that the action of chlorine was to reduce the amount of chlorophyll, and this action was more clearly shown on the plants in field plots than on potted plants.

Chlorine in many cases has a morphological effect upon the plants growing in the medium to which it is applied. Harter (1908) found that plants of wheat, oats, and barley grown from seeds in soil containing 0.7, 1.0, and 1.4 per cent, respectively, of sodium chloride developed a waxy bloom upon the leaf surface and a thickening of the cuticle, and this thickening increased with the concentration of the solution. The size of the epidermal cells decreased as the concentration of the salt in the soil increased, but when the amount of sodium chloride present was much below the minimum concentration that is injurious under field conditions no perceptible modification of the plant structure occurred. Hendry (1918) grew Windsor, kidney, lima, Red Mexican, Terpary, and other varieties of beans in sand to which were added 0.04, 0.16, 0.3, 0.5, and 1.0 per cent, respectively, of sodium chloride based on the dry weight of the sand. The visible effects of the sodium chloride upon the development of the plants were retardation of height, a reduction in the number and size of the leaves, a retardation of the blooming period, and a reduction in the number and size of the nodules formed. It was noted by Hendry and also by others that there is a distinct difference in the ability of different species of plants to withstand the toxic effects of chlorides when they are applied in excessive amounts to the medium in which the plant is growing, and Harrison and King (1925) have shown that the age of the plant is an important factor in determining the degree of toxicity of sodium chloride.

Asparagus is an example of a plant that can tolerate a large amount of sodium chloride and the salting of asparagus beds with sodium chloride has long been a traditional practice among gardeners. Rudolfs (1927) applied sodium chloride at the rate of 150. 300, and 500 lb. per acre to asparagus beds for 2 years. The application of 150 lb. did not increase the number of stalks over plants on unsalted plots. Doubling the amount of salt resulted in an increase in average number of stalks from six to seven, while an application of 500 lb. per acre brought the yield per plant up to seven and one-half stalks.

The manner in which the chlorides may bring about the responses that have been mentioned is not known. All the explanations that have been put forth are purely theoretical. Tottingham (1919) considered that the apparent relation of the chlorine content to the accumulation of starch in the potato tuber is a suggestion that enzyme activity might be the main controlling condition for this effect. The fact that chlorides have a stimulating effect upon the enzymes obtained from plant sources would seem to lend strength to this general theory. Since many of the vital activities of the plant are apparently controlled by enzyme action, the physiological responses might be brought about in this manner. Rudolfs (1919) from his observations on the effects of sodium chloride on trees considered that it is possible that the chlorine increases the acidity of the cell, which could accelerate the vital activities or enzymatic actions to a certain extent, but that by a too large increase of acidity the acceleration would change into a toxic action.

Our knowledge of the influence of chlorine upon the metabolism of the green plant is extremely meager. Tottingham (1919) is of the opinion that the most valuable general conclusion that can be drawn from the review of the literature is that the influence of chlorine upon plants seems to be impossible of any general statement. The facts observed indicate that the effect of chlorine will depend upon the nature of the plant under consideration and upon the nature of the soil or other medium in which the plant is growing, other than the chlorine content. Some consider that the influence of chlorine is in many cases indirect in that the reactions of the plant are due to changed conditions in the mineral content of the soil induced by the applied chlorides and not to their direct action upon the plant cells (Wheeler and Hartwell, 1902; Prianischnikow, 1905). Trelease (1917, 1920) reported extensive experiments which greatly illuminate this subject. He found in growing wheat in water cultures that if the solutions were properly balanced, the same growth was obtained regardless of the addition of chlorine. If the cultural solutions, however, were not properly balanced, the addition of chlorine apparently diminished to some extent the retarding effect produced by the unbalanced solution, and a better growth was obtained. The addition of an unessential element may thus improve the properties of the medium that contains the essential ones in improper proportions.

The action of chlorides on plants is evidently influenced also by the general climatic conditions that prevail during growth. This fact is suggested by Tottingham (1917) and also by Lipman, Davis, and West (1926), who obtained evidence in the study of the tolerance of wheat, barley, and peas to sodium chloride that led them to state that the environmental conditions are a very important determinant of the kind of results that may be obtained in studying the influence of chlorine upon plant life.

L. Aluminum

1. Occurrence.—Aluminum has been found in the ash of all plants that have been examined. The amount of this element varies greatly for different plants, and Kratzmann (1914), who analyzed 130 species of plants, considered that they possess specific selective powers toward aluminum. As a rule, the percentage of aluminum in plants is very small, but, in a few cases, as certain species of club mosses, aluminum approximates 22 to 39 per cent of the ash. Latshaw and Miller (1924) found that aluminum made up 0.07, 0.013, 0.023, and 0.98 per cent of the dry weight of the leaves, stem, grain, and roots, respectively, of corn plants that had reached their full vegetative growth and development. Berthelot and André (1895) also stated that the amount of aluminum in the roots is much higher than that in the leaves. There is some question, however, in regard to the relatively high amounts of aluminum reported in the roots, since the adhering particles of soil even in exact preparation of material could exert a marked influence on the results. Meyers and Voegtlin (1914) analyzed several of the grain and vegetable foods for aluminum compounds with the results shown in the table on page 357.

At least 65 per cent of the aluminum in corn and carrots was soluble in water, while the total aluminum content was completely soluble in 0.3 per cent of hydrochloric acid. Only a trace of the aluminum present in wheat flour and in the sunflower seed was soluble in water. McLean and Gilbert (1927) found in the case of corn, cabbage, peppers, redtop, lettuce, and timothy that the greatest accumulation of aluminum

Food	Percentage of aluminum on dry basis	Food	Percentage of aluminum on dry basis
Corn	0.120	Sunflowers (hulled)	0.401
Corn oil cake	0.810	Cotton seed meal	0.996
Rye	0.273	Carrots	0.301
Millet	0.428	Irish potato	0.126
Wheat flour	0.045		

is in the cortex of the roots. In the cell they observed that it accumulates in the protoplasm rather than in the cell walls or vacuoles and that it is especially abundant in the nucleus. Little or nothing is known concerning the form in which aluminum occurs in plants.

Levy (1931) examined over 75 species of plants belonging to 37 families and found aluminum in all the phanerogams examined. The accumulation of this element was more rapid during early than during later growth. It was more abundant in leaves than elsewhere and more plentiful in green than in pale and etiolated leaves. As compared with iron and zinc, the amount of aluminum was low in seeds. In the vegetative parts of plants, aluminum was present in about the same quantity as iron.

2. Influence upon Plant Life.—The role of aluminum has been studied along three general lines: (*a*) the effect of this element upon growth and production, (*b*) the effect upon the soil relationships, and (*c*) the role of this element if any in the functions of the plant cell.

a. Growth and Production.—Yamano (1905) found that 0.2 per cent of ammonium alum injured wheat and rye plants in nutrient solutions after 3 weeks and that 0.8 per cent killed the plants in a few days. Prianischnikow (1911) grew wheat, oats, barley, peas and buckwheat in sand cultures fertilized with aluminum phosphate alone and with calcium carbonate alone. He concluded that aluminum phosphate is assimilated and that calcium carbonate has no appreciable depressing effect on the assimilability of aluminum phosphate. Baguley (1912) made comparative tests with the orthophosphates of iron, calcium, and aluminum on oats, peas, and Swedish turnips grown in sand and chalk. He obtained better results with the iron and aluminum phosphate than with calcium phosphate. Truog (1913) working at Wisconsin with 10 different kinds of plants concluded that precipitated ferric and aluminum phosphates produced, with but few exceptions, as good growth as, and in a few cases better growth than, the acid phosphate. Kratzmann (1914) found that salts of aluminum in a concentration of 0.005 per cent hindered the growth of corn, horse beans, lentils, and sunflowers, while a solution of 0.0001 per cent somewhat increased growth. Ruprecht (1915) found that a solution of aluminum sulphate greater than 40 p.p.m. had a very toxic action on clover seedlings. The effect of the aluminum salts was localized in the cells near the root hairs, and he attributed the death of the plants grown in aluminum salts to starvation incident upon the inability of the plant to obtain nutrient salts for normal metabolism. Miyake (1916) showed that aluminum chloride is toxic to rice seedlings in concentrations greater than 0.000133 N and that the toxicity of the salt is not due to the hydrogen formed by the hydrolysis of the salt solution. Mirasol (1920) reported that in the case of barley growing in aluminum sulphate the solution is toxic up to 0.0001 N but with a concentration 0.000033 N it seemed to have a stimulating effect. McLean and Gilbert (1927) found that lettuce, beets, timothy, and barley are the crops most sensitive to aluminum poisoning. The medium-sensitive crops are radishes, sorghum, cabbage, oats, and rye, while corn, turnips, and redtop are noticeably resistant to the toxic effects of aluminum, redtop being the most resistant of the three. Later (1928) they observed that very low concentrations of aluminum (3 to 13 p.p.m.) were stimulatory to plants, while higher concentrations were toxic. Nondialyzing forms of aluminum that could not pass through a collodion membrane in sufficient amount to cause toxicity were decidedly toxic when placed in contact with barley roots. High acidity is not necessarily an accompaniment of toxicity, since aluminum citrate is toxic at pH 6.5 or less acid.

Gilbert and Pember (1931) found that seedlings of Cos lettuce, growing in culture solutions with a pH range of 3.2 to 7.5, showed no appreciable differences in the production of dry weight, but if aluminum sulphate was added to the solutions, the yields in dry matter were considerably reduced. The yields in dry weight of barley plants, grown in soils of various and widely distributed types, were more closely correlated with the active aluminum content of the soil than with its acidity. The same authors (1935) showed that the various weed species commonly found in competition with lawn grasses vary greatly in their sensitivity to aluminum when grown in nutrient cultures. They suggested that the toxicity of aluminum in the soil solution might be one of the causes of the inhibition of the growth of weeds on acid soils. Eisenmenger (1935) noted that organic aluminum compounds exerted a distinct toxic effect on plants of corn, soybeans, and buckwheat and found that this toxicity could be most effectively overcome by the application of calcium hydroxide. He believed that the presence of phosphate lowers the amounts of ionic aluminum in the soil. He considered that low concentrations of aluminum served to increase slightly the rate of growth of these plants.

b. Relationship of the Soil to the Plant.—It is not the intention here to enter into any detailed discussion of the chemical reactions that concern aluminum in the soil but to mention only a few observations in which the beneficial results of lime or fertilizers are apparently due to their action upon the aluminum compounds of the soil. Hartwell and Pember (1918) noted that in treating acid soil with lime the growth of rye was very little benefited but that the yield of barley was increased from two- to threefold. In an acid nutrient solution, however, these two plants were affected equally, while aqueous extract of the acid soil like the soil itself affected the two kinds of plants very differently. These facts indicated that the toxicity of the so-called "acid" soils is not due only to the acid but to some ingredient not present in the ordinary nutrient solution. After a search for the cause of this toxicity they decided that aluminum is the causative agent under conditions such as have been described. The application of phosphoric acid or acid phosphate, while it increases the acidity of the soil, has a beneficial effect on the crop, since the solubility of the aluminum in the soil is reduced by the phosphate. Their results indicate that the practical advantage of phosphating and liming may often be due to the precipitation of active aluminum, which is toxic to many plants quite as much as supplying phosphorus as a nutrient or lime as a reducer of acids. Conner and Sears (1922) likewise considered that the toxicity of aluminum salts is due to the aluminum ion more than it is to the hydrogen ion on such plants as barley, and that this toxicity is reduced when much phosphate is used in the nutrient solution. They considered that acid soils are toxic to many plants largely because they obtain easily soluble aluminum salts. Magistad (1925) noted the difference in plants in the reaction to aluminum and considered that the benefits obtained in practice through the use of lime in the case of clover, alfalfa, oats, and rye result usually from a decrease in acidity and not from a decrease of aluminum present. In corn and barley, however, over the same range of acidity this benefit results from a decrease in both acidity and soluble aluminum. Hoffer and

Carr (1923) and Hoffer and Trost (1923) in investigations on corn diseases reported on the accumulation of iron and aluminum compounds in the nodal tissues of the plant, which causes a purplish brown discoloration of these tissues and finally disintegration. They found that the accumulation of aluminum in these tissues is affected by the genetic composition of the strain of corn and by conditions of the soil. When abundant aluminum injury occurs in the corn plant it is an indication that the soil is deficient in available phosphates. This injury from aluminum can be corrected by the application of lime and phosphates to the soil. Arndt (1922) likewise observed that the composition of the nutrient solution had a marked influence in determining the toxicity of aluminum salts. Coville (1923) obtained striking results in the growth of plants of *Rhododendron catawbiense* by the addition of aluminum sulphate to a rich garden soil in which the plants were growing. One set of plants was treated with distilled water, another with distilled water and magnesium sulphate, and a third with distilled water and aluminum sulphate. The increase in the diameter of the rosettes of leaves was 0, 30, and 250 per cent, respectively. Potted plants that for more than a year had made sickly growth in rich garden soil were revived in an 11-week period by two mild applications of aluminum sulphate. The leaves of the treated plants became a bright green color, while those of the controls remained sickly and dull. Coville believed that the aluminum sulphate acts to replace lime with aluminum, the released lime leaching away in the form of calcium sulphate, and as a result the soil reaction is changed from neutral or alkaline to acid, and this is a favorable medium for this plant, which thrives best on an acid soil.

c. Function in the Cell.—The physiological role of aluminum in plants is not definitely known, although marked reactions of plants due to this element have frequently been observed. Fluri (1909) found that aluminum salts cause the disappearance of starch from certain water plants. He used the salts: aluminum nitrate, sulphate, chlorate, and bichromate and observed that these salts caused a disappearance of starch even when the plants were well illuminated. The aluminum found in the cell was small and the results could not be attributed to chemical reaction. Fluri considered that the disappearance of starch under these conditions is probably due to three reactions induced by the aluminum salts: (1) increased permeability of the protoplasm causes a more rapid loss of sugar, (2) the diastatic action is increased, and (3) the action of photosynthesis is slowed down. Meurer (1909) considered that aluminum forms some chemical compound with the pectic bodies of the middle lamella, since he found that tissues killed with chloroform absorbed aluminum as readily as living tissue. Szücs (1913) believed that aluminum salts cause the protoplasm to set or harden, while Stocklasa (1911, 1918, 1922) considered that they combine with the cell wall to form colloidal salts, an action that renders the cell less permeable. He suggested that aluminum may act as a catalytic agent performing a function in the assimilation of carbon by increasing the rate of photosynthetic action. Mazé (1915) considered that it is an essential element in the growth of corn.

The influence of aluminum upon the color of the flowers of *Hydrangea hortensis* was discovered by Molisch (1897). It had long been known that this plant when grown in certain soils had reddish flowers and in other soils blue flowers. Cultures of soil treated with potassium sulphate produced plants that bore the ordinary reddish flowers, while the soil treated with aluminum sulphate produced plants that bore blue flowers.

Later work by various investigators indicates that the coloring of these flowers is dependent upon the reaction of the medium in which they grow as well as upon its composition. Thus Connors (1923) noted that plants of *Hydrangea*, growing in soils with a pH of 5.0, or lower, consistently produced blue flowers while those in a soil of pH 6.6, or higher, produced pink flowers. He was unable to determine whether the color of the flowers was influenced by the presence of iron or aluminum. Allen (1931) found in spraying the flowers of *Hydrangea* at various stages of development with dilute solutions of aluminum sulphate that a blue coloration of the flowers resulted. In general, the same results were obtained when the solutions were injected into the stem, the flowers being most affected on the side of the injection. He obtained no effects on the coloration of the flowers when they were sprayed with various salts of iron, copper, magnesium, or potassium. Later (1934) he attempted to determine the most efficient method of maintaining the aluminum in an available form in the soil, and to determine the effects of various materials upon the acidity. The best results were obtained from the use of aluminum sulphate, but it was difficult to determine the rate of application to a given soil in order to lower the pH sufficiently for the optimum solubility of the aluminum. Chouard (1933) reported that iron, chromium, and aluminum could produce the bluing agent in the flowers of *Hydrangea*. Poesch (1935) applied aluminum sulphate to the soil at the rate of 1 lb. of salt to 5 gal. of water, or in the dry form by using 0.5 teaspoon of salt per 6-in. pot. When the soil gave a reaction of pH 6.0 or lower, bluing of the flower resulted, the intensity of the color varying with the variety of plant used.

M. Sodium

Although sodium is not generally considered essential to the growth and development of the plant, its presence apparently has a marked effect upon the growth of some plants under certain conditions. Hellriegel (1898) found that sodium salts always gave increases in yield even when potassium salts were present in quantity. The same fact was also observed by Breazeale (1906) for wheat plants growing in culture solutions. Wheeler, Hartwell, Pember, and others (1904 to 1913), however, obtained no increase in growth by the addition of sodium when an optimum amount of potassium was present, but they observed, as did Breazeale (1906, 1916), that a larger amount of potassium was left in the solution by the growing plants when the potassium was supplemented by sodium. Sodium thus seems to retard the absorption of potassium or to conserve it in the metabolism of the plant. Wheeler, Hartwell, and others found that when the deficiency of potassium was great enough to bring about a 30 per cent depression in the green weight of plant produced, the addition of sodium gave an increase in growth amounting to 10 to 25 per cent within a period of from 2 to 3 weeks. Hartwell and Wessels (1913) observed no decrease in the percentage of sugar in the onion when the potassium deficiency was supplemented with sodium, although there was a decrease in the crop corresponding to the scarcity of potassium. In the case of mangel roots there was a tendency toward depression in both the total and the reducing sugars as a result of the addition of sodium to supplement a potassium deficiency. Pfeiffer, Rippel, and Pfotenhauer (1920) stated that sodium salts may to some extent replace potassium salts as a nutritive material for oats. In the production of dry matter sodium replaced potassium in the molecular ratio of 0.25:1. When relatively small quantities of potassium and relatively large quantities of sodium are present, this ratio is increased to 0.38:1. When such a replacement occurs, the sodium appears in the grain as well as in the leaves and stems.

According to Vincent and Herviaux (1929) the soluble chlorides in the leaves of barley at the blooming stage were 80 per cent potassium chloride and 20 per cent sodium chloride, while in the grain the ratio of potassium chloride to sodium chloride was 1.2:1. In young cabbage leaves, 60 per cent of the soluble chlorides was sodium chloride. They concluded that the quantity of sodium present in cultivated plants is largest in the organs of greater physiological activity. In beets the use of sylvanite, rich in sodium chloride, resulted in higher yields than the use of potassium chloride.

Apparently sodium is sometimes extremely toxic to certain plants. Thus Beaumont (1932) found in fertilizing "run out" pastures in Massachusetts that sodium nitrate was toxic to *Polytrichum commune.* He considered that toxicity was linked with anionic and cationic effects, since the nitrates were more toxic than the corresponding chlorides. The effects were found to be both immediate and cumulative since several smaller applications gave the same result as a larger application.

Sodium salts are apparently necessary for numerous marine algae, according to Osterhout (1912). His experiments show that the replacement in sea water of sodium by ammonium, calcium, magnesium, barium, and others is decidedly injurious. Sodium is apparently not needed for osmotic pressure in this case, since an equivalent concentration of other compounds does not satisfy the plants. Osterhout could not determine the value of sodium as a nutrient but considered that it might serve to antagonize the toxic action of other salts that are present in the sea water. From the experimental evidence at hand it seems that sodium may be able to perform one or more of the following roles depending, however, upon the environmental conditions and the type of plant under consideration: (1) It appears to be a conserver of potassium, since less is absorbed by the plant when sodium is added. (2) It may to a limited extent replace potassium as a plant nutrient (Schulze, 1913). It can apparently perform certain functions of potassium but by no means all of them. It may delay potassium starvation for the plant but cannot keep it off entirely. (3) When applied to a soil that is deficient in potassium it may liberate a certain amount of potassium which was adsorbed by the soil and make it available for the plant, thus bringing about an increase in growth. The experimental evidence with water cultures, however, makes it appear doubtful whether sodium functions in this manner, although it is doubtful whether water cultures would indicate anything in regard to base exchange that might occur in the soil. (4) In certain specific cases, it evidently acts as an antidoting agent against certain toxic salts present in the medium in which the plant is growing.

N. Copper

Brenchley (1914) concluded from a careful survey of the literature to that date that copper compounds under the most typical conditions act as poisons to the higher plants and that it is only under particular and peculiar conditions and in very great dilution that copper shows any stimulating action. It is, however, very widely distributed in plants and in considerable quantity. McHargue (1925) found that the copper content of various plants and plant parts ranged from a trace to 46 p.p.m. of dry material, while Maquenne and Demoussy (1920) found 0.25 mg. of it per liter of expressed sap of the potato and as high as 40 mg. per kilogram of dry leaf substance of lettuce.

Webster and Jansma (1929) examined wheat from 14 states and found that the amount of copper in the various samples of grain varied from 281 to 497 mg. per kilogram of ash. Pitcaithly and Worley (1933) reported that the amounts of copper in the leaves, bark, kernel, and fresh fruit of the Karaka tree (*Corynocarpus laevigata*) were 7.0, 2.7, 4.2, and 2.7 mg., respectively, per kilogram of fresh material. The response of plants to copper has been noted in its effects on increasing vigor and yield, on quality, on the control of chlorosis, exanthema and "frenching," and on the prevention of "salt sick" in live stock.

Marquenne and Demoussy (1920) found copper in the greatest abundance in cells that are active in growth and considered that its translocation is controlled by processes accompanying metabolism. They further found that the addition of other solutions of copper sulphate to the extent of 0.02 to 0.2 mg. per liter increased not only the length of life of various plants but also their dry weight (Forbes, 1917). A wide range of plants has been found by Allison, Bryan, and Hunter (1927) to respond in growth and production in a remarkable way to the application of 30 to 50 lb. per acre of copper sulphate to the raw peat soils of the Florida Everglades.

Sommer (1931) stated that amounts of copper as low as 0.06 mg. per liter of nutrient solution produced a marked effect on the dry weight of the plants growing therein as shown by the following table:

Plant	Weight of plants, in grams	
	Without copper	With copper
Sunflower	0.16	0.70
Tomato	0.30	2.60
Flax	1.40	4.50

Lipman and Mackinney (1931) reported that barley plants were unable to form seed without having copper in the root medium. Amounts as low as $\frac{1}{16}$ to $\frac{1}{8}$ p.p.m. were sufficient to produce seed.

Felix (1927), Knott (1932) in New York, and Harmer (1932) in Michigan have obtained marked results in increasing the yield and quality of onions and lettuce by the application of copper sulphate to muck soils. The amount of this salt necessary to produce the desired results varied from 100 to 300 lb. per acre, depending upon the kind of crop and the type of soil. Painting the leaves of lettuce with a dilute solution of copper sulphate or dusting them with 20–80 copper-lime dust at the rate of 75 lb. per acre had similar effects to the application of copper sulphate to the soil. Andressen (1932) reported that deciduous fruit trees affected with chlorosis, rosetting, and dieback were cured by the application of copper sulphate to the soil in amounts ranging from ¼ to 2 lb. per tree. The application of this salt was reflected in the increased copper in the leaves. It was also found that leaves that had been dipped in very dilute solutions of copper sulphate became green within 2 weeks. The treatment of the soil with potassium, magnesium, manganese, sulphur, iron, or manure had no effect on these maladies. This gave further evidence that the lack of copper was the cause of the disturbances. The ash content of the chlorotic parts of these plants was always much higher than that of the healthy material. Oserkowsky and Thomas (1933) found that Bartlett pear trees affected with xanthema could be cured or markedly improved by spraying with Bordeaux mixture or by the introduction of soluble copper-salts into the trunks. Similar treatments with iron citrate, manganese chloride, and zinc sulphate, respectively, were without effects. The amount of copper in the diseased leaves was lower than that of healthy leaves. It was found by Burge, Wickwire, and Orth (1933) and Orth, Wickwire, and Burge (1934) in Florida that the surface application of 4 lb. of copper sulphate around each of several 3-year-old orange trees, whose leaves were spotted, yellow, or frenched, greatly improved the condition of the trees. Chlorophyll determinations made 4 months after the application show that the leaves of the treated plants contained 4.6 times more of this pigment than did the leaves of the control trees. It was observed also that spraying with Bordeaux mixture greatly increased the vigor of the diseased plants. Willis and Piland (1934) noted that the application of copper sulphate to newly limed soils corrected the excessive absorption of iron by the corn plant. Haas and Quayle (1935) studied the copper content of the leaves and fruit of citrus and found that the amount of this element necessary for healthy growth varies with the soil and climate, although the solubility of the copper may be of considerable importance. Hopkins (1933) with cultures of *Lemna* and *Chlorella* could find no increase in growth that he could attribute to addition of copper.

Neal, Becker, and Shealy (1931) found that the forage plants from certain regions of Florida contained a small amount of copper due primarily to the scarcity of this element in the soils in which they grew. The cattle, sheep, or swine that consumed these plants developed salt sickness. This disease may be prevented or controlled by the application of copper to the soils in which the plants are growing or by the addition of copper to the ration of the animals. There is no definite knowledge concerning the role of copper in plants, and all ideas relative to its uses are based purely upon theory and not fact.

O. Zinc

Mazé (1915) considered that zinc is one of the elements essential for the best development of the corn plant, and his statement was verified to some extent by Lipman, Davis, and West (1926). Nakamura (1904), however, could observe no stimulatory effect of zinc upon the growth of rye, peas, onion, or mustard. Conner (1920) observed that the zinc salts, formed by the action of acid soils on galvanized iron containers, were very toxic to oats and buckwheat growing therein. Finch, Albert, and Kinnison (1933) reported that the pecan tree is susceptible to zinc chlorosis.

Sommer and Lipman (1926) furnished evidence that boron is indispensable to the development of Windsor bean, buckwheat, flax, mustard, castor bean, cotton, and barley and that zinc is essential to the dwarf sunflower and barley. Sommer (1928) noted in the case of buckwheat that the plants grown without zinc were much smaller than the controls. The average dry weight per culture grown without zinc was 1.94 g., while that with zinc was 8.45 g. In *Vicia faba* there was no visible difference between plants grown with and without zinc until the flowering stage was reached. Then most of the flower buds fell off, and if pods developed no seeds were formed when zinc was lacking, but where it was supplied, the plants bloomed freely and produced seed. The plants of buckwheat and sunflower that survived in the absence of zinc had the appearance of plants that had grown on poor soil.

Roberg (1932) observed in cultures of *Chlorella vulgaris* that 0.001 mg. of zinc per 100 cc. of nutrient solution stimulated growth but that 0.005 mg. per 100 cc. of solution retarded growth under all conditions. In an examination of the leaves of 11 different species of plants, Bertiand and Andreitcheva (1933) found that the zinc content ranged from 0.36 mg. per kilogram of fresh tissue in endive to 8.9 mg. in the same weight of cress. They considered that a high zinc content in leaves is correlated with a high chlorophyll content.

It was found by Chandler, Hoagland, and Hibbard (1933) that the spraying of peaches and apricots in the spring with zinc-lime when the symptoms of "little leaf" were most pronounced, improved the fruit but did not increase the amount of fruit set. The injection of zinc sulphate into holes 1.5 in. deep and 3 to 4 in. apart around the trunk or branch gave the most dependable results of any method of application used except spraying in the case of citrus and grape. These benefits seldom last longer than through the second summer. Little leaf is apparently a symptom of a deficiency of zinc. Hoagland, Chandler, and Hibbard (1935) reported that if zinc was withheld from nutrient solutions in which apricot, tobacco, squash, corn, mustard, tomato, and other plants were growing, these plants developed symptoms of little leaf.

Mowry and Camp (1934) found that the bronzing of tung trees can be overcome by a treatment with zinc sulphate applied either as a spray or directly to the soil. The fact that the spraying of the aerial parts of the tree corrects the trouble would indicate that the zinc itself is responsible for the recovery and that it is not a case of this element replacing or rendering soluble in the soil some other ion in which the plant is deficient.

It was observed by Reed and Dufrenoy (1935) that there was renewed activity in the cells of the mottled leaves of citrus and an accelerated growth of new shoots subsequent to the application of a spray of a very dilute solution of zinc sulphate. In the green leaves of new shoots whose growth had been promoted by this spray, neither calcium deficiency nor phloem necrosis was evident, and the chloroplasts developed to a fair size and formed starch. Barnette and Warner (1935) and Barnette, Camp, Warner, and Gall (1936) found in Florida that the chlorotic condition of corn plants, termed "white bud," could be prevented or overcome by the application of 20 lb. of zinc sulphate per acre mixed with an inorganic fertilizer and an alkaline peat. Following this application, the chlorotic plants regained their green color, renewed growth, and produced grain. They further found that peanuts, oats, velvet bean, cowpeas, millet, sugar cane, and Napier grass responded favorably to the application of zinc sulphate.

P. Iodine

According to Wynd (1934) the presence of iodine in the higher plants was first verified by Chatin (1850). It occurs in the plant apparently in both the inorganic and organic forms. Masuda and Nishida (1934) found in *Lamnaria japonica* that the amount of organic iodine varied from 13 to 85 per cent of the total iodine present. It has been shown by Stocklasa (1924, 1930), Scharrer and Strobel (1927), Scharrer and Schwaibold (1927), Hiltner (1928), Orr, Kelly, and Stuart (1928), McHargue, Roy, and Pelphrey (1930), Wendt (1930), Conner (1931), Mitchell (1931), Beaumont and Karns (1932), and McHargue, Young, and Calfee (1935) that plants absorb iodine more or less proportionately to the amount in the medium in which they grow, although because of reactions in the soil this fact at times is not very evident. Some of the plants that have been shown by these investigators to increase their content of iodine when this element was applied to the soil are beets, lettuce, peas, spinach, radishes, bean, turnip, red clover, oats, and various forage crops. The iodine content of these plants was increased from two to sixteen times over that of the controls. The amount of iodine in the soil has been increased by the application of potassium iodide or potassium iodate, or of raw phosphate rock, Chile nitrate, or limestone, which contain considerable quantities of iodine as impurities.

Frear (1934) analyzed 135 samples of potatoes for iodine from as many localities in Pennsylvania. The minimum iodine content on a dry basis was 10 p.p.b., the maximum 216 p.p.b., and the mean content 77.8 p.p.b. McClendon, Barrett, and Canniff (1934) found, from the analysis of 55 samples of potatoes in Minnesota, that the average iodine content in the eastern section was 0.085 mg. per kilogram and in the western section 0.226 mg. per kilogram. The cases of simple goiter per 1,000 drafted men during the World War was 19 for the eastern and 7.5 for the western portion of that state. This indicates an inverse relation between the amount of iodine in potatoes and the proportion of goiter in the drafted men.

It was found by Suzuki and Aso (1903) that an application of potassium iodide at the rate of 0.5 g. to 20 sq. m. of soil surface increased the production of oats, and a dosage of 0.05 gm. for the same surface had a beneficial influence on radishes. Brenchley (1924), however, could not find any significant response of mustard and barley to the application of iodine. She suggested that it might have some specific and effective action if means could be found to prevent it from being so rapidly converted from the elemental state to the form of compounds, which ordinarily occurs when it is applied to the soil. Stewart (1930) in Hawaii showed that iodine in the presence of copper was beneficial to the production of sugar cane. Wendt (1930) noted that root systems of various garden crops were larger, and that the flowering and fruiting of these plants were earlier when iodine was applied to the soil at the rate of 6 to 40 g. per hectare. It was observed by Cotton (1930) that buckwheat plants perished within 3 weeks when growing in nutrient solutions to which 40 to 127 p.p.m. of iodine had been added. The growth of these plants was markedly depressed when this element was added to the solution in quantities as low as 1.2 to 12 p.p.m. Wynd (1934) added potassium iodide to Shive's three-salt solution in portions of 1, 5, 10, and 20 p.p.m. of iodine, and found that all these concentrations had a depressing effect upon the growth of plants as indicated by a loss of green color and a progressive dropping of the lower leaves.

Stocklasa (1929) reported that certain halophytes growing in iodine-free and in iodine-containing mediums showed marked differences in growth. It appears that iodine and chlorophyll are closely associated, and that the presence of this element creates a condition favorable to the production of greener leaves. It was observed by Marine, Baumann, and Webster (1930) that the iodine-absorbing materials are present in the greatest concentration in the rapidly growing portions of plants but decrease during maturation. These iodine-absorbing materials decreased rapidly during drying and slowly during storage. It is likely that they belong to the oxidation system. At a concentration of 1 p.p.m. of potassium iodide in the nutrient solution, Wynd (1934) observed that the amount of peroxidase and invertase present in the plant and the rate of respiration decreased. These greatly increased, however, at higher concentrations, while oxidase activity was progressively lowered.

Q. Selenium

Selenium occurs in soils in varying amounts over wide areas, and in certain regions of Nebraska, South Dakota, Wyoming, and probably in other regions it is present in relatively large quantities. Some of the plants growing on these soils absorb a considerable amount of this element and are toxic or poisonous to the animals that consume them. According to Byers and Knight (1935), Byers (1935), Hurd-Karrer (1935), and Robinson (1936) the quantity of selenium absorbed by plants and their subsequent toxicity depend upon numerous factors. Chief among these are: the composition of the soil, the moisture relations of the soil, the concentration of this element in the soil, the species of plant, the stage of its development, the available sulphur, and the crops previously grown.

Hurd-Karrer (1933, 1934) found that wheat plants grown in soil to which sodium selenate had been added at the rate of 15 p.p.m. or less became characteristically chlorotic. The young leaves frequently became almost snow white, with green tips and green midveins. When the selenate was added to the pots containing older plants, white chlorosis appeared only on the leaves emerging subsequent to the addition of selenium. This investigator (Hurd-Karrer) noted that the toxicity of sodium selenate is determined by the amount of sulphur available to the plants. Thus in cultural solutions, concentrations of selenium as low as 0.1 p.p.m. produced distinct injury after a few weeks if no sulphate was present, whereas a concentration of 18 p.p.m. was required for this degree of injury if the solution contained 192 p.p.m. of sulphur. There was no visible injury to the plants when the proportion of selenium to sulphur was 1:12 or less. When the ratio was 1:8, or greater, the plants were chlorotic and stunted, while when the ratio was as high as 1:2, growth was almost completely inhibited. Hurd-Karrer (1935) grew 17 different crop plants in the greenhouse in Keyport clay loam to which had been added 5 p.p.m. of selenium in the form of sodium selenate. Under these conditions the mustard plants absorbed the greatest amount of the element, 1,240 p.p.m., while sorgo absorbed only 130 p.p.m.

It was found by Beath, Eppson, and Gilbert (1935) in Wyoming that plants of *Astragalus racemosus*, *A. bisulcatus*, *A. scobinatulus*, *A. pectinatus*, and *A. flaviflorus*, among others, were absorbers of selenium when growing in the proper soil. A group of plants, including *Atriplex*, *Solidago*, and several native grasses, sometimes gave negative and sometimes positive tests for selenium, while a large number of plants including white and purple loco, wheat grass, soapweed, four species of *Astragalus*, foxtail, wild flax, and Canada thistle did not absorb a sufficient quantity of selenium to be detected. Trelease and Martin (1936) in reviewing the literature on the absorption of selenium stated that western wheat grass growing on various soils accumulated 1 to 60 p.p.m. of selenium, while *Astragalus bisulcatus* growing on the same soils accumulated 200 to 4,300 p.p.m. The highest accumulation of this element, 8,840 p.p.m., was found in the latter plant. The difference in the ability of plants to absorb and store selenium is illustrated by *Astragalus bisulcatus* and *A. missouriensis* growing on a soil containing 2.1 p.p.m. of selenium. The former plant accumulated 1,250 p.p.m., while the latter under identical conditions stored only 3.1 p.p.m. On one type of soil containing 2.5 p.p.m. of selenium, wheat plants accumulated 45 p.p.m., while in a different soil containing 3.0 p.p.m. the plants absorbed only 0.5 p.p.m.

According to Levine (1925), selenium compounds in concentrations of 0.01 per cent and above exerted a detrimental influence on the growth of white lupine and timothy seedlings. However, he obtained increased growth of these plants in solutions of selenium dioxide and selenic acid with concentrations of 0.0001 and 0.001 per cent, respectively.

R. Other Elements

The influence upon plant growth of barium, strontium, titanium, thallium, vanadium, arsenic, fluorine, rubidium, cobalt, and nickel has been observed by numerous investigators. McHargue (1919) in his experiments with cowpeas, soybeans, oats, wheat, and corn grown in sand cultures found that neither barium nor strontium compounds can be considered important plant nutrients, but the presence of a small amount of the carbonate of each of these elements in the presence of calcium carbonate gives increased yields that are noteworthy. Both the barium and strontium carbonate accelerate especially the growth of the roots of plants under consideration. The compounds of both these elements are toxic to plants in the absence of calcium carbonate. He also found, as did Suzuki (1900), that neither can replace calcium in any degree in the metabolism of the plant. Osterhout (1916) noted that *Spirogyra* underwent a peculiar cytological reaction when placed in a 0.0001 *M* solution of barium chloride. The reaction observed was a characteristic contraction of the chloroplasts, while the protoplasm remained intact in its normal position. Numerous other salts including calcium chloride, magnesium chloride, manganese chloride, nickel chloride, cobalt chloride, sodium chloride, and potassium chloride were tried in the same manner but none produced this peculiar result. This reaction suggests that there is some specific constituent of the protoplasm of *Spirogyra* that reacts in this peculiar manner to barium.

The reaction of plants to titanium has been studied by Nemec and Kas (1923), who obtained increased crop yields by manuring *Sinapis alba, Pisum sativum*, and *Medicago sativa* with titanium as the insoluble titanate and the soluble titanium sodium citrate. The amount of phosphorus, silicon, and aluminum increased and decreased with the titanium content of the plants. Since the iron content decreased with increasing applications of titanium, it was considered that iron might be replaced by this element.

Sideris (1930) found that pineapple plants, growing in a nutrient solution lacking iron but containing 5 p.p.m. of titanium, formed chlorophyll as readily as when iron was added in the form of ferric chloride. Inman, Barclay, and Hubbard (1935), however, in using titanium chloride in amounts of 15 to 20 p.p.m. found in the plants with which they worked that titanium could not be substituted for iron in the formation of chlorophyll.

Arsenic in the form of disodium arsenate showed a beneficial influence in low concentrations upon beans, peas, wheat, and radish, according to the observations of Stewart (1922). The beneficial influence of arsenic was not manifest by any increase in dry weight but by the more vigorous and healthy appearance of the plant. He considered that the accumulation of arsenic in the soil from sprays might be beneficial rather than injurious to plants. Beans, peas, and wheat were stimulated by concentrations of disodium arsenate up to 75 p.p.m. Radishes were stimulated by concentrations up to and including 240 p.p.m. Beans were the most sensitive to this element. When the above plants were killed by the absorption of arsenic, they contained the following parts per million of dry weight of plant: wheat, 269; potato, 524; bean, 678; pea, 1,190; and radish, 940. Brenchley (1914) stated that the toxic effect of arsenic upon higher plants is much more marked with arsenious acid and its compounds than with arsenic acid and its derivatives.

Albert and Paden (1931) believed that the addition of calcium arsenate to light, sandy-loam soils may be expected eventually to interfere seriously with the growing of such arsenic-sensitive crops as cowpeas, oats, cotton, and various grasses. The effect of the amount of arsenic in the soil upon its content in the plant is shown in the following table:

Soil, arsenic content, p.p.m.	Plants, arsenate content, p.p.m.			
	Oats		Cowpeas	
	Roots	Tops	Roots	Tops
30	180	6	40	11
8	80	3	10	3

Gautier and Clausman (1919) reported that the application of 100 p.p.m. of fluorine, as calcium fluoride, to field plots definitely increased the yield in some cases. Mazé (1919) considered that fluorine is a necessary element for corn, but Wilson (1933) and Wöher (1920) found that it had a toxic effect upon the plants they observed. The former worker found that the tobacco plant responds to sodium fluoride by transforming free water into bound water in such quantities as to indicate that this compound induces a state of physiological drought. Bartholomew (1935) found that concentrations of fluorine as high as 50 p.p.m. did not significantly decrease the germination of Sudan grass, cowpeas, soybean, or red clover. The addition of soluble fluorides to the amount of 10 p.p.m. did not decrease the amount of dry matter produced by cowpeas growing in a nutrient solution. The fluorine was found mostly in the roots of these plants.

Scharrer and Schropp (1935) reported that the application of extremely small amounts of vanadium stimulated slightly the later growth of all the plants studied by them but that it exerted a detrimental influence when applied in larger amounts. The roots of corn were especially injured by this element even when applied in the smallest amounts.

McMurtrey (1932) found that a nutrient solution containing 1 p.p.m. of thallium was toxic to tobacco plants. This toxicity was manifested by a series of effects that are more or less the characteristic symptoms of "frenching," although these two effects are not identical. The first effect is a slowing-down of the growth rate and the development of a lighter green color along the veins of the upper leaves of the plant. As the younger leaves develop, they show at first a chlorosis which progresses along the smallest branches of the vascular system. The subsequent growth produces leaves which are decidedly distorted and may often consist essentially of only the midrib. This is followed by a proliferation of the lateral buds resulting in the so-called "witches-broom" effect. McCool (1933) noted that the presence of 2.1 p.p.m. of thallium sulphate in sandy loam slightly retarded the development of the roots and tops of soybean, wheat, buckwheat, alfalfa, corn, tobacco, and tomato plants, while an application of 8.5 p.p.m. was very injurious.

It was found by Loew (1878) that rubidium chloride permitted normal growth of buckwheat until flowering, after which time injury became progressively extended until the plants died. Arndt (1922) found that buckwheat, grown with rubidium salts instead of potassium salts showed premature dying of the roots, curling of the leaves, and deterioration of the chlorophyll, but that these symptoms could be allayed or removed by the addition of potassium salts. It was suggested by Blanck, Giesecke, and Henkeshoven (1933) that rubidium interferes with potassium assimilation although it may not have a specific toxic effect. Brenchley (1934) studied the action of rubidium sulphate and palladium chloride in nutrient solutions on the growth of barley, wheat, oats, peas, and beans and found that, over a wide range of concentrations, rubidium sulphate had no effect on the growth of these plants. A relatively low concentration of palladium chloride, however, was injurious, the tolerance varying with the species. Barley appeared to be the least and oats the most sensitive of the cereals.

Nakamura (1904) could observe no stimulatory effect of cobalt or nickel upon the growth of rye, peas, onion, or mustard. Bishop and Lawrenz (1932) found that the presence of cobalt in plants depends on the nature of the soil and the type of plant. Scharrer and Schropp (1933) reported that corn and barley were stimulated by a low concentration of cobalt but that wheat, oats, beans, and peas were injured. Cotton (1930) found that concentrations of 5.8 to 58.7 p.p.m. of nickel were very toxic to plants, while a concentration as low as 1.8 p.p.m. caused the spotting and chlorosis of leaves. According to Nakamura (1904), lithium in the form of lithium carbonate in amounts ranging from 10 to 100 mg. per kilogram of soil exerted a stimulatory action upon barley and peas. He also found that cesium chloride at the rate of 100 mg. per kilogram of soil increased the height of rice plants.

V. ELEMENTAL REQUIREMENTS AND THE AGE OF THE PLANT

The amount of nutrients necessary for the optimum growth of plants at the different stages of their growth is an important factor from the standpoint of crop production and is a subject that has received considerable experimental attention.

McCall and Richards (1918) studied the salt requirements of wheat plants at three stages of growth when grown in a three-salt solution composed of monopotassium phosphate, calcium nitrate, and magnesium sulphate. Of these salts, 36 different proportions were used, which necessitated the growing of 36 different cultures for each of the three growing periods. The three stages of development of the plant that were studied were: (*a*) the first 30 days from germination, (*b*) the second 30 days of growth, and (*c*) the period extending from the close of the second 30-day period to the maturity of the plant. For the early growth period, the solutions that gave the highest yield of tops were characterized by a high calcium nitrate content and a low proportion of magnesium sulphate, while the lowest yield of tops was associated with low calcium nitrate and a high proportion of magnesium sulphate. For this period the effect of monopotassium phosphate appeared to be overshadowed by the other two salts. The results for the second 30-day period showed results very similar to those of the first period. This indicates that the elemental requirements of the plants during the second 30-day period were substantially the same as for the first period. For the third and final growth period, the solutions that gave the highest yielding plants were characterized by a relatively high concentration of calcium nitrate and a low proportion of both magnesium sulphate and monopotassium phosphate, while the solutions producing low yields were characterized by a high proportion of monopotassium phosphate without regard to the ratio of the other two salts. Hibbard and Gershberg (1924) in the case of Marquis wheat using a cultural solution high in magnesium sulphate and relatively low in both calcium nitrate and monopotassium phosphate produced the best growth during the vegetative period between the time the plants were 3 weeks old and the time of blooming. Shive and Martin (1918) found in buckwheat that during the 4 weeks following germination the highest yields of tops and roots were produced by a solution of monopotassium phosphate 0.0144 M, calcium nitrate 0.0052 M, and magnesium sulphate 0.0200 M. During the second 4-week period, including seed production and ripening, the best solution was 0.0108 M

monopotassium phosphate, 0.0130 *M* calcium nitrate, and 0.0100 *M* magnesium sulphate.

In corn, Hornberger (1882) noted that when the tassels were beginning to form there was an abrupt slowing down of the rate of absorption of the various elements. This was followed by a period of rapid absorption, which was succeeded by the ripening period in which there was some loss of practically all the elements, due perhaps to diffusion outwardly of solutes and to the loss of leaves and roots. Jones and Huston (1914) found that for 8 weeks after the germination of corn a very rapid absorption of potassium occurred. This was followed by a period of relatively slow absorption and then, at the time of greatest starch formation in the grain, a period of rapid absorption. They observed no striking difference in the rate of phosphorus absorption during the growing season. The rate of nitrogen absorption reached its highest point about the end of the 8-week period. This was followed by a decreased rate, which again became high at the 16-week period. In corn grown in nutrient solutions, Duley and Miller (1921) found that an optimum nutrient during the last 30-day period largely determined ear production, but, when there was a copious supply of mineral elements at the end of the vegetative stage, the leaves and stems contained enough material to produce fair ears even if there was a small amount of nutrient supply during the last 30-day period of the plant. Burd (1919) studied the rate of absorption of certain elements from the soil by the barley plant at different stage of its growth under conditions that precluded leachings or gains to the soil or the removal of elements from the leaves by rain. During the first period of growth, ending 8 or 9 weeks from planting or about the time the heads begin to form, the increase of nitrogen and potassium conformed very closely to the gain in total weight and water content of the plant. The increase of calcium, magnesium, and phosphorus closely paralleled the formation of dry matter up to 8 to 9 weeks, respectively, after which they lagged behind. At this time the absolute amounts of potassium and nitrogen contained in the plant very closely approximated the amount present at maturity, while the potassium content may be even greater than at maturity. During the second period, which extended from the beginning of head formation to the beginning of the ripening of the grain, there was not only a decreased rate of absorption, as has been noted in corn, but there was a decided loss of potassium and nitrogen from the aerial parts of the plant. This loss took place concurrently with the migration of these same constituents into the developing head. The plant, however, had a tendency toward the close of the period to absorb again the soil constituents previously lost. The third period or stage occurring at the time of ripening of the grain was characterized by a practically complete cessation of absorption of all constituents and an actual loss of some of these. The striking thing in this experiment is the fact that the absorption of potassium and nitrogen during the first period of growth is approximately proportional to the growth attained and that in the succeeding periods the final dry-matter content of the crop more than doubles without any very substantial increase in nitrogen content and with an actual loss of potassium. Burd concluded from his work that it is generally desirable to have the large amounts of solutes in the soil solution at the commencement of the plant's growth cycle but that it is unnecessary and may be undesirable to maintain this condition during certain later stages of growth. Pember (1917) and Pember and McLean (1924) found that potassium was utilized most successfully by barley, oats, and wheat when supplied early in the growth period. Wheat was able to make a more effective use of a limited amount of potassium if supplied during a 4-week period from the fourth to seventh week after planting than if the same amount were supplied over a period of 12 weeks. The time of the application of phosphorus made little difference, but deficient amounts of nitrogen were most beneficial if supplied gradually over the entire growth period.

The tendency toward a delayed rate of absorption or the absolute loss of some constituents at a comparatively early age is probably characteristic of certain types of plants, as indicated by the work of Wilfarth, Römer, and Wimmer (1906) with wheat, barley, and potatoes. They found evidence of potassium and nitrogen losses from wheat and barley at comparatively early stages of development, but in the case of the potato they found no losses of potassium and nitrogen at any stages of development, nor did they observe any outstanding changes in the rate of absorption of these elements during its growing period.

Rice grown in Porto Rico (Gile and Carrero, 1915) showed a decrease in the percentage of potassium, phosphorus, and sulphur in the ash and of nitrogen in the dry matter with the age of the plant, which is a common observation for most plants. The amount of magnesium, phosphorus, iron, sulphur, and nitrogen in 103-day-old plants when the panicles were just out was the same as found in 123-day-old plants when the seed was ripe. The amount of potassium, however, was approximately twice as much in the mature as in the younger stage. The amount of iron in the ash of the green leaves and straw decreased regularly with the age of the plant, but the amount in the whole plant, including stem, leaves, and grain, remained constant. According to Chizhov (1926) all the necessary ash and nitrogenous substances are accumulated by winter and spring crops about the period of blooming or grain forming. In the sunflower this accumulation is complete at about the time of ripening, while in beans and potatoes it is completed at the end of the vegetative period.

Gericke (1924 to 1925) grew wheat plants in complete nutrient solutions for varying periods and then transferred them to cultural solutions lacking one of the following elements: potassium, calcium, iron, magnesium, phosphorus, nitrogen, and sulphur, and allowed them to complete their growth. The following is a summary of the results obtained:

After plants had grown for 4 weeks in a complete nutrient solution and had attained approximately one-seventh of the dry weight obtainable in such a medium and were transferred to solutions devoid of magnesium, phosphorus, or sulphur, they produced more grain and straw than did the plants grown to maturity in complete nutrient solutions which were three times renewed. The removal of plants grown for 4 weeks in a complete nutrient solution to a potassium-free medium resulted in a production of dry matter equal to that obtained from cultures grown in a complete nutrient solution. The removal of plants grown 4 to 10 weeks in complete nutrient solutions to solutions lacking nitrogen decreased the production of both grain and straw, but otherwise the plants appeared normal. The maximum development among all cultures was obtained from the plant grown 4 weeks in a complete nutrient solution and then transferred into a solution lacking phosphorus. A better growth of plants was obtained from the cultures grown to maturity in a complete nutrient solution which was not renewed than was obtained from cultures grown in a complete nutrient solution which was three times renewed. The elements required longest in available form in the growth mediums for the normal development of wheat were calcium and iron.

Brenchley (1929) tested the effect of depriving barley plants of phosphorus after varying initial periods during which it had been supplied and also supplying phosphorus after initial periods of deprivation. The providing of phosphorus for the first 6 weeks or longer permitted normal growth to be made as judged by appearance, increase in dry weight, and grain produced. If phosphorus was withheld for the first 4 weeks, tiller production was not affected, but no heads were produced. With longer periods of initial deprivation, growth was steadily depressed in all respects. The amount of phosphorus absorbed by the plant increased steadily in more or less direct proportion to the length of time phosphorus was given at the beginning of growth.

Sufficient phosphorus, however, was taken up in the first 6 weeks to enable the plant to make its maximum dry weight. The percentage of phosphorus in the dry matter rapidly increased from this time onward.

In connection with the elemental requirements of plants at different stages of their growth, it is important to consider the availability to the seedling of the elements stored in the seed and to note how the amounts thus supplied correspond to the total amounts found in the young seedling. Buckner (1915, 1919) studied the translocation of the mineral constituents of the bean and corn seed and of the potato tuber during germination. He found that approximately 47 per cent of the minerals stored in the cotyledons of garden beans was retained in them when they were no longer functional. The cotyledon of corn contained about 39 per cent of its original mineral content when it had ceased to supply any nutrients to the young seedling. In the case of the potato tuber, the young stem removed only about 50 per cent of the minerals stored in the fresh tuber. The percentage distribution of the various mineral constituents in the seedlings studied is found in the table below.

It was observed that there were no very striking differences in the quantities of the minerals translocated and no marked selective influences shown by the root, stem, or leaves of the seedlings for any particular mineral reserve contained in the seeds or tuber.

The intake of nitrogen, potassium, and phosphorus by wheat plants grown in nutrient solutions for fractional parts of a day was studied by Breazeale (1923). The absorption of these elements for 1 hr. or other fractional parts of a day was out of proportion to what might be expected if time alone governed absorption. Thus during the first hour the plants absorbed 0.0520 g. of potassium. During a 24-hr. period the same plants absorbed 0.2157 g. of this element. If the rate of absorption of the

PERCENTAGE DISTRIBUTION OF THE MINERAL CONSTITUENTS IN SEEDLINGS
Adapted from Buckner, 1915

Part	P_2O_5	CaO	MgO	K_2O	SiO_2
Beans:					
Exhausted cotyledons	47.2	54.5	45.7	45.1	40.8
Leaves	24.3	10.5	27.3	27.8	23.5
Stems	20.7	22.4	20.8	18.4	15.4
Roots	7.7	13.7	6.1	8.7	19.5
Corn:					
Exhausted cotyledons	43.2	43.9	47.0	20.0	58.6
Leaves	26.4	30.0	28.7	35.7	19.8
Stem	17.4	13.2	16.8	21.2	6.4
Root	12.9	12.9	7.4	23.1	15.1
Potato:					
Exhausted tuber	67.1	42.0	65.7	64.4	12.4
Sprouts	17.8	13.1	15.8	12.7	5.1

plants kept in the nutrient solution 1 hr. had been maintained for 24 hr., the plants would have absorbed 1.248 instead of 0.2157 g. of potassium. Thus over five times as much potassium was absorbed during the first-hour period as might be expected if time alone governed absorption.

Davidson (1926) observed that 18-day corn seedlings had absorbed from the soil eight to ten times the amount of potassium contained in the seed. The amount absorbed during 1 week after the plants had been thinned in the hill was about twice that which had been absorbed during the previous 18 days. The phosphorus content of the seedlings was lower at both stages of growth in the seedling than in the seed. During the first 18 days the amount of nitrogen absorbed by the seedling was about equal to that in the original seed, but during the next 7 days it absorbed twice as much.

Gerdel (1931) working with corn noted that, with a continuous abundance of nitrogen, phosphorus, and potassium throughout the season, marked differentiation of the vegetative and reproductive cycles occurred associated with a narrow silking range, earliness of silking, and higher yield. With a low supply of nitrogen, phosphorus, and potassium throughout the season, he found that an overlapping and mergence of the two growth cycles occurred associated with a wide silking range, lateness of silking, and low yields.

Knowles and Watkin (1931) observed in wheat that the plant attained its maximum quantities of the nutrients in the following order: potassium, calcium, and nitrogen in 7, 5, and 3 weeks, respectively, and phosphorus, carbon, and silicon, 2 weeks before harvest. During the 6 weeks preceding harvest marked losses of calcium, chlorine, and potassium occurred. These losses could not have been caused by loss of parts or by leaching. It was considered that as maturity approaches there is a downward movement of these elements to the roots.

Échevin (1927, 1931) in studying the leaves of the hardwoods in autumn found that the various elements of the ash might increase or decrease prior to leaf fall. He concluded that no general conclusion can be drawn in regard to the movement of the various elements to and from the leaves previous to their fall from the tree. He found in young beech trees that the total amount of phosphorus in the entire plant remained constant throughout the year. In the spring, however, practically 50 per cent of the phosphorus of the stem and root migrated into the leaves. When the leaves had completed their growth this amount of phosphorus gradually returned to the stem and roots.

Davis and Moore (1934) noted in pear leaves that large amounts of calcium accumulated throughout the season so that at the time of leaf fall there was as much as 7.0 mg. per leaf. There was a seasonal increase of magnesium that reached its maximum 3 months prior to leaf fall. The amount of potassium and phosphorus remained more or less constant during the season until after September 20. Between that date and leaf fall the loss of potassium amounted to between 27 and 43 per cent. The amount of phosphorus remained constant until the last 2 or 3 weeks before leaf fall when its loss was approximately 25 per cent. Davis (1934) determined the amount of calcium, magnesium, and potassium in alternate-bearing prune trees. The production of a crop affects the amount of the various elements differently. A crop of fruit reduces the potassium and phosphorus in the wood, bark, and spurs, but in the leaves only the amount of potassium is diminished. There is no difference in the amount of phosphorus in the leaves of bearing and nonbearing trees. The calcium and magnesium content is higher in the leaves of bearing trees than in the leaves of nonbearing ones.

Bibliography

Acton, E. H., 1889: The assimilation of carbon by green plants from certain organic compounds, *Proc. Roy. Soc.*, **47**: 150–175.

Addoms, R. M., and F. C. Mounce, 1931: Notes on the nutrient requirements and the histology of the cranberry (*Vaccinium macrocarpon*) with special reference to mycorrhiza, *Plant Physiol.*, **6**: 653–666.

AGULHON, H., 1910: "Recherches sur la présence et le rôle du bore chez les végétaux," *thèse*, 163 pp., Paris.

ALBERT, W. B., and W. R. PADEN, 1931: Calcium arsenate and unproductiveness in certain soils, *Sci.*, **73**: 622.

ALBRECHT, W. A., and F. L. DAVIS, 1929: Physiological importance of calcium in legume inoculation, *Bot. Gaz.*, **88**: 310–321.

ALLEN, R. C., 1931: Factors influencing the flower color of *Hydrangea, Proc. Am. Soc. Hort. Sci.*, **28**: 410–412.

———, 1934: Controlling the color of greenhouse hydrangeas (*Hydrangea macrophylla*) by soil treatments with aluminum sulfate and other materials, *Proc. Am. Soc. Hort. Sci.*, **32**: 632–634.

ALLISON, R. V., O. C. BRYAN, and J. H. HUNTER, 1927: The stimulation of plant response on the raw peat soils of the Florida Everglades through the use of copper sulphate and other chemicals, *Fla. Sta. Bull.* 190.

ALLYN, W. P., 1927: The relation of lime to the absorption of iron by plants, *Proc. Ind. Acad. Sci.*, **43**: 405–409; *Exp. Sta. Rec.*, **66**: 121.

ALTEN, F., and R. GOTTWICK, 1933: Ein Beitrag zur Frage der Vertretbarkeit des Kaliums durch Rubidium und Caesium für die Pflanzenernährung, *Ernähr. der Pflanze*, **29**: 393–399.

AMES, J. W., and G. E. BOLTZ, 1912: Nitrogen and mineral constituents of the alfalfa plant, *Ohio Sta. Bull.* **247**: 755–773.

———, 1923: Sulfur experiments, *Ohio Sta. Mo. Bull.* **8**: 85–90.

——— and K. KITSUTA, 1932: Availability of rock phosphate as indicated by phosphorus assimilation of plants, *J. Am. Soc. Agron.*, **24**: 103–123.

——— and ———, 1933: Assimilation of phosphorus and potassium by barley plants grown according to Neubauer procedure and in undiluted soil, *Soil Sci.*, **35**: 197–207.

ANDERSON, P. J., T. R. SWANBACK, and O. E. STREET, 1932: Potash requirements of the tobacco crop, *Conn. Sta. Bull.*, **334**: 133–217.

ANDERSSEN, F. G., 1932: Chlorosis of deciduous fruit trees due to a copper deficiency, *J. Pom. and Hort. Sci.*, **10**: 130–146.

ANDRÉ, G., and E. DEMOUSSY, 1927: Sur la répartition du potassium et du sodium chez les végétaux, *C. R. Acad. Sci. Paris*, **184**: 1501–1503.

ANDREWS, J. S., and C. H. BAILEY, 1932: Distribution of organic phosphorus in wheat, *Ind. and Eng. Chem.*, **24**: 80–82.

ARCHIBALD, J. G., and E. BENNETT, 1935: Yield and chemical composition of certain species of grass, *J. Agr. Res.*, **50**: 711–715.

ARENDT, R., 1859: Untersuchungen über einige Vorgänge bei der Vegetation der Haferpflanze, *Landw. Versuchs-Stat.*, **1**: 31–66.

ARNDT, C. H., 1922: The growth of field corn as affected by iron and aluminum salts, *Am. J. Bot.*, **9**: 47–72.

ARNDT, J., 1922: New investigations of the problem of the substitution of potassium in the plant cell by rubidium, as well as the chlorine requirements of buckwheat and barley, *Ernähr. der Pflanze*, **18**: 177–180; 191–194.

ASHBY, E., 1929: The interaction of factors in the growth of *Lemna*. IV, The influence of minute quantities of organic matter upon growth and reproduction, *Ann. Bot.*, **43**: 805–816.

ASO, K., 1907: On the continuous application of manganous chloride in rice culture. II, *Col. Agr. Imp. Univ. Tokyo Bull.*, **7**: 449–453.

AVERILL, H. P., and C. G. KING, 1926: The phytin content of food stuffs, *J. Am. Chem. Soc.*, **48**: 724–728.

BAGULEY, A., 1912: The phosphate nutrition of plants, *J. Agr. Sci.*, **4**: 318–322.

BAMFORD, R., 1931: Changes in root tips of wheat and corn grown in nutrient solutions deficient in calcium, *Bull. Torrey Bot. Club*, **58**: 149–179.

BARNETTE, R. M., 1921: The influence of volume upon the rate of change in the hydrogen-ion concentration of nutrient solutions in contact with plant roots and the effect of this change upon iron availability, *N. J. Sta. Rept.*, **1921**: 345–347.

——— and J. W. SHIVE, 1923: The influence of solution volume upon plant growth in relation to reaction change and iron availability in culture solutions, *Soil Sci.*, **15**: 413–425.

——— and J. D. WARNER, 1935: A response of chlorotic corn plants to the application of zinc sulfate to the soil, *Soil Sci.*, **39**: 145–160.

———, J. P. CAMP, J. D. WARNER, and O. E. GALL, 1936: The use of zinc sulphate under corn and other field crops, *Fla. Sta. Bull.* 292.

BARRENSCHEEN, H. K., and W. ALBERS, 1928: Ueber die Rolle der Phosphorylierung im intermediären kohlenhydrat Stoffwechsel der Pflanzen, *Biochem. Z.*, **197**: 261–277.

BARTHOLOMEW, R. P., and G. JANSSEN, 1929: Luxury consumption of potassium by plants and its significance, *J. Am. Soc. Agron.*, **21**: 751–765.

——— and ———, 1929: The relation between concentrations of potassium in culture solutions and optimum plant growth, *Soil Sci.*, **27**: 189–203.

——— and ———, 1931: Rate of absorption of potassium by plants and its possible effect upon the amount of potassium remaining in soils from application of potassium fertilizers, *Ark. Sta. Bull.* 265.

———, 1935: Fluorine, its effect on plant growth and its relation to the availability to plants of phosphorus in phosphate rocks, *Soil Sci.*, **40**: 203–218.

BASSLAVSKAYA, S., and M. SYROESHKINA, 1936: Influence of the chloride ion on the content of chlorophyll in the leaves of potatoes, *Plant Physiol.*, **11**: 149–157.

BAUER, F. C., 1923: The foraging power of plants for phosphate rock, *J. Am. Soc. Agron.*, **15**: 99–110.

BEATH, O. A., H. F. EPPSON, and C. S. GILBERT, 1935: Selenium and other toxic minerals in soils and vegetation, *Wyo. Sta. Bull.* 206.

BEAUMONT, A. B., 1932: Toxicity of sodium nitrate for a species of moss, *Sci.*, **75**: 312–313.

——— and G. M. KARNS, 1932: Effect of an iodine fertilizer on iodine content of a food plant, *Sci.*, **78**: 567.

——— and R. F. CHANDLER, 1933: A statistical study of the effect of potassium fertilizers upon the firmness and keeping quality of fruits, *Proc. Am. Soc. Hort. Sci.*, **30**: 37–44.

——— and M. E. SNELL, 1935: The effect of magnesium deficiency on crop plants, *J. Agr. Res.*, **50**: 553–562.

BENNETT, J. P., and J. OSERKOWSKY, 1933: Copper and iron in the tracheal sap of deciduous trees, *Am. J. Bot.*, **20**: 632–637.

BERTHELOT, M., and G. ANDRÉ, 1895: Sur la présence de l'alumine dans les plantes et sur sa répartition, *C. R. Acad. Sci. Paris*, **120**: 288–290.

BERTRAND, G., 1897: Sur l'intervention du manganèse dans les oxydations provoquées par la laccase, *C. R. Acad. Sci. Paris*, **124**: 1032–1035.

———, 1897: Sur l'action oxydante des sels manganeux et sur la constitution chimique des oxydases, *C. R. Acad. Sci. Paris*, **124**: 1355–1358.

———, 1905: Sur l'emploi favorable du manganèse comme égrais, *C. R. Acad. Sci. Paris*, **141**: 1255–1257.

———, 1912: Sur le rôle des infiniment petits chimiques en agriculture, *Ann. Inst. Pasteur*, **26**: 852–867.

——— and M. ROSENBLATT, 1921: Recherches sur la présence du manganèse dans le règne végétal, *Ann. Inst. Pasteur*, **35**: 815–819.

——— and ———, 1921: Sur la présence générale du manganèse dans le règne végétal, *C. R. Acad. Sci. Paris*, **173**: 333–336.

——— and ———, 1922: Sur la répartition du manganèse dans l'organisme des plantes supérieures, *Ann. Inst. Pasteur*, **36**: 230–232.

——— and ———, 1922: Recherches sur la variations de la teneur en manganèse des feuilles avec l'âge, *Ann. Inst. Pasteur*, **36**: 491–501.

——— and ———, 1932: The unequal manganese content in green leaves and etiolated leaves, *C. R. Acad. Sci. Paris*, **194**: 1405–1408.

——— and M. ANDREITCHEVA, 1933: Sur la teneur comparée en zinc des feuilles vertes et des feuilles étiolées, *C. R. Acad. Sci. Paris*, **197**: 1374–1376.

BISHOP, E. R., and M. LAWRENZ, 1932: Cobalt in plant ash, *Sci.*, **75**: 264–265.

BISHOP, W. B. S., 1928: The distribution of manganese in plants and its importance in plant metabolism, *Australian J. Exp. Biol. and Med. Sci.*, **5**: 125–141; *Exp. Sta. Rec.*, **67**: 118.

BLAIR, A. W., and B. E. BROWN, 1921: The influence of fertilizers containing borax on the yield of potatoes and corn—season 1920, *Soil Sci.*, **11**: 369–385.

——— and A. L. PRINCE, 1932: The influence of phosphates on the phosphoric acid content of the plant, *J. Agr. Res.*, **44**: 579–590.

BLANCK, E., F. GIESECKE, and W. HEUKESHOVEN, 1933: Vegetationsversuch mit Hafer zur Klärung der Frage nach der Vertretbarkeit des Kaliums durch Rubidium, *J. landw. Wiss.*, **81**: 205–222.

BOBKO, E. V., and M. A. BELVOUSSOV, 1933: The importance of boron for the sugar beet, *Ann. Agron.*, **3**: 493–504; *Chem. Abst.*, **27**: 5374.

BOEHM, J., 1875: Ueber den vegetabilischen Nährwerth der Kalksalze, *Sitzungsber. Akad. Wiss. Wien*, **71**: 287–384.

BOLIN, D. W., 1934: The manganese content of grasses and alfalfa from grazed plots, *J. Agr. Res.*, **48**: 657–663.

BOLIN, P., 1916: Sodium chloride as a fertilizer, *Chem. Abst.*, **10**: 2268–2269.

BORROWNICKA-ORDZYWOLSKA, A., 1926: Effect of K and Na salts on metabolism in barley seedlings grown in the dark, *Bull. Int. Acad. Polonaise Sci. et Lett. B. Sci. Nat.*, **1925**: 80–846.

BRANDENBURG, E., 1931: Die Herz- und Trockenfäule der Rüben als Bormangel-Erscheinung, *Phytopath. Z.*, **3**: 499–517.

BRANNON, J. M., 1923: Influence of certain sugars on higher plants, *Bot. Gaz.*, **75**: 370–390.

BREAZEALE, J. F., 1906: The relation of sodium to potassium in soil and solution cultures, *J. Am. Chem. Soc.*, **28**: 1013–1025.

———, 1916: Effect of sodium salts in water cultures on the absorption of plant food by wheat seedlings, *J. Agr. Res.*, **7**: 407–416.

———, 1923: Nutrition of plants considered as an electrical phenomenon, *J. Agr. Res.*, **24**: 41–54.

BRENCHLEY, W. E., 1910: The influence of copper sulphate and manganese sulphate upon the growth of barley, *Ann. Bot.*, **24**: 571–583.

———, 1914: "Inorganic Plant Poisons and Stimulants," 110 pp., Cambridge University Press.

———, 1914: On the action of certain compounds of zinc, arsenic and boron on the growth of plants, *Ann. Bot.*, **28**: 283–302.

———, 1924: The effect of iodine on soils and plants, *Ann. App. Biol.*, **11**: 86–112.
——— and H. G. Thornton, 1925: The relation between the development, structure and functioning of the nodules on *Vicia faba* as influenced by the presence or absence of boron in the nutrient medium, *Proc. Roy. Soc.*, B, **98**: 373–400.
———, E. J. Maskell, and K. Warington, 1927: The inter-relation between silicon and other elements in plant nutrition, *Ann. App. Biol.*, **14**: 45–82.
——— and K. Warington, 1927: The rôle of boron in the growth of plants, *Ann. Bot.*, **41**: 167–187.
———, 1929: The phosphate requirement of barley at different periods of growth, *Ann. Bot.*, **43**: 89–110.
———, 1932: The action on the growth of crops of small percentages of certain metallic compounds when applied with ordinary artificial fertilizers, *J. Agr. Sci.*, **22**: 704–735.
———, 1934: The effect of rubidium sulphate and palladium chloride on the growth of plants, *Ann. App. Biol.*, **21**: 398–417.
Briggs, G. E., 1922: Experimental researches on vegetable assimilation and respiration. XVI, The characteristics of subnormal photosynthetic activity resulting from deficiency of nutrient salts, *Proc. Roy. Soc.*, B, **94**: 20–35.
Brown, B. A., 1932: The effects of fertilization on the chemical composition of vegetation in pastures, *J. Am. Soc. Agron.*, **24**: 129–145.
Brown, B. E., 1922: Effect of borax in fertilizer on the growth and yield of potatoes, *U. S. Dept. Agr. Bull.* 998.
———, V. C. Houghland, O. Smith, and R. L. Carolus, 1933: The influence of magnesium on different potato soil types, *Am. Potato J.*, **10**: 55–65.
Brown, J. W., 1927: Chemical studies in the physiology of apples. VI, Correlation in the individual apple between the mineral constituents and other properties, *Ann. Bot.*, **41**: 127–139.
———, 1929: Chemical studies in the physiology of apples. XI, The relation between the mineral constitution of apples and the soil on which they are grown, *Ann. Bot.*, **43**: 817–831.
Brown, P. E., and G. A. Minges, 1916: The effect of some manganese salts on ammonification and nitrification, *Soil Sci.*, **2**: 67–87.
Bruce, O. C., 1925: The relation of sulfur to alfalfa production, *J. Agr. Res.*, **30**: 937–949.
Bruch, P., 1902: Zur physiologischen Bedeutung des Kalciums in der Pflanze, *Landw. Jahrb.*, **30**: 127–144.
Bruno, A., 1928: Hypothèses nouvelles sur le mode d'action de potasse, *Rev. Viticulture*, **69**: 60–62; *Biol. Abst.*, **5**: 1623.
Bryan, O. C., 1929: The stimulating effect of external applications of copper and manganese on certain chlorotic plants of the Florida Everglades, *J. Am. Soc. Agron.*, **21**: 923–933.
——— and R. B. Becker, 1935: The mineral content of soil types as related to "salt sick" of cattle, *J. Am. Soc. Agron.*, **27**: 120–127.
Buckner, G. D., 1915: Translocation of mineral constituents of seeds and tubers of certain plants during growth, *J. Agr. Res.*, **5**: 449–459.
———, 1919: The translocation of the mineral constituents of the jack bean, *J. Am. Chem. Soc.*, **41**: 282–287.
———, 1919: The composition of the ash of crab grass (*Digitaria sanguinalis*) as affected by the soil in which it is grown, *J. Am. Chem. Soc.*, **41**: 1384–1385.
Burd, J. S., 1919: Rate of absorption of soil constituents at successive stages of plant growth, *J. Agr. Res.*, **18**: 51–72.

BURGE, W. E., G. C. WICKWIRE, and O. S. ORTH, 1933: The stimulating effect of copper on chlorophyll formation, *abst. Am. J. Bot.*, **20**: 669.

BURKE, E., 1932: Chlorosis of trees, *Plant Physiol.*, **7**: 329–334.

BYERS, H. G., 1935: Selenium occurrence in certain soils in the United States with a discussion of related topics, *U. S. Dept. Agr. Tech. Bull.* 482.

——— and H. G. KNIGHT, 1935: Selenium in soils, *Ind. and Eng. Chem.*, **27**: 902–904.

CANALS, E., J. CANAYE, and M. E. CABANES, 1930: Recherches physiochimiques sur les sucs végétaux, *Bull. Soc. Chim. Biol.*, **12**: 1022–1024; *Biol. Abst.*, **5**: 2654.

CARMIN, J., 1931: Growth and variability of wheat seedlings in magnesium sulphate solutions, *Bull. Torrey Bot. Club*, **58**: 179–190.

CAROLUS, R. L., 1933: Tomato fertilization. II, The effect of different fertilizer ratios on the chemical composition of tomatoes, *Va. Truck Exp. Sta. Bull.* 81.

———, 1935: Experiences with rapid chemical tests for the determination of nutrient deficiencies in vegetable crops, *Proc. Am. Soc. Hort. Sci.*, **33**: 579–583.

———, 1935: The relation of potassium, calcium, and sodium to magnesium deficiency, *Proc. Am. Soc. Hort. Sci.*, **33**: 595–599.

CATTLE, M., 1933: Studies of the physiological importance of the mineral elements in plants. V, The distribution of diastase, invertase and catalase in normal and potassium-starved bean plants, *New Phytol.*, **32**: 364–381.

CHANDLER, W. H., D. R. HOAGLAND, and P. L. HIBBARD, 1933: Little-leaf or rosette of fruit trees. III, *Proc. Am. Soc. Hort. Sci.*, **30**: 70–86.

CHAPMAN, G. W., 1931: The relation of iron and manganese to chlorosis in plants, *New Phytol.*, **30**: 266–283.

CHAPMAN, R. E., 1934: The composition of the salts of the leaves of some desert plants, *Ann. Bot.*, **48**: 777–780.

CHATIN, A., 1850: Existence de l'iode dans les plantes d'eau douce. Conséquences de ce fait pour la géognosie, la physiologie végétale, la thérapeutique et peut-être pour l'industrie, *C. R. Acad. Sci. Paris*, **30**: 352–354.

CHATTERJEE, U. N., 1933: Studies on the effect of phosphates on respiration of green leaves. 1, *Eugenia jambolana;* 2, *Allium tuberosum*, *Bull. Acad. Sci. United Prov. Agra and Oudh, Allahabad, India*, **2**: 179–192; *Biol. Abst.*, **9**: 1013.

CHIZHOV, B. A., 1926: The accumulation of dry matter, ash and nitrogenous substances in cultivated plants under field conditions, *J. Exp. Landw. Südosten Russ.*, **3**: 40–75.

CHOUARD, P., 1933: Preliminary studies of bluing and chlorosis of *Hydrangea opuloides*, *Bull. Mens. Natl. Hort. France*, **5**: 289–291. Quoted by Poesch, 1935.

CLARK, E. R., and H. K. WILSON, 1933: Lodging in small grains, *J. Am. Soc. Agron.*, **25**: 561–572.

CLARK, N. A., and C. L. FLY, 1930: The role of manganese in the nutrition of *Lemna*, *Plant Physiol.*, **5**: 241–248.

———, 1933: Manganese and the growth of *Lemna*, *Plant Physiol.*, **8**: 157–161.

COCKEFAIR, E. A., 1931: The role of phosphorus in the metabolism of plants, *Am. J. Bot.*, **18**: 582–598.

COLBY, H. L., 1933: Effect of starvation on distribution of mineral nutrients in French prune trees grown in solution cultures, *Plant Physiol.*, **8**: 357–394.

COLLA, S., 1929: Untersuchungen über Plasma und Plasmaströmung bei *Characeen.* II, Die Wirkung verschiedener Salze auf die Plasmaströmung, *Protoplasma*, **6**: 438–448.

COLLINGS, G. H., 1927: The influence of boron on the growth of the soybean plant, *Soil Sci.*, **23**: 83–107.

CONNER, S. D., 1920: The effect of zinc in soil tests with zinc and galvanized iron pots, *J. Am. Soc Agron.*, **12**: 61–65.

——— and O. H. SEARS, 1922: Aluminum salts and acids at varying hydrogen-ion concentrations, in relation to plant growth in water cultures, *Soil Sci.*, **13**: 23–43.

———, 1932: Factors affecting manganese availability in soils, *J. Am. Soc. Agron.*, **24**: 726–733.

CONNER, W. H., 1931: Study of the iodine content of Florida grown crops, *Fla. Sta. Ann. Rept.*, **45**: 65.

CONNORS, C. H., 1923: Color control in forcing *Hydrangeas*, *Proc. Am. Soc. Hort. Sci.*, **20**: 282–288.

COOK, F. C., 1916: Boron: its absorption and distribution in plants and its effect on growth, *J. Agr. Res.*, **5**: 877–891.

——— and J. B. WILSON, 1917: Effect of three annual applications of boron on wheat, *J. Agr. Res.*, **10**: 591–599.

——— and ———, 1918: Boron: its effect on crops and its distribution in plants and soil in different parts of the United States, *J. Agr. Res.*, **13**: 451–471.

CORSON, G. E., and A. L. BAKKE, 1917: The use of iron in nutrient solution for plants, *Proc. Iowa Acad. Sci.*, **24**: 477–482.

COTTON, M., 1930: Toxic effects of iodine and nickel on buckwheat grown in solution cultures, *Bull. Torrey Bot. Club*, **57**: 127–141.

COVILLE, F. V., 1923: The effect of aluminum sulphate on *Rhododendron* seedlings. *Am. Hort. Soc. Bull.* **1**: 6.

COWELL, S. J., 1932: A note on the calcium content of cabbage, *Biochem. J.*, **26**: 1422–1423.

CROCKER, W., 1923: The necessity of sulfur carriers in artificial fertilizers, *J. Am. Soc. Agron.*, **15**: 129–141.

DADSWELL, I. W., 1935: Some observations on the inorganic composition of wheat grown in eastern Australia, *Australian J. Exp. Biol. and Med. Sci.*, **13**: 33–42.

DAMIENS, A., and S. BLAIGNAN, 1931: Sur le brome normal (règne végétal) graines comestible, blé, pain, *C. R. Acad. Sci. Paris*, **193**: 1460–1462.

——— and ———, 1932: Sur le brome normal (règne végétal) plantes et fruits comestible, *C. R. Acad. Sci. Paris*, **194**: 2077–2080.

DANIEL, H. A., 1934: The calcium, phosphorus, and nitrogen content of grasses and legumes and the relation of these elements in the plant, *J. Am. Soc. Agron.*, **26**: 496–503.

——— and H. J. HARPER, 1934: The relation between total calcium and phosphorus in mature prairie grass and available plant food in soil, *J. Am. Soc. Agron.*, **26**: 981–986.

———, 1935: The magnesium content of grasses and legumes and the ratios between this element and the total calcium, phosphorus and nitrogen in these plants, *J. Am. Soc. Agron.*, **27**: 922–927.

DASTUR, R. H., and R. E. COOPER, 1932: The osmotic and suction pressures of the rice plant, *Oryza sativa*, when treated with different salts: a method of determining the salt requirements of plants, *Indian J. Agr. Sci.*, **2**: 99–137; *Chem. Abst.*, **26**: 5122.

DAVIDSON, J., and J. A. LECLERC, 1923: Effect of various inorganic nitrogen compounds, applied at different stages of growth on the yield, composition, and quality of wheat, *J. Agr. Res.*, **23**: 55–69.

———, 1926: Removal of plant food in thinning corn, *J. Am. Soc. Agron.*, **18**: 962–967.

——— and C. E. CHAMBLISS, 1932: Chemical composition of rice and its relation to soil fertility in China and Japan, *Sci.*, **75**: 294.

DAVIS, A. R., D. R. HOAGLAND, and C. B. LIPMAN, 1923: The feeding power of plants, *Sci.*, **57**: 299–301.

DAVIS, L. D., 1934: Some ash constituents of alternate-bearing sugar prune trees, *Proc. Am. Soc. Hort. Sci.*, **32**: 125–130.

——— and N. P. MOORE, 1934: Seasonal changes in Bartlett pear leaves, *Proc. Am. Soc. Hort. Sci.*, **32**: 131–137.

DAVIS, M. B., 1930: Investigations on the nutrition of fruit trees. Some effects of deficiencies of nitrogen, potassium, calcium and magnesium with special reference to the behavior of certain varieties of apple trees, *J. Pom. and Hort. Sci*, **8**: 316–344.

———, H. HILL, and F. B. JOHNSON, 1934: Nutritional studies with *Fragaria*. II, *Sci. Agr.*, **14**: 411–432.

DAY, D., 1928: Some effects on *Pisum sativum* of a lack of calcium in the nutrient solution, *Sci.*, **68**: 426.

———, 1929: Some effects of calcium deficiency on *Pisum sativum*, *Plant Physiol.*, **4**: 463–506.

———, 1935: Some chemical aspects of calcium deficiency effects on *Pisum sativum*, *Plant Physiol.*, **10**: 811–816.

DEMOLON, M. A., 1912: Sur l'action fertilisante du soufre, *C. R. Acad. Sci. Paris*, **154**: 524–526.

DENIGES, G., 1920: Réaction de coloration extrêmement sensible des phosphates et des arséniates. Ses applications, *C. R. Acad. Sci. Paris*, **171**: 802–804.

DE PERALTA, F., 1927: Influence upon the development of young rice plants of sodium chloride added to a complete solution, *Philippine Agr.*, **15**: 471–479; *Exp. Sta. Rec.*, **61**: 719.

DE TURK, E. E., J. R. HOLBERT, and B. W. HOWK, 1933: Chemical transformation of phosphorus in the growing corn plant with results on two first generation crosses, *J. Agr. Res.*, **46**: 121–142.

DEUBER, C. G., 1926: Can a pyrrole derivative be substituted for iron in the growth of plants? *Am. J. Bot.*, **13**: 276–285.

———, 1926: Potassium ferrocyanide and ferric ferrocyanide as sources of iron for plants, *Soil Sci.*, **21**: 23–27.

DHEIN, A., 1929: Untersuchungen über Art und Wesen einiger Kalisalzwirkungen, *Landw. Jahrb.*, **70**: 255–309.

DICKSON, J. G., 1918: The value of certain nutritive elements in the development of the oats plant, *Am. J. Bot.*, **5**: 301–325.

DOMONTOVITSCH, M. K., and P. A. SCHELESNOW, 1930: Feeding plants via the leaves with salts of potassium and of magnesium, *J. landw. Wiss.*, **7**: 134–143; *Exp. Sta. Rec.*, **67**: 22.

DOVE, W. F., 1934: A study of the cause of nutritional deficiency diseases in the livestock and inhabitants of Maine with possible corrective methods, etc., *Maine Sta. Bull.* 375.

DOWDING, E. S., 1925: The regional and seasonal distribution of potassium in plant tissues, *Ann. Bot.*, **39**: 459–474.

DULEY, F. L., 1916: Relation of sulphur to soil productivity, *J. Am. Soc. Agron.*, **8**: 154–160.

——— and M. F. MILLER, 1921: The effect of a varying supply of nutrients upon the character and composition of the maize plant at different periods of growth, *Mo. Sta. Tech. Bull.* 42.

DYMOND, T. S., F. HUGHES, and C. W. C. JUPE, 1905: The influence of sulphates as manure upon the yield and the feeding value of crops, *J. Agr. Sci.*, **1**: 217–229.

EATON, F. M., 1930: The effect of boron on powdery mildew and spot blotch of barley, *Phytopath.*, **20**: 967–973.

———, 1932: Boron requirements of cotton, *Soil Sci.*, **34**: 301–306.

——— and G. Y. Blair, 1935: Accumulation of boron by reciprocally grafted plants, *Plant Physiol.*, **10**: 411–424.

Eaton, S. V., 1922: Sulphur content of soils and its relation to plant nutrition, *Bot. Gaz.*, **74**: 32–59.

———, 1935: Influence of sulphur deficiency on the metabolism of the soybean, *Bot. Gaz.*, **97**: 68–100.

Échevin, R., 1927: Variations de la matière minérale des feuilles au cours du jaunissement automnal, *Rev. Gén. Bot.*, **39**: 405–419; 488–514; *Biol. Abst.*, **5**: 2660.

———, 1931: Nitrogen, phosphorus and sulfur in woody plants with caducous leaves, *Rev. Gén. Bot.*, **43**: 517–677; *Chem. Abst.*, **26**: 1318.

Eckerson, S., 1929: Influence of phosphorus deficiency on the metabolism of tomato plants, *abst. Am. J. Bot.*, **16**: 852.

———, 1931: Influence of phosphorus deficiency on metabolism of the tomato (*Lycopersicum esculentum*), *Contr. Boyce Thompson Inst.*, **3**: 197–217.

Eidelman, Z. M., 1930: The influence of phosphorus and of day length on physiological functions in plants, *J. landw. Wiss.*, **7**: 387–402; *Exp. Sta. Rec.*, **67**: 116–117.

Eisenmenger, W. S., 1935: Toxicity of aluminum salts on tobacco plants, *J. Agr. Res.*, **51**: 917–924.

———, 1935: Toxicity of aluminum on seedlings and action of certain ions in the elimination of the toxic effects, *Plant Physiol.*, **10**: 1–25.

Emmert, E. M., 1930: A method for the rapid determination of phosphate in fresh plant tissue, *Plant Physiol.*, **5**: 411–417.

———, 1932: Field method for estimating nitrate, phosphate, and potassium in plants, *Plant Physiol.*, **7**: 315–321.

——— and F. K. Ball, 1933: The effect of soil moisture on the availability of nitrate, phosphate and potassium to the tomato plant, *Soil Sci.*, **35**: 295–307.

———, 1934: Tests for phosphate, nitrate and soluble nitrogen in conducting tissue of tomato and lettuce plants, as indicators of availability and yield, *Ky. Sta. Circ.* 43.

———, 1934: Tests for nutrients in conducting tissue as indicators of the nutritional status in horticultural crops, *Proc. Am. Soc. Hort. Sci.*, **32**: 604–609.

———, 1935: New methods for the determination of the availability of nitrogen and phosphorus to plants, *J. Am. Soc. Agron.*, **27**: 1–7.

Fagundes, A. B., 1934: O boro em nutriçâo vegetal, *Arq. Inst. Bio. Veg.* (*Rio de Janeiro*), **1**: 3–26; *Biol. Abst.*, **9**: 1234.

Felix, E. L., 1927: Correction of unproductive muck by the addition of copper, *Phytopath.*, **17**: 49–50.

Finch, A. H., D. W. Albert, and A. F. Kinnison, 1933: A chlorotic condition of plants in Arizona related to iron deficiency, *Proc. Am. Soc. Hort. Sci.*, **30**: 431–434.

———, ———, and ———, 1934: Progress of the control of citrus chlorosis or decline, *Proc. Am. Soc. Hort. Sci.*, **32**: 20–23.

Fisher, P. L., 1935: Responses of the tomato in solution cultures with deficiencies and excesses of certain essential elements, *Md. Sta. Bull.* 375.

Fluri, M., 1909: Der Einfluss von Aluminiumsalzen auf das Protoplasma, *Flora*, **99**: 81–126.

Fonder, J. F., 1929: Variations in potassium content of alfalfa due to stage of growth and soil type and the relationship of potassium and calcium in plants grown upon different soil types, *J. Am. Soc. Agron.*, **21**: 732–750.

Forbes, R. H., 1917: Certain effects under irrigation of copper compounds upon crops, *Univ. Cal. Agr. Sci.*, **1**: 395–494.

Fraps, G. S., 1931: How reliable are existing chemical methods for determining soil deficiencies in ash constituents of plants? *J. Am. Chem. Soc.*, **23**: 337–352.

FRAZER, J., 1935: Sulphur content of plants, *Plant Physiol.*, **10**: 529–535.

FREAR, D. E. H., 1934: A study of the iodine content of Pennsylvania potatoes, *J. Agr. Res.*, **48**: 171–181.

FREY-WYSSLING, A., 1930: Über die Ausscheidung der Kieselsäure in der Pflanze, *Ber. bot. Ges.*, **48**: 179–183.

———, 1930: Vergleich zwischen der Ausscheidung von Kieselsäure und Kalziumsalzen in der Pflanze, *Ber. bot. Ges.*, **48**: 184–191.

FUNCHESS, M. J., 1916: The effect of certain organic compounds on plant growth, *Ala. Sta. Bull.* 191.

FURNSTAL, A. H., and B. JOHNSON, 1936: Preparation of sintered pyrex glass aerators for use in water culture experiments with plants, *Plant Physiol.*, **11**: 189–194.

GARNER, W. W., J. E. MCMURTREY, C. W. BACON, and E. G. MOSS, 1923: Sand drown, a chlorosis of tobacco due to magnesium deficiency, and the relation of sulphates and chlorids of potassium to the disease, *J. Agr. Res.*, **23**: 27–41.

———, ———, J. D. BOWLING, and E. G. MOSS, 1930: Magnesium and calcium requirements of the tobacco crop, *J. Agr. Res.*, **40**: 145–169.

———, ———, ———, and ———, 1930: Role of chlorine in nutrition and growth of the tobacco plant and its effect on the quality of the cured leaf, *J. Agr. Res.*, **40**: 627–649.

———, C. W. BACON, J. D. BOWLING, and D. E. BROWN, 1934: The nitrogen nutrition of tobacco, *U. S. Dept. Agr. Tech. Bull.* 414.

GASSNER, G., 1910: Beobachtungen und Versuche über den Anbau und die Entwicklung von Getreidepflanzen in subtropischem Klima, *Jahres. Ver. Angew. Bot.*, **8**: 95–163. Quoted by Plitt, 1935.

——— and G. GOEZE, 1932: The influence of potash nutrition on the rate of assimilation of wheat leaves, *Ber. bot. Ges.*, **50a**: 412–482.

GAUTIER, A., and P. CLAUSMAN, 1919: Action des fluorures sur la végétation, *C. R. Acad. Sci. Paris*, **169**: 115–122.

GEIGEL, A. R., 1935: Effect of boron on the growth of certain green plants, *J. Agr. Univ. Puerto Rico*, **19**: 5–28; *Biol. Abst.*, **10**: 312.

GERDEL, R. W., 1931: Effect of fertilizer and date of planting on the physiological development of the corn plant, *Plant Physiol.*, **6**: 695–714.

GERICKE, W. F., 1922: "Magnesia injury" of plants grown in nutrient solutions, *Bot. Gaz.*, **74**: 110–113.

———, 1924: The beneficial effect to plant growth of the temporary depletion of some of the essential elements in the soil, *Sci.*, **59**: 321–324.

———, 1924: The beneficial effect to wheat growth due to depletion of available phosphorus in the cultural media, *Sci.*, **60**: 297–298.

———, 1925: Salt requirements of wheat at different growth phases, *Bot. Gaz.*, **80**: 410–425.

———, 1925: Effect of light on availability of iron to wheat plants in water cultures, *Bot. Gaz.*, **79**: 106–109.

GERMAR, B., 1934: Some functions of silicic acid in cereals with special reference to resistance to mildew, *Pflanzenernähr. Düng. u. Bodenk.* **A35**: 102–115; *Chem. Abst.*, **29**: 1461.

GILBERT, B. E., 1926: The adaptation of certain colorimetric methods to the estimation of nitrates, phosphates, and potassium in plant solutions, *Plant Physiol.*, **1**: 191–199.

——— and L. J. HARDIN, 1927: The current mineral nutrient content of the plant solution as a possible means of chemical control of optimum fertilization, *J. Agr. Res.*, **35**: 185–192.

———, F. T. McLean, and W. L. Adams, 1927: The current mineral nutrient content of the plant solution as an index of metabolic limiting conditions, *Plant Physiol.*, **2**: 139–153.

——— and J. B. Smith, 1929: Nitrates in soil and plant as indexes of the nitrogen needs of a growing crop, *Soil Sci.*, **27**: 459–468.

——— and F. R. Pember, 1931: The sensitivity of red clover (*Trifolium pratense*) to small amounts of boron and manganese, *Plant Physiol.*, **6**: 727–729.

——— and ———, 1931: Further evidence concerning the toxic action of aluminum in connection with plant growth, *Soil Sci.*, **31**: 267–273.

———, 1934: Normal crops and the supply of available soil manganese, *R. I. Sta. Bull.* 246.

——— and F. R. Pember, 1935: Tolerance of certain weeds and grasses to toxic aluminum, *Soil Sci.*, **39**: 425–430.

——— and ———, 1935: The use of seedling plants to determine soil nutrient deficiencies, *R. I. 47th Ann. Sta. Rept.*, 80–83.

Gildehaus, E. J., 1931: The relation of nitrogen to potassium in the nutrition of fruit trees, *Bot. Gaz.*, **92**: 384–395.

Gile, P. L., 1911: Relation of calcareous soils to pineapple chlorosis, *Porto Rico Sta. Bull.* 11.

——— and J. O. Carrero, 1914: Assimilation of colloidal iron by rice, *J. Agr. Res.*, **3**: 205–210.

——— and ———, 1915: Ash composition of upland rice at various stages of growth, *J. Agr. Res.*, **5**: 357–364.

——— and ———, 1916: Assimilation of iron by rice from certain nutrient solutions, *J. Agr. Res.*, **7**: 503–528.

——— and ———, 1916: Immobility of iron in the plant, *J. Agr. Res.*, **7**: 83–87.

——— and ———, 1917: Absorption of nutrients as affected by the number of roots supplied with the nutrient, *J. Agr. Res.*, **9**: 73–95.

——— and ———, 1920: Cause of lime-induced chlorosis and availability of iron in the soil, *J. Agr. Res.*, **20**: 33–63.

——— and ———, 1921: Assimilation of nitrogen, phosphorus and potassium by corn when nutrient salts are confined to different roots, *J. Agr. Res.*, **21**: 545–575.

——— and J. G. Smith, 1925: Colloidal silica and the efficiency of phosphates, *J. Agr. Res.*, **31**: 247–260.

Gines, F. G., 1930: Relative effects of different iron salts upon growth and development of young rice plants, *Philippine Agr.*, **19**: 43–52; *Exp. Sta. Rec.*, **67**: 23.

Ginsburg, J. M., 1925: Composition and appearance of soybean plants grown in cultural solutions each lacking a different essential element, *Soil Sci.*, **20**: 1–15.

Godden, W., 1926: Report of the effect of fertilizers on the mineral content of pastures, *J. Agr. Sci.*, **16**: 98–104.

Gössl, J. V., 1905: Ueber das Vorkommen des Mangans in der Pflanze und über seinen Einfluss auf Schimmelpilze, *Beih. bot. Cent.*, I, **18**: 119–132.

Grafe, V., and L. Portheim, 1906: Untersuchungen über die Rolle des Kalkes in der Pflanze, *Sitzungsber. Akad. Wiss. Wien*, **18**: 324.

Greaves, J. E., and D. H. Nelson, 1925: The iron, chlorine and sulfur contents of grains and the influence of irrigation water upon them, *Soil Sci.*, **19**: 325–330.

——— and C. T. Hirst, 1929: The mineral content of grains, *Utah Sta. Bull.* 210.

Greenhill, A. W., 1930: The availability of phosphatic fertilizers as shown by an examination of the soil solution and of plant growth, *J. Agr. Sci.*, **20**: 559–572.

Gregory, F. G., and F. Crowther, 1928: A physiological study of varietal differences in plants. I, A study of the comparative yields of barley varieties with different manurings, *Ann. Bot.*, **42**: 757–770.

——— and ———, 1931: A physiological study of varietal differences in plants. II, Further evidence for the differential response in yield of barley varieties to manurial deficiencies, *Ann. Bot.*, **45**: 579–592.

Gris, E., 1844: Nouvelles expériences sur l'action des composés ferrugineux solubles appliqués à la végétation et spécialement en traitement de la chlorose et de la débilité des plantes, *C. R. Acad. Sci. Paris*, **19**: 1118–1119.

Grizzard, A. L., 1935: Effects of soil type and soil treatments on the chemical composition of alfalfa plants, *J. Am. Soc. Agron.*, **27**: 81–99.

Groom, P., 1896: Preliminary note on the relation between calcium and the conduction of carbohydrates in plants, *Ann. Bot.*, **10**: 91–96.

Haas, A. R. C., and H. S. Reed, 1927: Significance of traces of elements not ordinarily added to culture solutions for growth of young orange trees, *Bot. Gaz.*, **83**: 77–85.

———, 1928: Water solubility of dry matter in relation to calcium nutrition of normal orange and lemon leaves, *Bot. Gaz.*, **85**: 334–341.

———, 1929: Toxic effect of boron on fruit trees, *Bot. Gaz.*, **88**: 111–131.

———, 1930: Boron as an essential element for healthy growth of citrus, *Bot. Gaz.*, **89**: 410–413.

——— and L. J. Klotz, 1931: Some anatomical and physiological changes in citrus produced by boron deficiency, *Hilgardia*, **5**: 175–190.

——— and ———, 1931: Further evidence of the necessity of boron for health in citrus, *Bot. Gaz.*, **92**: 94–100.

———, 1932: Injurious effects of manganese and iron deficiencies on the growth of citrus, *Hilgardia*, **7**: 181–206.

———, 1933: Walnut-yellows in relation to ash composition, manganese, iron and other ash constituents, *Bot. Gaz.*, **94**: 495–511.

——— and H. J. Quayle, 1935: Copper content of citrus leaves and fruit in relation to exanthema and fumigation injury, *Hilgardia*, **9**: 143–177.

———, 1936: Phosphorus content of citrus and factors affecting it, *Soil Sci.*, **41**: 239–258.

———, 1936: Phosphorus relations of lemon cuttings grown in solution cultures, *Bot. Gaz.*, **97**: 794–807.

Haley, D. E., J. B. Longenecker, and O. Olson, 1931: Composition and quality of Pennsylvania cigar-leaf tobacco as related to fertilizer treatment, *Plant Physiol.*, **6**: 177–182.

Hall, A. D., and C. G. T. Morison, 1905: On the function of silica in the nutrition of cereals. I, *Proc. Roy. Soc.*, B, **77**: 455–477.

Hall, D. M., 1934: The relationship between certain morphological characters and lodging in corn, *Minn. Sta. Tech. Bull.* 103.

Hall, E. H., 1922: Sulfur and nitrogen content of alfalfa grown under various conditions, *Bot. Gaz.*, **73**: 401–412.

Halma, F. F., and A. R. C. Haas, 1931: Solubility changes of inorganic constituents in citrus cuttings, *Bot. Gaz.*, **91**: 213–218.

Harmer, P. M., 1932: Muck soil management for onion production, *Mich. Sta. Ext. Bull.*, **123**: 1–23.

Harper, H. J., and H. A. Daniel, 1934: Chemical composition of certain aquatic plants, *Bot. Gaz.*, **96**: 186–189.

———, ———, and H. F. Murphy, 1934: The total nitrogen, phosphorus, and calcium content of common weeds and native grasses in Oklahoma, *Proc. Okla. Acad. Sci.*, **14**: 36–44.

Harrison, G. T., and C. J. King, 1925: Age of seedlings as a factor in the resistance of maize to sodium chloride, *J. Agr. Res.*, **31**: 633–641.

HART, E. B., and W. H. PETERSON, 1911: Sulphur requirements of farm crops in relation to the soil and air supply, *Wis. Sta. Res. Bull.* 14.

——— and W. E. TOTTINGHAM, 1915: Relation of sulfur compounds to plant nutrition, *J. Agr. Res.*, **5**: 233–251.

HARTER, L. L., 1908: The influence of a mixture of soluble salts, principally sodium chloride, upon the leaf structure and transpiration of wheat, oats, and barley, *U. S. Plant. Ind. Bull.* 134.

HARTT, C. E., 1929: Potassium deficiency in sugar cane, *Bot. Gaz.*, **88**: 229–261.

———, 1934: Some effects of potassium upon the growth of sugar cane and upon the absorption and migration of ash constituents, *Plant Physiol.*, **9**: 399–451.

———, 1934: Some effects of potassium upon the amounts of protein and amino forms of nitrogen, sugars and enzyme activity of sugar cane, *Plant Physiol.*, **9**: 452–490.

HARTWELL, B. L., H. J. WHEELER, and F. R. PEMBER, 1907: The effect of the addition of sodium to deficient amounts of potassium upon the growth of plants in both water and sand cultures, *R. I. Sta. Rept.*, **20**: 299–357.

——— and F. R. PEMBER, 1908: The relative toxicity of ferrous sulfate to barley and rye seedlings, *R. I. Sta.* 21*st Rept.*, 286–294.

——— and F. S. HAMMETT, 1911: The effect of phosphorus manuring on the amount of inorganic phosphorus in flat turnip roots, *J. Ind. Eng. Chem.*, **3**: 831–832.

———, 1913: The percentage of total phosphorus in flat turnips as influenced by the amount available in soils, *R. I. Sta. Bull.* 154.

——— and P. H. WESSELS, 1913: Effect of sodium manuring on the composition of plants, *R. I. Sta. Bull.* 153.

———, 1916: Starch congestion accompanying certain factors which retard plant growth, *R. I. Sta. Bull.* 165.

———, F. S. HAMMETT, and P. H. WESSELS, 1917: Reactions of the phosphorus of the thickened root of the flat turnip, *J. Agr. Res.*, **11**: 359–371.

——— and F. R. PEMBER, 1918: The presence of aluminum as a reason for the difference in the effect of so-called acid soil on barley and rye, *Soil Sci.*, **6**: 259–281.

——— and ———, 1918: Aluminum as a factor influencing the effect of acid soils on different crops, *J. Am. Soc. Agron.*, **10**: 45–47.

———, 1927: Relative crop response to potash, *J. Am. Soc. Agron.*, **19**: 479–482.

——— and S. C. DAMON, 1927: The degree of response of different crops to various phosphorus carriers, *R. I. Sta. Bull.* 209.

HASELHOFF, E., 1913: Über die Einwirkung von Borverbindungen auf das Pflanzenwachstum, *Landw. Versuchs-Stat.*, **79–80**: 399–429.

HEADDEN, W. P., 1915: Occurrence of manganese in wheat, *J. Agr. Res.*, **5**: 349–355.

———, 1921: Titanium, barium, strontium, and lithium in certain plants, *Colo. Sta. Bull.* 267.

HEATH, H. C., 1930: Potash shale as a source of potassium for growing plants, *Bot. Gaz.*, **90**: 121–151.

HECK, A. F., 1934: Some indications of a relation of soil fertility and plant nutrition to cane diseases in Hawaii, *J. Am. Soc. Agron.*, **26**: 381–389.

HELLER, K., and C. L. WAGNER, 1931: Zur Anreicherung von Kaliumisotopen in Pflanzen, *Z. Anorgan. u. Allgem. Chem.*, **200**: 105–112; *Biol. Abst.*, **6**: 2206.

HELLRIEGEL, H., and H. WILFARTH, 1898: Vegetationsversuche über den Kalibedarf einiger Pflanzen, *Arb. landw. Ges.*, **34**: 1–101.

HENDRY, G. W., 1918: Relative effect of sodium chloride on the development of certain legumes, *J. Am. Soc. Agron.*, **10**: 246–250.

HESTER, J. B., 1934: The influence of soil acidity and soil type upon the growth and composition of the lima bean plant, *Proc. Am. Soc. Hort. Sci.*, **32**: 600–603.

Hibbard, R. P., and S. Gershberg, 1924: The salt requirements of Marquis wheat in water culture for the vegetative phase of development, *Mich. Sta. Tech. Bull.* 64.

——— and B. H. Grigsby, 1934: Relation of light, potassium and calcium deficiencies to photosynthesis, protein synthesis and translocation, *Mich. Sta. Tech. Bull.* 141.

Hill, H., and E. P. Grant, 1935: The growth of turnips in artificial cultures, *Sci. Agr.*, **15**: 652–659.

Hiltner, E., 1928: Ist eine wesentlich Jodanreicherung in Pflanzen möglich? *Fortschr. Landw.*, **3**: 1–4.

Hoagland, D. R., 1919: Relation of nutrient solution to composition and reaction of cell sap of barley, *Bot. Gaz.*, **68**: 297–304.

———, 1926: Some phases on the inorganic nutrition of plants in relation to the soil solution, *Sci. Agr.*, **6**(**5**): 141–151, 177–189.

——— and J. C. Martin, 1933: Absorption of potassium by plants in relation to replaceable, non-replaceable and soil solution potassium, *Soil Sci.*, **36**: 1–34.

——— and W. C. Snyder, 1933: Nutrition of strawberry plant under controlled conditions: (*a*) effects of deficiencies of boron and certain other elements; (*b*) susceptibility to injury from sodium salts, *Proc. Am. Soc. Hort. Sci.*, **30**: 288–296.

———, W. H. Chandler, and P. L. Hibbard, 1935: Little-leaf or rosette of fruit trees. V, Effect of zinc on the growth of plants of various types in controlled soil and water culture experiments, *Proc. Am. Soc. Hort. Sci.*, **33**: 131–141.

Hobart, F. G., 1924: Calcium in plant metabolism, *Pharm. J.*, **113**: 500; *Bot. Abst.*, **14**: 569.

Hoffer, G. N., and R. H. Carr, 1920: Iron accumulation and mobility in diseased corn stalks, *Phytopath.*, **10**: 56.

——— and ———, 1923: Accumulation of aluminum and iron compounds in corn plants and its probable relation to root rots, *J. Agr. Res.*, **23**: 801–825.

——— and J. F. Trost, 1923: The accumulation of iron and aluminum compounds in corn plants and its probable relation to root rots, *J. Am. Soc. Agron.*, **15**: 323–331.

———, 1926: Testing corn stalks chemically to aid in determining their plant food needs, *Ind. Sta. Bull.* 298.

Holynski, S., 1929: Assimilation of phosphorus and of potassium by sprouts and by young seedlings of rye, *Mém. Inst. Natl. Polon. Econ. Rurale Pulawy.*, **10**: 163–174; *Exp. Sta. Rec.*, **67**: 118.

Hopkins, E. F., and F. B. Wann, 1926: Relation of hydrogen ion concentration to growth of *Chlorella* and of the availability of iron, *Bot. Gaz.*, **81**: 353–376.

——— and ———, 1927: Iron requirement for *Chlorella*, *Bot. Gaz.*, **84**: 407–427.

———, 1930: The necessity and function of manganese in the growth of *Chlorella sp.*, *Sci.*, **72**: 609–610.

———, 1930: Manganese as an essential element for a green alga, *abst. Am. J. Bot.*, **17**: 1047.

———, 1930: Iron-ion concentration in relation to growth and other biological processes, *Bot. Gaz.*, **89**: 209–240.

———, 1931: Manganese and the growth of *Lemna minor*, *Sci.*, **74**: 551–552.

———, 1933: The effect of small amounts of copper on the growth of *Chlorella* and *Lemna*, *abst. Am. J. Bot.*, **20**: 681.

———, 1934: Manganese an essential element for green plants, *N. Y. Sta.* (*Cornell*) *Mem.* 151.

Hornberger, R., 1882: Chemische Untersuchungen über das Wachsthum der Maispflanze, *Landw. Jahrb.*, **11**: 359–523.

Houghland, G. V. C., and J. A. Schricker, 1933: The effect of potash on starch in potatoes, *J. Am. Soc. Agron.*, **25**: 334–340.

HUMPHREY, H. B., 1934: Rapport sur les recherches Américaines relatives aux maladies végétales dues aux carences minérales, *Deuxième Cong. int. Path. comparée*, 278–295.

HURD-KARRER, A. M., 1933: Inhibition of selenium injury to wheat plants by sulfur, *Sci.*, **78**: 560.

———, 1934: Selenium injury to wheat plants and its inhibition by sulfur, *J. Agr. Res.*, **49**: 343–358.

———, 1935: Factors affecting the absorption of selenium from soil by plants, *J. Agr. Res.*, **50**: 413–428.

———, 1935: Plant physiology involved in the problem of the selenium diseases of live stock, *J. Wash. Acad. Sci.*, **25**: 335–336.

———, 1935: Selenium absorption by plants and their resulting toxicity to animals, *Smithsonian Pub.* 3361.

———, 1936: Inhibition of arsenic injury to plants by phosphorus, *J. Wash. Acad. Sci.*, **26**: 180–181.

——— and M. H. KENNEDY, 1936: Inhibiting effect of sulphur in selenized soil on toxicity of wheat to rats, *J. Agr. Res.*, **52**: 933–942.

——— and F. W. POOS, 1936: Toxicity of selenium-containing plants to aphids, *Sci.*, **84**: 252.

INGALLS, R. A., and J. W. SHIVE, 1931: Relation of H-ion concentration of tissue fluids to the distribution of iron in plants, *Plant Physiol.*, **6**: 103–125.

INMAN, O. L., G. BARCLAY, and M. HUBBARD, 1935: Effect of titanous chloride on the formation of chlorophyll in *Zea mays*, *Plant Physiol.*, **10**: 821–822.

INOZEMTZEV, S. L., 1930: Nature of potassium compounds in plants, *Ergeb. Veg. Lab. arb. Pryanishnikov*, **15**: 85–101; *Chem. Abst.*, **27**: 5368.

JACOB, K. D., R. P. BARTHOLOMEW, B. E. BROWN, W. H. PIERRE, F. R. REID, and J. W. TIDMORE, 1935: Nutrient value of the phosphorus in calcined phosphate as determined by growth of plants in greenhouse experiments, *J. Agr. Res.*, **50**: 837–848.

JACOBSON, H. G. M., and T. R. SWANBACK, 1929: Manganese toxicity in tobacco, *Sci.*, **70**: 283–284.

——— and ———, 1932: Manganese content of certain Connecticut soils and its relation to the growth of tobacco, *J. Am. Soc. Agron.*, **24**: 237–245.

JAMES, W. O., 1930: Studies on the physiological importance of the mineral elements in plants. I, The relation of potassium to the properties and functions of the leaf, *Ann. Bot.*, **44**: 173–198.

———, 1931: Studies of the physiological importance of the mineral elements in plants. II, Potassium: its distribution, movement and relation to growth in the potato, *Ann. Bot.*, **45**: 425–442.

——— and N. L. PENSTON, 1933: Studies of the physiological importance of the mineral elements in plants. IV, The quantitative distribution of potassium in the potato plant, *Ann. Bot.*, **47**: 279–293.

——— and M. CATTLE, 1935: Studies on the physiological importance of the mineral elements in plants. VII, The effects of potassium and chloride ions on the diastase of broad bean leaves, *New Phytol.*, **34**: 283–295.

JANSSEN, G., and R. P. BARTHOLOMEW, 1929: The translocation of potassium in tomato plants and its relation to their carbohydrate and nitrogen distribution, *J. Agr. Res.*, **38**: 447–465.

——— and ———, 1930: The influence of the potash concentration in the cultural medium on the production of carbohydrates in plants, *J. Agr. Res.*, **40**: 243–261.

——— and ———, 1932: The effect of potassium on the production of proteins, sugar and starch in cowpea and in sugar beet plants and the relation of potassium to plant growth, *J. Am. Soc. Agron.*, **24**: 667–679.

———, ———, and V. M. Watts, 1934: Some effects of nitrogen, phosphorus and potassium on the composition and growth of tomato plants, *Ark. Sta. Bull.* 310.

Jashnova, N. V., 1930: Nitrification in podzol soil, *Trans. Sci. Inst. Fertilizers*, 76, No. 415. Quoted by Nicol, 1934.

Javillier, M., and S. Imas, 1926: Le manganèse dans le grain de blé et dans les produits de mouture, *C. R. Acad. Agr. France*, **12**: 721–727.

——— and ———, 1926: Le zinc dans le grain de blé et dans les produits de mouture, *C. R. Acad. Agr. France*, **12**: 727–732.

Jennings, D. S., 1919: The effect of certain colloidal substances on the growth of wheat seedlings, *Soil Sci.*, **7**: 201–215.

Jodin, V., 1883: Du rôle de la silica dans la végétation du maïs, *Ann. Chim. et Phys.*, **30**: 485–494.

Johnson, M. O., 1916: The spraying of yellow pineapple plants on manganese soils with iron sulphate solutions, *Hawaii Sta. Press Bull.* 51.

———, 1917: Manganese as a cause of the depression of the assimilation of iron by pineapple plants, *J. Ind. Eng. Chem.*, **9**: 47–49.

———, 1924: Manganese chlorosis of pineapples: its cause and control, *Hawaii Sta. Press Bull.* 52.

———, 1928: Control of chlorosis of the pineapple and other plants, *Ind. and Eng. Chem.*, **20**: 724–725.

Johnston, E. S., 1917: Seasonal variations in the growth rates of buckwheat plants under greenhouse conditions, *Johns Hopkins Univ. Circ.*, n.s., **3**: 211–217.

——— and W. H. Dore, 1928: The relation of boron to the growth of the tomato plant, *Sci.*, **67**: 324–325.

——— and ———, 1929: The influence of boron on the chemical composition and growth of the tomato plant, *Plant Physiol.*, **4**: 31–62.

——— and D. R. Hoagland, 1929: Minimum potassium level required by tomato plants grown in water cultures, *Soil Sci.*, **27**: 89–109.

——— and P. L. Fisher, 1930: The essential nature of boron to the growth and fruiting of the tomato, *Plant Physiol.*, **5**: 387–392.

Jones, H. W., 1920: The distribution of inorganic iron in plants and animal tissues, *Biochem., J.*, **14**: 654–659.

Jones, J. P., 1929: Deficiency of magnesium the cause of a chlorosis in corn, *J. Agr. Res.*, **39**: 873–892.

———, 1931: The diagnostic value of plant symptoms in determining nutrient deficiencies of soils, *J. Am. Soc. Agron.*, **23**: 352–357.

Jones, J. S., and D. E. Bullis, 1921: Manganese in commonly grown legumes, *J. Ind. Eng. Chem.*, **13**: 524–525.

Jones, L. H., 1920: The effect of ammonium sulfate upon the availability of iron in nutrient solutions, *N. J. Sta. Rept.*, **1920**: 405–409.

——— and J. W. Shive, 1921: The influence of iron in the forms of ferric phosphate and ferrous sulphate upon the growth of wheat in a nutrient solution, *Soil Sci.*, **11**: 93–101.

——— and ———, 1921: Effect of ammonium sulfate upon plants in nutrient solutions supplied with ferric phosphate and ferrous sulphate as sources of iron, *J. Agr. Res.*, **21**: 701–729.

——— and ———, 1923: Influence of ammonium sulphate on plant growth in nutrient solutions and its effect on hydrogen-ion concentration and iron availability, *Ann. Bot.*, **37**: 355–377.

JONES, R. L., and F. R. PEMBER, 1925: The fertilizer nutrients required by barley, wheat, and oats as shown by both soil and water cultures, *Soil Sci.*, **19**: 169–201.

JONES, W. J., and H. A. HUSTON, 1914: Composition of maize at various stages of its growth, *Ind. Sta. Bull.* 175.

JORDAN, W. H., 1913: Studies in plant nutrition. I, *N. Y. Sta.* (*Geneva*) *Bull.* **358**: 11–30.

———, 1913: Studies in plant nutrition. II, *N. Y. Sta.* (*Geneva*) *Bull.* **360**: 53–77.

JUNG, J., 1920: Über den Nachweis und die Verbreitung des Chlors im Pflanzenreiche, *Sitzungsber. Akad. Wiss. Wien*, **129**: 297–340; *J. Chem. Soc.*, **121, 122**: 1098.

KARRAKER, P. E., and C. E. BORTNER, 1933: Symptoms of phosphorus deficiency in Turkish tobacco, *J. Am. Soc. Agron.*, **25**: 556–558.

——— and ———, 1934: Studies of frenching of tobacco, *Ky. Sta. Res. Bull.* 349.

KATAYMA, T., 1906: On the degree of stimulating action of manganese and iron salts on barley, *Bull. Col. Agr. Tokyo*, **7**: 91–93.

KELLEY, W. P., 1912: The function and distribution of manganese in plants and soils, *Hawaii Sta. Bull.* 26.

———, 1914: The function of manganese in plants, *Bot. Gaz.*, **57**: 213–227.

KIMBROUGH, W. D., 1931: The effect of fertilizer on the quality and keeping quality of watermelons, *Plant Physiol.*, **5**: 373–384.

KING, E. J., and V. DAVIDSON, 1933: The biochemistry of silicic acid. IV, Relation of silica to the growth of phytoplankton, *Biochem. J.*, **27**: 1015–1021.

KLEIN, G., 1930: Changes in the nitrogen of green plants, *Ergeb. Agr.-Chem.*, **2**: 143; *Chem. Abst.*, **26**: 2484.

———, K. TAUBÖCK, and H. LINSER, 1930, 1931: Urea and ureides in higher plants. I, Occurrence of urea in the vegetable kingdom and its transformation during the vegetative period. II, Appearance and disappearance of urea in plants, *Jahrb. Wiss. Bot.*, **73**: 193–225; **74**: 429–458.

KNOP, W., 1861–1862: Quantitativ-analytische Arbeiten über den Ernährungsprocess der Pflanzen, *Landw. Versuchs-Stat.*, **3**: 295–324; **4**: 173–185.

KNOTT, J. E., 1932: The effect of certain mineral elements on the color and thickness of onion scales, *N. Y. Sta.* (*Cornell*) *Bull.* 552.

KNOWLES, F., and J. E. WATKIN, 1931: The assimilation and translocation of plant nutrients in wheat during growth, *J. Agr. Sci.*, **21**: 612–637.

——— and ———, 1932: The amounts and distribution of some phosphorus and nitrogen compounds in wheat during growth, *J. Agr. Sci.*, **22**: 755–766.

KNUDSON, L., 1915: Toxicity of galactose for certain of the higher plants, *Ann. Mo. Bot. Garden*, **2**: 659–666.

———, 1916: Influence of certain carbohydrates on green plants, *N. Y. Sta.* (*Cornell*) *Mem.* 9.

———, 1917: The toxicity of galactose and mannose for green plants and the antagonistic action of other sugars toward these, *Am. J. Bot.*, **4**: 430–437.

——— and E. W. LINDSTROM, 1919: Influence of sugars on the growth of albino plants, *Am. J. Bot.*, **6**: 401–406.

KOSTYTSCHEW, S., and P. ELIASBERG, 1920: Über die Form der Kaliumverbindungen in lebenden Pflanzengeweben, *Z. physiol. Chem.*, **111**: 228–235.

KRATZMANN, E., 1913: Der microchemische Nachweis und die Verbreitung des Aluminiums im Pflanzenreich, *Sitzungsber. Akad. Wiss. Wien*, **122**: 311–336.

———, 1914: The physiological action of the salts of aluminum upon plants, *Sitzungsber. Akad. Wiss. Wien*, **123**: 211–233; *Exp. Sta. Rec.*, **34**: 525.

KRAYBILL, H. R., 1930: Plant metabolism studies as an aid in determining fertilizer requirements. I, *Ind. and Eng. Chem.*, **22**: 275–276.

KREUZHAGE, C., and E. WOLFF, 1884: Bedeutung der Kieselsäure für die Entwicklung der Haferpflanze, *Landw. Versuchs-Stat.*, **30**: 161–197.

LANHAM, W. B., 1927: Effect of potash fertilizer on carrying quality of tomatoes, *Tex. Sta. Bull.* 357.

LARSON, H. W. E., 1928: The relation of the concentration of calcium ion required by alfalfa to the amount present in the soil solution, *Soil Sci.*, **25**: 399–408.

LATSHAW, W. L., and E. C. MILLER, 1924: Elemental composition of the corn plant, *J. Agr. Res.*, **27**: 845–861.

LAURENT, J., 1904: Recherches sur la nutrition carbonée des plantes vertes à l'aide de matières organiques, *Rev. Gén. Bot.*, **16**: 14–48, 66–80, 96–119, 120–128, 155–166, 188–202, 231–241.

LAURIE, A., and E. W. MCELWEE, 1934: A method of studying nutrient deficiencies in greenhouse crops, *Proc. Am. Soc. Hort. Sci.*, **32**: 617–619.

LAURITZEN, C. W., 1934: Displacement of soil solubles through plant roots by means of air pressure as a method of studying soil fertility problems, *J. Am. Soc. Agron.*, **26**: 807–819.

———, 1934: A method for studying soil-plant nutrient relationships, *Sci.*, **79**: 187.

LAUSBERG, T., 1935: Determination of the cuticular excretion of leaves, *Jahrb. wiss. Bot.*, **81**: 769–806; *Chem. Abst.*, **29**: 7387.

LEEPER, G. W., 1934: Relationship of soils to manganese deficiency of plants, *Nature*, **134**: 972–973.

LEMMERMANN, O., 1922: Die ertragssteigernde Wirkung der Kieselsäure bei unzureichender Phosphorsäuredüngung, *Z. pflanz. Düng.*, 8vo., 72 pp.

———, H. WIESSMANN, and K. SAMMETT, 1925: Untersuchungen über die Ursache der ertragssteigerenden Wirkung der Kiesselsäure, *Z. pflanz. Düng.*, A, **4**: 265–315.

LEVINE, V. E., 1925: The effect of selenium compounds upon growth and germination in plants, *Am. J. Bot.*, **12**: 82–91.

LEVY, G., 1931: The presence, distribution and role of aluminium in plants, thesis, Paris; *Exp. Sta. Rec.*, **73**: 452.

LEYDHECKER, A., 1886: Ueber die physiologische Bedeutung des Chlor in der Buchweizenpflanze, *Landw. Versuchs-Stat.*, **8**: 177–187.

LIGON, W. S., and W. H. PIERRE, 1932: Soluble aluminum studies. II, Minimum concentrations of aluminum found to be toxic to corn, sorghum and barley in culture solutions, *Soil Sci.*, **34**: 307–317.

LILLELAND, O., 1930: Does potassium increase the sugar content of prunes? *Proc. Am. Soc. Hort. Sci.*, **27**: 15–18.

LIPMAN, C. B., 1916: A critique of the hypothesis of the lime-magnesium ratio, *Plant World*, **19**: 83–105, 119–135.

———, A. R. DAVIS, and E. S. WEST, 1926: The tolerance of plants for NaCl, *Soil Sci.*, **22**: 303–322.

———, 1928: The chemical elements indispensable to plants, *Sci. Month.*, **26**: 289–294.

——— and G. MACKINNEY, 1931: Proof of the essential nature of copper for higher green plants, *Plant Physiol.*, **6**: 593–599.

LLOYD, J. W., and E. P. LEWIS, 1935: Substitution of commercial fertilizers for manure in vegetable production, *Ill. Sta. Bull.* 421.

LOEW, O., 1878: Kann das Rubidium die physiologische Function des Kaliums in der Pflanzenzelle übernehmen? *Landw. Versuchs-Stat.*, **21**: 389–395.

———, 1891: Ueber die physiologische Function der Phosphorsäure, *Biol. Cent.*, **11**: 269–281.

——— and S. SAWA, 1902: On the action of manganese compounds on plants, *Bull. Col. Agr. Tokyo*, **5**: 161–172.

———, 1903: The physiological role of mineral nutrients in plants, *U. S. Dept. Agr. Bull.* 45.
——— and S. Honda, 1904: Über den Einfluss des Mangans auf Waldbäume, *Bull. Col. Agr. Tokyo*, **6**: 125–130.
———, 1934: The physiological importance of potassium and calcium, *Ernähr. der Pflanze*, **30**: 141–144; *Chem. Abst.*, **28**: 4448.
Lomanitz, S., 1923: A preliminary study of the effects of sodium chloride upon alfalfa grown in solution cultures, *Soil Sci.*, **16**: 183–195.
———, 1924: The influence of sodium chloride upon alfalfa grown in solution cultures, *Soil Sci.*, **18**: 353–371.
Loomis, W. E., and J. J. Wilson, 1933: Boron-iron relationship in the growth of tomato, *Proc. Iowa Acad. Sci.*, **40**: 53–56.
Loring, F. H., and J. G. F. Druce, 1930: Potassium: its radioactivity, its isotopes, and its association with the phenomena of life, *Chem. News*, **140**: 34–36.
Lukaszewicz, J., 1926: Über das Vorkommen von Kalium in manchen pflanzlichen Gebilden, *extr. du Bull. del'Académie Polonaise des Sci. et Lettres*, **1926**: 559–576.
Lutman, B. F., and N. L. Walbridge, 1929: The role of magnesium in the aging of plants, *Vt. Sta. Bull.* 296.
———, 1934: Cell size and structure in plants as affected by various inorganic elements, *Vt. Sta. Bull.* 383.
Lyon, C. J., 1923: The effect of phosphates on respiration, *J. Gen. Physiol.*, **6**: 299–306.
———, 1927: The role of phosphate in plant respiration, *Am. J. Bot.*, **14**: 274–284.
MacGillivray, J. H., 1926: The importance of phosphorus in the production of seed and non-seed portions of a tomato fruit, *Proc. Am. Soc. Hort. Sci.*, **22**: 374–379.
———, 1927: Effect of phosphorus on the composition of the tomato plant, *J. Agr. Res.*, **34**: 97–129.
Magistad, O. C., 1925: The aluminum content of the soil solution and its relation to soil reaction and plant growth, *Soil Sci.*, **20**: 181–227.
Magistris, H., 1932: Zur Biochemie und Physiologie organischer Phosphorverbindungen in Pflanze und Tier. III, Über den Austritt von anorganischem und organischem Phosphor aus Pflanzenzellen unter verschiedenen Bedingungen, *Biochem. Z.*, **253**: 64–80.
Maiwald, K., 1923: Wirkung hoher Nährstoffgaben auf den Assimilationsapparat, *Angew. Bot.*, **5**: 33–74.
———, 1930: Nährstoffaufnahme und Stofferzeugung beim Buchweizen (*Fagopyrum esculentum*); ein Beitrag zum t-Ertragsgesetz der höheren Pflanze, *Z. Pflanzenernähr. Düng. u. Bodenk.*, **17**: 12–42; *Biol. Abst.*, **5**: 2684.
Malyschev, N., 1932: Das Wachstum des isolierten Wurzelmeristems auf sterilem Nährboden, *Biol. Zent.*, **52**: 257–205.
Mameli, E., 1918: Sulla influenza del magnesis sopra la formazione della chlorofilla, *Univ. di Pavia. Atti Ist. Bot.*, **15**: 151–205.
Mann, C. E. T., 1924: The physiology of the nutrition of fruit trees. I, Some effects of calcium and potassium starvation, *Univ. Bristol Agr. Hort. Res. Sta. Ann. Rept.*, **1924**: 36–45.
——— and T. Wallace, 1925: The effects of leaching with cold water on the foliage of the apple, *J. Pom. and Hort. Sci.*, **4**: 146–161.
Maquenne, L., and E. Demoussy, 1920: Sur la distribution et la migration du cuivre dans les tissus des plantes vertes, *C. R. Acad. Sci. Paris*, **170**: 87–93.
——— and ———, 1920: Sur l'absorption du calcium par les racines des plantes et ses propriétés antitoxiques vis-à-vis du cuivre, *C. R. Acad. Sci. Paris*, **170**: 420–425.

——— and ———, 1920: Un cas d'action favorable du cuivre sur la végétation, *C. R. Acad. Sci. Paris*, **170**: 1542–1545.

Marine, D., E. J. Baumann, and B. Webster, 1930: The relative amounts of iodine-absorbing (reducing) material in various plants, *J. Biol. Chem.*, **89**: 213–219.

Marsh, R. P., and J. W. Shive, 1925: Adjustment of iron supply to requirements of soybean in solution culture, *Bot. Gaz.*, **79**: 1–28.

———, 1935: A modified culture jar, *Sci.*, **82**: 256.

Marsh, R. S., 1922: Preliminary studies on the sulfur content of the tomato, *Proc. Am. Soc. Hort. Sci.*, **19**: 83–84.

———, 1923: Seasonal variation of sulphur content in certain tissues of the apple tree, *Bot. Gaz.*, **75**: 400–413.

Martin, A. L., 1936: Toxicity of selenium to plants and animals, *Am. J. Bot.*, **23**: 471–483.

Masuda, E., and K. Nishida, 1934: Iodine of sea weeds, *J. Pharm. Soc. Japan*, **54**: 243–245.

Mather, T. H., 1929: The effect of fertilizers upon the forms of phosphorus and the amounts of phosphorus, nitrogen and silica in hays, *Sci. Agr.*, **10**: 35–63.

Maume, L., and J. Dulac, 1934: Varietal differences in the absorption of nitrogen, phosphoric acid and potash by wheat at the same physiological age and under the same environmental conditions, *C. R. Acad. Sci. Paris*, **198**: 199–202.

Mayer, G. G., 1929: Der Einfluss verschiedener Nährstoffzuführung auf das Längenwachstum isolierter Wurzeln, *diss. Hessischen Ludwigs-Univ.*

Mazé, P., and A. Perrier, 1904: Recherches sur l'assimilation de quelques substances ternaires par les végétaux à chlorophylle, *Ann. Inst. Pasteur*, **18**: 721–747.

——— and M. Lemoigne, 1912: Recherches sur la chlorose végétale provoquée par le carbonate de calcium, *C. R. Acad. Sci. Paris*, **155**: 435–437.

———, M. Ruot, and M. Lemoigne, 1913: Chlorose calcaire des plantes vertes. Rôle des excrétions des racines dans l'absorption du fer des sols calcaires, *C. R. Acad. Sci. Paris*, **157**: 495–498.

———, 1914: Influences respectives des éléments de la solution minérale sur le développement du maïs, *Ann. Inst. Pasteur*, **28**: 21–68.

———, 1915: Détermination des éléments minéraux rares nécessaires au développement du maïs, *C. R. Acad. Sci. Paris*, **160**: 211–214.

———, 1919: Recherche d'un solution purement minérale capable d'assurer l'évolution complète du maïs cultivé à l'abri des microbes, *Ann. Inst. Pasteur*, **33**: 139–173.

McCall, A. G., and P. E. Richards, 1918: Mineral food requirements of the wheat plant at different stages of its development, *J. Am. Soc. Agron.*, **10**: 127–134.

——— and J. R. Haag, 1921: The relation of the hydrogen-ion concentration of nutrient solutions to growth and chlorosis of wheat plants, *Soil Sci.*, **12**: 69–77.

McCalla, A. G., 1934: The effect of potassium supply on the composition and quality of wheat. I, *Can. J. Res.*, **11**: 687–700.

——— and E. K. Woodford, 1935: The effect of potassium supply on the composition and quality of wheat. II, *Can. J. Res.*, **13**: 339–354.

McCallum, A. B., 1905: On the distribution of potassium in animal and vegetable cells, *J. Physiol.*, **32**: 95–128.

McCance, R. A., and E. M. Widdowson, 1935: Phytin in human nutrition, *Biochem. J.*, **29**: 2694–2699.

McClelland, C. K., 1931: Effects of various plant foods on growth activities and development of oats, *J. Am. Soc. Agron.*, **23**: 304–311.

McClendon, J. F., E. Barrett, and T. Canniff, 1934: The iodine content of potatoes, *Biochem. J.*, **28**: 1209–1211.

McCool, M. M., 1913: The toxicity of manganese and the antidotal relations between this and various other cations with respect to green plants, *N. Y. Sta.* (*Cornell*) *Mem.* 2.

——— and M. D. Weldon, 1930: The effect of sodium nitrate on the composition of the expressed sap of the small grains, *J. Am. Soc. Agron.*, **22**: 434–438.

——— and R. L. Cook, 1930: Rate of intake, accumulation and transformation of nitrate nitrogen by small grains and Kentucky blue grass, *J. Am. Soc. Agron.*, **22**: 757–765.

———, 1933: Effect of thallium sulphate on the growth of several plants and on nitrification in soils, *Contr. Boyce Thompson Inst.*, **5**: 289–296.

———, 1935: Effect of light intensity on the manganese content of plants, *Contr. Boyce Thompson Inst.*, **7**: 427–438.

McDonnell, C. C., and R. C. Roark, 1917: Occurrence of manganese in insect flowers and insect flower stems, *J. Agr. Res.*, **11**: 77–83.

McGeorge, W. T., 1923: The chlorosis of pineapple plants grown on manganiferous soils, *Soil Sci.*, **16**: 269–275.

McHargue, J. S., 1919: The effect of manganese on the growth of wheat: a source of manganese for agricultural purposes, *J. Ind. Eng. Chem.*, **11**: 332–335.

———, 1919: Effect of certain compounds of barium and strontium on the growth of plants, *J. Agr. Res.*, **16**: 183–195.

———, 1921: The occurrence and distribution of manganese in plants, *Ky. Sta. Rept.*, **1921**: 36–37; **1922**: 35–36.

———, 1922: The role of manganese in plants, *J. Am. Chem. Soc.*, **44**: 1592–1598.

———, 1923: Effect of different concentrations of manganese sulphate on the growth of plants in acid and neutral soils and the necessity of manganese as a plant nutrient, *J. Agr. Res.*, **24**: 781–795.

———, 1923: Common earthenware jars a source of error in pot experiments, *J. Agr. Res.*, **26**: 231–233.

———, 1923: Iron and manganese content of certain species of seeds, *J. Agr. Res.*, **23**: 395–401.

———, 1925: The occurrence of copper, manganese, zinc, nickel and cobalt in soils, plants and their possible function as vital factors, *J. Agr. Res.*, **30**: 193–196.

———, 1925: The significance of the occurrence of copper, manganese and zinc in forage crops and foods, *J. Am. Soc. Agron.*, **17**: 368–372.

———, 1926: Mineral constituents of the cotton plant, *J. Am. Soc. Agron.*, **18**: 1076–1083.

———, 1927: Significance of the occurrence of manganese, copper, zinc, nickel and cobalt in Kentucky bluegrass, *Ind. and Eng. Chem.*, **19**: 274–276.

———, W. R. Roy, and J. G. Pelphrey, 1930: Iodine in some fertilizer materials, *Am. Fertilizer*, **73**: 40, 42, 63.

——— and O. M. Shedd, 1930: The effect of manganese, copper, zinc, boron and arsenic on the growth of oats, *J. Am. Soc. Agron.*, **22**: 739–747.

——— and R. K. Calfee, 1931: Effect of manganese, copper and zinc on the growth of yeast, *Plant Physiol.*, **6**: 559–566.

——— and ———, 1931: Effect of manganese, copper, and zinc on growth and metabolism of *Aspergillus flavus* and *Rhizopus nigricans*, *Bot. Gaz.*, **91**: 183–193.

——— and W. R. Roy, 1931: The mineral, nitrogen and fat content of some varieties of mature bean seed and of string beans, *J. Nutrition*, **3**: 479–481.

——— and R. K. Calfee, 1932: Effect of boron on the growth of lettuce, *Plant Physiol.*, **7**: 161–164.

——— and ———, 1932: Manganese essential for the growth of *Lemna minor*, *Plant Physiol.*, **7**: 697–703.

———, W. R. Roy, and J. G. Pelphrey, 1932: The iron, manganese, copper, zinc and iodine content of some important forage crops, *J. Am. Soc. Agron.*, **24**: 562–565.

——— and R. K. Calfee, 1933: Further evidence that boron is essential for the growth of lettuce, *Plant Physiol.*, **8**: 305–313.

———, D. W. Young, and R. K. Calfee, 1935: The effect of certain fertilizer materials on the iodine content of important foods, *J. Am. Soc. Agron.*, **27**: 559–565.

McKibben, R. R., 1928: The effect of sulphur on soils and on crop yields, *Md. Sta. Bull.* 296.

McLean, F. T., and B. E. Gilbert, 1925: Manganese as a cure for chlorosis of spinach, *Sci.*, **61**: 636.

———, 1927: Feeding plants manganese through the stomata, *Sci.*, **66**: 487–489.

——— and B. E. Gilbert, 1927: The relative aluminum tolerance of crop plants, *Soil Sci.*, **24**: 163–177.

——— and ———, 1928: Aluminum toxicity, *Plant Physiol.*, **3**: 293–302.

McLean, R. C., and W. L. Hughes, 1936: The quantitative distribution of boron in *Vicia faba* and *Gossypium herbaceum*, *Ann. App. Biol.*, **23**: 231–244.

McMurtrey, J. E., 1929: The effect of boron deficiency on the growth of tobacco plants in aerated and unaerated solutions, *J. Agr. Res.*, **38**: 371–380.

———, 1932: Effect of thallium on growth of tobacco plants, *Sci.*, **76**: 86.

———, 1932: Relation of calcium and magnesium to the growth and quality of tobacco, *J. Am. Soc. Agron.*, **24**: 707–716.

———, 1933: Distinctive effects of the deficiency of certain essential elements on the growth of tobacco plants in solution cultures, *U. S. Dept. Agr. Tech. Bull.* 340.

———, 1935: Boron deficiency in tobacco under field conditions, *J. Am. Soc. Agron.*, **27**: 271–273.

Meurer, R., 1909: Ueber die regulationische Aufnahme anorganischer Stoffe durch die Wurzeln von *Beta vulgaris* und *Daucus carota*, *Jahrb. wiss. Bot.*, **46**: 503–568.

Mevius, W., and I. Dikussar, 1930: Nitrites as sources of nitrogen for the higher plants, *Jahrb. wiss. Bot.*, **73**: 633–703; *Chem. Abst.*, **26**: 3005.

Meyer, A. H., 1931: Some neglected soil factors in plant growth, *J. Am. Soc. Agron.*, **23**: 606–625.

Meyers, C. N., and C. Voegtlin, 1914: Soluble aluminum compounds, *U. S. Health Rept.*, **29**: 1625–1629.

Michiels, H., and P. DeHeen, 1906: Note au sujet de l'action stimulante du manganèse sur la germination, *Bull. de l'Acad. Roy. Belg.*, **5**: 288–289.

Miller, H. G., 1919: The relation of sulfates to plant growth and composition, *J. Agr. Res.*, **17**: 87–103.

———, 1921: Further studies on relation of sulfates to plant growth and composition, *J. Agr. Res.*, **22**: 101–111.

Miller, L. P., 1933: Effect of manganese deficiency on the growth and sugar content of plants, *Am. J. Bot.*, **20**: 621–631.

Mirasol, J. J., 1920: Aluminum as a factor in soil acidity, *Soil Sci.*, **10**: 153–219.

Mitchell, H. L., 1934: Pot culture tests of forest soil fertility, *Black Rock Forest Bull.* 5.

———, 1935: A method for determining the nutrient needs of shade trees with special reference to phosphorus, *Black Rock Forest Papers*, Vol. 1, No. 1.

Mitchell, J. H., 1931: A study of relationships between the mineral content of the soil and plants grown on the soil, *S. C. Sta. Ann. Rept.*, **44**: 56.

Miyake, K., 1916: The toxic action of soluble aluminum salts upon the growth of the rice plant, *J. Biol. Chem.*, **25**: 23–28.

MOLISCH, H., 1892: "Die Pflanzen in ihren Beziehungen zum Eisen. Eine physiologische Studie," 8vo., 119 pp., G. Fischer, Jena; *Bot. Cent.*, **50**: 370.

———, 1897: Der Einfluss des Bodens auf die Blütenfarbe der *Hortensien*, *Bot. Ztg.*, **55**: 49–61.

———, 1902: Ueber vorübergehende Rothfärbung der Chlorophyllkörner in Laubblättern, *Ber. bot. Ges.*, **20**: 442–448.

MOLLIARD, M., 1907: Action morphogénique de quelques substances organiques sur les végétaux supérieurs, *Rev. Gén. Bot.*, **19**: 241–291, 329–349, 357–391.

MOORE, H. I., and A. P. JONES, 1936: Iron and plant growth, *Gard. Chron.*, **99**: 199–200.

MORRIS, H. S., 1931: Physiological effects of boron on wheat, *Bull. Torrey Bot. Club*, **58**: 1–30.

MORRIS, V. H., 1933: Form of potassium in corn tissues, *Ohio Sta. 51st Ann. Rept. Bull.*, **516**: 35–36.

——— and J. D. SAYRE, 1935: Solubility of potassium in corn tissue, *Plant Physiol.*, **10**: 565–568.

MORSE, F. W., 1927: Relation between water and potash in plant production, *J. Agr. Res.*, **35**: 939–946.

MOSER, F., 1933: The calcium magnesium ratio in soils and its relation to crop growth, *J. Am. Soc. Agron.*, **25**: 365–376.

MOTHES, K., 1932: Natural regulation of the vegetable protein metabolism, *Naturwiss.*, **20**: 102–103; *Chem. Abst.*, **26**: 3820.

MOWRY, H. and A. F. CAMP, 1934: A preliminary report on zinc sulfate as a corrective for bronzing of tung trees, *Fla. Sta. Bull.* 273.

MURNEEK, A. E., and E. J. GILDEHAUS, 1931: Leaf diagnosis and the interpretation of fertilizer requirements of plants, *Sci.*, **74**: 39–41.

MURPHY, H. F., 1933: Recovery of phosphorus from prairie grasses growing on central Oklahoma soils treated with superphosphate, *J. Agr. Res.*, **47**: 911–917.

———, 1936: The nitrogen, phosphorus and calcium content of the cotton plant at pre-blooming to early boll stages of growth, *J. Am. Soc. Agron.*, **28**: 52–57.

NAGAOKA, M., 1903, 1904: On the stimulating action of manganese upon rice, *Bull. Col. Agr. Tokyo*, **5**: 467–472; **6**: 135–136.

———, 1906: On the stimulating action of manganese upon rice. III, *Bull. Col. Agr. Tokyo*, **7**: 77–81.

NAKAMURA, M., 1904: Can lithium and cesium salts exert any stimulant action on phanerogams? *Bull. Col. Agr. Tokyo*, **6**: 153–160; *Bot. Cent.*, **95**: 669.

———, 1904: Can salts of zinc, cobalt and nickel in high dilution exert a stimulating action on agricultural plants? *Bull. Col. Agr. Tokyo*, **6**: 147–153; *Bot. Cent.*, **96**: 109.

NEAL, W. M., R. B. BECKER, and A. L. SHEALY, 1931: A natural copper deficiency in cattle rations, *Sci.*, **74**: 418–419.

NELLER, J. R., and W. J. MORSE, 1921: Effects upon the growth of potatoes, corn and beans resulting from the addition of borax to the fertilizer used, *Soil Sci.*, **12**: 79–132.

———, 1935: Phosphorus content and buffer capacity of plant sap as related to the physiological effect of phosphorus fertilizers in fibrous low-moor peat, *J. Agr. Res.*, **51**: 287–300.

NEMEC, A., and V. KAS, 1923: Studien über die physiologische Bedeutung des Titans im Pflanzenorganismus, *Biochem. Z.*, **140**: 583–590.

——— and M. GRAČANIN, 1930: Der Einfluss des Lichtes auf die Resorption von Kali und Phosphorsäure bei Neubaueruntersuchungen, *Z. Pflanzenernähr. Düng. u. Bodenk.*, **16**: 102–110.

NEWELL, H. J., 1930: Range of boron concentration in nutrient solutions for plants, *master's thesis, Univ. Md.*

NEWTON, J. D., 1928: The selective absorption of inorganic elements by various crop plants, *Soil Sci.*, **26**: 85–91.

NICOLAS, G., 1921: Contribution à l'étude du mécanisme de l'action fertilisante du soufre, *C. R. Acad. Sci. Paris*, **172**: 85–87.

NIGHTINGALE, G. T., L. G. SCHERMERHORN, and W. R. ROBBINS, 1930: Some effects of potassium deficiency on the histological structure and nitrogenous and carbohydrate constituents of plants, *N. J. Sta. Bull.* 499.

———, R. M. ADDOMS, W. R. ROBBINS, and L. G. SCHERMERHORN, 1931: Effects of calcium deficiency on nitrate absorption and on metabolism in tomato, *Plant Physiol.*, **6**: 605–631.

———, L. G. SCHERMERHORN, and W. R. ROBBINS, 1932: Effect of sulphur deficiency on metabolism in tomato, *Plant Physiol.*, **7**: 565–596.

NOBBE, F., and T. SIEGERT, 1862, 1863: Über das Chlor als specifischer Nährstoff der Buchweizenpflanze, *Landw. Versuchs-Stat.*, **4**: 318–340; **5**: 116–136.

——— and ———, 1864: Ueber das Chlor als Pflanzennährstoff, *Landw. Versuchs-Stat.*, **6**: 108–120.

———, 1865: Ueber die physiologische Function des Chlor in der Pflanze, *Landw. Versuchs-Stat.*, **7**: 371–386.

———, J. SCHRÖDER, and R. ERDMANN, 1871: Ueber die organische Leistung des Kalium in der Pflanze, *Landw. Versuchs-Stat.*, **13**: 321–423.

NOLL, C. F., 1923: The effects of phosphate on early growth and maturity, *J. Am. Soc. Agron.*, **15**: 87–99.

ODDO, B., and G. POLLACCI, 1920: Influenza del nucleo pirrolico sulla formazione della chlorofilla, *Gaz. Chim. Italiana*, **50**: 54–70.

ODLAND, T. E., and F. K. CRANDALL, 1932: The effect of the lack of available manganese in the soil on crop yields, *J. Am. Chem. Soc.*, **24**: 622–626.

OLSON, G. A., and J. L. ST. JOHN, 1921: An investigation of sulfur as a plant food, *Wash. Sta. Bull.* 165.

ORR, J. B., F. C. KELLY, and G. L. STUART, 1928: The effect of iodine manuring on the iodine content of plants, *J. Agr. Sci.*, **18**: 159–161.

ORTH, O. S., G. C. WICKWIRE, and W. E. BURGE, 1934: Copper in relation to chlorophyll and hemoglobin formation, *Sci.*, **79**: 33–34.

OSERKOWSKY, J., 1932: Hydrogen-ion concentration and iron content of tracheal sap from green and chlorotic pear trees, *Plant Physiol.*, **7**: 253–259.

———, 1933: Quantitative relation between chlorophyll and iron in green and chlorotic pear leaves, *Plant Physiol.*, **8**: 449–468.

——— and H. E. THOMAS, 1933: Exanthema in pears and its relation to copper deficiency, *Sci.*, **78**: 315–316.

OSTERHOUT, W. J. V., 1912: Plants which require sodium, *Bot. Gaz.*, **54**: 532–536.

———, 1916: Specific action of barium, *Am. J. Bot.*, **3**: 481–483.

OVERHOLSER, E. L., and F. L. OVERLEY, 1933: Fertilizers as related to leaf area in apple production, *Proc. Am. Soc. Hort. Sci.*, **30**: 52–54.

OWEN, O., 1931: The analysis of tomato plants. II, The effect of manurial treatment on the composition of tomato foliage, *J. Agr. Sci.*, **21**: 442–451.

PALLADIN, V. I., 1927: "Plant Physiology," Eng. trans. Livingston, 86 pp., P. Blakiston's Son & Company, Philadelphia.

PARBERY, N. H., 1935: Mineral constituents in relation to chlorosis of orange leaves, *Soil Sci.*, **39**: 35–45.

PARKER, F. W., and E. TRUOG, 1920: The relation between the calcium and the nitrogen content of plants and the function of calcium, *Soil Sci.*, **10**: 49–56.

——— and W. H. PIERRE, 1928: The relation between the concentration of mineral elements in a culture medium and the absorption and utilization of these elements by plants, *Soil Sci.*, **25**: 337–343.

PARKER, M. M., J. B. HESTER, and R. L. CAROLUS, 1933: The effect of soil conditions on the growth and composition of certain vegetable crop plants as influenced by soil reaction, *Proc. Am. Soc. Hort. Sci.*, **30**: 452–457.

PEMBER, F. R., 1917: Studies by means of both pot and solution culture of the phosphorus and potassium requirements of the barley plant during its different periods of growth, *R. I. Sta. Bull.* 169.

——— and F. T. MCLEAN, 1925: Economical use of nitrogen, phosphorus and potassium by barley, oats, and wheat in solution cultures, *R. I. Sta. Bull.* 199.

PENSTON, N. L., 1931: Studies on the physiological importance of the mineral elements in plants. III, A study by microchemical methods of the distribution of potassium in the potato plant, *Ann. Bot.*, **45**: 671–692.

———, 1935: Studies on the physiological importance of the mineral elements in plants. VIII, The variation in potassium content of potato leaves during the day, *New Phytol.*, **34**: 296–309.

PETERSON, W. H., 1914: Forms of sulphur in plant materials and their variation with the soil supply, *J. Am. Chem. Soc.*, **36**: 1290–1300.

——— and C. B. PETERSON, 1926: The water-soluble content of calcium and phosphorus in cabbage, *J. Agr. Res.*, **33**: 695–699.

——— and C. W. LINDOW, 1928: Variations in the manganese content of certain vegetables, *Soil Sci.*, **26**: 149–153.

PETRIE, A. H. K., 1934: The drift of the content of potassium and calcium with age on plants, *Australian J. Exp. Biol. Med. Sci.*, **12**: 99–110; *Chem. Abst.*, **29**: 2575.

PETTINGER, N. A., 1931: The expressed sap of corn plants as an indicator of nutrient needs, *J. Agr. Res.*, **43**: 95–119.

———, R. G. HENDERSON, and S. A. WINGARD, 1932: Some nutritional disorders in corn grown in sand cultures, *Phytopath.*, **22**: 33–51.

PFEFFER, W., 1900: "Physiology of plants," Eng. trans. by Ewarts, Vol. I, 416, 435 pp.; Vol. II, 33, 55 pp., Clarendon Press, London.

PFEIFFER, T., E. FRANK, O. LEMMERMANN, and H. SCHILLBACH, 1898: Über die Wirkung verschiedener Kalisalze auf die Zusammensetzung und den Ertrag der Kartoffeln, *Landw. Versuchs-Stat.*, **49**: 349–385.

———, A. RIPPEL, and C. PFOTENHAUER, 1919: Über den Verlauf der Nährstoffaufnahme und Stofferzeugung bei der Gerstenpflanze, *Fühlung landw. Ztg.*, **68**: 81–101.

———, ———, and ———, 1920: The action of sodium in the presence of potassium as a plant nutrient, *J. Landw.*, **68**: 255–284; *Exp. Sta. Rec.*, **45**: 624.

PHILLIPS, M., J. DAVIDSON, and H. D. WEIHE, 1931: Studies of lignin in wheat straw with reference to lodging, *J. Agr. Res.*, **43**: 619–627.

PHILLIPS, T. G., T. O. SMITH, and R. B. DEARBORN, 1934: The effect of potassium deficiency on the composition of the tomato plant, *N. H. Sta. Tech. Bull.* 59.

PIERRE, I., 1866: La silice et la verse des blés, *C. R. Acad. Sci. Paris*, **63**: 374–377.

PITCAITHLY, N. P., and F. P. WORLEY, 1933: Distribution of copper in the Karaka tree (*Corynocarpus laevigata*), *J. Agr. Sci.*, **23**: 204–207.

PITZ, W., 1916: Effect of elemental sulfur and of calcium sulfate on certain of the higher and lower forms of plant life, *J. Agr. Res.*, **5**: 771–781.

PLAGGE, H. H., T. J. MANEY, and B. S. PICKETT, 1935: Functional diseases of the apple in storage, *Iowa Sta. Bull.* 329.

PLIMMER, R. H. A., 1913: The hydrolysis of organic compounds by dilute acid and by dilute alkali, *Biochem. J.*, **7**: 72–80.

POEHLMAN, J. M., 1935: Some limitations on plant juice analyses as indicators of the nutrient needs of plants, *J. Am. Soc. Agron.*, **27**: 195–207.

POESCH, G. H., 1935: Coloring and fertilizing hydrangeas, *Ohio Sta. Mo. Bull.*, **22**: 92–94.

POHLMAN, G. G., and W. H. PIERRE, 1933: The phosphorus concentration of the exuded sap of corn as a measure of the available phosphorus of the soil, *J. Am. Soc. Agron.*, **25**: 160–171.

POLLACCI, G., and B. ODDO, 1915: Influenza del nucleo pirrolico sulla formazione della chlorofilla. Nota preliminare, *Roy. Accad. Lincei*, **24**: 37–39.

———, 1935: Influence of the pyrrole nucleus on the formation of chlorophyll, *Ber. bot. Ges.*, **53**: 540–542; *Chem. Abst.*, **29**: 6275.

POSTERNAK, S., 1903: Sur la matière phospho-organique de réserve des plantes à chlorophylle. Précédé de préparation, *C. R. Acad. Sci. Paris*, **137**: 202–206.

POWERS, W. L., 1923: Progress of sulfur investigations with Oregon soils, *J. Am. Soc. Agron.*, **15**: 158–160.

———, 1930: The role of sulphur in plant nutrition, *J. Am. Soc. Agron.*, **22**: 371–374.

PRIANISCHNIKOW, D. N., 1905: Results of vegetation experiments in the years 1901–1903, *Ann. Inst. Agron. Moscow*, **11**: 155–219; *Exp. Sta. Rec.* **18**: 320–321.

———, 1911: Über den Einfluss von kohlensaurem Kalk auf die Wirkung von verschiedenen Phosphaten, *Landw. Versuchs-Stat.*, **75**: 357–376.

———, 1923: Zur Frage über die Bedeutung des Calciums für die Pflanzen, *Ber. bot. Ges.*, **41**: 138–144.

PROEBSTING, E. L., 1933: Absorption of potassium by plants as affected by decreased exchangeable potassium in the soil, *J. Pom. and Hort. Sci.*, **11**: 199–204.

PRUTZKOVA, M., M. LEBEDEVA, A. MELNIKOV, and S. OSTANIN, 1932: Lodging of wheat, *Bull. App. Bot. Genetics and Plant Breeding*, Ser. A., No. 3: 67–87; *Exp. Sta. Rec.*, **72**: 325.

QUARTAROLI, A., 1929: Copper as an element necessary to plant life, *Ann. Chim. Applicata*, **19**: 467–519; *Chem. Abst.*, **24**: 2773.

RAVIN, P., 1913: Nutrition carbonée des plantes à l'aide des acides organiques libres et combinés, *Ann. Sci. Nat. Bot.*, ser. IX, **18**: 289–446.

REED, H. S., 1907: The value of certain nutritive elements to the plant cell, *Ann. Bot.*, **21**: 501–543.

——— and A. R. C. HAAS, 1923: The pseudo-antagonism of sodium and calcium in dilute solutions, *J. Agr. Res.*, **24**: 753–759.

——— and ———, 1923: Studies on the effects of sodium, potassium, and calcium on young orange trees, *Calif. Sta. Tech. Paper* **11**: 1–23.

——— and ———, 1924: Nutrient and toxic effects of certain ions on citrus and walnut trees with especial reference to the concentration and pH of the medium, *Calif. Sta. Tech. Paper* **17**: 1–75.

——— and ———, 1924: Iron supply in a nutrient medium, *Bot. Gaz.*, **77**: 290–300.

——— and J. DUFRENOY, 1935: The effects of zinc and iron salts on the cell structure of mottled orange leaves, *Hilgardia*, **9**: 113–137.

——— and ———, 1935: The effects of zinc salts on the oxidation process in plant cells, *Sci.*, **82**: 249–250.

REIMER, F. C., and H. V. TARTAR, 1919: Sulphur as a fertilizer for alfalfa in southern Oregon, *Ore. Sta. Bull.* 163.

RICHARDS, F. J., 1932: Physiological studies in plant nutrition. III, Further studies on the effect of potash deficiency on the rate of respiration in leaves of barley, *Ann. Bot.*, **46**: 367–388.

RICHARDSON, W. D., 1920: The ash of dune plants, *Sci.*, **51**: 546–551.

Rippel, A., 1923: Über die durch Mangan verursachte Eisenchlorose bei grünen Pflanzen, *Biochem. Z.*, **140**: 315–323.

———, G. Behr, and R. Meyer, 1933: Zur Kenntnis der Wirkung des Kaliums auf höhere Pflanzen, *Z. Pflanzenernähr. Düng. u. Bodenk.*, **32**: 95–109; *Biol. Abst.*, **9**: 748.

Roach, B. M. B., 1926: On the relation of certain soil algae to some soluble carbon compounds, *Ann. Bot.*, **40**: 149–203.

———, 1927: On the carbon nutrition of some algae isolated from soil, *Ann. Bot.*, **41**: 509–517.

Robbins, W. J., 1918: Direct assimilation of organic carbon by *Ceratodon purpureus*, *Bot. Gaz.*, **65**: 543–551.

———, 1922: Cultivation of excised root tips and stem tips under sterile conditions, *Bot. Gaz.*, **73**: 376–390.

——— and W. E. Maneval, 1924: Effect of light on growth of excised root tips under sterile conditions, *Bot. Gaz.*, **78**: 424–432.

———, 1929: A study of the metabolism of roots, *Mo. Sta. Bull.*, **272**: 41.

Robbins, W. R., 1927: Nutrient requirements of the sweet potato plant when grown in sand culture and effect of various nutrient salt proportions on the composition of the roots, *Ann. Rept. N. J. Sta.*, **48**: 178–179.

———, G. T. Nightingale, L. G. Schermerhorn, and M A. Blake, 1929: Potassium in relation to the shape of the sweet potato, *Sci.*, **70**: 558.

Roberg, M., 1932: Ein Beitrag zur Stoffwechselphysiologie der Grünalgen. II, Über die Wirkung von Eisen-, Zink-, und Kupfersalzen, *Jahrb. wiss. Bot.*, **76**: 311–332; *Biol. Abst.*, **9**: 536–537.

Robinson, W. O., L. A. Steinkoenig, and C. F. Miller, 1917: The relation of some of the rarer elements in soils and plants, *U. S. Dept. Agr. Bull.* 600.

———, 1936: Selenium content of wheat from various parts of the world, *Ind. and Eng. Chem.*, **28**: 736–738.

Rodriguez, A. G., 1933: Effect of boron on the availability of iron, *abst. Am. J. Bot.*, **20**: 679.

Rogers, C. H., and J. W. Shive, 1932: Factors affecting the distribution of iron in plants, *Plant Physiol.*, **7**: 227–252.

Rudolfs, W., 1919: Influence of sodium chloride upon the physiological changes of living trees, *Soil Sci.*, **8**: 397–427.

———, 1927: Influence of salt on the growth rate of asparagus, *Bot. Gaz.*, **83**: 94–98.

Ruprecht, R. W., 1915: Toxic effect of iron and aluminum salts on clover seedlings, *Mass. Sta. Bull.*, **161**: 125–129.

Russell, E. J., 1915: "Soil Conditions and Plant Growth," p. 35, Longmans, Green & Company, New York.

Saeger, A., 1933: Manganese and the growth of *Lemnaceae*, *Am. J. Bot.*, **20**: 234–245.

Salm-Horstmar, F., 1856: Versuche und Resultate über die Nahrung der Pflanzen, *diss., Braunschweig.*

Salmon, S. C., 1931: An instrument for determining the breaking strength of straw and a preliminary report on the relation between breaking strength and lodging, *J. Agr. Res.*, **43**: 73–82.

Salter, R. M., and J. W. Ames, 1928: Plant composition as a guide to the availability of soil nutrients, *J. Am. Soc. Agron.*, **20**: 808–837.

Samuel, G., and C. S. Piper, 1929: Manganese as an essential element for plant growth, *Ann. App. Biol.*, **16**: 493–523.

Sayre, J. D., 1930: Accumulated iron in the nodes of corn plants, *Plant Physiol.*, **5**: 393–399.

Scharrer, K., and J. Schwaibold, 1927: Zur Kenntnis des Jods als biogenes Element. X, Mitteilung: Untersuchung einiger Kulturpflanzen auf ihren natürlichen Jodgehalt und dessen Steigerung durch Joddüngung, *Biochem. Z.*, **185**: 405–413.

——— and A. Strobel, 1927: Die Jodanreicherung der Pflanzen durch Jodzufuhr, *Angew. Bot.*, **9**: 187–199.

——— and W. Schropp, 1933: Sand- und Wasserkulturversuche mit Nickel und Kobalt, *Z. Pflanzenernähr. Düng. u. Bodenk.*, **31**: 94–113.

——— and ———, 1935: The action of vanadium upon cultivated plants, *Z. Pflanzenernähr. Düng. u. Bodenk.*, **37**: 196–202; *Chem. Abst.*, **29**: 6624.

Schimper, A. F. W., 1888: Ueber Kalkoxalatbildung in den Laubblättern, *Bot. Ztg.*, **46**: 65–69, 81–89, 97–107, 113–123, 129–139, 145–153.

———, 1890: Zur Frage der Assimilation der Mineralsalze durch die grüne Pflanze, *Flora*, **73**: 207–278.

Schmidt, E. W., 1915: The effects of chlorides on plant growth, *unpub. thesis*, *Univ. Wis.* Quoted by Tottingham, 1919.

Schmucker, T., 1934: The influence of boric acid on plants, especially germinating pollen grains, *Planta*, **23**: 264–283; *Chem. Abst.*, **29**: 2201.

Schollenberger, C. J., 1922: Silica and silicates in relation to plant growth and composition, *Soil Sci.*, **14**: 347–363.

Schreiner, O., 1923: Organic phosphorus in soils, *J. Am. Soc. Agron.*, **15**: 117–125.

——— and P. R. Dawson, 1927: Manganese deficiency in soils and fertilizers. *Ind. and Eng. Chem.*, **19**: 400–404.

Schropp, W., and K. Scharrer, 1933: Water culture investigations with Hoagland's "A-Z" solution, *Jahrb. wiss. Bot.*, **78**: 544–563; *Exp. Sta. Rec.*, **71**: 171–172.

Schulze, B., 1913: Beitrag zur Frage der Düngung mit Natronsalzen, *Landw. Versuchs-Stat.*, **79-80**: 431–448.

Schuster, G. L., 1927: Potash in relation to quality of crop, *J. Am. Soc. Agron.*, **19**: 506–517.

Scofield, C. S., and L. V. Wilcox, 1930: The boron content of oranges, *Sci.*, **71**: 542.

——— and ———, 1931: Boron in irrigation waters, *U. S. Dept. Agr. Tech. Bull.* 264.

Scott, S. G., 1929: Phosphorus deficiency in forage feeds of range cattle, *J. Agr. Res.*, **38**: 113–130.

Shaw, J. K., 1934: Some experience with the Thornton test for potash in apple trees, *Proc. Am. Soc. Hort. Sci.*, **32**: 56–60.

Shedd, O. M., 1913: The sulphur content of some typical Kentucky soils, *Ky. Sta. Bull.*, 174.

———, 1914: The relation of sulphur to soil fertility, *Ky. Sta. Bull.* 188.

———, 1917: Effect of sulphur on different crops and soils, *J. Agr. Res.*, **11**: 91–105.

Shive, J. W., and W. H. Martin, 1918: A comparative study of salt requirements for young and for mature buckwheat plants in solution cultures, *J. Agr. Res.*, **14**: 151–176.

———, 1918: A comparison of salt requirements for young and for mature buckwheat plants in water cultures and sand cultures, *Am. J. Bot.*, **5**: 186–191.

———, 1918: A comparative study of salt requirements for young and mature buckwheat plants in sand cultures, *Soil Sci.*, **6**: 1–32.

Shull, C. A., 1920: Sulphur as a fertilizer, *Sci.*, **52**: 376–378.

Shulov, I. S., 1909: Various smaller experiments with fertilizers and soils, *Ann. Inst. Agron. Moscow*, **15**: 116–125; *Exp. Sta. Rec.* **22**: 223–224.

Sideris, C. P., 1930: Titanium replacing the iron essential in plant growth, *Pineapple News*, **4**: 98–100.

Sidorin, M. I., 1914: On the assimilation of iron by plants, *Rec. trav. lab. Agron. Moskow, Selsk. Khoz. Inst.*, **10**: 241–257; *Exp. Sta. Rec.* **36**: 633.

Skinner, J. J., and F. R. **Reid**, 1921: Nutrient requirements of clover and wheat in solution cultures, *Soil Sci.*, **12**: 287–301.

——— and F. E. **Allison**, 1923: Influence of fertilizers containing borax on the growth and fruiting of cotton, *J. Agr. Res.*, **23**: 433–445.

———, B. E. **Brown**, and F. R. **Reid**, 1923: The effect of boron on the growth and yield of crops, *U. S. Dept. Agr. Bull.* 1126.

——— and R. W. **Ruprecht**, 1930: Fertilizer experiments with truck crops, *Fla. Sta. Bull.*, **218**: 1–65.

———, 1931: Influence of potash sources and chlorine content of fertilizers on yield of cotton, *J. Am. Soc. Agron.*, **23**: 13–21.

Smith, J. B., F. K. **Crandall**, and D. E. **Frear**, 1932: The relative effect of single and fractional applications of soluble nitrogen on nitrates in soil and plant and on the yield of certain vegetable crops, *J. Am. Soc. Agron.*, **24**: 203–221.

Smith, T. O., and O. **Butler**, 1921: Relation of potassium to growth in plants, *Ann. Bot.*, **35**: 189–227.

Sommer, A. L., 1926: Studies concerning the essential nature of aluminum and silicon for plant growth, *Univ. Calif. Pub.*, **5**: 57–81.

——— and C. B. **Lipman**, 1926: Evidence of the indispensable nature of zinc and boron for higher green plants, *Plant Physiol.*, **1**: 231–249.

———, 1927: The search for elements essential in only small amounts for plant growth, *Sci.*, **66**: 482–484.

———, 1928: Further evidence of the essential nature of zinc for the growth of higher green plants, *Plant Physiol.*, **3**: 217–223.

——— and H. **Sorokin**, 1928: Effect of the absence of boron and of some other essential elements on the cell and tissue structure of the root tips of *Pisum sativum*, *Plant Physiol.*, **3**: 237–261.

———, 1931: Copper as an essential for plant growth, *Plant Physiol.*, **6**: 339–345.

———, 1936: The relationship of the phosphate concentration of solution cultures to the type and size of root systems and the time of maturity of certain plants, *J. Agr. Res.*, **52**: 133–148.

Sorokin, H., and A. L. **Sommer**, 1929: Changes in the cells and tissues of root tips induced by the absence of calcium, *Am. J. Bot.*, **16**: 23–40.

Sreenivasan, A., 1934: The role of silicon in plant nutrition, *Current Sci.*, **3**: 193–197; *Chem. Abst.*, **29**: 2205.

Steward, F. C., 1935: Mineral nutrition of plants, *Ann. Rev. Biochem.*, **4**: 519–544.

Steward, W. D., and J. M. **Arthur**, 1933: Some effects of radiation from a quartz mercury arc upon the mineral composition of plants, *abst. Am. J. Bot.*, **20**: 673.

——— and ———, 1934: Some effects of radiation from a quartz mercury vapor lamp upon the mineral composition of plants, *Contr. Boyce Thompson Inst.*, **6**: 225–245.

Stewart, G. R., 1930: Report on chemistry, *Proc. Hawaiian Sugar Planters' Assoc.*, **49**: 328–351; *Chem. Abst.*, **24**: 5404.

Stewart, J., 1922: Some relations of arsenic to plant growth. I and II, *Soil Sci.*, **14**: 111–127.

Stewart, R., 1920: Sulphur in relation to soil fertility, *Ill. Sta. Bull.* 227.

Stiebeling, H. K., 1932: The iron content of vegetables and fruits, *U. S. Dept. Agr. Circ.* 205.

Stocklasa, J., and J. **Pitra**, 1901: Ueber die Einwirkung der Kalisalze auf die Entwicklung der Gerste, *Z. landw. Versuch. im Oesterr.*, p. 567.

———, 1908: Beiträge zur Kenntniss der physiologischen Funktion des Kalis im Pflanzenorganismus, *Z. landw. Versuch.*, **11**: 52–61.

———, 1911: Catalytic fertilizers for sugar beets, *Bl. Zuckerrühenbau*, **18**: 193–197; *Exp. Sta. Rec.* **26**: 225.

——— and E. Senft, 1912: Ist das Kalium an dem Auf- und Abbau der Kohlenhydrate bei höheren Pflanzen Beteiligt? *Z. landw. Verwesen Öesterr*, **15**: 711–736.

———, 1916: Ist das Kaliumion an der Eiweisssynthese in der Pflanzenzelle beteiligt? *Biochem. Z.*, **73**: 107–160.

———, 1918: Über den Einfluss des Aluminiumions auf die Keimung des Samens und die Entwicklung der Pflanzen, *Biochem. Z.*, **91**: 137–223.

———, 1918: Über die Verbreitung des Aluminiumions in der Pflanzenwelt, *Biochem. Z.*, **88**: 292–322.

———, 1920: Der Mechanismus der physiologischen Wirkung der Radiumemanation und der Radioaktivität des Kaliums auf die biochemischen Vorgänge bei dem Wachstumsprocess der Pflanzen. II, *Biochem. Z.*, **108**: 140–172.

———, 1920: Bedeutung der Radioaktivität des Kaliums bei der Photosynthese, *Biochem. Z.*, **108**: 173–184.

———, 1920: Ueber die Radioaktivität des Kaliums und ihre Bedeutung in der chlorophyllosen und cholorophyllhaltigen Zelle. I, *Biochem. Z.*, **108**: 109–139.

———, 1922: Ueber die Resorption des Aluminiumions durch das Wurzelsystem der Pflanzen, *Biochem. Z.*, **128**: 35–47.

———, 1922: "Über die Verbreitung des Aluminiums in der Natur und seine Bedeutung beim Bau- und Betriebsstoffwechsel der Pflanzen," 8vo., x + 500 pp., Fischer, Jena.

———, 1924: De la fonction physiologique de l'iode dans organisme de la betterave à sucre, *C. R. Acad. Sci. Paris*, **178**: 120–122.

———, 1929: Über den Einfluss des Jodions auf das Wachstum und die Zellvermehrung der Halophyten, *Biochem. Z.*, **211**: 213–228; *Biol. Abst.*, **5**: 1616.

———, 1930: Die Resorption des Jods durch das Wurzelsystem der Pflanze, *Protoplasma*, **8**: 199–214.

Stokes, W. E., W. A. Leukel, and R. M. Barnette, 1930: Effect of irrigation with sewage effluent on the yields and establishment of Napier grass and Japanese cane, *Fla. Sta. Bull.* 215.

Storey, H. H., and R. Leach, 1933: A sulfur-deficiency disease of the tea bush, *Ann. App. Biol.*, **20**: 23–56.

Street, O. E., 1934: Carbohydrate-nitrogen and base element relationships of peas grown in water culture under various light exposures, *Plant Physiol.*, **9**: 301–322.

Suzuki, V., 1900: Can strontium and barium replace calcium in phanerogams? *Bull. Col. Agr. Tokyo*, **4**: 69–77.

——— and K. Aso, 1903: On the stimulating action of iodine and fluorine compounds on agricultural plants, *Bull. Col. Agr. Tokyo*, **5**: 4; *Bot. Cent.*, **93**: 143.

———, 1919: The translocation of the mineral constituents of the jack bean, *J. Am. Chem. Soc.*, **41**: 282–287.

Swain, R. E., and A. B. Johnson, 1936: Effect of sulfur dioxide on wheat development, *Ind. and Eng. Chem.*, **28**: 42–47.

Swanback, T. R., 1927: The effect of boric acid on the growth of tobacco plants in nutrient solutions, *Plant Physiol.*, **2**: 475–486.

Swanson, C. O., and R. W. Miller, 1917: The sulfur content of some typical Kansas soils and the loss of sulphur due to cultivation, *Soil Sci.*, **3**: 139–148.

Szücs, J., 1913: Über einige charakteristische Wirkungen des Aluminiumions auf das Protoplasma, *Jahrb. wiss. Bot.*, **52**: 269–332.

Tanaka, I., 1931: Nutrition of higher plants with organic compounds, *Japan J. Bot.*, **5**: 323–350; *Chem. Abst.*, **28**: 2388; *Exp. Sta. Rec.*, **68**: 458.

Teakle, L. J. H., 1929: The absorption of phosphate from soil and solution cultures, *Plant Physiol.*, **4**: 211–232.

Thatcher, R. W., and G. A. Olson, 1912: Experiments in fertilizing alfalfa, *Wash. Sta. Pop. Bull.* 49.

———, 1932: Some new aspects of plant nutrition, *Sci.*, **76**: 281–285.

———, 1934: A proposed classification of the chemical elements with respect to their functions in plant nutrition, *Sci.*, **79**: 463–466.

Thoday, D., and H. Evans, 1932: Studies in growth and differentiation. III, The distribution of calcium and phosphate in the tissues of *Kleinia articulata* and some other plants, *Ann. Bot.*, **46**: 781–807.

——— and ———, 1933: Studies in growth and differentiation. IV, The distribution of some solutes in the tissues of *Kleinia articulata*, *Ann. Bot.*, **47**: 1–20.

Thomas, W., 1930: The conception of balance with respect to the absorption of nitrogen, phosphorus and potassium by plants and the influence of the balance of nutrition, *Sci.*, **72**: 425–427.

———, 1932: The reciprocal effects of the nitrogen, phosphorus, and potassium as related to the absorption of these elements by plants, *Soil Sci.*, **33**: 1–21.

———, 1934: Misconceptions relative to the mineral composition of plants, *Sci.*, **80**: 587.

Thornton, S. F., 1932: A field and laboratory test on plant material for diagnosing phosphorus deficiencies, *Ind. Sta. Bull.* 355.

———, 1933: A simple and rapid chemical test on plant material as an aid in determining potassium needs, *J. Am. Soc. Agron.*, **25**: 473–481.

———, S. D. Conner, and R. R. Frasure, 1934: The use of rapid chemical tests on soil and plants as aids in determining fertilizer needs, *Ind. Sta. Circ.* 204.

Tidmore, J. W., 1930: The phosphorus content of the soil solution and its relation to plant growth, *J. Am. Soc. Agron.*, **22**: 481–488.

———, 1930: Phosphate studies in solution cultures, *Soil Sci.*, **30**: 13–33.

Tincker, M. A. H., 1932: The influence of the daily period of light and the supply of potassium on the rate of growth of tubers and other storage organs, *J. Roy. Hort. Soc.*, **57**: 326–331.

——— and F. V. Darbishire, 1933: Studies on the formation of tubers and other storage organs. The influence upon translocation of the period of light and the supply of potassium, *Ann. Bot.*, **47**: 27–55.

Tobler, F., 1929: Zur Kenntnis der Wirkung des Kaliums auf den Bau der Bastfaser, *Jahrb. wiss. Bot.*, **71**: 26–51.

Tottingham, W. E., 1915: The role of chlorine in plant nutrition, *Sci.*, **42**: 68.

——— and A. J. Beck, 1916: Antagonism between manganese and iron in the growth of wheat, *Plant World*, **19**: 359–370.

———, 1917: On the role of chlorine to plant growth, *Johns Hopkins Univ. Circ.*, **3**: 217–221.

———, 1918: The sulphur requirement of the red clover plant, *J. Biol. Chem.*, **36**: 429–438.

———, 1919: A preliminary study of the influence of chlorides on the growth of certain agricultural plants, *J. Am. Soc. Agron.*, **11**: 1–33.

——— and E. J. Rankin, 1923: The availability of iron in nutrient solutions for wheat, *Am. J. Bot.*, **10**: 203–211.

Trelease, S. F., 1917: A study of salt proportions in a nutrient solution containing chloride as related to the growth of young wheat plants, *Johns Hopkins Univ. Circ.*, **3**: 222–225.

———, 1920: The relation of salt proportions and of concentrations to the growth of young wheat plants in nutrient solutions containing a chloride, *Philippine J. Sci.*, **17**: 527–603.

——— and H. M. TRELEASE, 1931: Magnesium injury of wheat, *Bull. Torrey Bot. Club*, **58**: 127–149.

——— and A. L. MARTIN, 1936: Plants made poisonous by selenium absorbed from the soil, *Bot. Rev.*, **2**: 373–396.

TRUE, R. H., and R. B. HARVEY, 1918: The absorption of calcium salts by squash seedlings, *Brooklyn Bot. Gard. Mem.*, **1**: 502–512.

———, O. F. BLACK, and J. W. KELLY, 1919: Ash absorption by spinach from concentrated soil solutions, *J. Agr. Res.*, **16**: 15–27.

———, 1921: The function of calcium in the nutrition of seedlings, *J. Am. Soc. Agron.*, **13**: 91–108.

———, 1922: The significance of calcium for higher green plants, *Sci.*, **55**: 1–6.

TRUOG, E., 1913: Availability of phosphates to various crops, *Wis. Sta. Bull.* 240.

———, 1916: The utilization of phosphates by agricultural crops, including a new theory regarding the feeding power of plants, *Wis. Sta. Res. Bull.* 41.

———, 1922: The feeding power of plants, *Sci.*, **56**: 294–298.

TUBBS, F. R., 1930: Physiological studies in plant nutrition. II, The effect of manurial deficiency upon the mechanical strength of barley straw, *Ann. Bot.*, **44**: 147–160.

TURNER, T. W., 1929: Effect of mineral nutrients upon seed plants. II, Phosphates, *Bot. Gaz.*, **88**: 85–95.

TYNER, E. H., 1935: The feeding power of plants for the potassium in feldspar, exchangeable form, and dilute solution, *Soil Sci.*, **39**: 405–424.

VANDECAVEYE, S. C., and G. O. BAKER, 1932: The effect of fertilizers on crop yields of different soils and on the composition of certain crops, *Wash. Sta. Bull.* 274.

VAN HAARLEM, J. R., 1935: Some results of mineral deficiency studies, *Ann. Rept. Mich. State Hort. Soc.*, **65**: 73–75.

VAN SCHREVEN, D. A., 1935: External and internal symptoms of boron deficiency in tomato, *Tijdschr. Plantenziekten*, **41**: 1–26; *Exp. Sta. Rec.*, **74**: 466–467.

VAN WIJK, D. J. R., 1932: Leaching of phosphorus compounds and proteins from grasses by rain water, *J. South Africa Chem. Inst.*, **15**: 54–59.

VILLE, G., 1889: Recherches sur les relations qui existent entre la couleur des plantes et la richesse des terres en agents de fertilité, *C. R. Acad. Sci. Paris*, **109**: 397–400.

VINCENT, V., and J. HERVIAUX, 1929: Le sodium dans les végétaux cultivés, *Ann. Sci. Agron.*, **46**: 444–460; *Biol. Abst.*, **5**: 2662.

VOELCKER, J. A., 1915: The influence of boron compounds on wheat and barley, *J. Roy. Agr. Soc. England*, **76**: 347–351.

VON EULER, H., and T. PHILIPSON, 1932: Water-soluble growth factors, *Biochem. Z.*, **245**: 418–430; *Chem. Abst.*, **26**: 3281.

VON SACHS, J., 1887: "Lectures on the Physiology of Plants," 288 pp., Clarendon Press, London.

VON WRANGELL, M., 1922: Laws governing the phosphoric acid nutrition of plants, *Landw. Jahrb.*, **57**: 1–78; *Exp. Sta. Rec.*, **48**: 426.

———, 1927: Das Jod als Pflanzennährstoff, *Natur. Wiss.*, **15**: 70–73.

WALLACE, T., 1930: Experiments on the manuring of fruit trees. III, The effects of deficiencies of potassium, calcium and magnesium, respectively, on the contents of these elements and of phosphorus in the shoot and trunk regions of apple trees, *J. Pom. and Hort. Sci.*, **8**: 23–43.

———, 1931: Chemical investigations relating to potassium deficiency of fruit trees, *J. Pom. and Hort. Sci.*, **9**: 111–121.

——— and E. L. PROEBSTING, 1933: The potassium status of soils and fruit plants in some cases of potassium deficiency, *J. Pom. and Hort. Sci.*, **11**: 120–148.

———, 1935: Investigations on chlorosis of fruit trees. V, The control of lime-induced chlorosis by injection of iron salts, *J. Pom. and Hort. Sci.*, **13**: 54–67.

WARBURG, O., 1925: Iron, the oxygen-carrier of respiration-ferment, *Sci.*, **61**: 575–582.

WARE, J. O., and V. H. YOUNG, 1934: Control of cotton wilt and "rust," *Ark. Sta. Bull.* 308.

WARINGTON, K., 1923: The effect of boric acid and borax on the broad bean and certain other plants, *Ann. Bot.*, **37**: 629–672.

———, 1926: The changes induced in the anatomical structure of *Vicia faba* by the absence of boron from the nutrient solution, *Ann. Bot.*, **40**: 27–42.

———, 1933: The influence of length of day on the response of plants to boron, *Ann. Bot.*, **47**: 429–457.

———, 1934: Studies in the absorption of calcium from nutrient solutions with special reference to the presence or absence of boron, *Ann. Bot.*, **48**: 743–776.

WARNE, L. G. G., 1934: The distribution of potassium in normal and scorched foliage, *Ann. Bot.*, **48**: 57–67.

WATSON, D. J., 1936: The effect of potassium chloride on the diurnal changes of the carbohydrates of the potato leaf, *Ann. Bot.*, **50**: 59–77.

WEBBER, I. E., 1935: Histological characteristics of plants grown in toxic concentrations of boron, *J. Agr. Res.*, **50**: 189–194.

WEBSTER, J. E., 1928: Phosphorus distribution in grains, *J. Agr. Res.*, **37**: 123–125.

———, 1929: Phosphorus in plants, *Bull. Univ. Okla.* (*Proc. Okla. Acad. Sci.*), **9**: 47–49.

——— and F. JANSMA, 1929: A comparison of the copper content of Oklahoma wheat with those of other states, *Sci.*, **70**: 174.

——— and C. DALBLOM, 1930: Changes in the phosphorus content of growing mung beans, *J. Agr. Res.*, **41**: 819–824.

WEEVERS, T., 1911: Untersuchungen über die Lokalisation und Funktion des Kaliums in der Pflanze, *Rec. Trav. Bot. Néerland.*, **8**: 289–332; *Exp. Sta. Rec.* **26**: 823.

———, 1917: Die physiologische Bedeutung des Kaliums in den Pflanzen, *Biochem. Z.*, **78**: 354–357.

WEIHE, A., 1931: Ein Beitrag zur Kieselsäuredüngung, *diss., Univ. Kiel; Biol. Abst.*, **9**: 44.

WEISZMANN, H., 1923: Ueber den Einfluss des Kaliums auf die Entwicklung d[illegible] Pflanzen und ihren morphologischen und anatomischen Bau, bei besonderer Berücksichtigung der landwirtschaftlichen Kulturpflanzen, *Z. Pflanzendüng.*, **2**: 1–79.

WELTON, F. A., 1928: Lodging in oats and wheat, *Bot. Gaz.*, **85**: 121–151.

——— and V. H. MORRIS, 1931: Lodging in oats and wheat, *Ohio Sta. Bull.* 471.

WENDT, G. V., 1930: Stimulationsversuche mit Jod, *Gartenflora*, **79**: 213–218; *Biol. Abst.*, **7**: 791–792.

WESTER, D. H., 1921: Über den Mangangehalt von (holländischen) Samen, *Biochem. Z.*, **118**: 158–163.

WHEELER, H. J., and B. L. HARTWELL, 1902: Conditions determining the poisonous action of chlorides, *R. I. Sta. 15th Ann. Rept.*, 287–304.

——— and ———, 1904: Magnesium as a manure, *R. I. Sta. 17th Ann. Rept.*, **1904**: 221–260.

———, 1905: Plant peculiarities as shown by the influence of sodium salts, *R. I. Sta. Bull.* 104.

——— and G. E. ADAMS, 1905: Concerning the agricultural value of sodium salts, *R. I. Sta. Bull.* 106.

WHERRY, E. T., and R. BUCHANAN, 1926: Composition of the ash of Spanish moss, *Ecology*, **7**: 303–307.

WHITE, H. L., 1936: The interaction of factors in the growth of *Lemna*. VII, The effect of potassium on growth and multiplication, *Ann. Bot.*, **50**: 175–196.

WILCOX, E. V., and W. P. KELLEY, 1912: The effect of manganese on pineapple plants and the ripening of the pineapple fruit, *Hawaii Sta. Bull.* 28.

WILFARTH, H., H. RÖMER, and G. WIMMER, 1906: Über die Nährstoffaufnahme der Pflanzen in verschiedenen Zeiten ihres Wachstums, *Landw. Versuchs-Stat.*, **63**: 1–70.

WILKINS, L. K., 1917: The high calcium content of some cucurbit vines, *N. J. Sta. Bull.* 310.

WILLIS, L. G., and J. O. CARRERO, 1921: Rice investigations, *Porto Rico Sta. Rept.*, **1921**: 8–9.

——— and J. R. PILAND, 1934: The influence of copper sulfate on iron absorption by corn plants, *Soil Sci.*, **37**: 79–84.

———, 1936: Bibliography of references on the literature on the minor elements and their relation to the science of plant nutrition, Chilean Nitrate Ed. Bur., New York, N. Y.

WILLSTÄTTER, R., 1906: Zur Kenntniss der Zusammensetzung des Chlorophylls, *Ann. Chem.*, **350**: 48–82.

——— and M. BENZ, 1908: Untersuchungen über Chlorophyll. VI, Ueber Krystallisirtes Chlorophyll, *Annalen*, **358**: 267–287.

——— and A. STOLL, 1913: "Untersuchung über Chlorophyllen, Methoden, und Ergebnisse," Julius Springer, Berlin.

WILSON, L. B., 1933: Effects of chlorine, bromine and fluorine on the tobacco plant, *J. Agr. Res.*, **46**: 889–899.

WINSTON, J. R., 1935: Reducing decay in citrus fruits with borax, *U. S. Dept. Agr. Tech. Bull.* 488.

WOHACK, F., 1930: Zur Frostschutzwirkung der Kalisalze, *Ernähr. der Pflanze*, **26**: 318–319; *Biol. Abst.*, **5**: 2692.

WÖHER, A., 1920: Uber die Giftwirkung von Arsen-, Antimon-, und Fluorverbindungen auf einige Kulturpflanzen, *Angew. Bot.*, **2**: 161–178.

WOLFF, J., 1913: De l'influence du fer dans le développement de l'orge et sur la spécificité de son action, *C. R. Acad. Sci. Paris*, **157**: 1022–1024.

WOODARD, J., 1922: Sulphur as a factor in soil fertility, *Bot. Gaz.*, **73**: 81–110.

WYATT, F. A., 1916: Influence of calcium and magnesium compounds on plant growth, *J. Agr. Res.*, **6**: 589–621.

WYND, F. L., and H. J. FULLER, 1931: Effects of ultra-violet radiation on the calcium and phosphorus of plants, *abst. Am. J. Bot.*, **18**: 895.

———, 1933: Sources of carbohydrate for germination and growth of orchid seedlings, *Ann. Mo. Bot. Gard.*, **20**: 569–581.

———, 1934: The effects of increasing the iodine content of the tomato plant on respiration and enzymatic activity, *Ann. Mo. Bot. Gard.*, **21**: 367–432.

YAMANO, Y., 1905: Can aluminum salts enhance plant growth? *Bull. Col. Agr. Tokyo*, **6**: 429–432.

YANOVSKY, E., E. K. NELSON, and R. M. KINGSBURY, 1932: Berries rich in calcium, *Sci.*, **75**: 564–566.

YOUNG, R. S., 1935: Certain rarer elements in soils and fertilizers and their role in plant growth, *N. Y. Sta.* (*Cornell*) *Mem.* 174.

YOUNG, V. H., G. JANSSEN, and J. O. WARE, 1932: Cotton wilt studies. IV, Effect of fertilizers on cotton wilt, *Ark. Sta. Bull.* 272.

ZAITSEVA, A. A., 1929: Influence of magnesium on chlorophyll accumulation in algae and higher plants, *Bull. Inst. Sci. Lesshaft.*, **15**: 137–175; *Biol. Abst.*, **5**: 2664.